J/3
10—

NATURAL HISTORY OF VERMONT

Zadock Thompson

NATURAL HISTORY OF VERMONT

by ZADOCK THOMPSON

with numerous engravings and an appendix

CHARLES E. TUTTLE COMPANY
Rutland, Vermont

Representatives
Continental Europe: BOXERBOOKS, INC., *Zurich*
British Isles: PRENTICE-HALL INTERNATIONAL, INC., *London*
Australasia: PAUL FLESCH & CO., PTY. LTD., *Melbourne*
Canada: M. G. HURTIG LTD., *Edmonton*

*Published by the Charles E. Tuttle Company, Inc.
of Rutland, Vermont & Tokyo, Japan
with editorial offices at
Suido 1-chome, 2-6, Bunkyo-ku, Tokyo, Japan*

Copyright in Japan, 1971, by Charles E. Tuttle Co., Inc.

All rights reserved

Library of Congress Catalog Card No. 77-152112

International Standard Book No. 0-8048-0983-6

*First edition published 1853 by the author
First Tuttle edition published 1972*

PRINTED IN JAPAN

TABLE OF CONTENTS

                Zadock Thompson *vii*

CHAPTER I: Descriptive and Physical Geography of Vermont *1*

CHAPTER II: Quadrupeds of Vermont *23*

CHAPTER III: Birds of Vermont *56*

CHAPTER IV: Reptiles of Vermont *112*

CHAPTER V: Fishes of Vermont *127*

CHAPTER VI: Invertebral Animals of Vermont *151*

CHAPTER VII: Botany of Vermont *173*

                Appendix to Thompson's Vermont *225*

                Index to the Appendix to Thompson's Vermont *284*

                Index to Natural History *285*

ZADOCK THOMPSON
SCIENCE MASTER FOR THE COMMON MAN

T. D. Seymour Bassett

Why be interested in a natural history of Vermont compiled over a century and a quarter ago by a rank amateur, self-confessed to be no botanist and no geologist? Thompson made a poor living as a country schoolmaster, doing astronomical observations for almanacs, editing two magazines that failed, supplying an occasional sermon, or performing a wedding or funeral to eke out his income. Yet the diffident clergyman who prepared this work was one of the half dozen most learned men of his time in Vermont. He produced a bird's-eye view of lasting value.

The first observations of Vermont nature were by explorers, soldiers, and travelers. Samuel de Champlain noted snow on Mount Mansfield in July 1609 and chestnuts on the shore of Lake Champlain. Swedish Peter Kalm's name was attached to plants he saw near the lake when he traversed it in the 1740's. Surveyors like Ira Allen recorded a random sampling of the original forest cover by using trees as landmarks. Samuel Williams's *History of Vermont* (1794) depicted the biological situation after a decade's flood of settlement. And in Vermont's first fifty years of statehood many began collecting rocks and shells, drying plants, stuffing birds and beasts, and pigeon-holing these pieces into Linnaean lists of Latin names.

Thompson built on these beginnings. He marshaled the available talent to publish the best summary possible in 1842, and the 1853 appendix went further; but it is obviously not the last word. You can't look here to find out what mushrooms are poisonous because he doesn't even mention fungi. For all the invertebrates except Professor Adams's mollusks, Thompson makes "a few remarks" in three and a half pages, with 14 lines on spiders and a page on the cecropia moth, called a butterfly (pp. 169-73).

Thompson's glory is in his attitude toward man and nature. He wrote for the common man more than the scientist—for the Vermonter who had only a few terms of common schooling and was not interested in taxonomy and nomenclature. He described "the specimen before me" in plain language, but also pointed out the history, habitat, and use of each member of each "tribe."

Like his friend George Perkins Marsh after him, Thompson always thought of man *and* plants *and* animals on a particular surface of soil, water, and rock, and the effects of the shifting balance. Marsh probably developed out of suggestions from Thompson, as well as from his own experience, his thesis that man may harmfully disturb his environment. See how frequently in the following pages Thompson counsels conservation by emphasizing how species help man, even when man calls them vermin. Repeatedly he will tell a Just So Story: "the deer, Best Beloved, were not always as scarce as they are now." In other words, Thompson kept the history in natural history—history with a moral.

Thompson and Marsh reached these insights via their pilgrim curiosity. Traveling wide awake, they pieced together what they found en route. On his father's Bridgewater, Vermont, farm, cleared only a few years before he was born in 1796, Thompson saw nature little disturbed by man. He heard wolves howl; he ate the wild black currant; he was scared by the screech owl. A farm accident nearly killed him at twelve, his friend Pliny H. White recalled, and the long convalescence bent him toward study.

He was twenty-seven before he could scrape through the University of Vermont with what he could earn selling almanacs. The only natural history he could have learned at college was outside of class. William Paddock was lecturing on botany and materia medica to medical students during Thompson's upper class years, and James Dean filled the slot of mathematics and the physical sciences but was infectiously curious about all phenomena. In the fall of his senior year Thompson was elected president of Phi Sigma Nu, a student literary society. He must surely have voted with the large affirmative majority on two of the weekly debate questions: "Is the study of the Mathematics and Natural Philosophy more beneficial than the study of the Languages?" (18–4, yes) "Is the science of Geography more beneficial to man than Grammar?" (11–5, yes)

Within a year of graduation he had married and had published his first *Gazetteer* (1824), replacing the 1808 compilation of his mentor, James Dean. A schoolmaster for the next decade in Burlington and the Eastern Townships of Lower Canada, Thompson included botany and natural history (listed separately) in his curriculum. (See his advertisement as principal of the Burlington High School for Young Ladies in the March 5, 1830, Burlington *Free Press*.)

Thompson saw life as an articulated whole. Trained like most college students of the day on William Paley's *Natural Theology: or, Evidences of the Existence and Attributes of the Deity, Collected from the Appearances of Nature* (1802), he was prepared to see natural design reflecting the Creator. "We have . . . paid a bounty for the destruction of crows, while in consequence of that destruction our fields were suffering the ravages of grubs, which the crows are designed to check" (p. 172). Again, invertebrates "all afford eminent manifestations of the wisdom and skill of the Creator" (p. 151).

Since Thompson assumed all nature to be interrelated, he was concerned with every kind, from angleworm to yellowbird. If someone else appeared to know more, he turned the section over to the specialist. Hence we have the "Catalogue of Vermont Plants" by William Oakes of Ipswich, Massachusetts, and "Fresh Water and Land Shells" collected by Professor Charles B. Adams of Middlebury College.

But how dry the specialists are compared to Thompson! Like them, he describes from direct observation but more often adds "vulgar" English names and easy terms. Besides, his histories relate each kind to its surroundings, to its past, and to man. Thompson keeps arguing for birds even if they steal seed corn or chickens or cherries—and describes the common scarecrow, not like ours. His synopsis of domestic animals (pp. 51–56) gives a view of livestock farming and the importation of breeds. We learn something of hunting and trapping, of bounties and the price of pelts. Thompson reflects the rural habit, perhaps derived from the Indians, of taming mink and martin, woodchuck and squirrel, robin and crow, and teaching the birds to talk (pp. 31, 32, 45, 72, 79).

He adds to Oakes's technical list what is important to people about trees—what nuts and fruits can be eaten and how the wood and bark are used (pp. 209–21). He explains the scarcity of red cedar because it makes the best fence posts; the scarcity of white pine "in consequence of the indiscriminate havoc . . . by our early settlers" (p. 216), who found it best for just about everything made of wood. We learn the fate of the Lombardy poplar and locust as shade trees (pp. 172, 218–19) and read a properly intertwined history of apple growing, cider brandy, and the temperance movement.

Keeping the common reader always in mind, Thompson records folklore, sometimes contradicting it from observation. Spruce beer prevents scurvy; extract of butternut bark is used as a cathartic; slippery elm bark, "macerated in water . . . makes a refreshing drink much used in colds, coughs and fevers." The balsam fir or "Balm of Gilead . . . is of some celebrity as a medicine" but apparently not yet used as a Christmas tree (pp. 215–17). Rattlesnakes do not charm birds and squirrels to hop into their mouths (p. 119); hair snakes do not originate from cow or horse hairs animated in the water (p. 170).

Thompson appears to have shifted careers in 1835, being ordained Episcopal deacon, probably by Bishop John Henry Hopkins, and thereafter earning travel money as substitute preacher for absent rectors. But he was still basically a teacher, helping out in the bishop's enlarged family. Hopkins's school had no vacations, but the pupils had plenty of fun learning about nature on hikes or rowing on the lake. Thompson's partipation in this kind of "progressive" education qualifies him as the professor about whom the "field trip" story is told. President Calvin Pease, finding one of his teachers fishing for specimens with his students, reprimanded him with "Why aren't you hearing recitations in your classroom instead of wasting your students' time like this?"

The bishop went bankrupt, and his instructor turned to "a larger work" than anything he had done hitherto: new editions of his 1824 *Gazetteer* and 1833 *Geography and History of Vermont*, combined with this natural history, "almost wholly the result of original investigations" (Preface).

Thompson's monumental accomplishment was ignored by big-city professionals, or belittled, if the carping of botanists Oakes and John Carey of New York was typical. "I am disappointed [in] ... the Vt. Cat., so different ... from what it ... would have been under your eye and correction," Carey wrote Oakes November 14, 1842 (Oakes Papers in American Antiquarian Society, Worcester, Massachusetts). The *American Journal of Science* did not notice the *History* until its September 1848 issue. To be sure, Thompson wrote Oakes, "my book is designed for general circulation in this state" (Dec. 7, 1841, *ibid.*) and not abroad.

Even at home, however, recognition was limited and informal. The 1842 legislature ordered 100 copies and voted Thompson a $250 bonus. Hunters and fishermen continued to bring him their curious kill for identification. A decade later, however, too much of the 5,000-copy edition remained unsold to warrant publishing a revision of the whole.

When in 1844 the Assembly finally authorized a state geological survey, Middlebury Governor Slade appointed Middlebury Professor Adams State Geologist. Adams, conchologist and teacher of physical sciences, may have been thought the best qualified man available, but he left the state after three seasons' efforts, with three annual reports and some rock collections, partly labeled, to show for the $6,000 appropriation. Field work by Thompson and Samuel R. Hall of Craftsbury had covered the larger northern half of the state, but no "Final Report" like those of New York and Massachusetts emerged. Appropriations were not renewed until after Adams's death in January 1853 and after Thompson had published his 18-page summary in the appendix to this work, when he was appointed State Naturalist.

The railroad was strangely responsible for his belated recognition. It uncovered while preparing its roadbed a fossil "elephant" at Mount Holly in 1848 and a fossil whale in Charlotte in 1849. The bones were brought to Thompson, who published descriptions in the *American Journal of Science* and the *Proceedings* of the Boston Society of Natural History. He published two more articles and an annual Burlington weather summary in the same journals between 1850 and 1855.

On June 5, 1850, Thompson delivered the annual address before the Boston naturalists. Their *Proceedings* (3: 300–301) devoted less attention to his paper, "Natural History of Vermont," than to Louis

Agassiz's subsequent remarks on mollusks; but his faithful neighbor, Chauncey Goodrich, published it in full.

The essay, after a dozen summary pages, enumerated the advantages and drawbacks facing the country naturalist. He can follow species through their life cycles, "in their natural relations to their localities" (p. 22) and note the "original distribution of plants and animals, and . . . the manner in which that distribution is affected in consequence of the changes wrought by human agency" (p. 18). But he lacks books and instruments. Thompson himself had no microscope and got along on two or three dozen volumes in biology while preparing the 1842 edition. In 1840, he claimed, "a respectable library for the use of a naturalist could not have been culled from all the public and private libraries and all the bookstores in Vermont" (p. 29).

Worst of all, Thompson complained, the country pupil is warped from his natural inclination by the grammarians: "To spell words without knowing their meaning, to read sentences with fluency, without understanding them, to recite the geography of the countries of the world, while their thoughts ranged no farther than the maps before them—exercises like these have usually absorbed nearly the whole time of children in the school-room, and, practically, deprived them of the means and motives for understanding, appreciating, and enjoying what is real, and valuable, and beautiful in the productions of the natural world around them" (p. 25).

Thompson was not against literature. He practiced it with precision, compression and compassion. Seeing a frog about to be swallowed by a snake, he wrote, "it has afforded me real satisfaction to destroy the cruel aggressor and liberate his wretched victim" (p. 115). He did not conceal his dislike for bluejays (p. 22), defended the cedar waxwing (p. 74), and mourned with the widower phoebe (p. 76). He has collected an anthology of vignettes. He sees sermons in stones and beauty even in the hogfish.

Imagine him in his little white cottage on the college green. The rooms are littered with specimens, filled with pens, cages, and tubs of living animals. He is telling that kindred spirit, his wife Phoebe, that the peonies are up, the peachbuds are opening, or a cowbird has invaded the warbler's nest. At another season he is discussing with Professor Joseph Torrey from next door the additions Torrey is making to Oakes's list of plants, for the appendix. Or he is setting out on a walk with another neighbor, J. H. Hills, who engraved the cuts for the *History*. He is writing, in his clear, neat hand, his *Journal of a Trip to London, Paris, and the Great Exhibition* (1851). At any time during his last five years, he might be preparing for his University of Vermont classes by reading Faraday on electricity, Jacob Bigelow on technology or Thomas Say's *Entomology*.

One day in January, 1856, he died of a heart attack. The stores closed for his funeral at St. Paul's Church. Not much was said about his life—could he have heard eulogy, it would have embarrassed him. Ex-President John Wheeler wrote to George Perkins Marsh in Washington that it was "a very very great loss to us . . . the geological survey is again ended. We need some one who has Thompson's love in making observations. . . ." Published obituaries repeated this thought. He was indeed an independent observer, a one-man chapter of the Society for the Propagation of Useful Knowledge—*ad maiorem gloriam dei.*

THOMPSON'S VERMONT.

Part First.

NATURAL HISTORY OF VERMONT.

CHAPTER I.

DESCRIPTIVE AND PHYSICAL GEOGRAPHY OF VERMONT.

SECTION I.

Situation, Boundaries, Extent and Divisions.

Situation.—Vermont is situated in the northwestern corner of New England, and lies between the parallels of 42° 44' and 45° of north latitude, and between 3° 35' and 5° 29' of east longitude from the Capitol of the United States at Washington, or between 71° 33' and 73° 25' of west longitude from Greenwich Observatory.* The most eastern extremity of Vermont is in the township of Canaan, and the most western in the township of Addison. This state lies nearly in the middle of the north temperate zone. The longest day at the south line of the state, is 15h. 9m. 9s., and at the north line, 15h. 25m. 50s.

Boundaries.—Vermont is bounded on the north by the province of Canada, on the east by New Hampshire, on the south by Massachusetts, and on the west by New York. The north line of the state runs upon the parallel of latitude 45° north. This line was first surveyed by commissioners appointed by the provinces of New York and Canada, in the year 1767. It was afterwards run, but very erroneously, by I. Collins and I. Carden. in 1772. In 1806, Dr. Samuel Williams made some observations with the view of ascertaining the true north line of the state, and still further observations were made in 1818, by Messrs. Hassler and Tiarks, surveyors under the treaty of Ghent. Ac-

---

* Where it is not otherwise specified, the longitudes given in this work are in all cases reckoned from the Capitol of the United States. The longitude of the Capitol from Greenwich, according to the most recent observations, is 77° 1' 48". It is very much to be lamented that the longitude of places in Vermont is so imperfectly known. We are not aware that a single point within the state has been determined with any pretensions to accuracy. True, a few solar eclipses have been observed and some calculations have been made, for the purpose of deducing from them the longitude of the places; but the only observations within our knowledge, which have hitherto been regarded as entitled to any degree of confidence, were those of the solar eclipse of 1811, made at Burlington by Prof. James Dean and John Johnson, Esq., and at Rutland by Dr. Williams. The longitude of the University of Vermont, deduced from these observations by Dr. Bowditch, was 73° 14' 34", and of Rutland court house 72° 57' 27" west from Greenwich observatory, and in accordance with these has the longitude of the different parts of the state been laid down upon our maps. In 1838, the author prepared, with much care, for observing the large solar eclipse of that year, for the purpose of determining the longitude of the University. But the opportunity proved unfavorable, the sun being hid by clouds during the greater part of the eclipse. Of the beginning he had a tolerable observation, and from this alone he carefully calculated the longitude by Dr. Bowditch's precepts, and the result was 73° 10' 36" for the longitude of the University, or about 4m. less than was obtained from the preceding observations; and, as he is inclined, from other circumstances, to think it as near an approximation to the true longitude as any yet obtained, he has adopted it in this work.

cording to the latter, the 45th parallel lies a little to the southward of the line previously established, but it is not yet finally settled. The eastern boundary was established by a decree of George III, July 20th, 1764, which declared the western bank of the Connecticut river to be the western boundary of New Hampshire. The southern boundary is derived from a royal decree of March 4th, 1740, and was surveyed by Richard Hazen, in February and March, 1741. This line, which was the divisional line between Massachusetts and New Hampshire, was to run due west from a point three miles to the northward of Patucket falls, till it reached the province of New York. It was run by the compass, and ten degrees allowed for westerly variation of the magnetic needle. This being too great an allowance, the line crossed the Connecticut river 2' 57" to the northward of a due west line. In consequence of this error, New Hampshire lost 59,873 acres, and Vermont 133,897 acres, and the south line of the state is not parallel with the north line. The western boundary was settled by the governments of Vermont and New York at the close of their controversy, in 1790. This line passes along the western boundaries of the townships of Pownal, Bennington, Shaftsbury, Arlington, Sandgate, Rupert, Pawlet, Wells and Poultney, to Poultney river; thence along the middle of the deepest channel of said river, East bay and lake Champlain to the 45th degree of north latitude, passing to the eastward of the islands called the Four Brothers, and to the westward of Grand Isle and Isle la Motte. The portion of this line between the southwest corner of the state and Poultney river, was surveyed in 1813 and 1814, and the report and plan of the survey are in the office of the Secretary of State at Montpelier.

Extent and Area.—The length of Vermont from north to south is 157½ miles, and the average width from east to west 57½ miles, which gives an area of 9,056¼ square miles, or 5,795,960 acres. The length of the north line of the state is 90 miles, and of the south line 41 miles, but, on account of the great bend of the Connecticut to the westward, the mean width of the state is considerable less than the mean between these two lines, as above stated. The width of the state from Barnet to Charlotte through Montpelier, which is 50 miles nearer to the northern than to the southern boundary, is only about 60 miles. On account of the irregularities in the western and eastern boundaries, both these lines are longer than the mean length of the state, the former being about 175 miles, and the latter, following the course of the Connecticut, 215 miles.* The state is divided into two equal parts by the parallel of 44d. 9m. north latitude, and also by the meridian in 4d. 19m. of east longitude. These two lines intersect each other near the western line of Northfield, and about 10 miles south westerly from Montpelier, and the point of intersection is the *geographical centre of the state.*

Divisions.—The Green Mountains extend quite through the state from south to north, and, following the western range, divide it into two very nearly equal parts. These form the only natural division, with the exception of the waters of lake Champlain, which divide the county of Grand Isle from the counties of Franklin and Chittenden, and the several islands which compose that county, from each other, and from the main land. For civil purposes the state is divided into 14 counties, which are sub-divided into 245 townships, and several small gores of land, which are not yet annexed to, or formed into, townships. The names of the counties, the date of their incorporation, the shire towns, and the number of towns in each county at the present time (1842,) are exhibited in the following table:

Counties.	Incorporated.	Shire Towns.	No
Addison,	Feb. 27, 1787	Middlebury,	22
Bennington	Feb. 11, 1779	Bennington Manchester,	17
Caledonia,	Nov. 5, 1792	Danville,	18
Chittenden,	Oct. 22, 1782	Burlington,	15
Essex,	Nov. 5, 1792	Guildhall,	17
Franklin,	Nov. 5, 1792	St. Albans,	14
Grand Isle,	Nov. 9, 1802	North Hero,	5
Lamoille	Oct. 26, 1835	Hydepark,	12
Orange,	Feb. 1781	Chelsea,	17
Orleans,	Nov. 5, 1792	Irasburgh,	19
Rutland,	Feb. 1781	Rutland,	26
Washington	Nov. 1, 1810	Montpelier,	17
Windham,	Feb. 11, 1779	Newfane,	23
Windsor,	Feb. 1781	Woodstock,	23

* Dr. Williams (vol. I, p. 24) seems to have, inadvertently, taken the mean of the two ends of the state for its mean width and thus computed the area at 10,237 1-4 square miles, or 1181m. too much; but this is the area which has usually been given in our geographies and other works respecting Vermont. As the area of countries forms the basis of statistical tables, it is a matter of some consequence that it should be correctly stated. Suppose for example, we wish to know how Vermont compares with the other states in density of population, we divide the population of each state by its area and the quotient is the average number of persons to each square mile in the states respectively. Now if we take the last census and the area at 10,237, the population is only about 28 to a square mile, but if we take the true area, 9,056, it is 32 to the square mile, which would effect very materially its relation to the other states. According to the census of 1820, Vermont was set down as the 10th state in density

FACE OF THE COUNTRY.   PRINCIPAL SUMMITS.

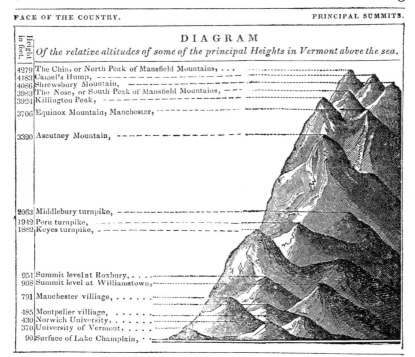

DIAGRAM
Of the relative altitudes of some of the principal Heights in Vermont above the sea.

Height in feet.

4279	The Chin, or North Peak of Mansfield Mountains, ...
4183	Camel's Hump,
4086	Shrewsbury Mountain,
3983	The Nose, or South Peak of Mansfield Mountains,
3924	Killington Peak,
3706	Equinox Mountain, Manchester,
3320	Ascutney Mountain,
2063	Middlebury turnpike,
1942	Peru turnpike,
1882	Keyes turnpike,
951	Summit level at Roxbury,
908	Summit level at Williamstown,
791	Manchester villiage,
485	Montpelier villiage,
430	Norwich University,
370	University of Vermont,
90	Surface of Lake Champlain,

SECTION II.

Face of the country.

Mountains.—The surface of Vermont is generally uneven. A few townships along the margin of lake Champlain may be called level; but with these exceptions, the whole state consists of hills and valleys, alluvial flats and gentle acclivities, elevated plains and lofty mountains. The celebrated range of Green Mountains, which give name to the state, extends quite through it from south to north, keeping nearly a middle course between Connecticut river on the east and lake Champlain on the west. From the line of Massachusetts to the southern part of Washington county, this range continues lofty, and unbroken through by any considerable streams; dividing the counties of Windham, Windsor and Orange from the counties of Bennington, Rutland and Addison. In this part of the state, the communication between the eastern and western sides of the mountain was formerly difficult, and the phrase, *going over the mountain,* denoted an arduous business. But on account of the great improvement of the roads, more particularly in their more judicious location near the streams, the difficulty of crossing the mountain has nearly vanished. In the southern part of Washington county, the Green Mountains separate into two ranges. The highest of these ranges, bearing a little east of north, continues along the eastern boundaries of the counties of Chittenden and Franklin, and through the county of Lamoille to Canada line; while the other range strikes off much more to the east through the southern and eastern parts of Washington county, the western part of Caledonia county and the north western part of Essex county to Canada. This last is called the *height of lands,* and it divides the waters, which fall into Connecticut river, in the north part of the state, from those which fall into lake Champlain and lake Memphremagog. This branch of the Green Mountains, though it no where rises so high as many points of the western branch, is much more uniformly elevated; yet the acclivity is so gentle as to admit of easy roads over it in various places. The western range, having been broken through by the rivers Winooski, Lamoille and Missisco, is divided into several sections, these rivers having opened passages for good roads along their banks, while

---
of population, whereas, if the true area had been used in the computation, she would have ranked as the eighth.

MOUNTAINS.

the intervening portions are so high and steep as not to admit of roads being made over them, with the exception of that portion lying between the Lamoille and Missisco. This part of the Green Mountains presents some of the most lofty summits in the state ; particularly the Nose and Chin in Mansfield, and Camel's Hump in Huntington. These, together with other important mountains and summits in the state, are exhibited in the foregoing table and cut, and will also be described in the Gazetteer, under their respective names. The sides, and, in most cases, the summits of the mountains in Vermont, are covered with evergreens, such as spruce, hemlock and fir. On this account the French, being the first civilized people who visited this part of the world, early gave to them the name of *Verd Mont*, or Green Mountain ; and when the inhabitants of the New Hampshire Grants assumed the powers of government, in 1777, they adopted this name, contracted by the omission of the letter *d*, for the name of the new state.*

---

* This name is said to have been adopted upon the recommendation of Dr. Thomas Young—(see part 2d, page 106.) The following account of the christening of the Green Mountains, is given by the Rev. Samuel Peters in his life of the Rev. Hugh Peters, published at New York in 1807.

" Verd-Mont was a name given to the Green Mountains in October, 1763, by the Rev. Dr. Peters, the first clergyman who paid a visit to the 30,000 settlers in that country, in the presence of Col. Taplin, Col. Willes, Col. Peters, Judge Peters and many others, who were proprietors of a large number of townships in that colony. The ceremony was performed on the top of a rock standing on a high mountain, then named Mount Pisgah because it provided to the company a clear sight of lake Champlain at the west, and of Connecticut river at the east, and overlooked all the trees and hills in the vast wilderness at the north and south. The baptism was performed in the following manner : Priest Peters stood on the pinnacle of the rock, when he received a bottle of spirits from Col. Taplin ; then haranguing the company with a short history of the infant settlement, and the prospect of its becoming an impregnable barrier between the British colonies on the south and the late colonies of the French on the north, which might be returned to their late owners for the sake of governing America by the different powers of Europe, he continued, ' We have here met upon the rock Etam, standing on Mount Pisgah, which makes a part of *the everlasting hill,* the spine of Asia, Africa and America, holding together the terrestrial ball, and dividing the Atlantic from the Pacific ocean—to *dedicate* and *consecrate* this extensive wilderness to God manifested in the flesh, and to give it a new name worthy of the Athenians and ancient Spartans,—which new name is *Verd Mont*, in token that her mountains and hills shall be ever green and shall never die.'

RIVERS AND STREAMS.

Rivers and Streams.—The rivers and streams lying within the state of Vermont are very numerous, but small. They, in most cases, originate among the Green Mountains, and their courses are short and generally rapid. Connecticut river washes the whole eastern border of the state, but belongs to New Hampshire, the western margin of that stream forming the boundary line between New Hampshire and Vermont. The Connecticut receives the waters from 3,700 square miles of our territory. It receives from Vermont, besides numerous smaller streams, the waters of the eleven following rivers, viz: Wantasticook, or West, Saxton's, Williams', Black, Ottaquechy, White, Ompompanoosuc, Wait's, Wells', Passumpsic, and Nulhegan. Clyde, Barton and Black river run northerly into Memphremagog lake. Missisco, Lamoille, Winooski and Poultney river and Otter creek flow westerly into lake Champlain, and the Battenkill and Hoosic westerly into Hudson river. Deerfield river runs southerly from Vermont and falls into the Connecticut in Massachusetts ; and the Coatacook and Pike river head in the north part of the state and run northerly into Canada, the former uniting with Massuippi river at Lenoxville and the latter falling into the head of Missisco bay. All these streams and many smaller ones will be described in the Gazetteer under their respective names.

No country in the world is better supplied with pure and wholesome water than Vermont. There are scarcely any farms in the state which are not well watered by springs, or brooks ; and none, with the exception of those upon the islands in lake Champlain, which are not in the vicinity of one, or more, considerable mill stream. But while Vermont is so abundantly supplied with water, there is, probably, no part of our country in which so little stagnant water is found. The waters of the lakes and ponds are usually clear and transparent, and nearly all the springs and streams are brisk and lively. It is a common remark that the streams in this state have diminished very much in size, since the country began to be cleared and settled, and it is doubtless true to some extent. Many mills, which

---

He then poured out the spirits and cast the bottle upon the rock Etam."

There is no doubt that the name *Verd Mont* had been applied to this range of mountains long previous to the above transaction, (if, indeed, it ever took place;) but we do not find that the name *Verd Mont*, or *Vermont*, was ever applied to the territory generally known as the New Hampshire Grants, previous to the declaration of the independence of the territory in January, 1777.

formerly had an abundance, have ceased to receive the necessary supply of water during a considerable portion of the year; and many mill sites, which were once thought valuable, have, from the same cause, become entirely useless. One of the principal causes of this diminution of our streams is supposed to be the cutting down of the forests, which formerly threw off immense quantities of vapor into the atmosphere, which was again precipitated upon the earth in rain and snow. But it is believed that the quantity of water which annually passes off in our streams is not so much less than formerly as is generally imagined. Before the country was cleared, the whole surface of the ground was deeply covered with leaves, limbs, and logs, and the channels of all the smaller streams were much obstructed by the same. The consequence was, that, when the snows dissolved in the spring, or the rains fell in the summer, the waters were retained among the leaves, or retarded by the other obstructions, so as to pass off slowly, and the streams were kept up, nearly uniform as to size, during the whole year. But since the country has become settled, and the obstructions, which retarded the water, removed by freshets, when the snows melt or the rains fall, the waters run off from the surface of the ground quickly, the streams are raised suddenly, run rapidly, and soon subside. In consequence of the water being thus carried off more rapidly, the streams would be smaller than formerly during a considerable part of the year, even though the quantity of water be the same. It is a well known fact that the freshets in Vermont are more sudden and violent than when the country was new.

The waters of the lakes, ponds and streams are universally soft, miscible with soap, and in general free from foreign substances. And the same may be said of most of the springs, particularly on the Green Mountains, and in that portion of the state lying east of these mountains. The waters of most of the springs and wells in the western part of the state are rendered hard and unsuitable for washing by the lime they hold in solution, and there are many springs which are highly impregnated with Epsom salts, and others containing iron, sulphuretted hydrogen, &c. These mineral springs will be described in another place.

Lakes and Ponds. Small lakes and ponds are found in all parts of Vermont, but there are no large bodies of water which lie wholly within the state. Lake Champlain lies between this state and the state of New York, and more than half of it within the limits of Vermont. It extends in a straight line from south to north, 102 miles along the western boundary, from Whitehall to the 45th degree of latitude, and thence about 24 miles to St. Johns in Canada, affording an easy communication with that province and with New York. This lake is connected with Hudson river, at Albany, by a canal 64 miles in length; so that the towns lying on the shores of Lake Champlain have direct communication by water with the cities of Troy, Albany, Hudson, and New York, and, by means of the great western canal, with the great western lakes. The length of this lake from south to north, measured in a straight line from one extremity to the other, and supposing it to terminate northerly at St. Johns, is 126 miles. Its width varies from one fourth of a mile to 13 miles, and the mean width is about 4½ miles. This would give an area of 567 square miles, two thirds of which lie within the limits of Vermont. The waters, which this lake receives from Vermont, are drained, by rivers and other streams, from 4088 miles of its territory. Its depth is generally sufficient for the navigation of the largest vessels. It received its present name from Samuel Champlain, a French nobleman, who discovered it in the spring of 1609, and who died at Quebec in 1635, and was not drowned in its waters, as has been often said.* One of the names given to this lake by the aborigines is said to have been *Caniaderi-Guarunte*, signifying the mouth or door of the country.† If so, it was very appropriate, as it forms the gate-way between the country on the St. Lawrence and that on the Hudson. The name of this lake in the Abenàqui tongue was *Petawâ-bouque*, signifying alternate land and water, in allusion to the numerous islands and projecting points of land along the lake. Previous to the settlement of the country by Europeans, this lake had long been the thorough-fare between hostile and powerful Indian tribes, and its shores the scene of many a mortal conflict. And after the settlement, it continued the same in reference to the French and English colonies, and subsequently in reference to the English in Canada and the United States. In consequence of this peculiarity of its location, the name of Lake Champlain stands connected with some of the most interesting events in the annals of our country; and the transactions associated with the names of Ticonderoga, and Crown Point,

---

* See Part II, p. 2. †Spafford'sGaz.ofN.Y., p. 98.

and Plattsburgh, and many other places, united with the variety and beauty of the scenery, the neatness and accommodation of the steamboats, and the unrivalled excellency of their commanders, render a tour through this lake one of the most interesting and agreeable to the enlightened traveller. A historical account of the most important transactions upon Lake Champlain, together with some account of the navigation of the lake, and particularly of the steamboats which have been built upon it, will be found in part second, and a much more minute description of the lake under its name in part third.

Memphremagog lake is situated on the north line of the state, and about midway between lake Champlain and Connecticut river. It extends from south to north, and is nearly parallel with lake Champlain. It is 30 miles long, and the average width about two miles. One third part of this lake lies in Vermont; the other two thirds in Canada. The name of this lake in the Abenâqui tongue was *Mem-plow-bouque*, signifying a large expanse of water. This, together with numerous small lakes and ponds, which lie wholly within the state, will be described in part third, either under their names, or in the account of the towns in which they are situated. There is abundant evidence that most of our lakes and ponds were formerly much more extensive than they are at present, and that they have been diminished, both by the deposit of earthy matter brought in by the streams, and by the deepening of the channels at their outlets; and there is also sufficient proof of the former existence of many ponds in this state, which have long since become dry land by the operation of the same causes. Several of these will be pointed out in the descriptions of the rivers in part third, particularly in the description of Winooski river, Barton river, &c.

Bays.—The shores of Lake Champlain are indented by numerous bays, most of which are small and of little consequence. Missisco bay is the largest of these, and belongs principally to Vermont, lying between the townships of Alburgh and Highgate, and extending some distance into Canada. The other bays of most consequence, lying along the east shore of the lake and belonging to Vermont, are M'Quam bay in Swanton, Belamaqueen bay lying between St. Albans and Georgia, Mallets bay in Colchester, Burlington bay between Appletree point and Red Rocks point, Shelburne bay between Red Rocks point and Pottier's point, Button bay in Ferrisburgh, and East bay between Westhaven and Whitehall. Besides these there are several smaller bays lying along the east shore of Lake Champlain, and a considerable bay at the south end of Lake Memphremagog, called South bay. Most of these bays will be more particularly described under their names in part third, and also some of the most important bays lying along the west shore of Lake Champlain, and belonging to New York.

Swamps.—These are hardly of sufficient importance to deserve a separate notice. Though considerably numerous, they are, in general, of small extent, and, in many cases, have been, or may be drained and converted into excellent lands. They are most common in the northern and northeastern parts of the state. In the county of Essex are several unsettled townships, which are said to be made up of hills and mountains with swamps lying between them, which render them to a great extent incapable of settlement. There is a considerable tract of swampy land at the south end of Memphremagog lake, and another in Highgate about the mouth of Missisco river. When the country was new, there were many stagnant coves along the margin and among the islands of Lake Champlain, which, during the hotter parts of the summer, generated intermittent and bilious fevers. But, since the clearing of the country, these have been, to a considerable extent, filled up, and, with the causes which produced them, those disorders have nearly disappeared.

Islands.—The principal islands belonging to Vermont, are South Hero, North Hero, and La Motte. South Hero, called also Grand Island, is 13 miles long, and is divided into two townships, by the name of South Hero and Grand Isle. North Hero is about 11 miles long, but very narrow, and constitutes a township bearing the same name as the island. Isle la Motte lies westward of North Hero, and constitutes a township by the same name. A more particular account of these islands, and also a description of Juniper island and several others lying in Lake Champlain, will be found under their names in part third.

Soil and Productions.—The soil of Vermont is generally a rich loam, but varies considerably according to the nature and compositions of the rocks in the different parts of the state. Bordering our lakes, ponds, and rivers, are considerable tracts of rich and beautiful intervale*

---

* *Intervale.* This word has not yet found a place in our dictionaries, and there has been much carping about it by Dr. Dwight, Mr. Kendall, and other travellers and critics. But we use it, notwithstand-

lands, which consist of a dark, deep and fertile alluvial deposit. These intervales are level tracts lying but little higher than the ordinary height of the water in the streams, and are in most cases subject to being flooded, when the water is very high. They were, while in a state of nature, covered with a heavy growth of forest trees, such as oak, butternut, elm, buttonwood, walnut, ash, and some other kinds. Back of these flats were frequently others, elevated a few feet higher, and covered with white pine. Still further back, the land rises, in most cases very gradually, into hills and upland plains, and the soil becomes harder and more gravelly, but very little diminished in richness and fertility. The timber upon these lands, which constitute the greater part of the state, was principally sugar maple, beech and birch, interspersed with bass, ash, elm, butternut, cherry, hornbeam, spruce and hemlock. And still further back the lands rise into mountains, which are in general timbered with evergreens, such as spruce, hemlock and fir. The loftiest mountains are generally rocky and the summits of some few of them consist of naked rock, with no other traces of vegetation than a few stinted shrubs and mosses; but they are, in general, thickly covered with timber to their very tops. Along the western part of the state, and bordering upon Lake Champlain, are extensive tracts of light sandy soil, which were originally covered with white, pitch and Norway pine, and in the northern part of the state, swamps are numerous, which were well stored with tamarack and white cedar. A more full account of the native vegetables found in this state will be given in a subsequent chapter. Since the country has been cleared, the soil has, in general, been found sufficiently free from stone to admit of easy cultivation, and to be very productive in corn, grain and grass. Without manuring the intervales usually produce large crops, and are easily cultivated, but these crops are liable, occasionally, to be destroyed by floods—the same agency which produces the fertility of the soil on which they grow. All parts are, however, sufficiently fertile amply to reward the labors of the husbandman, and the farmer who is saving and industrious seldom fails of having his barn filled with fodder for his horses, cattle and sheep, his granary with corn, wheat, rye, oats, peas and beans, and his cellar with potatoes, apples, and other esculent vegetables. A sufficient quantity of grain for the supply of the inhabitants might easily be raised in all parts of the state, yet the greater part of the lands are better adapted for grazing than for tillage. The hills and mountains, which are not arable on account of their steepness, or rocks, afford the best of pasturage for cattle and sheep. Of the fruits, nuts, berries, &c., which grow in Vermont, both wild and cultivated, a more particular account will be given in a subsequent chapter on the botany of the state.

Medicinal Springs.—There are in Vermont springs which are more or less impregnated with mineral, or gaseous substances, but none which have yet acquired a very general or permanent celebrity for their curative properties. Along the shore of Lake Champlain, in the counties of Addison and Rutland, the waters generally are impregnated with Epsom salts, (*sulphate* of magnesia). Some of the springs are so highly charged with these salts, in the dryer parts of the year, that a pail full of the water will produce a pound of the salts. They have been manufactured, for medicinal purposes, in some quantities, and, did the price of the article make it an object, they might be made here to almost any extent.

The medicinal properties of most of the waters in this state, which have acquired any notoriety, are derived from gaseous and not from mineral substances. In different towns in the northeastern part of the state, are springs of cold, soft and clear water, which are strongly impregnated with sulphuretted hydrogen gas, and said to resemble the Harrow-Gate waters in England, and those of Ballcastle and Castlemain in Ireland. These waters are found to be efficacious in scrofulous and many other cutaneous complaints, and the springs at Newbury, Tunbridge, Hardwick, &c., have been much resorted to by valetudinarians in their vicinity.

Of medicinal springs on the west side of the Green Mountains, those of Clarendon and Alburgh have acquired the greatest notoriety. It is now about 16 years since the springs at Clarendon began to be known beyond their immediate neighborhood. Since that time their reputation has been annually extending, and the number of visiters increasing, till they have at length become a place of considerable resort for the afflicted from various

---

ing, because it will express our meaning more briefly and intelligibly to the greater part of our readers, than any other we could employ. It may be derived from *inter*—within, and *vallis*—a vale, or valley; and in its specific signification, it denotes those alluvial flats, lying along the margins of streams, which have been, or occasionally are overflowed in consequence of the rising of the water. For the use of the word in this sense, we have the authority of Dr. Belknap and Dr. Williams, the historians of New Hampshire and Vermont, and other good writers.

CLARENDON SPRINGS.

parts of the country. They are situated in a picturesque and beautiful region, 7 miles southwest from Rutland, and have, in their immediate vicinity, good accommodations for 500 visiters. The waters are found to be highly efficacious in affections of the liver, dispepsia, urinary and all cutaneous complaints, rheumatism, inveterate sore eyes, and many others, and they promise fair to go on increasing in notoriety and usefulness. These waters differ in their composition from any heretofore known, but resemble most nearly the German Spa water. For their curative properties they are believed to be indebted wholly to the gases they contain. They have been analyzed by Mr. Augustus A. Hayes, of Roxbury, Mass., with the following results. One gallon, or 235 cubic inches of the water contained,

Carbonic acid gas	46.16 cubic inch.
Nitrogen gas	9.63 " "
Carbonate of Lime	3.02 grains.
Murate of Lime	}
Sulphate of Soda	} 2.74 grs.
Sulphate of Magnesia	}

One hundred cubic inches of the gas which was evolved from the water, consisted of

Carbonic acid gas	0.05	cubic inches.
Oxygen gas	1.50	" "
Nitrogen gas	98.45	" "

The Alburgh springs do not differ materially from the springs at Newbury, Tunbridge, and other places in the northeastern part of the state, owing their medicinal properties principally to the sulphuretted hydrogen gas, which they contain.

Caves. There are no caves in Vermont which will bear comparison with some of the caverns found in other parts of the world, and yet we have several, which are deserving the attention of the curious. Those at Clarendon, Plymouth and Danby are the most interesting. The Clarendon cave is situated on the southeasterly side of a mountain in the westerly part of that town. The descent into it is through a passage 2½ feet in diameter and 31 feet in length, and which makes an angle of 35 or 40° with the horizon. It then opens into a room 20 feet long, 12½ wide, and 18 or 20 feet high. The floor, sides and roof of this room are all of solid rock, but very rough and uneven. From the north part of this room is a passage about 3 feet in diameter and 24 feet in length, but very rough and irregular, which leads to another room 20 feet wide, 30 feet long and 18 feet high. This room, being situated much lower than the first, is usually filled with water in the spring of the year, and water stands in the lowest parts of it at all seasons.*

CLARENDON AND PLYMOUTH CAVES.

The Plymouth caves are situated at the base of a considerable mountain, on the southwest side of Black river, and about 50 rods from that stream. They are excavations among the lime rock, which have evidently been made by running water. The principal cave was discovered about the first of July, 1818, and on the 10th of that month was thoroughly explored by the Author, who furnished the first description of it, which was published shortly after in the Vermont Journal at Windsor. The passage into this cavern is nearly perpendicular, about the size of a common well, and 10 feet in depth. This leads into the first room which is of an oval form, 30 feet long, 20 wide, and its greatest height about 15 feet. It appears as if partly filled up with loose stones, which had been thrown in at the mouth of the cave. From this to the second room is a broad sloping passage. This room is a little more than half as large as the first. The bottom of it is the lowest part of the cave, being about 25 feet below the surface of the ground, and is composed principally of loose sand, while the bottoms of all the other rooms are chiefly rocks and stones. The passage into the third room is 4 feet wide and 5 high, and the room is 14 feet long, 8 wide, and 7 high. The fourth room is 30 feet long, 12 wide, and 18 high, and the rocks, which form the sides, incline towards each other and meet at the top like the ridge of a house. The fifth room, very much resembling an oven in shape, is 10 feet long, 7 wide, and 4 high, and the passage into it from the third room is barely sufficient to admit a person to crawl in. At the top of this room is a conical hole, 10 inches across at the base and extending 2 feet into the rock. From the north side of the second room are two openings leading to the sixth and seventh, which are connected together, and each about 15 feet long, 7 wide, and 5 high. From the seventh room is a narrow passage which extends northerly 15 or 16 feet into the rocks, and there appears to terminate. When discovered, the roof and sides of this cavern were beautifully ornamented with stalactites, and the bottom with corresponding stalagmites, but most of these have been rudely broken off and carried away by the numerous visiters. The temperature, both in winter and summer, varies little from 44½°, which is about the mean temperature of the climate of Vermont in that latitude. A few

---

* Williams' History of Vermont, vol. I, p. 29.

rods to the westward of this cavern there is said to be another which is about two thirds as large.

Section III.

Climate and Meteorology.

Temperature.—Though situated in the middle of the north temperate zone, the climate of Vermont is subject to very considerable extremes both of heat and cold, and the changes of temperature are often very sudden. The usual annual range of the thermometer, in the shade, is from about 92° above to 22° below zero on Farenheit's scale, though it is sometimes known to rise as high as 100°, and at other times to sink as low as 36°, and even to 39° or 40° below zero. But so great a degree of cold as that last mentioned, which is the freezing point of mercury, has not, to our knowledge, been experienced but twice since the means of measuring temperature have been in use in the state, and these were both in the year 1835; the first on the 4th of January, and the second on the morning of the 18th of December. The temperature of the 4th of January, as noted at several places in this state, was as follows: Montpelier —40°, White River —40°, Bradford —38°, Newbury —36°, Norwich —36°, Windsor —34°, Hydepark —36°, Rutland —30°, and Burlington —26°; and the temperature varied but little from the above at those places on the 18th of December. For some time after the first settlement of Vermont the thermometer was hardly known in this part of the country; and since that instrument has become common, very few meteorological journals have been kept, and those few have not, in general, been kept with sufficient care to render them of much value, nor have many of them been preserved in a condition to be accessible to those who may wish to consult them. And hence we possess few accurate data, either for determining the mean annual temperature of the different sections of the state, or for settling the mooted question with regard to a change of climate corresponding to the clearing and cultivating of the country. The results of the principal observations, to which we have access, and which have been made in this state, to ascertain the temperature of the months and the mean annual temperature, are contained in the following tables:

MONTHS.	Rutland. Williams. 1789.	Burlington. Sanders. 1803–8.	Windsor. Fowler. 1806.	Burlington.						
				Thompson.						
				1828.	1832.	1833.	1838.	1839.	1840.	1841.
January,	18.0°	14.4°	22.0°	25.0°	19.7	22.8	26.1	18.6	12.2	25.3
February,	18.5	18.9	26.5	31.1	19.3	15.3	12.3	24.2	28.4	19.6
March,	32.0	28.5	30.3	32.4	30.8	28.2	22.9	36.6	31.4	25.3
April,	41.0	39.5	38.1	39.2	39.4	46.1	35.8	46.3	47.0	39.1
May,	50.0	56.3	57.1	57.6	52.4	57.0	51.7	53.3	57.2	52.8
June,	64.0	66.6	66.4	69.7	61.3	59.6	68.1	60.7	65.6	67.1
July,	67.5	68.2	68.5	70.1	68.5	66.2	71.8	71.5	71.6	68.9
August,	67.5	67.6	64.3	70.2	68.3	63.3	67.5	68.3	72.5	70.5
September	57.0	57.1	62.1	60.8	58.7	57.2	60.5	60.6	58.3	61.9
October,	41.0	45.2	49.5	46.7	47.7	44.9	46.8	50.8	48.0	45.0
November,	37.0	33.5	36.2	38.9	35.6	34.5	31.3	34.0	35.6	35.3
December,	30.0	24.7	24.6	29.3	23.6	24.7	19.1	26.2	21.1	26.4
	43.6	43.4	45.6	47.6	43.8	43.3	43.6	45.5	45.7	44.8

Meteorological observations at Williamstown by Hon. Elijah Paine.

MONTHS.	1829	1830	1831	1832	1833	1834	1835	1836	1837	1838	1839	1840	1841
January,		11.4	10.9	17.1	19.3	12.5	17.9	17.3	9.7	23.9	15.3	9.0	21.6
February,	10.9	14.3	14.6	14.6	13.5	26.5	12.6	10.5	16.7	9.9	20.8	23.7	15.8
March,	23.5	26.4	26.4	25.4	23.5	27.2	25.1	22.9	23.6	30.9	25.8	26.0	24.1
April,	36.6	44.6	39.8		41.2	41.7	36.1	34.5	36.5	31.2	41.2	40.7	34.7
May,	54.8	49.6	53.2		54.7	48.9	48.0	51.6	45.9	48.5	48.7	51.7	47.7
June,	58.7	58.9	64.8	59.3	55.4	57.4	59.4	58.8	60.6	63.0	54.9	58.5	63.1
July,	60.2	64.1	64.4	63.3	62.3	68.2	64.6	63.4	61.2	66.2	65.2	64.8	62.6
August,	60.7	60.7	63.6	63.5	59.5	60.5	60.9	57.0	59.8	61.6	61.4	64.6	63.9
September,	47.9	51.4	53.0	53.9	52.7	55.4	50.0	53.3	52.0	54.6	54.2	52.5	57.9
October,	42.6	44.4	44.6	43.9	41.2	39.7	47.8	34.5	39.0	39.7	45.4	41.9	38.5
November,	29.7	38.2	30.9	31.7	29.5	28.9	29.8	30.6	25.8	28.1	25.3	30.2	29.4
December,	27.3	24.9	7.1	19.7	21.1	16.0	13.1	17.8	14.4	14.1	21.4	16.2	21.7
		40.7	39.4		39.5	40.2	38.8	37.7	37.5	39.1	40.2	39.9	40.0

With the exception of the first three columns in the *first* of the two preceding tables, the particulars of which are not known, all the means for the months have been deduced from *three* daily observations, taken at sun-rise, 1 o'clock, P. M. and 9 in the evening. Now, as the three daily observations at Burlington synchronize for several years with those at Williamstown, the two tables enable us to make a very accurate comparison of the mean temperature of the two places; and the comparison shows that the mean temperature of Burlington, although situated 22' farthest north, is about 5° warmer than that of Williamstown, that of the former being 44.6° and the latter 39.4°. But the cause of this difference is obvious in the location of the two places, Burlington being situated on the margin of lake Champlain, and the place of observation elevated only 250 feet above it, while Williamstown lies among the Green Mountains near the geographical centre of the state, and, the place of Judge Paine's observation, elevated 1500 feet above the lake.*

The mean annual temperature of Burlington, deduced from all of the 12 years observations in the preceding table, is 44.1°, and from the seven years observations by the author 44.9°, but, as the year 1828 was very remarkably warm, that should, perhaps, be set aside, and the mean of the other six, 44.4°, taken as probably a fair statement of the mean annual temperature of Burlington. The mean annual temperature of Williamstown, deduced from the whole of Judge Paine's observations, is 40.3°.

Many perennial springs, and deep wells are found to continue nearly of the same temperature, both in summer and winter, and to be but very little affected by the changes of temperature which are constantly going on at the surface of the earth; the temperature of these may, therefore, be regarded as a pretty fair indication of the mean annual temperature of the climate. The temperature of a well 40 feet deep, belonging to Mr. Samuel Reed, in Burlington, has been observed and noted during the year 1841 as follows, the first number after the day of the month being the depth in feet to the surface of the water at the time of the observation: Jan. 1, 14—46°, Feb. 12, 18—44½°, April 14, 16—44°, June 1, 10—44°, July 20, 10—46½°, and Dec. 8, 20—45½°, giving a mean of 45.1°, or .3° higher than that deduced from the daily observations.

Winds.—For small sections of country the prevailing winds usually take their direction from the position of the mountains and valleys. That is very much the case in Vermont. Through the valley of the Connecticut and of lake Champlain the winds usually blow in a northerly or southerly direction, while easterly and westerly winds are comparatively of rare occurrence. In the valley of lake Champlain east winds are exceedingly rare, as will be seen by the following tables.* Along our smaller rivers, particularly the Winooski and the Lamoille, the prevailing winds are from the northwest. The following tables contain the result of observations made at Burlington, for eleven years, and at Rutland for one year. In the journal kept by the author at Burlington, and from which the tables on the following page were copied, three observations of wind and weather were entered each day, which synchronize with the observations of temperature for the same years in the preceding table, on the ninth page.

The following table contains the results of five years observation at Burlington, by Dr. Saunders, and one year at Rutland, by Dr. Williams.

Place.	Time.	No.Obs.	N	NE	E	SE	S	SW	W	NW	fair.	clody	rain	snw	fog	thun	au
Burlington	1803—8	1682	739	11	19	1	826	25	43	18	1025	676	289	127	19	45	27
Rutland	1789		1095	153	13	16	76	272	182	125	258	452	643	89	41	37	15	21

---

* The author has in his possession a meteorological journal kept at Hydepark by Dr. Ariel Huntoon, for a period of 9 years, of which he had intended to insert an abstract; but, finding the three daily observations to have been made too near the warmest part of the day to furnish the true mean temperature of the 24 hours, and consequently unsuitable for comparison with the other tables, he concluded not to insert it. In order to render meteorological observations of service in determining the relative temperature of places, uniformity in the method of making them seems to be indispensable, and a want of this renders a great part of the journals which have been kept nearly useless.

* Although, at Burlington, we seldom have a wind from the east sufficiently strong to turn the vanes upon our churches, it is not uncommon, during the latter part of the night and early in the morning, when the weather is fair, to have a light breeze from the east, which is doubtless occasioned by the rolling down of the cold air from the mountains to supply the rarefaction over the lake. In other words, it is strictly a *land breeze*, similar to what occurs between the tropics. That these breezes are local and limited is evident from the fact, that, at the same time, the general motion of the air is in a different direction, as indicated by the motion of clouds in higher regions of the atmosphere.

CHAP. 1. DESCRIPTIVE GEOGRAPHY. 11

METEOROLOGICAL TABLE.—WINDS AND WEATHER AT BURLINGTON.

1832	Whole No. observations	WINDS.								WEATHER.			
		N	ne	E	SE	S	sw	W	NW	fair	cldy	r'n	sno
Jan	93	16	2	0	2	40	4	1	18	49	29	3	12
Feb	87	35	1	1	0	37	11	4	8	42	24	3	18
Mar	93	37	0	0	2	44	2	4	5	59	23	2	9
Apr	90	40	1	1	7	24	1	6	9	54	25	8	0
May	93	42	1	0	4	31	6	2	12	54	20	19	0
Jne	90	4	1	1	4	38	0	5	4	62	20	8	0
July	93	42	0	0	1	47	0	4	5	59	23	11	0
Aug	93	21	4	0	0	64	1	2	7	68	20	5	0
Sept	90	14	0	0	0	52	3	0	9	70	15	5	0
Oct	93	22	0	0	0	53	12	2	2	53	30	10	0
Nov	90	38	0	3	3	38	3	2	13	36	41	7	0
Dec	93	45	3	0	0	32	0	3	5	34	50	0	9
Total	1098	364	16	6	21	500	26	57	107	640	329	81	57

1833	Obs.	N	ne	E	SE	S	sw	W	NW	fair	cldy	r'n	s
Jan	93	36	2	0	0	43	3	3	7	45	37	3	8
Feb	84	41	1	1	1	26	0	3	12	51	19	0	14
Mar	93	34	0	0	2	47	1	4	6	65	22	3	3
Apr	90	43	0	1	8	39	0	5	4	60	23	7	0
May	93	34	0	0	3	40	1	10	3	63	19	11	0
Jne	90	27	0	2	3	38	2	5	8	62	20	8	0
July	93	29	1	1	0	58	0	1	0	65	25	3	0
Aug	93	38	0	2	0	50	0	2	2	55	28	10	0
Sept	90	45	0	0	0	41	0	0	3	66	20	4	0
Oct	93	47	0	0	0	40	4	4	3	58	26	6	0
Nov	90	35	0	0	1	46	5	5	6	42	41	7	0
Dec	93	34	18	0	0	18	3	1	8	28	61	0	4
Total	1095	413	23	6	25	486	11	46	55	660	341	65	29

1838	Whole No. observations	N	ne	E	SE	S	sw	w	nw	fair	cldy	r'n	sno
Jan	93	31	0	1	1	51	1	7	1	43	39	6	5
Feb	84	60	3	0	0	14	0	4	0	56	24	0	4
Mar	93	54	0	2	3	32	0	2	1	56	33	1	3
Apr	90	42	0	0	2	31	0	15	2	59	27	4	0
May	93	30	0	0	0	36	0	2	0	66	20	7	0
Jne	90	53	0	0	0	51	1	2	0	62	20	7	0
July	93	32	0	1	0	46	1	2	3	73	14	6	0
Aug	93	33	6	0	1	37	0	10	2	75	13	5	0
Sept	90	33	3	1	6	35	2	0	6	72	15	3	0
Oct	93	45	2	0	3	44	1	2	2	50	36	7	0
Nov	90	39	2	0	0	46	0	3	1	50	31	6	0
Dec	93	24	4	0	4	55	4	1	4	68	23	0	2
Total	1095	482	24	5	25	478	17	55	20	731	295	52	17

1839	Obs.	N	ne	E	SE	S	sw	w	nw	fair	cldy	r'n	sno
Jan	93	41	2	0	1	42	0	2	2	45	40	5	3
Feb	84	21	1	0	3	57	0	1	2	42	37	2	2
Mar	93	40	3	3	0	40	0	4	3	66	23	2	2
Apr	90	50	0	0	0	36	0	3	1	65	16	6	0
May	93	39	3	0	1	47	1	2	0	69	16	8	0
Jne	90	47	0	1	0	35	4	1	4	60	19	11	0
July	93	13	1	0	5	65	1	3	1	71	14	8	0
Aug	93	21	0	0	4	38	4	9	5	84	6	3	0
Sept	90	8	2	1	5	54	0	8	11	64	21	5	0
Oct	93	23	2	0	9	49	3	0	4	73	17	3	0
Nov	90	18	9	0	3	31	6	14	3	39	44	4	3
Dec	93	40	10	0	1	23	3	0	12	51	32	6	4
Total	1095	361	38	13	47	563	23	57	43	729	285	64	17

1840	Whole No. observations	N	ne	E	SE	S	sw	W	N W	N W	fair	cldy	r'n	sno
Jan	93	35	3	0	6	30	10	8	2	1	52	33	6	6
Feb	87	24	0	1	1	54	4	2	2	1	47	36	6	0
Mar	93	26	3	1	7	38	4	2	2	11	55	20	15	0
Apr	90	24	0	2	3	42	5	2	5	7	63	20	5	0
May	93	30	0	0	13	33	0	1	6	5	72	15	3	0
Jne	90	24	1	2	5	48	1	6	3	0	64	13	3	0
July	93	17	1	1	5	54	0	2	2	5	79	11	3	0
Aug	93	7	9	1	0	54	1	2	13	6	69	18	6	0
Sept	90	19	7	3	5	52	1	2	6	11	63	20	6	1
Oct	93	24	1	4	5	37	3	3	9	10	49	39	4	1
Nov	90	10	0	3	3	27	3	3	9	9	36	49	2	3
Dec	93	27	8	5	6	27	9	9	9	9	48	32	3	10
Total	1098	278	42	36	70	479	37	72	84		697	307	68	26

1841	Obs.	N	ne	E	SE	S	sw	W	NW	NW	fair	cldy	r'n	sno
Jan	93	21	2	1	0	42	5	12	3	3	52	25	6	10
Feb	84	18	1	0	1	34	1	3	13	11	50	28	2	6
Mar	93	16	6	4	4	21	3	5	15	7	64	19	2	3
Apr	90	45	1	1	0	28	6	6	3	9	63	10	9	0
May	93	33	0	0	0	33	5	10	5	8	65	16	8	0
Jne	90	20	4	3	3	36	5	4	2	4	60	28	12	0
July	93	8	0	1	1	36	1	3	10	1	77	12	0	0
Aug	93	28	3	1	1	43	4	2	6	6	78	10	1	0
Sept	90	34	3	0	0	40	3	0	4	4	60	23	0	0
Oct	93	33	2	1	0	34	0	3	3	12	44	42	6	0
Nov	90	28	4	0	1	19	6	15	2	6	30	45	4	6
Dec	93	25	2	1	0	40	5	5	14	5	40	37	6	10
Total	1095	343	29	13	43	410	37	93	107		678	288	77	52

ANNUAL QUANTITY OF RAIN.   ANNUAL FALL OF SNOW.

Rain.—The quantity of water, which falls in rain and snow in any one year, does not probably differ very considerably in the different sections of the state, but observations are too few to enable us to speak with much confidence on this point. The quantity of water, however, which falls at the same places in different years, varies very considerably, as will appear from the following table:

MONTHS.	RUTLAND. *Williams.* 1789.	WINDSOR. *Fowler* 1806.	BURLINGTON. *Thompson.*							
			1828.	1832.	1833.	1838.	1839.	1840.	1841.	
	Inches.	Inches.	Inches.	Inches.	Inches.	Inches.	Inches.	Inches.	Inches.	
January,	3.50	2.90	1.30	3.56	1.26	2.52	0.85	1.26	3.49	Mean quantity at B. for 7 years, 37.28 in's
February,	2.78	2.44	2.10	3.22	2.63	1.32	1.20	2.19	0.80	
March,	3.10	0.48	1.35	2.31	1.48	1.10	1.43	3.05	3.23	
April,	3.01	2.78	2.75	1.96	1.28	1.34	1.60	4.69	3.54	
May,	4.72	2.06	2.45	5.71	9.85	4.51	2.43	2.46	2.28	
June,	3.91	2.73	3.70	3.41	4.28	5.37	3.70	2.84	5.16	
July,	2.31	4.34	5.95	3.52	7.54	3.25	6.26	4.18	2.87	
August,	2.11	0.95	4.30	4.76	7.34	2.41	1.91	3.51	1.40	
September,	2.48	4.57	9.85	1.81	4.17	1.33	2.91	4.71	3.62	
October,	5.66	1.40	1.65	4.05	6.01	2.98	0.45	3.76	0.83	
November,	4.10	2.17	6.25	3.01	1.91	3.78	2.57	2.22	2.47	
December,	3.49	2 36	1.65	2.27	1.59	0.92	2.68	2.41	3.02	
Total,	41.17	29.18	43.30	39.59	49.24	30.83	27.99	37.28	32.71	

The depth of water, which falls during a rain storm or thunder shower, is much less than people generally suppose. A fall of 4 or 5 inches during a severe thunder shower would not be thought at all extravagant by persons who have paid no attention to the accurate measurement of the quantity which fell. But during the seven years observations at Burlington contained in the above table, the depth of water which fell in one shower has never exceeded two inches, and the whole amount in 24 hours has, in only one instance, exceeded three inches, and that was on the 13th of May, 1833, when the fall of water was 3.54 inches.

Snow.—For more than three months of the year the ground is usually covered with snow, but the depth of the snow, as well as the time of its lying upon the ground, vary much in the different parts of the state. Upon the mountains and high lands, snows fall earlier and deeper, and lie later in the Spring than upon the low lands and valleys, and it is believed that they fell much deeper in all parts of the state, before the country was much cleared, than they have for many years past. As little snow falls at Burlington, probably, as at any place in the state. The following table exhibits the amount at this place for the last five winters:

Fall of Snow at Burlington in the winters of

1837–'8.	Inc.	1838–'9.	Inc.	1839–'40.	Inc.	1840–'1.	Inc.	1841–'2.	Inc.
Nov. 9,	2	Oct. 29,	1	Nov. 6,	2	Oct. 26,	2½	Oct. 8,	2
" 26,	5	Nov. 7,	3½	" 9,	1½	Nov. 22,	7	" 26,	3½
Dec. 10,	3	" 19,	2	Dec. 11,	3	" 26,27,	3½	" 29,	3
" 11,	1	" 28,	2	" 16,	9	Dec. 7,	6	Dec. 2,	1
" 18,	3	Dec. 7,	¼	" 17,	1	" 22,	3	" 14,	1½
" 28,	1	" 17,	1	" 28,	5	" 27,	8	" 18,	15
Jan. 15,	1	" 18,	4	" 29,	4	Jan. 2,	10	Jan. 5,	2
" 19,	2	" 23,	6	Jan. 5,	4	" 6, 11,	5	" 9,	2
" 28,	12	" 29,	1	" 15,	1½	" 22,25,	8½	" 27,	3
Feb. 11,	5	Jan. 4,	1	" 23,	6	" 30,	2	Feb. 17,	15
" 13,	3	" 5,	1½	Feb. 26,	1	Feb. 2,	2½	" 22,	1
" 17,	8	" 28,	1	March 7,	1	" 6, 10,	4½	" 26,	4
" 22,	1	Feb. 2,	1	" 10,	2	" 17,27,	7	March 7,	5
March, 6,	6	" 8,	2	" 24,	7	March 7,	5	" 15,	1
" 21,	1	" 27,	4			" 9,	4	" 26,	5
" 28,	2	March 3,	1			" 29,	7		
" 30,	3	" 19,	5			Apr. 6,13,	2		
April, 2,	1	April 13,	3½			" 22,	5		
	60		41		48		92½		64

CHAP. 1.   DESCRIPTIVE GEOGRAPHY.   13

SLEIGHING.   SEASONS.   APPEARANCES OF BIRDS AND BLOSSOMS.

In 1838–'9, sleighs run from December 23, to January 8, but there was no good sleighing during the winter. In 1839–'40 sleighing was excellent from December 16, to February 5, *fifty one days.* In 1840–'41, sleighs run from November 22, to November 29, and from December 7, to December 12, but the sleighing was not good. From December 27, the sleighing was good till the 8th of January, after which there was no good sleighing, although sleighs continued to run till the 20th of March. In 1841–'2, sleighing tolerable from December 18, to January 20, after that no good sleighing though sleighs run at several periods for a few days at a time.

The deepest snows, which fall in Vermont, are usually accompanied by a north or northeasterly wind, but there is sometimes a considerable fall of snow with a northwesterly, or southeasterly wind. A long continuance of south wind usually brings rain, both in winter and summer. Although snows are frequent in winter and rains in summer, storms are not of long continuance, seldom exceeding 24 hours. Storms from the east, which are common on the sea board, do not often reach the eastern part of this state, and on the west side of the Green Mountains they are wholly unknown, or rather, they come to that portion of the country from a northeastern, or southeastern direction. Thunder showers are common in the months of June, July and August, but seldom at other seasons. They usually come from the west, or southwest, but are not often violent or destructive, and very little damage is ever done by hurricanes or hail. The crops oftener suffer from an excess, than from a deficiency, of moisture, though seldom from either.

Seasons.—During the winter the ground is usually covered with snow, seldom exceeding one or two feet deep on the low lands, but often attaining the depth of three or four feet on the high lands and mountains. The weather is cold, and, in general, pretty uniformly so, with occasional snows and driving winds, till the beginning of March, when with much boisterous weather there begin to appear some slight indications of spring. About the 20th of that month the snows begin to disappear, and early in April the ground is usually bare. But the snows fall some weeks earlier and lie much later upon the mountains than upon the low lands. The weather and state of the ground is usually such as to admit of sowing wheat, rye, oats, barley and peas, the latter part of April. Indian corn is commonly planted about the 20th of May, flowers about the 20th of July, and is ripe in October. Potatoes are planted any time between the 1st of May and the 10th of June. Frosts usually cease about the 10th of May and commence again the latter part of Sept., but in some years slight frosts have been observed, at particular places, in all the summer months, while in others, the tenderest vegetation has continued green and flourishing till November. The observations contained in the following table will afford the means of comparing the springs of a few years past. They are gathered from the Meteorological journal kept by the author at Burlington:

Year.	Robins seen.	Song Sparrows seen.	Barn Swallows seen.	Currants Blossom.	Red Plum Blossom.	Plums and Cherries Blossom.	Crab Apple Blossom.	Common Apple Blossom.
1828			April 28	May 9		May 12		May 16
1829			" 23	" 9	May 12	" 16		" 22
1832	Mar. 25	Mar. 28	" 26	" 12	" 14	" 20	May 24	June 3
1833	" 23	" 28	" 21	" 4	" 7	" 12	" 15	May 18
1837	" 20	" 23	" 30	" 16	" 19	" 28	" 30	June 2
1838	" 23	" 31	May 2	" 19	" 22	" 26	June 1	" 2
1839	" 25	" 25	April 26	" 4	" 12	" 14	May 22	May 26
1840	" 15	" 21	" 21	" 3	" 12	" 17	" 20	" 23
1841	" 27	" 27	" 27	" 23	" 25	" 26	" 29	" 31

Vegetation, upon the low lands and along the margin of the lakes and large streams, is, in the spring, usually, a week or ten days in advance of that upon the high lands and mountains; but frosts usually occur, in the fall, earliest upon the low lands, allowing to each nearly the same time of active vegetation. The low lands, however, enjoy a higher tempera- ture, and bring fruits and vegetables to maturity which do not succeed well upon the high lands. To the above remark, with regard to early frosts, there are several exceptions. On the low islands and shores of lake Champlain, vegetation is frequently green and flourishing long after the frosts have seared it in other parts of the state, and, along several of the rivers, vegetation is protected by the morning fogs for some time after its growth has been stopped upon the uplands. The early part of the autumn is usually pleasant and agreeable and the cold advances gradually, but as it proceeds the changes become more considerable and frequent, and the great contrast between the temperature of the day and night at this season render much precaution necessary in order to guard against its injurious effects upon health. The ground does not usually become much frozen till some time in November, and about the 25th of that month the ponds and streams begin to be covered with ice, and the narrow parts of lake Champlain become so much frozen as to prevent the navigation from Whitehall to St. Johns, and the line boats go into winter quarters, but the broad portions of the lake continue open till near the first of February, and the ferry boats from Burlington usually cross till the first of January. The following table contains the times of the closing and the opening of the broad lake opposite to Burlington, and when the steamboats commenced and stopped their regular trips through the lake from Whitehall to St. Johns, for several years past:

Year.	Lake Champl'n closed.	Lake Champl'n opened.	Lineboats comenc'd running.	Line Boats stopped.
1816	Feb. 9			
1817	Jan. 29	Apr. 16		
1818	Feb. 2	Apr. 15		
1819	Mar. 4	Apr. 17	Apr. 25	
1820	{ Feb. 3 { Mr. 8	Feb. Mar. 12		
1821	Jan. 15	Apr. 21		
1822	Jan. 24	Mar. 30		
1823	Feb. 7	Apr. 5	Apr. 15	
1824	Jan. 22	Feb. 11		
1825	Feb. 9			
1826	Feb. 1	Mar. 24		
1827	Jan. 21	Mar. 31		
1828	not clos'd			
1829	Jan. 31	Apr.	Apr. 6	
1830				
1831			Apr. 11	
1832	Feb. 6	Apr. 17	Apr. 23	
1833	Feb. 2	Apr. 6	Apr. 8	
1834	Feb. 13	Feb. 20	Apr. 4	Dec. 5
1835	{ Jan 10 { Feb 7	Jan. 23 Apr. 12	Apr. 21	Nov. 29
1836	Jan. 27	Apr. 21	Apr. 25	Nov. 29
1837	Jan. 15	Apr. 26	Apr. 29	Dec. 10
1838	Feb. 2	Apr. 13	Apr. 19	Nov. 26
1839	Jan. 25	Apr. 6	Apr. 11	Nov. 28
1840	Jan. 25	Feb. 20	Apr. 11	
1841	Feb. 18	Apr. 19	Apr. 28	Dec. 1
1842	not clos'd		Apr. 13	

It frequently happens that the ice continues upon the lake for some time after the snows are gone in its neighborhood and the spring considerably advanced. In such seasons the ice often disappears very suddenly, instances having been observed of the lake being entirely covered with ice on one day and the next day no ice was to be seen, it all having disappeared in a single night. People in the neighborhood, being unable to account for its vanishing thus suddenly in any other way, have very generally supposed it to sink. This opinion is advanced in the account of this lake contained in Spafford's Gazetteer of New York, and the anomaly is very gravely attempted to be accounted for on philosophical principles. But the true explanation of this phenomenon does not require the absurdity of the sinking of a lighter body in a heavier. It is a simple result of the law by which heat is propagated in fluids. That bodies are expanded, or contracted, according to the increase or diminution of the heat they contain, is a very general law of nature. Fresh water observes this law, when its temperature is above 40°, but below 40° the law is reversed, and it expands with the reduction of temperature.

When winter sets in, the waters of the lake are much warmer than the incumbent atmosphere. The surface, therefore, of the water communicates its heat to the atmosphere, and, becoming heavier in consequence, sinks, admitting the warmer water from below to the surface. Now since heat is propagated in fluids almost entirely by the motion of the fluids, this circulation will go on, if the cold continues, till all the water from the surface downward to the bottom is cooled down to the temperature of 40°. It will then cease. The colder water now being lighter than that below, will remain at the surface and soon be brought down to the freezing point and congealed into ice. This accounts for the ice taking soonest where the water is most shallow, and also for the closing of the broad parts of the lake earliest in those winters in which there is most high wind, the process of cooling being facilitated thereby.

After the ice is formed over the lake, and during the coldest weather, the great mass of water, after getting a few inches below the ice, is of a temperature 8° above the freezing point. While the cold is severe, the ice will continue to increase in thickness, but the mass of water below the ice will be unaffected by the temperature of the atmosphere above. Now the mean annual temperature of the climate in the neighborhood of lake Champlain does not vary much from 45°, and this is about the uniform temperature of the earth at some distance below the surface. While then the mass of the waters of the lake is at 40°, and ice is forming at the top, the earth, beneath the water, is at the temperature of 45°, or 5° warmer than the water. Heat will, therefore, be constantly imparted to the water from beneath, when the temperature of the water is less than 45°. The only effect of this communication of heat to the water from beneath, during the earlier and colder parts of the winter, is to retard the cooling of the lake and the formation of ice upon its surface. But after the cold abates in the end of winter and beginning of spring, so that the lower parts of the ice are not affected by the frosts from above, the heat, which is communicated from below, acts upon the under surface of the ice, and, in conjunction with the sun's rays, which pass through the transparent surface and are intercepted by the more opaque parts below,* dissolves the softer portions, rendering it porous and loose like wet snow, while the upper surface of the ice, hardened by occasional frosts, continues comparatively more compact and firm. In this state of things, it often happens that, by a strong wind, a rent is made in the ice. The waters of the lake are immediately put in motion, the rotten ice falls into small fragments, and, being violently agitated, in conjunction with the warmer water beneath, it all dissolves and vanishes in the course of a few hours.

There is one phenomenon, which is of common occurrence in many of our streams, during the coldest part of winter, and which may not at first appear reconcilable with what has been said above, and that is, the formation of ice upon the stones at the bottom of the streams, usually called *anchor ice.* Anchor ice is formed at falls and places where the current is so rapid that ice is not formed upon the surface. In the case of running water, and particularly where the water is not deep and the current rapid, over a rough bottom, the temperature of the whole mass is probably reduced nearly or quite to the freezing point before any ice is formed ; and then, where the current is so rapid that the ice cannot form at the surface, the ice-cold waters of the surface, in their tumultuous descent, are successively brought in contact with the stones at the bottom, which, themselves, soon become ice-cold, after which they serve as nuclei upon which the waters are crystilized and retained by attraction, forming anchor ice.

Smoky Atmosphere.—From the earliest settlement of this country there have been observed a number of days, both in spring and autumn, on which the atmosphere was heavily loaded with smoke. The smoke has generally been supposed to result wholly from extensive burnings in some unknown part of the country. There is no doubt but that much of the smoke often is produced in this way, but it has appeared to us, that, since smoke is not a product, but a defect, of combustion, it may be possible for it to be produced even where there is no fire. We have been led to this conclusion by observing that the amount of smoke has not always been greatest in those years in which burnings were known to be most extensive ; and by observing, moreover, that the atmosphere was usually most loaded with smoke in those autumns and springs which succeeded warm and productive summers. These circumstances have led us to the opinion that the atmosphere may, by its solvent power, raise and support the minute particles of decaying leaves and plants, with no greater heat than is necessary to produce rapid decomposition. When, by the united action of the heat and moisture of autumn and spring, the leaves are separated into minute particles, we suppose these particles may be taken up by the atmosphere, before they are entirely separated into their original elements, or permitted to form new compounds. This process goes on insensibly, until, by some atmospheric change, a condensation takes place, which renders the effluvia visible, with all the appearance and properties of smoke.

Dark Days.—It sometimes happens that the atmosphere is so completely filled with smoke as to occasion, especially when accompanied by clouds, a darkness, in the day-time, approaching to that of night. The most remarkable occurrences of this kind, within our own recollection, were in the fall of 1819, and in the spring of 1820. At both of these seasons, the darkness was so great, for a while near the middle of the day, that a book of ordinary print could not be read by the sun's light. The darkness in both cases was occasioned principally by smoke, and without any known extensive burnings; but the summer of 1819, is known to have been remarkable for the abundant growth of vegetation. But the most remarkable

---

* A remarkable phenomenon attending this disintegration of the ice by the influence of the sun's rays, and one which we think worthy of investigation, is its separation into parallel icicles, or candles, as they are sometimes called, extending perpendicularly from the upper to the lower surface of the ice, giving the mass, particularly the lower portions, somewhat the appearance of a honey comb.

darkness of this nature, which has occurred since the settlement of this country, was on the memorable 19th of May, 1780, emphatically denominated the *dark day.* The darkness at that time is known to have covered all the northern parts of the United States and Canada, and to have reached from lake Huron eastward over a considerable portion of the Atlantic ocean. It was occasioned chiefly by a dense smoke, which evidently had a progressive motion from southwest to northeast. In some places it was attended with clouds and in some few with rain. The darkness was not of the same intensity in all places, but was so great through nearly the whole of this extensive region as to cause an entire suspension of business during the greater part of the day, where the country was settled, and in many places it was such as to render candles as necessary as at midnight. Several hypotheses have been advanced to account for this remarkable darkness, such as an eruption of a volcano in the interior of the continent, the burning of prairies, &c., but by the one advanced in the preceding article, it receives an easy explication. The regions at the southwest are known to be extremely productive, and to have been, at that period, deeply covered with forest sand plants, whose leaves and perishable parts would be sufficient, during their decay, to fill the atmosphere to almost any extent; and nothing more would be necessary for the production of the phenomenon, than a change of atmospheric pressure, which should produce a sudden condensation, and a southwesterly wind.

Indian Summer.—It has been said, though we do not vouch for its truth, that it was a maxim with the aborigines of this country, which had been handed down from time immemorial, that there would be 30 smoky days both in the spring and autumn of each year; and their reliance upon the occurrence of that number in autumn was such that they had no fears of winter setting in till the number was completed. This phenomenon occurred between the middle of October and the middle of December, but principally in November; and it being usually attended by an almost perfect calm, and a high temperature during the day, our ancestors, perhaps in allusion to the above maxim, gave it the name of *Indian Summer.* But it appears that from the commencement of the settlement of the country, the Indian Summers have gradually become more and more irregular and less strikingly marked in their character, until they have almost ceased to be noticed. Now upon the hypothesis advanced in the preceding articles, this is precisely what we should expect. When our ancestors arrived in this country, the whole continent was covered with one uninterrupted, luxuriant mantle of vegetation, and the amount of leaves and other vegetable productions, which were then exposed to spontaneous dissolution upon the surface of the ground, would be much greater than after the forests were cut down and the lands cultivated. Every portion of the country being equally shielded by the forest, the heat, though less intense, on account of the immense evaporation and other concurring causes, would be more uniformly distributed, and the changes of wind and weather would be less frequent than after portions of the forests had been removed, and the atmosphere, over those portions, subjected to sudden expansions from the influence of the sun upon the exposed surface of the ground. It is very generally believed, that our winds are more variable, our weather more subject to sudden changes, our annual amount of snow less and our mean annual temperature higher than when the settlement of the country was commenced. And causes, which would produce these changes, would, we believe, be sufficient to destroy, in a great measure, the peculiar features of our Indian Summers. The variableness of the winds, occasioned by cutting down large portions of the forests, would of itself be sufficient to scatter and precipitate those brooding oceans of smoke, and prevent the long continuance of those seasons of dark and solemn stillness, which were, in ages that are past, the unerring harbingers of long and dreary winters and deluges of snow.

Meteors and Earthquakes.—Upon these subjects Vermont affords nothing peculiar. The common phenomenon of shooting stars is witnessed here as in other parts of the country, and those uncommon displays which have several times occurred about the 13th of November, have been observed from various parts of the state. In addition to these, several of those rare meteors, from which meteorolites or meteoric stones are thrown, have been noticed, but the records of them are few and meagre. These meteors make their appearance so unexpectedly and suddenly, and continue visible for so short a period of time, that it is hardly possible to make observations sufficiently accurate to furnish data for calculating their velocity, distance or magnitude. That most remarkable meteor which passed over New England in a southerly direction in the morning of the 14th of December, 1807, and from which fell large quantities of meteoric stones in Weston, Connecticut, was seen from Rutland in this state, and the observation there made formed one of the elements in Dr. Bowditch's calculations of its velocity, distance and size. A meteor of the same kind passed over New England and New York in a southwesterly direction a little before 10 o'clock in the evening of the 23d of February, 1819, and was seen from many parts of Vermont. We had the pleasure of witnessing it at Bridgewater in this state. The meteor there made its appearance about 10° south of the zenith, and, descending rapidly towards the southwest, it disappeared when about 25° above the horizon. Indeed, its velocity was such over Windsor and Rutland counties as to give to all, who observed it, though at the distance of 10, 20 and even 30 miles from each other, along the line of its course, the impression that its fall was nearly perpendicular; and each observer supposed that it fell within a few hundred yards of himself. Now as this meteor was probably moving nearly parallel to the horizon, the deception must have arisen from the rapid diminution of the visible angle between the meteor and the horizon, occasioned by the great horizontal velocity of the meteor in its departure from the zenith of the observer. These facts should teach us to guard against the illusions of our own senses and to admit with caution the testimony of others respecting phenomena of this nature.

According to the best of our judgment, the meteor was visible three or four seconds, in which time it passed through an arc of near 50° of the heavens. Its apparent diameter was about 20′, or two thirds that of the moon, and the color of its light was very white and dazzling, like that of iron in a furnace in a state of fusion. It left a long train of light behind it, and just at the time of disappearance a violent scintillation was observed, and the fragments detached continued luminous at considerable distance from the main body of the meteor, but no meteoralites are known to have fallen. Five or six minutes after the disappearance of the meteor, a very distinct report was heard accompanied by a jarring of the earth, like the report of a cannon at the distance of five or six miles. Now, assuming the correctness of the above data, and that the report was given at the time of the scintillation, the distance of the meteor was then between 70 and 80 miles, and its diameter about one third of a mile.

Another, and still more remarkable meteor, was seen from this state as well as from the rest of New England, and from New York and Canada, about 10 o'clock in the evening of the 9th of March, 1822. From observations made at Burlington and Windsor, Prof. Dean computed its course to be S. 35° W., its distance from Burlington 59 miles and from Windsor 83 miles, and its height above the earth about 37 miles when it first appeared, and when it disappeared its distance from Burlington was 144 miles and its distance from Windsor 133 miles and its height 29 miles. According to these computations, at the first appearance of the meteor, it was vertical over the unsettled parts of Essex county in the state of New York, and at its disappearance, it was over the western part of Schoharie county in the same state.

Several other meteors of this kind have been observed, the most remarkable of which was seen from the northern part of the state and from nearly the whole of Lower Canada, about 4 o'clock in the morning of the 28th of May, 1834. It being a time when people generally were in bed and asleep, comparatively few had the opportunity of seeing it. Many, however, were awakened by its light, and still more by its report. Residing then at Hatley in Canada, which is 15 miles north of the north line of Vermont at Derby, we were suddenly awakened by a noise resembling that of a large number of heavy carriages driven furiously over a rough road or pavement, and by a shaking of the house, which caused a rattling of every door and window. Supposing it to be an earthquake, we sprung out of bed and reached the door two seconds at least before the sound ceased. The atmosphere was calm and the sky was perfectly clear, with the exception of a narrow train of cloud or smoke, extending from southwest to northeast, and at considerable distance to the northward of the zenith. It was nearly motionless, and was apparently at a vastly greater height than clouds usually lie. Indeed there was something so peculiar in its appearance as to make it the subject of remark and careful observation till after sunrise, when it gradually vanished, although at this time we had no reason to suspect its connexion with the noise and shaking of the earth, which had awakened us. We, however, soon learned that a remarkable meteor had been seen, and that its course lay along the very line occupied by the remarkable cloud above mentioned. From an intelligent young man, who was fishing at the time on Massuippi lake in Hatley, and who had a full view of the meteor during the whole time it was visible, we learned that it made its

NEW ENGLAND EARTHQUAKES.

appearance at a point a little north of west at an elevation of about 35°, passed the meridian at a considerable distance north of the zenith and disappeared in the northeast with an altitude of about 25°. He thought its apparent magnitude to be 8 or 10 times that of the moon, and that it was visible about 10 seconds. It was of a fiery red color, brightest when it first appeared, and gradually decreased in brilliancy, all the time throwing off sparks, till it disappeared. About 4 minutes after the vanishing of the meteor, a rumbling or rattling sound, which sensibly agitated the surface of the lake, commenced in the point where the meteor was first seen, and following the course of the meteor died away at the point where the meteor vanished. This meteor was vertical on a north and south line, about 50 miles to the northward of Derby in this state, or nearly over Shipton in Canada, and its altitude must have been at least 30 miles, and still the agitation it produced in the atmosphere was such as to break considerable quantities of glass in the windows at Shipton, Melbourne and some other places. The course of this meteor was mostly over an unsettled country. The most remarkable circumstances attending this meteor were the train of smoke which it left behind, and the long continued noise and shaking of the earth.

Since the settlement of New England, there have been recorded a considerable number of earthquakes, and several have been noticed in Vermont. The sound accompanying these is usually described as having a progressive motion; and that, and the shaking of the earth have been supposed to be produced by the rushing of steam through the cavities in the interior of the earth, but the effect known to have been produced by the meteor last described, furnishes strong reasons for suspecting that the cause of many, and perhaps of all the earthquakes which have occurred in New England, has been in the atmosphere above instead of the earth beneath. Had this meteor passed without being seen, the sound and shaking of the earth, which it produced, would have been regarded as a real earthquake, and its origin in the atmosphere would not have been suspected.

Aurora Borealis.—This meteor has been very common in Vermont, ever since the first settlement of the state; but in some years it is of more frequent occurrence, and exhibits itself in a more interesting and wonderful manner than in others. Its most common appearance is that of streams of white light shooting up from near the

AURORA BOREALIS.

horizon towards a point not far from the zenith; but at times it assumes forms as various and fantastic as can well be imagined, and exhibits all the colors of the rainbow. It is not uncommon that it takes the form of concentric arches spanning the heavens from west to east, usually at the north, but sometimes passing through the zenith, or even at considerable distance to the south of it. At times the meteor is apparently motionless, but it is not an uncommon thing for it to exhibit a violent undulating motion like the whipping of a flag in a brisk wind. But it is so variable in its appearance, that it is vain to attempt its description. We will, however, mention a few of the remarkable occurrences of this meteor which have fallen under our own observation, and some of the attending circumstances.

On the 12th of October, 1819, at about 7 o'clock in the evening, the Aurora Borealis assumed the form of three luminous resplendant arches, completely spanning the heavens from west to east. The lowest arch was in the north a little below the pole star, the second about midway between the pole star and the zenith, and the third 10° or 15° to the southward of the zenith. These belts gradually spread out till they became blended with each other, and the whole concave heavens was lit up with a soft and beautiful glow of white light. It would then concentrate to particular points whose brightness would equal that of an ordinary parhelion, and around them would be exhibited the prismatic colors melting into each other in all their mellow loveliness. The motions of the meteor were rapid, undulatory and from north to south varying a little towards the zenith. The sky was clear and of a deep blue color where it was not overspread by the meteor. It was succeeded in the morning of the 13th by a slight fall of snow with a northwest wind. The aurora exhibited itself in a manner very similar to the above in the evening of the 3d of April, 1820, and several times since.

But the most remarkable exhibition of this meteor, which has fallen under our own observation, was in the evening of the 25th of January, 1837. It first attracted our attention at about half past 6 o'clock in the evening. It then consisted of an arch of faint red light extending from the northwest and terminating nearly in the east, and crossing the meridian 15 or 20° north of the zenith. This arch soon assumed a bright red hue and gradually moved towards the south. To the northward of it, the sky was nearly black, in which but few stars could be seen. Next

AURORA BOREALIS.

to the red belt was a belt of white light, and beyond this in that direction, the sky was much darker than usual, but no clouds were any where to be seen. The red belt, increasing in width and brightness, advanced towards the south and was in the zenith of Burlington about 7 o'clock. The light was then equal to the full moon, and the snow and every other object from which it was reflected, was deeply tinged with a red or bloody hue. Between the red and white belts, were frequently exhibited streams of beautiful yellow light, and to the northward of the red light were frequently seen delicate streams of blue and white curiously alternating and blending with each other. The most prominent and remarkable belt was of a blood-red color, and was continually varying in width and intensity. At eight o'clock, the meteor, though still brilliant, had lost most of its unusual properties.

MAGNETIC VARIATION.

This meteor, when very brilliant, is usually regarded as an indication of an approaching storm, but, like other signs, it often fails. It is most common in the months of March, September and October, but it is not unusual in the other months.

Magnetic Variation.—Very few observations have hitherto been made in Vermont for the purpose of determining the variation of the magnetic needle, and these few have generally been made with a common surveyor's compass, and, probably, in most cases, without a very correct determination of the true meridian; and hence they cannot lay claim to very minute accuracy. But since such observations may serve to present a general view of the amount and change of variation, since the settlement of the state, we have embodied those to which we have had access, in the following table.

Magnetic Variation in Vermont.

Place of Observation.	Date.	Vari. west.	Latitude.	Lon.w.G'h	Authorities.
Burlington,	1793	7° 38'	44° 28'	73°	Dr. S. Williams.
"	1818	7 30	"	"	J. Johnson, Esq.
"	1822	7 42	"	"	"
"	1830	8 10	"	"	"
"	1831	8 15	"	"	"
"	1832	8 25	"	"	"
"	1834	8 50	"	"	"
"	1837	8 45	"	"	Prof. Benedict.
"	1840	9 42	"	"	J. Johnson, Esq.
Rutland,	1789	7 3	43 37	72	Dr. S. Williams.
"	1810	6 4	"	"	"
"	1811	6 1	"	"	"
Ryegate,	1801	7 0	44 10	72	Gen. J. Whitelaw,
Holland,	1785	7 40	45 0	71	"
St. Johnsbury,	1837	9 16	44 26	71	Prof. A. C. Twining.
Barton,	1837	10 51	44 44	"	"
Montpelier,	1829	12 25	44 17	72	Exec. Documents.
Pownal,	1786	5 52	42 46	72	Dr. S. Williams.
Canaan,	1806	9 00	45 0	71	"

From repeated observations and from a careful examination of the lines of the original surveys, John Johnson, Esq. was of the opinion that in 1785, the westerly variation at Burlington was about 7° 12' and that it diminished till the year 1805 when it was about 6° 12". From 1805 the variation has been increasing up to the present time, 1842; and is now 9° 54. This would give a mean annual change of variation of 6' since 1805, and of 3' previous to that time. And although he thought the change of variation may not have been perfectly uniform, yet he was of opinion that a table constructed with the above variation would not differ materially from the truth. The following is such a table.

Magnetic Variation at Burlington.

Year	Var.w	Year.	Var.w	Year.	Var.w	Year.	Var.w
1785	7°12'	1800	6°27'	1815	7°12'	1830	8°42'
1786	7 9	1801	6 24	1816	7 18	1831	8 48
1787	7 6	1802	6 21	1817	7 24	1832	8 54
1788	7 3	1803	6 18	1818	7 30	1833	9 0
1789	7 0	1804	6 15	1819	7 36	1834	9 6
1790	6 57	1805	6 12	1820	7 42	1835	9 12
1791	6 54	1806	6 18	1821	7 48	1836	9 18
1792	6 51	1807	6 24	1822	7 54	1837	9 24
1793	6 48	1808	6 30	1823	8 0	1838	9 30
1794	6 45	1809	6 36	1824	8 6	1839	9 36
1795	6 42	1810	6 42	1825	8 12	1840	9 42
1796	6 39	1811	6 48	1826	8 18	1841	9 48
1797	6 36	1812	6 54	1827	8 24	1842	9 54
1798	6 34	1813	7 0	1828	8 30	1843	10 0
1799	6 30	1814	7 6	1829	8 36	1844	10 6

Remarkable Seasons.—Although the mean temperature of Vermont has not usually varied much from year to year, yet seasons have occasionally occurred, which became, for a time, proverbial on account of their unusual coldness, or heat, or on account of an excess or deficiency of snow or rain. Of the years, which were remarkable on any of these accounts in early times, we have no accurate records. But it is universally conceded that the year 1816, was the coldest, and perhaps the dryest during the early part of summer, ever known in Vermont, although we have no meteorological observations for that year, and are therefore unable accurately to compare the temperature of its seasons with other years. Snow is said to have fallen and frosts to have occurred at some places in this State in every month of that year. On the 8th of June, snow fell in all parts of the State, and upon the high lands and mountains, to the depth of five or six inches. It was accompanied by a hard frost, and on the morning of the 9th, ice was half an inch thick on shallow, standing water, and icicles were to be seen a foot long. The weather continued so cold that several days elapsed before the snow disappeared. The corn, which was up in many places, and other vegetables, were killed down to the ground, and, upon the high lands, the leaves of the trees, which were about two thirds grown, were also killed and fell off. The summer was not only excessively cold, but very dry. Very little Indian corn came to maturity, and many families suffered on account of the scarcity of bread stuffs and their consequent high prices.

The year, 1828, was nearly as remarkable for warmth as 1816 was for cold. The mean temperature of all the months of this year, with the exception of April, was higher than their average mean, and the temperature of the year 3° higher than the mean of the annual temperatures which have been observed. The broad parts of lake Champlain were not frozen over during the winter.

The year 1830 was distinguished on account of the great quantity of water which fell in rain and snow, and especially for one of the most extensive and destructive freshets ever known in Vermont. Up to the 15th of July, the weather was exceedingly cold as well as wet. It then changed, and became suddenly and excessively warm. The following table shows the height to which the thermometer rose in the shade, on each day from the 15th of July to the 21st, inclusive.

July 15.	Thursday,	. . . 94°
" 16.	Friday,	. . . 92
" 17.	Saturday,	. . . 92½
" 18.	Sunday,	. . . 92
" 19.	Monday,	. . . 90
" 20.	Tuesday,	. . . 91
" 21.	Wednesday,	. . . 94

Nor was the heat much diminished in the absence of the sun. In some cases the thermometer stood as high as 80° during the whole night, and it sunk but little below 80° during any part of the time included in the above table. Another such succession of hot days and nights was perhaps never experienced in the state. From the 15th up to Saturday the 24th, the weather was for the most part clear and calm. On Saturday afternoon, the rain commenced and continued with only short intermissions, till Thursday following. During the 5 days from Saturday noon to Thursday noon, the fall of water at Burlington, exceeded 7 inches, and of this 3.85 inches fell on the 26th in the space of about 16 hours, and this is believed to be one of the greatest falls of water, in that length of time, ever known in Vermont. The Winooski, which was most affected of any of our large streams, was at its greatest height in the afternoon of Tuesday the 27th, and was then from 4 to 20 feet, according to the width of the channel, higher than had ever before been observed. Although the county of Chittenden, and the northern parts of the county of Addison, seemed to be the section upon which the storm spent its greatest force, yet its disastrous effects were felt with unusual severity throughout the valley of lake Champlain, and in all the northern and central parts of the state, and the destruction of property in bridges, mills, buildings and growing crops was great, almost beyond computation. But its most melancholly effect was the destruction of human life. By a change of the channel of New Haven river, in the town of New Haven, during the night, between the 26th and 27th, several buildings containing families were insulated, and afterwards swept away by the waters. Of 21 persons, who were thus surprised and washed away, 7 only escaped; the remaining 14 found a watery grave.*

The whole quantity of water which fell at Burlington, in 1830, measured 59.3 in. being half as much again as the mean annual quantity, and probably exceeding the amount in any other year since the state was settled.

Comparative view of the Climate.—As Vermont extends through 2° 16' of latitude, there is, as might be expected, a

---
* See part III. Article, New Haven.

sensible difference between the temperature of the northern and southern parts, and there is a difference still more marked between the elevated and mountainous parts and the lower country along our lakes and rivers; but observations are too limited to enable us to form any accurate comparison between the different sections of the state.* Between the climate of this state and that of those portions of other states, lying in the same latitude, there is no material difference, with the exception, perhaps, of the sea-coast of New Hampshire and Maine, whose mean annual temperature may be a little higher. But between Vermont and the countries of Europe, lying in the same latitude, there is a remarkable difference, the temperature of the latter being no less than $11\frac{1}{2}°$ higher than ours; and there is a like contrast, increasing towards the north, between the whole western coast of Europe and the eastern coast of North America.

This singular contrast was observed by the earliest navigators, who visited the coast of North America, and has since been confirmed by numerous meteorological observations. A comparison of the journals kept in this country with those kept in Europe shows us that the climate of Vermont, which lies in the latitude of the southern part of France, is as cold as that of Denmark, situated 11 or 12° further north. The following table exhibits pretty nearly the mean temperatures along the coasts of the two continents, with the differences, from the 30th to the 60th degree of latitude.

Table.

Latitude.	Europe. Mean Temp.	America. Mean Temp.	Differences.
30°	70.1°	66.8°	3.3°
35	66.5	60.5	6.0
40	63.1	54.2	8.9
45	56.8	45.0	11.8
50	50.8	37.9	12.9
55	46 0	28.0	18.0
60	40.0	18.0	22.0

A contrast so remarkable, as is exhibited in the preceding table, has been the source of much speculation, but, as it appears to us, without throwing much light upon the true cause of the phenomenon.

Among the earliest writers who attempted to account for it was Father Bresani, an Italian Jesuit, who spent most of his life in Canada. He says that "a certain mixture of dry and moist makes ice, and that in Canada there is a remarkable mixture of water and dry sandy soil; and hence the long duration of cold and great quantities of snow." To this he adds another cause, which is "the neighborhood of the northern sea, which is covered with monstrous heaps of ice, more than 8 months of the year." Father Charlevoix, who visited Canada in 1720, and from whose travels the forgoing opinions of Bresani are taken, says* that, in his opinion," " no person has explained the cause, why this country is so much colder than France in the same latitude." "Most writers," he continues, " attribute it to the snow lying so long and deep on the ground. But this only makes the difficulty worse. Whence those great quantities of snow ?" His own opinion is that the cold and snow are to be attributed to the mountains, woods and lakes. Many European writers have supposed the great lakes, which abound in the country, to be the cause of the coldness of our climate; while others have imagined that there must be a chain of very high mountains in the interior of the continent, running from southwest to northeast, which produce the coldness of our north westerly winds. Doct. Dwight supposes these

---

* As the extremes of heat and cold were not noted in the preceding meteorological tables, we have collected in the following table the extremes of cold which have been entered at sun-rise upon journals kept at three different places within the state since 1829. Degrees in all cases below zero.

Year.	Williamstown.		Burlington.		Hydepark.	
1829	Feb. 5.	11°				
1830	Jan. 31,	22				
1831	Dec. 22,	18	Dec.	14°		
1832	Feb. 24,	22	Jan. 26,	16		
1833	Jan. 19,	26	Jan. 19,	20	Dec. 15,	12°
1834	Dec. 15,	18			Jan. 24,	28
1835	Feb. 4,	24			Jan. 4,	36
1836	Feb. 2,	26			Feb. 18,	34
1837	Jan. 4,	16	Dec. 22,	15	Jan. 26,	34
1838	Dec. 13,	15	Jan. 21,	13	Feb. 2,	22
1839	Jan. 24,	24	Jan. 24,	16	Feb. 10,	22
1840	Jan. 16,	17	Jan. 18,	16		
1841	Feb. 9,	9	Jan. 4,	10		

It would appear from various observations and circumstances, that during calm weather, when the sun does not shine, the temperature of vallies and low situations is lower than that of the high lands, but in windy weather and when the sun shines, it is coldest on the high lands. In confirmation of this statement, in part, we give the following extract of a letter to the author from the Hon. Elijah Paine, of Williamstown, (see pages 9 and 10.) "I have found," says he, "that in extremely cold, still weather, the mercury in the thermometer at Burlington, Montpelier, at Northfield, on Dog river, on the low lands at the meeting-house in this town, at Woodstock, Hanover, N. H., and even at Albany, N. Y., has sometimes been 14 degrees lower than in mine. Sometimes, even in March, I have found the difference equally great, when the wind was light and the weather very cool for the season. But the reverse is the case in extremely cold, *windy* weather. I have known my thermometer in such weather 11 degrees lower than some of those I have mentioned."

* Charlevoix's Travels in America, Vol. I. p. 136.

winds to be descending currents from the higher regions of the atmosphere; and hence their coldness. Doct. Holyoke attributed the coldness of our climate to the extensive forests of evergreens. Doct. Williams, the able historian of Vermont, attributed it to the forest state of the country, and has endeavoured to prove that, eighteen centuries ago, the climate of Europe was even colder than that of America at the present time.* But other writers have, with equal plausibility, shown that no considerable change has taken place in the mean temperature of Europe within that period.† The fact, moreover, that the western coasts of America, which are wholly uncultivated, are very much warmer than the eastern coasts of Asia in the same latitude, which are cultivated to considerable extent, shows that these differences of temperature do not depend upon cultivation, nor, indeed, upon any of the causes which have been mentioned, but upon some more general cause. And this cause, we believe, is to be sought in the influence of the ocean upon the prevailing winds in high northern latitudes. We regard the ocean as the great equalizer of temperature upon the surface of our globe—as the instrument for distributing the heat of the equatorial regions towards the poles and bringing thence cold towards the equator, and thus meliorating the climate of both. We look upon it as a truth established both by theory and fact that there is a general circulation of the waters of the ocean between the equatorial and polar regions—that the warm water from the equator is flowing along the surface of the ocean towards the poles, while the colder water from the poles is advancing along the bottom of the ocean towards the equator. Such a motion of the waters might be inferred, as the result of the unequal distribution of heat through the oceanic mass, increased by the rotation of the earth on its axis. But independent of this, facts furnish indubitable proof of its existence. The temperature of the earth, at a distance below the surface, being a pretty correct index of the mean temperature of the climate, without the circulation we have supposed, the temperature of the ocean at considerable depths, ought, particularly in the warmer parts of the year, to be as high, at least, as the mean annual temperature. But on the contrary, observation proves it to be much lower. In latitude 67°, where the mean temperature is 39°, Lord Mulgrave found, on the 20th of June, when the temperature of the air was 48½°, that the temperature of the ocean at the depth of 4680 feet, was 26°, or 6° below the freezing point. On the 31st of August, in latitude 69° where the annual temperature is 38°, that of the air being 59½°, the temperature of the water at the depth of 4038 feet was 32°.* At the tropic where the temperature does not vary more than 7° or 8° during the year, at the depth of 3600 feet the temperature of the water was found to be only 53°, while that of the air was 84°, making a difference of 31°, and indicating a degree of cold in the lower parts of the ocean nearly 25° more intense than is ever experienced in the atmosphere in that latitude,† How else can we account for the coldness of these waters, but by supposing them to come from higher latitudes in the manner we have described?

Of the opposite motion of the warmer waters along the surface of the Atlantic ocean, from the equatorial towards the polar regions, the gulf stream, the currents setting along the western coasts of Norway, and the vast quantities of tropical productions, lodged upon the costs and islands of the northern ocean, afford abundant proof.

Now this transportation of the colder waters towards the equator and of the warmer waters towards the poles, serves, as already remarked, to mitigate the otherwise intolerable heat of the former, and the excessive cold of the latter; and affords an obvious manifestation of the wisdom and goodness of providence. And it is to the influence of the warm superficial waters of the ocean, which have come from tropical regions, upon the winds, or currents of the atmosphere, that we are to look for the cause of the difference of temperature in the climate of the eastern coasts of North America and the western coasts of Europe, and also in that of the eastern coasts of Asia and the western coasts of North America. If we observe the gulf stream, which is only a concentration by the trade winds of those warm waters which are flowing northerly along the surface of the ocean, we shall perceive it to be very narrow, presenting to the atmosphere only a small surface of its warm water, while near the American coast. But as it proceeds to the northeast its warm waters are spread out upon the surface of the ocean and are thrown directly along or upon the western coasts of Europe. Observation also shows that the prevailing winds in high northern latitudes, are from a north west-

---

* Williams' History of Vermont, Vol. 1, p. 475.
† Edinburgh Review, Vol. XXX, p. 25.

* Count Rumford's Essays, Vol. II. page 304.
† Phil. Transactions, 1752.

erly direction, or passing nearly at right angles across the great northeasterly current of the ocean, and we believe it to be the influence of these warm waters of the ocean upon the westerly and northwesterly winds, which produces the phenomenon in question. On the eastern coasts of North America, these winds come from mountainous, snowy regions, or from lakes and seas, which are covered with ice the greater part of the year; and hence they are excessively cold. In their progress over the Atlantic, they are gradually warmed by imbibing heat from the surface of the ocean, so that when they arrive upon the continent of Europe, their temperature is so much elevated as to produce the remarkable difference observed between the climates of the coasts of the two continents.*

CHAPTER II.

QUADRUPEDS OF VERMONT.

Preliminary Observations.

All animals are divided by Baron Cuvier, the celebrated French naturalist, whose arrangement we shall endeavor mainly to follow, into four general divisions, viz. I. *Vertebrated animals*, or such as have a spine, or back bone, II. *Moluscous animals*, or such as have no skeleton, III. *Articulated animals*, whose trunk is divided into rings, and IV. *Radiated animals*, or zoophytes. The *first* division embraces the mammalia, the birds, the reptiles and the fishes; the *second*, the shell fishes; the *third*, the insects, and the *fourth*, polypi. In this work we shall attempt but little beyond an account of our vertebrated and moluscous animals.

MAMMALIA.

The Mammalia are such animals as suckle their young, and are divided by Cuvier into the following orders:

I. *Bimana*—having two hands and three kinds of teeth. Man is the only species.

II. *Quadrumana*—animals having four hands and three kinds of teeth. Monkies and baboons belong to this order.

III. *Carnivora*—having three kinds of teeth and living principally upon animal food, as the dog, cat, &c.

IV. *Marsupialia*—producing their young prematurely and bringing them to perfection in an abdominal pouch, which incloses the teats, of which the opossum is an example.

V. *Rodentia*—have large incisory teeth suitable for gnawing, and grinders with flat or tuberculated crowns, but no canine teeth, as the rat, beaver, &c.

VI. *Edentata*—having no incisory teeth in either jaw, and in some genera no teeth at all, of which the sloth and ant eater are examples.

VII. *Pachydermata*—having either three or two kinds of teeth, toes variable in number and furnished with strong nails or hoofs, and the digestive organs not formed for ruminating, as the horse, elephant and hog.

VIII. *Ruminantia*—having no incisory teeth in the upper jaw, cloven hoofed feet, and four stomachs fitted for ruminating, or chewing the cud, as the ox, sheep, deer, &c.

IX. *Cetacea*—Aquatic animals having their bodies shaped like fishes, as the whale, dolphin, &c.

Of these nine orders of animals, only three are found in Vermont, in a wild state. These are the *Carnivora*, the *Rodentia* and the *Ruminantia*. We have one order more, the *Pachydermata*, among our domestic quadrupeds, including the horse, ass and hog.

---

* Mr. Daniels in his meteorological essays endeavors to account for the higher temperature of the western coasts of continents in a different manner. He supposes the northwesterly winds to arrive loaded with vapor and that the caloric, liberated by its condensation, raises the general temperature of the atmosphere on the western coast; but, as the winds proceed eastward, they become dryer and when they reach the eastern coasts contain little vapor to be condensed, and consequently do not produce an elevation of temperature. If this were the principal cause of the phenomenon under consideration, the quantity of rain on the western coasts should be greater than upon the eastern in proportion as the temperature is higher, but so far as observations extend the reverse of this seems to be true, the quantity of rain on the eastern coast being greatest.

QUADRUPEDS OF VERMONT.

The following is a catalogue of the native quadrupeds of Vermont, arranged in the order, in which they are described in the following pages:

ORDER CARNIVORA—*Carniverous Animals.*

Vespertilio subulatus,	Say's Bat.
" *pruinosus,*	Hoary Bat.
" *carolinensis,*	Carolina Bat.
" *noctivagans,*	Silver-haired Bat.
Sorex Forsteri,	Forster's Shrew.
" *brevicaudus,*	Short tail Shrew.
Scalops canadensis,	Shrew Mole.
Condylura macroura,	Star-nosed Mole.
Ursus americanus,	Black Bear.
Procyon lotor,	Raccoon.
Gulo luscus,	Wolverene.
Mustela vulgaris,	Weasel.
" *erminea,*	Ermine.
" *vison,*	Mink.
" *canadensis,*	Fisher Martin.
" *martes,*	Pine Martin.
Mephitis americana,	Skunk.
Lutra brasiliensis,	American Otter.
Canis lupus,	Wolf.
" *fulvus,*	Red Fox.
" *var. decussatus,*	Cross Fox.
" *var. argentatus,*	Black or SilverFox.
Felis canadensis,	Lynx.
" *rufa,*	Bay Lynx.
" *concolor,*	Catamount.
Phoca vitulina,	Common Seal.

ORDER RODENTIA—*Gnawing Animals.*

Castor fiber,	Beaver.
Fiber zibethicus,	Musk Rat.
Arvicola riparius,	Meadow Mouse.
Mus decumanus,	Norway Rat.
" *rattus,*	Black Rat.
" *musculus,*	Common Mouse.
Gerbillus canadensis,	Jumping Mouse.
Arctomys monax,	Woodchuck.
Sciurus cinereus,	Gray Squirrel.
" *niger,*	Black Squirrel.
" *hudsonius,*	Red Squirrel.
" *striatus,*	Stiped Squirrel.
Pteromys volucella,	Flying Squirrel.
Hystrix dorsata,	Hedge Hog.
Lepus americanus,	Rabbit.
" *virginianus,*	Hare.

ORDER RUMINANTIA—*Ruminating Animals.*

Cervus alces,	Moose.
" *canadensis,*	Elk.
" *virginianus,*	Common Deer.

ORDER CARNIVORA.

The animals of this order have three kinds of teeth, a simple, membranaceous stomach, and short intestines. They live principally on flesh, or animal food.

GENUS VESPERTILIO.—*Linnæus.*

Generic Characters.—Teeth from 32 to 36,—incisors $\frac{4}{6}$, canines $\frac{1-1}{1-1}$, grind. $\frac{4-4}{5-5}$, $\frac{5-5}{5-5}$ to $\frac{5-5}{6-6}$. Upper incisors in pairs, cylindrical and pointed; the anterior grinders simply conical, posterior having short points or prominences. Nose, simple, without grooves, or wrinkles; ears, with an auricule, lateral and more or less large; tongue smooth, and not protractile; index finger with but one phalanx, the middle with three, the annular and little finger with two; tail comprised in the interfemoral membrane; sebaceous glands under the skin of the face, which vary in different species.

The bats consist of a great number of species, but they agree very nearly in their general form and habits. They produce and nourish their young in the manner of other quadrupeds, but unlike them they are furnished with delicate membranous wings upon which they spend much of their time in the air, thus seeming to form the connecting link between the quadrupeds and birds. They are nocturnal in their habits, lying concealed during the day, but venturing abroad on the approach of evening, during the early part of which they may be seen flitting lightly and noiselessly through the air in quest of food, which consists chiefly of insects. At such times they often enter the open windows of our dwellings and sometimes commit depredations upon our larders, being exceedingly fond of fresh meat. Their nocturnal habits manifest themselves in the domesticated state as well as the wild, and it is with difficulty that they are made to mount upon their wings, or take food during the day, but in the evening they devour food voraciously and fly about the room without reluctance. On the approach of winter bats retire to dry caverns and hollow trees where they suspend themselves by the hooked nails of their hind feet, and thus remain in a torpid state during the winter. They void their excrement, which is found in abundance in these retreats, by reversing their position and suspending themselves by the hooks upon their thumbs till their object is accomplished, when they resume their former position. Bats produce their young in June or July, and have from one to three at a time. The teats of the female are situated on the chest and to these, as we are assured by Dr. Godman, (Nat. His. I. 56.), the young attach themselves so firmly as to be carried about by the mother in her flight, till they have attained a considerable size. The four following species are all that have hitherto been distinguished in Vermont. It is, however, probable that others may hereafter be detected.

SAY'S BAT.
Vespertilio subulatus.—SAY.

DESCRIPTION.—Head short, broad and flat; nose blunt with a small, flat, naked muzzle; eyes small, situated near the ears and covered with fur; ears longer than the head, thin ovate, obtuse and hairy at the base behind; tragus thin, broadly subulate below, tapering upwards and ending in an obtuse tip, at about two thirds the height of the ear; color of the back yellowish brown, the belly yellowish gray; fur soft and fine, and blackish towards the roots; head covered with fur, excepting about the nostrils; color blackish about the mouth; whiskers few, short and stiff; membrane between the hind legs broad, thinly covered with fur next the body, and tapering to a point near the extremity of the tail, which it envelopes; toes of the hind feet long; hooked thumb including the nail ¼ of an inch. Length of the specimen before me, from the nose to the insertion of the tail, 2 inches; tail 1½ inches; spread of the wings, 10 inches.

HISTORY.—This Bat seems to be distributed very generally through the continent. It was first described scientifically by Mr. Say, in the notes to the account of Long's expedition, from a specimen obtained at the foot of the Rocky Mountains. It was afterwards minutely described by Dr. Richardson from specimens obtained on the upper branches of the Saskatchewan and Peace rivers.* Specimens have since been obtained from Labrador, Georgia, Ohio, New Hampshire and Columbia river. It is one of the smallest, and, I think, the most common Bat found in Vermont, especially in the central mountainous parts, where it enters the houses in the evening and is easily captured. The specimen, from which my description was drawn was taken in Waterbury.

THE HOARY BAT.
Vespertilio pruinosus.—SAY.

DESCRIPTION.— Ears broad, shorter than the head, broadly emarginate behind, hairy on the outside more than half the length,

* Fauna Boreali Americana, part I. p. 4.

and at the central part of the inside, tragus bent, club-shaped and blunt at the tip. Canine teeth large and prominent; incisors in the upper jaw conical with a tubercle near the base, very near the canines, and nearly in a line with them; snout cartilaginous and moveable; nostrils wide apart. Eyes black and prominent. Fur on the body blackish brown at its base, then pale brownish yellow, then brownish and terminated with clear, delicate white, like hoar frost; fur on the throat, on and about the ears, and on the inside of the wings towards their base, fulvous; snout, chin, margin of the ears and the posterior part of the wing membrane, blackish; the anterior part of the wings and the base of the fur on the interfemoral membrane, dark chestnut. Tail, wholly embraced in the interfemoral membrane, which is thickly covered with fur, except at the very posterior extremity. Length of the specimen before me, from the snout to the extremity of the tail, 5½ inches; spread of the wings, when fully extended, 16½ inches.

HISTORY.—This bat was also first described by Say in Long's expedition and has since been minutely described by Richardson,* Cooper† and others. It has been found in most parts of the United States and was obtained by Dr. Richardson as far north as lat. 54°. It is not common in Vermont, but is occasionally met with. The only Vermont specimen, which I have examined, and that from which the preceding description was drawn, was sent me alive by my friend, David Reed, Esq., of Colchester. It was taken at his place in Colchester the latter part of October, 1841, and was kept alive for some time in a large willow basket with a flat cover of the same material. On opening the basket, he was almost invariably found suspended by his hind claws from the central part of the cover. When the basket was open, he manifested little fear, or disposition to fly, or get away, during the day time, but in the evening would readily mount on the wing and fly about the room, and on lighting always suspended himself by his hind claws with his head downward. He ate fearlessly and voraciously of fresh meat when offered to him, but could not be made to eat the common house fly.

CAROLINA BAT.
Vespertilio carolinensis.—GEOFFROY.

DESCRIPTION.—Ears rather large and naked, except on the back side near the

* Fauna Boreali Americana I. p. 1.
† Annals N. Y. Lyceum of Nat. His. Vol. IV. 54.

head, emarginate on the outer posterior edge, tragus shorter and less pointed than in Say's Bat. Head long and narrow; canine teeth very prominent; snout, interfemoral and wing membranes black and entirely naked; a few scattering hairs on the feet. Fur on the head and back long and color uniform bright ferruginous; beneath yellowish brown; last joint of the tail not enveloped in the membrane. Bones supporting the membrane very apparent. Length of the specimen before me, from the snout to the extremity of the tail 4 7 inches, head and body 3 inches, tail 1.7, fore arm 1.8, tibia .7, spread of the wings 11.5 inches.

HISTORY.—Of the history of this bat I know nothing. It is said to be quite common in the southern states particularly in the Carolinas and Georgia and also on Long Island near New York. The only specimen I have seen and that from which the above description was made, was taken in Burlington, and deposited in the museum of the college of Natural History of the University of Vermont by Mr. John H. Morse, a student of the University. A Vermont specimen of this species is also preserved in the museum of Nat. His. of Middlebury college.

SILVER-HAIRED BAT.
Vespertilio noctivagans.—LE CONTE.

DESCRIPTION.—Ears dusky black, rather large, naked on the anterior portion, somewhat ovate and obtuse, with two emarginations, on the outer posterior border, produced by two plaits; naked within, and with the tragus moderate, ovate and obtuse. Color above, a uniform dark dusky brown, approaching to black. On the back the fur is somewhat glossy and tipped with silvery white, forming an interrupted line across the shoulders, and thence irregularly mixed down the centre of the back. Interfemoral membrane thickly hairy on the upper part becoming thinner downward and naked near the border. Tip of the tail projecting about a line beyond the membrane. Feet hairy. Wing membrane entirely naked. Beneath very similar to the upper parts, though the light colored tips of the hairs are more yellowish. Total length 3.8 inches, tail 1.5, fore-arm 1.8, tibia .8, spread of the wings 11 inches.

HISTORY.—This Bat I have not seen in Vermont, but I am informed by my friend Prof. Adams that there is a specimen of it, which was taken in this state, in the museum of Natural History of Middlebury College. The above is Mr. Cooper's

*Annals N. Y. Lyceum Nat. His. Vol. IV. p. 9.

description of this Bat*, who says that "it was first described in 1831 by Major Le Conte and Dr. Harlan, and that it may be easily recognized by its dark black-brown fur tipped with white on the back." It was named *V. noctivagans* by Le Conte and *V. Audiboni,* by Harlan, and the former of these names is retained, because Le Conte's account was first published.

GENUS SOREX.—*Linnæus.*

Generic Characters.—Teeth variable from 26 to 34. The two middle upper incisors hooked and dentated at their base; the lower ones slanting and elongated; lateral incisors small, usually five on each side above, and two below; grinders, most commonly 4 on each side above, and 3 below. The body is covered with fine, short fur; toes, five on each foot, separate, furnished with hooked nails not proper for digging: head and nose elongated, the latter moveable; ears short and rounded; eyes small but visible.

FORSTER'S SHREW.
Sorex Forsteri.—RICHARDSON.

DESCRIPTION.—Color yellowish brown or dark olive above, bluish white or cinerous beneath; base of the fur plumbeous for two thirds its length both above and below; teeth white at the base and at their points, deep chestnut brown; tail long, four sided, covered with short hair and terminated in a fine pencil of hairs; feet small, light flesh-colored and nearly naked; nails slender and white; whiskers half an inch long, light brown. Length of the head and body 2 inches, tail 1.4, head .9, from the eye to the point of the nose .3.

HISTORY.—This little animal is occasionally met with in our pastures and fields, having their places of retreat in stone walls and under old fences and logs. The specimen from which the above description was made was taken in Bridgewater and is now in my possession. This shrew was first described by Dr. Richardson who says that it is common throughout the fur countries, even as far north as the 67° of latitude and that its delicate footsteps are often seen imprinted on the snow when the temperature is 40 or 50° below zero.* It is also found according to Dr. Bachman on Long Island in the vicinity of New York.†

*Fauna Boreali, vol. I. page 6.
†Journal Acad. Nat. Sci. of Phil. vol. VII. p. 386

THE SHORT-TAILED SHREW.
Sorex Brevicaudus.—SAY.

DESCRIPTION.—Color of the head, body and tail dark plumbeous brown above, a little lighter beneath ; lips naked fleshy and flesh-colored ; extremity of the snout brown, notched ; teeth tipped with dark chestnut brown at their points fading into white at their base ; feet flesh-colored, nearly naked and slender ; nails slender, white on the fore feet, and on the hind feet chestnut brown at the base and white at the tip. The inner toe on each foot is shortest, the outer a little longer and the other three nearly equal, the third being a little the longest. The tail is squarish, largest in the middle, slightly strangulated at the base and sparsely covered with short hairs ; whiskers whitish, sparse, half an inch long, situated between the eye and the snout and turned backwards. No external ear, opening large. Total length of the specimen before me 4-8 inches, to the origin of the tail 3-8, tail 1, head 1-1, hind foot to the point of the longest nail .6.

HISTORY.—This species of Shrew bears a very considerable resemblance to the Shrew mole in its general appearance, but is much inferior to it in size, and differs from it remarkably in the structure of its fore feet. As they seldom venture into cleared fields, very little is known of their habits, but in the woods they are often seen and heard rustling among the leaves and digging little holes into the ground, probably in quest of food. This and the preceding species are occasionally caught and brought in by cats ; but they will seldom attempt to eat them on account, probably, of their disagreeable musky odor. In addition to the foregoing we certainly have one other species, and probably more, but they require further examination.

GENUS SCALOPS.—*Cuvier.*

Generic Characters.—Teeth 36 to 44—Incisors $\frac{2}{4}$, canines $\frac{6-6}{3-3}$ or $\frac{6-6}{6-6}$, grinders $\frac{3-3}{3-3}$ or $\frac{4-4}{3-3}$, crowns of the grinders furnished with sharp tubercles ; nose long and pointed ; eyes very small ; no external ears ; fore feet very broad and strong, with long flattened nails fit for excavating the earth ; hind feet small and thin, with slender, arched nails ; tail short ; body thickly covered with fine, soft fur, which is perpendicular to the skin ; feet five toed.

THE SHREW MOLE.
Scalops aquaticus.—LINNÆUS.
Scalops canadensis.—Desmarest.

DESCRIPTION.—Color, grayish brown ; body, plump, cylindrical and tapering from the shoulders backward ; nose long, terminated by a button shaped cartilage ; eyes and ears concealed by the fur ; fore feet broad and strong, with the toes united up to the roots of the nails ; nails broad, flat and strong ; palms naked, bordered by small stiff hairs, above slightly covered with grayish down ; hind legs and feet slender and delicate, with slender, sharp, hooked nails ; tail short and covered with hair. Length of the specimen before me, from the nose to the insertion of the tail, 5.3 in. tail 1 in. head 1.3 in.

HISTORY.—The Shrew Mole inhabits fields and meadows, but seems to prefer the banks of rivers and other water courses. In its habits it resembles the other moles. Its large and powerful paws are well calculated for digging in the earth, and by their aid it is enabled to burrow with surprising quickness. They spend most of their time in the ground, where they form extensive and connected galleries, through which they can range at pleasure to considerable distances and in various directions, without coming to the surface. In excavating these galleries, they throw up, in a manner difficult to be explained, little mounds of loose earth, by which their burrows may be detected. These mounds occur at distances, from one to three feet, and are from three to six inches in height, but exhibit externally no appearance of passages into the burrows. The fur of this animal is exceedingly beautiful, being thick, fine, soft and even, with delicate glossy, or silvery reflections.

GENUS CONDYLURA.—*Illiger.*

Generic Characters.—Teeth, 40—Incisors $\frac{4}{4}$, canines $\frac{4-4}{5-5}$, grinders $\frac{4-4}{3-3}$— In the upper jaw are two large, triangular incisors, two very small ones, and on each side a large, strong canine. In the lower jaw the four incisors slant forward, and the canine on each side is small and pointed. Body cylindrical, clumsy, and covered with short thick fur, which is perpendicular to the skin ; nose elongated and sometimes furnished with a membranous crest disposed in the form of a star around the nostrils ; feet five-toed ; fore feet broad and strong, fitted for digging ; hind feet slender ; eyes very small ; no external ear.

THE STAR-NOSED MOLE.
Condylura macroura.—HARLAN.

DESCRIPTION.—Color dark brown approaching to black; body cylindrical; nose long, tapering and surrounded at the extremity by a fringed membrane, having twenty points; tail nearly as long as the body, strangulated at the base and then becoming suddenly enlarged as if swollen and thence tapering to a point. The tail is scaly and sparsely covered with stiff hairs. The fore legs very short; the paws large and naked, excepting the edges, which are fringed with stiff hairs; nails long and flat with cutting edges. The hind feet are naked, long and narrow, and the nails long, slender and sharp resembling birds claws; eyes concealed and very small; no external ear, 4 pectoral mammae; length from the nose to the insertion of the tail 4.7 inches, tail 2.8 inches, hand .7 inches, longest nail .3 inches, hind foot 1.

HISTORY.—This animal being rare, its habits are not well understood. They appear, however, from what is known of them, to be similar to those of the other moles. They are usually found about old buildings, fences and stone-walls, and they occasionally find their way into cellars of dwelling houses. I have two specimens of this animal, both of which were before me, while making out the foregoing description. The color of one is a little darker than the other, but they scarcely differ in any other respects. They were both caught in Burlington, one in 1830, in the cellar of the Rev. G. G. Ingersoll, and the other in 1840, on the surface of the ground in a door-yard. Their fore feet are so closely attached to their bodies, that they serve but little purpose except for digging, and their progress upon the surface of the ground, is extremely slow, labored and awkward. Like the shrew moles, they probably reside most of the time in the ground and venture abroad only in the night. On account of their clumsiness they are frequently drowned in cisterns and tubs of water and are sometimes brought in by cats; but cats are not fond of eating them on account of the musky odor which they have in common with the shrew and shrew mole. It proceeds, as in the other cases, from a white viscous fluid contained in a sack near the vent.

GENUS URSUS.—*Linnæus.*

Generic Characters.—Teeth, 32 to 44,—incisors $\frac{6}{6}$, canines $\frac{1-1}{1-1}$, grinders $\frac{4 \cdot 4}{4 \cdot 4}$ to $\frac{7 \cdot 7}{7 \cdot 7}$. Three of the grinders on each side in each jaw, are large, with square tuberculous crowns; the other are small, most of which appear late and are shed early. Body thick, covered with strong hair; ears long and slightly pointed; toes, five, furnished with strong, curved claws, calculated for climbing or burrowing; tail, short.

THE BLACK BEAR.
Ursus americanus.—PALLAS.

DESCRIPTION.—Color shining black; hair long and not curled; nose fawn colored, projecting, brightest about the angle of the mouth, and terminated by a naked black snout; forehead slightly arched; ears oval, rounded at the tip and far apart; palms and soles of the feet short in comparison with the brown bear; claws black and strong with the hairs of the feet projecting over them; tail short.

HISTORY.—The specimen from which our description is drawn was killed in Williston in 1838, and presented to the College of Natural History of the University of Vermont. It measures 6 feet from the nose to the tail; tail 2 inches; height of the ears 4 inches; height to the top of the shoulders 3 feet; rump 2 feet 4 inches. This Bear, which is found throughout all the woody parts of North America, was formerly very common in Vermont, and continues so plentiful at the present day, that our Legislature continue in force a law allowing a bounty of $5 each, for its destruction. It appears from our Treasurer's reports for several years past that the number of bears for which the bounty has been paid has varied from 40 to 50 annually. The black bear, under ordinary circumstances, is neither very carniverous nor very ferocious. Its favorite food consists of vegetables, such as Indian corn, nuts, berries and roots. But when these fail, it is compelled by necessity rather than choice to resort to animal food. In such cases, impelled by hunger, it will sometimes attack and destroy young cattle, sheep and hogs, but will seldom, if ever, attack a person except in defence of its cubs, or when provoked, or wounded. The early settlers of this State suffered most from them in consequence of their ravages upon their fields of Indian corn. They entered the fields in the night when the corn was in the milk and broke down and devoured the ears with great greediness; and it was a common business for the settlers to watch for them with guns and shoot them while committing their depredations; and in this way large numbers were annually killed. During the fall, when their food is abundant, bears usually become very fat, and, as the winter sets in, they retire to some natural den among the rocks, or uprooted trees, or into some hollow tree, where they remain in a torpid state and without food until the return of warm weather in the spring. The female produces her young during her hibernation and has from one to five at a litter, but the more common number is two. Their period of gestation is about 15 or 16 weeks, and during this time the females conceal themselves so effectually that we have no record of any being killed while pregnant though they are often discovered while the cubs are very small. When the bears first leave their winter quarters, they are said to be about as fat as when they retired in the fall, but with exercise they shortly lose their fat so as to appear in a few days much emaciated. When the bear is in high order he is valued for his flesh, his grease, and his skin. He is, with the exception of the moose, the largest native quadruped found in Vermont, and has been frequently killed weighing from 400 to 500 pounds. Their skins are worth from $2, to $4, or $5 according to their size and quality.

GENUS PROCYON.—*Storr.*

Generic Characters.—Teeth 40,—Incisors $\frac{6}{6}$, canines $\frac{1-1}{1-1}$, grinders $\frac{6-6}{6-6}$. The three first grinders on each side in each jaw, are pointed, the others are tuberculated. Body low set; nose pointed; external ears small, oval; tail long and pointed; feet five toed; nails sharp; mammæ six.

THE RACCOON.
Procyon lotor.—CUVIER.

DESCRIPTION.—General color blackish gray which results from the hairs being alternately ringed with black and dirty white; belly lighter; tail bushy, like that of the fox, but more tapering, surrounded by alternate rings of dark and yellowish white, about six of each; head roundish with the snout projecting beyond the upper jaw and terminating in a smooth black membrane through which the nostrils open; face whitish in front, with a black patch surrounding the eye and descending to the lower jaw, and a black line descending from the forehead between the eyes; pupils of the eyes round; the ears oval, rounded at the tip and the edges of a dirty white color; legs short; whiskers strong. Usual length of the head and body 22 inches, tail 9 inches; height 12 inches.

HISTORY.—Raccoons were very plenty in all parts of Vermont, when the country was new, and they exist in the mountainous and woody parts in considerable numbers at the present time. In the general aspect of this animal there is some resemblance of the fox, but in its movements it is more like the bear. It also like the bear subsists both upon animal and vegetable food and its destructive propensity is well known. It sleeps during the day in its nest in some hollow tree or among the rocks, and prowls for its prey during the night; and is said to destroy many more animals than it consumes, merely sucking their blood or eating their brain. It sometimes makes great havoc in the farmer's poultry-yard, and being an excellent climber scarcely any roost can be placed beyond his reach. But it probably does most mischief in the fields of Indian corn, of which it is extremely fond, while the corn is soft, or " in the milk." Here it breaks down and destroys much more than it eats. The Raccoon is said to be fond of dipping its food in water before it eats it, and hence, Linnæus gave it the specific name of *lotor*, which signifies *washer*. The price of the skin is variable, from 17 to 37½ cents. The largest of these animals in Vermont, weigh about 32 pounds, according to Dr. Williams, who says that its flesh is eaten and considered very excellent food.

GENUS GULO.—*Cuvier.*

Generic Characters.—Teeth 36 to 38—Incisors $\frac{6}{6}$, canines $\frac{1-1}{1-1}$, grinders $\frac{4-4}{6-6}$ or $\frac{5-5}{6-6}$. The three first grinders in the upper jaw, and four first in the lower are small, succeeded by a large carniverous or cutting tooth, and small tuberculous teeth further back. Body low; head moderately elongated; ears short and round; tail short; feet with five toes armed with crooked nails.

THE WOLVERENE.
Gulo luscus.—SABINE.

DESCRIPTION.—Head broad and rounded; jaws like the dog; ears low, rounded and much hidden by the fur: back arched; tail low and bushy; legs thick and short and the whole aspect of the animal indicates more strength than activity. Color dark brown, passing into almost black on the back in winter with a pale reddish brown band passing from each shoulder along the flanks and meeting on the rump. Fur similar to that of the bear, but not so long nor valuable. The tail is thickly covered with long black hair. Some white marking on the throat and between the fore legs; legs brownish black; claws strong and sharp. Length 2 feet 6 inches; tail (*vertebræ*) 7 inches; tail with the fur 10 inches.

HISTORY—This animal was occasionally found when the country was new, in all parts of the state, but was never very plentiful. For many years past, however, it has been known only in the most woody and unsettled districts, and in such places it is now extremely rare, none having been met with to my knowledge for several years. According to Dr. Richardson, from whose work the above description is abridged, this animal is quite common in the fur countries at the north, and is a great annoyance to the hunters, robbing their traps of game, or of the bait, which they do so dexterously as seldom to be caught themselves.* The Wolverene is represented as being very fierce and carniverous in its disposition, and many marvellous stories have been told of its cunning and artifice and gormandizing propensities, which are totally unfounded. Its food ordinarily consists of mice, moles, hares and other small animals, seldom meddling with larger ones, excepting such as have been previously killed or disabled It produces once a year from two to four cubs which are covered with a downy fur of a pale cream color. It is found throughout all the northern parts of North America, even as far north as the 75th degree of latitude.

GENUS MUSTELA.—*Linnæus.*

Generic Characters.—Teeth 34 or 38—Incisors $\frac{6}{6}$, canines $\frac{1-1}{1-1}$, grinders $\frac{4-4}{5-5}$, or $\frac{5-5}{6-6}$. Second inferior incisors on each side slightly receding; canines strong; grinders cutting; the anterior false grinders conical and compressed; true grinders trilobate, the last with a blunt crown. Body long and cylindrical; head small and oval; ears short and round; legs short; toes 5, armed with sharp, crooked claws, and glands producing a strong, fetid secretion.

THE WEASEL.
Mustela vulgaris.—LINNÆUS.
Putorius vulgaris.—Cuvier.

DESCRIPTION.—Color above, *in summer* dull yellowish brown deepening into hair brown on the upper part of the head and nose, and yellowish white beneath, the brown extending in a rounded spot into the white behind the angle of the mouth; tail next the body the same color as the back, but darker as it approaches the extremity, where it is quite black, and the hairs terminate in a point resembling that of a camel's hair pencil. Color *in winter* wholly white, excepting the posterior half of the tail, which is always black, or reddish brown. Forehead flatish; ears slightly pointed; eyes small, black and lively; body long and cylindrical; tail short, less than half the length of the body. Length of the head and body of the specimen before me 8 inches; tail (*vertebræ*) 2 inches.

HISTORY.—The Weasel, though nowhere greatly multiplied, is frequently met with in all parts of Vermont. It is generally seen in stone walls, old fences and heaps of bushes. When in sight it seems to be always in motion and its motions are very quick. When in a stone wall or heap of bushes he will sometimes show himself for an instant in half a dozen places in the course of half that number of minutes. The weasel feeds upon mice, young rats, young birds and birds eggs, and sometimes commits depredations upon the eggs and young of our domestic fowls. It is not uncommon for it to enter the barns and granaries and cellars of the farmers in quest of food, and particularly in pursuit of mice of which it destroys large numbers, and on which account it might be regarded as a public benefactor, were it not for its occasional depredations upon the poultry yard. The female produces her young several times in the course of the year and has from three to five at a litter. But notwithstanding their apparent fecundity, they never become very numerous.

---

*Fauna Boreali, I. 41.

THE ERMINE.
Mustela erminea.—LINN. GMEL.
Putorius erminea.—Cuvier.

DESCRIPTION.—Color, both in summer and winter, nearly the same as that of the Weasel, excepting that the upper parts of the Ermine are darker in summer and the under parts a clearer white than the same parts of the Weasel. The Ermine also grows to a larger size than the Weasel and is likewise more thick set, its forehead and nose more convex; its ears broader and more rounded, and its tail about twice as long in proportion to the length of the body. Length of the head and body of the specimen before me 8 inches; tail (vertebræ) 3.5. The tuft or pencil at the extremity extends about .7 inches beyond the vertebræ both in this and the Weasel.

HISTORY.—It has been a matter of dispute whether this and the preceding animal do or do not belong to the same species. Dr. Harlan describes them as two,* Dr. Godman, as one.† With these authorities before him, Dr. Richardson says that both these species are, indubitably, inhabitants of the American continent, the Ermine extending to the most remote arctic districts and the Weasel as far north, at least, as the Saskatchewan river.‡ Dr. Williams also describes the two as distinct species, and says that the Ermine, which he calls "one of the greatest beauties of nature" sometimes weighs 14 ounces, but that the Weasel is smaller.§ The skin of the Ermine, in its winter pelage of pure white, was formerly held in very high estimation, and was much worn by the nobility and high functionaries of Europe upon their robes and dresses, and particularly by judges. Thence it became the emblem of judicial purity, and the judge who was any way corrupted was said to have soiled his *Ermine*. The value of the skins at present is hardly sufficient to pay for collecting them. The Ermine in its summer dress is, in many places, called the Stoat.

* Fauna Americana p. 61. † Nat. His. I. p. 193.
‡ Fauna Boreali, I. p. 45. § His. Vt. I. p. 111.

THE MINK.
Mustela vison.—LINN. GMEL.
Putorius vison.—Cuvier.

DESCRIPTION.—The head is depressed and small; eyes small and far forward; ears low and rounded; neck and body long and slender; tail round and thick next the body and tapering towards the tip; legs short; toes connected by short hairy webs; claws nearly straight, sharp, white and concealed by the fur. The fur is of two sorts, a very dense down mixed with strong hairs; shortest on the head and increasing in length backwards; color of the down brownish gray; that of the hairs varying in different parts from chocolate brown to brownish black; occasional white spots about the throat; two oval glands which secrete a very fetid fluid. Length of the head and body 20 inches, tail 9 inches.

HISTORY.—The Mink is a common animal in Vermont. Its favorite haunts are along the banks of streams, where it dwells in holes near the water, or in the ruins of old walls, or in heaps of flood wood, or in piers and abutments of bridges. It does not venture far from the streams and when pursued betakes himself immediately to the water. It does not run well on land, but swims and dives admirably, and can remain a long time under water. When irritated it ejects a fluid, which diffuses a very unpleasant odor. Its fine short fur, Otter-like tail, short legs and webbed feet, all denote its aquatic habits. Its fur though not highly prized, is more valuable than that of the Musk rat. The food of the Mink consists of frogs, fishes, muscles and fish spawn; and also rats, mice, young birds and other small land animals. They sometimes enter the poultry yard, where they make great havoc among the fowls, by cutting off their heads and sucking their blood. It is not a very timid animal when in the water, but dives instantly at the flash of a gun, which makes it difficult to shoot them. It is easily tamed and in that state is very fond of being caressed, but, like the cat, is easily offended, and, on a sudden provocation, will sometimes bite its kindest benefactor. This animal is found throughout the United States and British America, but there has been some confusion with regard to its name. The Mink produces from three to six at a litter. When fully grown their weight is about four pounds. Mink skins are worth from 20 to 40 cents, according to quality.

THE FISHER MARTIN.
Mustela canadensis.—LINNÆUS.

DESCRIPTION.—Head, neck, shoulders and top of the back, mixed with gray and brown; nose, rump, tail and extremities, brownish black; sometimes a white spot under the throat, and also between the fore and hind legs; lower part of the fore legs, the fore feet and the whole of the hind legs, black; tail full, black, lustrous and tapering to a point; fur on the head short, but gradually increasing in length towards the tail; the head has a strong, roundish compact appearance; the ears are low semicircular and far apart, leaving a broad and slightly rounded forehead; fore legs short and strong; toes on all the feet connected at the base by a short web which is covered on both sides with hair. Length from the nose to the insertion of the tail, 23 inches; tail, including the fur, 16 inches.

HISTORY.—This animal is known in different places under a great variety of appellations, but in Vermont it is usually called the Fisher, or Fisher Martin. This name is, however, badly chosen, as it is calculated to deceive those unacquainted with the animal, with regard to its nature and habits. From its name the inexperienced would conclude that it led an aquatic mode of life, and that like the otter, it subsisted principally upon fishes But this is by means true; and they, who have had an opportunity to observe its habits, aver that it manifests as much repugnance to water as the domestic cat. It may, perhaps, sometimes devour fishes, which are thrown upon the shore, but it usually subsists by preying upon small quadrupeds, birds, eggs, frogs, &c. like the martin and other kindred species. It is said to kill the porcupine, by biting it on the belly, and then devour it. It lives in woods, preferring those which are low and damp. This animal is much valued for its fur, and considerable numbers are taken in the state, annually. The price of the skin varies from $1 to $2. It is sometimes called the *Pekan*, or the *Pekan Weasel*, or the *Fisher Weasel*.

THE PINE MARTIN.
Mustela martes.—LINNÆUS.

DESCRIPTION.—General color, fulvous brown, varying in different individuals, and at different seasons, from bright fulvous, to brownish black; bright yellow under the throat; hair of the tail longer, coarser and darker than that of the body; the color on all parts darker and more lustrous, and the fur more valuable in winter than in summer; nose and legs, at all seasons, dark, and the tip of the ears light. The fur of this animal is of two kinds, one coarse and the other fine and downy. The usual length of the head and body, 18 inches; tail, 9.

HISTORY.—In Vermont the name of Martin and Sable are indifferently applied to this animal, but the latter incorrectly, as the true sable is not found in this country. In works on natural history it is usually denominated the Pine Martin. This animal was formerly very plentiful in most parts of the state, but it is at present chiefly confined to the mountainous and woody portions. Though small it is much hunted for its fine and valuable fur, which, with the clearing and settling of the country, has very much reduced their numbers. Many are, however, still taken on the forest-clad mountains along the central part of the state. They are usually caught in traps baited with some kind of fresh meat. Their food consists of mice, hares, partridges, and other birds. They often rob birds nests of their eggs, or young, and will ascend trees for that purpose, or to escape pursuit. When its retreat is cut off, it will turn upon its assailant, arch its back, erect its hair and hiss and snarl like a cat. It will sometimes seize a dog by the nose and bite so hard, that, unless the latter is accustomed to the combat, it suffers the little animal to escape. It is sometimes tamed and will manifest considerable attachment to its master, but never becomes docile. Martins burrow in the ground. The female is smaller than the male. Her time of gestation is said to be only six weeks, and she brings forth from four to seven at a litter, about the last of April. A full grown martin weighs about four pounds. The price of prime skins is from $1, to $1.25.

Genus Mephitis.—*Cuvier.*

Generic Characters.—Teeth 34—incisors, $\frac{6}{6}$, canine $\frac{1-1}{1-1}$, grinders $\frac{4-4}{5-5}$; canines strong and conical; superior tuberculous teeth very large and as broad as they are long; the inferior grinders with two tubercles on the inside. Head short; nose projecting; feet five toed, hairs on the bottom, and furnished with nails suitable for digging; trunk of the tail of moderate length, or very short; hair of the body long, that of the tail very long; and glands, which secrete an excessively fetid liquor.

THE SKUNK.
Mephitis americana.—Desm.

Description.—General color black, with a white spot between the ears, which often extends along the sides towards the hips in the form of the letter V, and a narrow strip of white in the face; tail bushy, tipped with white; nails of the fore feet strong and about the length of the palm; hair on the head short, longer on the body and very long on the tail. Length from the nose to the insertion of the tail 16 inches, head $4\frac{1}{2}$ inches, body $11\frac{1}{2}$ inches, tail (trunk 10, tuft 4) 14 inches.

History.—The skunk is a very common animal in Vermont. It is not confined to the forests, nor to the thinly settled parts of the country, but frequently makes its residence in the midst of our villages. During the day he shelters himself in stone walls, or beneath barns, or out buildings, and prowls for his food during the night. This consists of eggs, young birds, mice and other small quadrupeds and reptiles. He frequently does considerable mischief in our poultry yards, by the destruction of eggs and fowls. What renders this animal most remarkable is its peculiar weapon of defence. When pursued, or attacked, it has the power of ejecting in the face of its enemy a fluid of the most nauseating and stifling scent, which exists in nature. This fluid is secreted by glands situated near the root of the tail, and seems to be designed wholly as a means of defence, being totally independent of the ordinary evacuations. When undisturbed the skunk has no disagreeable odor, and whole nests of them may lie under a barn floor for months, without betraying their presence by their scent. The flesh of the skunk when the odorous parts have been removed is well flavored and wholesome food.

Genus Lutra.—*Briss.*

Generic Characters.—Teeth 36—Incisors, $\frac{6}{6}$, canines $\frac{1-1}{1-1}$, grinders $\frac{5-5}{5-5}$; canines of moderate length and hooked; the first superior grinder small and blunt, the second and third cutting, the fourth with a strong spur on the inner side, the fifth with three external points and a broad internal spur: the inferior vary from five to six but resemble the superior. Head large and flattened; ears short and round; body very long, and low upon the legs; tail long, flattened horizontally and tapering; feet webbed; nails crooked and sharp; body covered with a fine fur mixed with long bristly hairs; two small oval glands secreting a fetid liquor.

THE AMERICAN OTTER.
Lutra brasiliensis.—Desm.

Description.—Color dark reddish glossy brown; pale or whitish about the throat and face; head globular; neck long; body long and cylindrical; tail depressed at the base; feet webbed, short and strong; 5 toes on the anterior feet, and 4 with the rudiment of a 5th on the posterior. Total length of one of the largest size, 4 feet; length of the head $4\frac{1}{2}$ inches, tail 17 inches, height 10 inches, circumference at the middle of the back 19 inches.

History.—The Otter lives in holes in the banks of creeks and rivers, and feeds principally upon fish, frogs and other small animals. They were formerly very common in this state, particularly along the streams which fall into lake Champlain and lake Memphremagog. *Otter Creek* derives its name from the great abundance of otter, which formerly inhabited its banks. They are now become scarce, but are occasionally taken at several places within the state.

The Otter is an active, strong and voracious animal. When attacked and unable to escape they fight with great fierceness, and when fully grown are more than a match for a common-sized dog. The teeth of the Otter are sharp and strong

THE COMMON WOLF.

and his bite very severe. His legs are very short and his feet webbed, on which account he seems to be better fitted for swimming than for running upon land; and he is so eminently aquatic in his habits that he is seldom seen at much distance from the water. This animal when fully grown measured according to Dr. Williams, 5 or six feet in length and weighed about 30 pounds, but the total length of those taken at present seldom exceeds 4 feet. The price of the skin is at present from 5 to 7 dollars, but it has been at times in such demand as to be worth 10 or 12 dollars.

Genus Canis.—*Linnæus.*

Generic Characters.—Teeth 42—Incisors, $\frac{6}{6}$, canine $\frac{1-1}{1-1}$, grinders $\frac{6-6}{7-7}$. The three first grinders in the upper jaw are small and edged, and are termed false molars, or grinders; the great carnivorous tooth above bicusped, with a small tubercle on the inner side, and two tuberculous teeth behind each of the carnivorous ones. Muzzle elongated, naked and rounded at the extremity; tongue smooth, ears pointed and erect in the wild species; fore feet with 5 toes and hind feet with 4, having robust nails.

THE COMMON WOLF.
Canis lupus.—Linnæus.

Description.—General color yellowish or reddish gray, blackish on the shoulders and rump, and yellowish white beneath, but varying much according to age and climate, being in some cases nearly black and in others almost white.* On the back and sides there is usually an intermixture of long black, and white hairs with a grayish wool, which partially appears, giving to those parts a grayish hue, which deepens along the back into black; hair on the back part of the cheeks, bushy; tail straight and bushy like that of the fox and nearly the color of the back; eyes oblique; ears erect; teeth very strong.

*Difference of colour has been the occasion of the division of this species into the following varieties:

Variety I. *Lupus griseus*, Common Gray Wolf.
" 2. *Lupus albus*, White Wolf.
" 3. *Lupus sticte*, Pied Wolf.
" 4. *Lupus nubilus*, Dusky Wolf.
" 5. *Lupus ater*, Black Wolf.

Length of the specimen in the collections of the College of Natural History of the Vermont University, from the nose to the tail 4 feet 3 inches, tail 17 inches; height at the shoulder 2 feet.*

History.—For some years after the settlement of this state was commenced, wolves were so numerous and made such havoc of the flocks of sheep, that the keeping of sheep was a very precarious business. At some seasons particularly in the winter they would prowl through the settlements by night in large companies, destroying whole flocks in their way, and, after merely drinking their blood and perhaps eating a small portion of the choicest and tenderest parts, would leave the carcases scattered about the enclosure and go in quest of new victims. Slaughter and destruction seemed their chief delight; and while marauding the country they kept up such horrid and prolonged howlings as were calculated, not only to thrill terror through their timorous victims, but to appall the hearts of the inhabitants of the neighborhood. Though the sheep seems to be their favorite victim, wolves sometimes destroy calves, dogs, and other domestic animals; and in the forest they prey upon deer, foxes, hares and such other animals as they can take. Impelled by hunger they have been known in this state to attack persons,* but they usually flee from the presence of man. The wolf bears a strong resemblance to our domestic dog; is equally prolific, and its time of gestation is said to be the same. It produces its young in the early part of summer, having from four to eight at a birth. Between the dog and the wolf prolific hybrids have often been produced, which however partake more of the nature of the wolf than of the dog.

Wolves have always been so great an annoyance that much pains have been taken for their extermination, but at present, their number is so much reduced that comparatively very little damage is done by them in this state. The legislature, however, continues in force a law, giving a bounty of $20 for the destruction of each grown wolf within the state, and $10 for each sucking whelp of a wolf; and the amount paid annually for wolf certificates is usually from one to two hundred dollars. The largest wolves killed in Vermont have weighed from 90 to 100 pounds. The only part of the wolf which is valuable is its skin, which affords a warm and durable fur.

* This specimen is distorted by too much stuffing. It was killed in Addison county about ten years ago.
*Williams Hist. I. 104.

THE RED FOX.

THE RED FOX.
Canis fulvus.—DESMAREST.

DESCRIPTION.—General color yellowish red, or straw yellow, less brilliant towards the tail; chin white; breast dark gray; belly whitish, tinged with red towards the tail; fronts of the legs and feet black; tail very bushy and less ferruginous than the body, the hairs being mostly terminated with black, giving it a dark appearance, with usually a few white hairs at the tip; eyes near to each other;—length of the head and body 28 inches; tail including the hair 16 inches; height of the shoulder 13 inches.

HISTORY.—The Fox has always been proverbial for slyness and cunning, and to illustrate these traits of character in the human species this animal has been largely taxed by fabulists, particularly by Æsop, who composed his fables 2400 years, ago. Foxes have their residence chiefly in holes, which they dig in the earth, or of which they get possession by ejecting the woodchuck from his. These burrows have two or more entrances and usually extend under ledges of rocks or roots of trees so that digging out the animal is often attended with considerable labor. Though sometimes seen skulking about in the day time, or basking in the sun, the Fox does not usually venture much abroad excepting in the night. He then prowls for his prey through the woods and fields and even among our out-buildings. His food consists of hares, rats, mice, small birds and poultry. He is said sometimes to feed upon frogs, snails and insects, and is fond of several kinds of berries and fruits. The fable of the fox and sour grapes, shows that the partiality of this animal for the fruit of the vine was understood in the days of Æsop. The Fox is a great annoyance in many parts of the state, sometimes destroying young lambs and often making great havoc among the poultry. A bounty of 25 cents each has been for several years paid for killing Foxes within the state; and the amount paid out of the treasury on this account has varied from $1000 to $2000 annually, showing that from 4000 to 8000 foxes have been annually destroyed. The law authorizing the bounty was repealed in 1841.

The red Fox is the common fox in Vermont, as well as in all the northern parts of the United States and Canada. Much doubt has existed with regard to the identity of this fox with the common fox of Europe, *Canis vulpes*, but it is at present regarded by the best naturalists as a distinct species. The particulars in which the two species differ are pointed out by Dr. Richardson in his Fauna Boreali Americana, Vol. I. p. 91. This fox is sometimes taken in traps, but he is so sly and suspicious that to trap for him successfully requires much skill. The best fox hunters attribute their success to the use of assafœtida or castoreum, with which they rub their traps, believing the foxes to be attracted by such perfumes. The fox is however more commonly taken in Vermont, by being shot under the pursuit of the hound. When the hound is put upon their track they do not retreat directly to their holes, nor lead off to any considerable distance in one direction, but take a circuit around the base of some hill which they will often encompass many times before they proceed to their burrows. The hunter, knowing this to be the habit of the fox, can judge of the course he will take and is enabled to place himself in a situation to shoot the animal as it passes. The skins of red foxes, if prime, are always valuable and the price for several years past has been from $1 to $1,25 and sometimes a little higher according to quality. The fox is a prolific animal. It produces its young usually in April and has from three to six at a litter.

THE AMERICAN CROSS FOX.
Canis fulvus.—Var. *decussatus.*

DESCRIPTION.—A blackish stripe passing from the neck down the back and another crossing it at right angles over the shoulders; sides ferruginous, running into gray on the back; the chin, legs and under parts of the body black, with a few hairs tipped with white; upper side of the tail gray; under side and parts of the body adjacent, pale yellow; tail tipped with white. The cross upon the shoulders is not always apparent even in specimens, which, from the fineness of the fur, are acknowledged to be Cross Foxes. Size the same as the common Fox.

HISTORY.—Instead of considering the Cross Fox a distinct species, as most American writers have done, I have concluded to adopt the opinion of Dr. Richardson, who regards it merely as a variety of the common fox. In form and size the Cross Fox agrees very nearly with the red fox, and differs from it chiefly in color, and perhaps a little in the fineness of its fur. The skin of the Cross Fox bears a much higher price than the red fox, which is owing almost entirely to the color. The price of a prime skin of this fox in Vermont is from $1,50 to $2,50.

THE BLACK, OR SILVER FOX.
Canis fulvus.—Var. *argentatus.*

DESCRIPTION.—Color sometimes entirely black and shining, with the exception of the tip of the tail, which is white; but more commonly hoary on some parts from an intermixture of hairs tipped with white; the nose, legs, sides of the neck, black, or nearly so; fur long and thick upon the body and tail, and short on the paws and face; soles of the feet covered with woolly fur. One of the largest of this variety measured from the nose to the insertion of the tail 31 inches, and the tail, including the hair, 18 inches.

HISTORY.—The Black or Silver Fox is regarded by Dr. Richardson as another variety of the common fox. It is much less common than the preceding variety and usually grows to a larger size. It has sometimes been taken in Vermont, but very seldom. Its fur is exceedingly valuable, prime skins being worth from $10 to $15 each.

There is another variety in Vermont, which is not uncommon, called the *Sampson Fox*. The fur is coarse resembling wool and of little value. The Gray Fox, *Canis virginianus*, is said to have been taken in this state, but as I have seen no Vermont specimen, it is here omitted. As we have before said, it is disputed whether our common red fox is, or is not identical with the common fox of Europe. Harlan, Godman, Richardson, and others, describe it as a distinct species. But Dr. McMurtrie, the translator of Cuvier's Animal Kingdom, says that the *Canis fulvus*, or American red fox, is identical with the European, and was introduced into the United States many years ago by some Englishmen, who thought they afforded better sport than the American species.*

GENUS FELIS.—*Linnæus.*

Generic Characters.—Teeth 30—Incisors $\frac{6}{6}$, canines $\frac{1-1}{1-1}$, grinders $\frac{4-4}{3-3}$. Inferior incisors forming a regular series; canines very strong; grinders, above, two conical ones on each side, one carnivorous one with three lobes and a small tuberculous one, below, two false compressed simple grinders and one carnivorous bicusped. Head round, jaws short, tongue aculeated; ears in general short and triangular; pupils of the eyes in some circular and in others vertically oval; fore feet with 5 toes, hind feet with 4, all furnished with long sharp retractile claws.

THE LYNX.
Felis canadensis.—LINNÆUS.

DESCRIPTION.—General aspect hoary, sometimes mottled; lighter and yellowish beneath, the extremity of the hairs being white, and below, yellowish brown; head rounded; ears erect, terminated with black pencils or tufts, 1½ inch long, black at the tip, with a black border on the posterior side. Anterior border yellowish. Base of the jaws surrounded by a fringe of long hair, intermixed with gray black and white; brownish around the mouth, white beneath; whiskers black and white; tail terminated with black; legs yellowish; toes 4 on each foot, much spread; nails sharp, white and concealed in long silky fur or hair. Total length 3 feet 4 inches; tail 5 inches. Height of the back 1 foot 4 inches; height of the ear 1¾ inches.

HISTORY.—The Lynx was never very greatly multiplied in Vermont, but when the country was new, it was frequently met with, and individuals have been taken occasionally, down to the present time. It resembles in fierceness and subtlety the other animals of the cat kind, preying upon hares, rabbits, mice and other small animals. Nor does it confine itself to small game, but sometimes destroys larger animals, such as deer, sheep, calves &c. This it is said to do by dropping upon them from branches of trees, clinging upon their necks with their sharp claws and opening their jugular veins and drinking their blood. Sheep and lambs have sometimes been destroyed by them in this state. This animal is found in large numbers in the vicinity of Hudson's Bay. Their skins are valuable and the Hudson Bay Company procure annually from seven to nine thousand of them. The flesh of the Lynx is used for food and is said to resemble that of the hare. It is a timid animal and makes but little defence when attacked. Its gait is by bounds but not swift. It swims well and will cross lakes 2 miles wide. It breeds once a year and has two young at a time.

---

* Cuvier's Animal Kingdom, Vol. 1, p. 433.

THE BAY LYNX.
Felis rufa.—GUILDENSTED.

DESCRIPTION.—Color yellowish, or reddish brown. Inferior parts of the throat white, or whitish. Eyes encircled with a whitish band. Front and portions about the upper lip striped with darkish; irides yellow. Ears short, tufted with black hair springing from the back of the ear, near the tip. Inside of the legs spotted with brown. Tail short, terminated with dark brown, and obscurely banded.—Fringe of hair longer than in other parts near the base of the jaw. Ears surrounded posteriorly with a black border, within which is a triangular patch of yellowish white. Length of the head and body, 2 ft. 3 inches; tail, 4 inches; height, 16 inches.

HISTORY.—This animal has been frequently met with in our woods, and has perhaps been most generally known by the name of *Wild Cat*. It is, however, to be distinguished from the smaller wild cats with long tails, which are met with, and which have probably sprung from the domestic cat. In its habits it resembles the preceding species, preying upon squirrels, birds, and other small animals. This animal is now very rare, being only occasionally seen, in the most unsettled parts of the State.

THE CATAMOUNT.
Felis concolor.—LINNÆUS.

DESCRIPTION.—General color, brownish red on the back, reddish gray on the sides, and whitish or light ash on the belly; tail, the same color as the back, excepting the extremity, which is brownish black, not tufted; chin, upper lip, and inside of the ears, yellowish white; the hairs on the back are short, thick, brownish, and tipped with red; on the sides and belly, longer, looser, lighter, and tipped with white; hairs of the face like the back, with whitish hairs intermingled, giving it a reddish gray tinge; body long, head round, jaws strong; teeth strong; canines conical; claws strong, retractile, and of a pearly white color. Dimensions of the specimen from which the above description is drawn—length from the nose to the tail, 4 ft. 8 inches; tail, 2 ft. 6 inches; from the top of the head to the point of the nose, 10 inches; width across the forehead, 8 inches; length of the fore legs, 1 ft. 2 inches; the hind legs, 1 ft. 4 inches.

HISTORY.—This ferocious American animal has been known in different places under a great variety of different names. In the southern and western parts of the United States it is called the Cougar, Painter, or American Lion; in New England it is known by the name of Catamount, or Panther; while in Europe it has more commonly borne the name of Puma. This is the largest and most formidable animal of the cat kind found in America. In form it bears considerable resemblance to the domestic cat, but when fully grown is about two-thirds the size of a lion. It, however, differs from the lion in not having the tail tufted, and the male being without a mane. These animals, though scattered over all the temperate and warmer parts of the continent, do not appear to have been any where very numerous. They were formerly much more common in Vermont than at the present day, and have at times done much injury by destroying sheep and young cattle. They usually take their prey, like the common cat, by creeping softly within proper distance, and then leaping upon it and seizing it by the throat. If the victim be a large animal, like a calf, sheep, or deer, they swing it upon their back, and dash off with great ease and celerity, into some retired place, where it is devoured at leisure. Some years ago one of these animals took a large calf out of a pen in Bennington, where the fence was four feet high, and carried it off on his back. With this load, he ascended a ledge of rocks, where one of the leaps was 15 feet in height.* During the day the Catamount usually lies concealed, but in the night prowls for his prey, and in early times his peculiar cry has often sent a thrill of horror through a whole neighborhood. When the country was new, much precaution was considered necessary, when travelling in the woods in this state, in order to be secure from the attacks of this ferocious beast.

---
* Williams' History, Vol. 1, p. 104.

Travellers usually went well armed, and at night built a large fire, which served to keep this cautious animal at a distance. Under such circumstances a catamount will sometimes approach within a few rods of the fire, and they have been thus shot in this state by aiming between the glaring eye-balls, when nothing else was visible. The Catamount will seldom attack a person in the day time, unless provoked or wounded. In the New York Museum is the skin of one of these animals, of which the following account is given in Dr. Godman's Natural History.* "Two hunters, accompanied by two dogs, went out in quest of game, near the Catskill mountains. At the foot of a large hill, they agreed to go round it in opposite directions, and when either discharged his rifle, the other was to hasten towards him to aid him in securing the game. Soon after parting, the report of a rifle was heard by one of them, who, hastening towards the spot, after some search, found nothing but the dog, dreadfully lacerated and dead. He now became much alarmed for the fate of his companion, and, while anxiously looking round, was horror struck by the harsh growl of a catamount, which he perceived on a large limb of a tree, crouching upon the body of his friend, and apparently meditating an attack on himself. Instantly he levelled his rifle at the beast, and was so fortunate as to wound it mortally, when it fell to the ground along with the body of his slaughtered companion. His dog then rushed upon the wounded catamount, which, with one blow of his paw, laid the poor creature dead by its side. The surviving hunter now left the spot, and quickly returned with several other persons, when they found the lifeless catamount extended near the dead bodies of the hunter and the faithful dogs." So recently as 1830, one of these animals sprang upon an unfortunate woman, as she was passing along a road in Pennsylvania, and killed her instantly.†

The weight of a full grown catamount is usually about 100 pounds. One of the largest taken in this State, to my knowledge, was killed in Roxbury, in December, 1821. It measured 7 feet from the nose to the extremity of the tail, and weighed 118 pounds. Under the name of panther, our legislature give a bounty of $20 each for the destruction of this animal within the state.

THE COMMON SEAL.
Phoca vitulina.—LINNÆUS.

But what! exclaims one, the Seal in Vermont—that inland mountain state?

* Vol. 1, p. 301. † Griff. Part V, p. 438.

Be not surprised, kind reader. It is even so, and there are living witnesses of the fact. While several persons were skating upon the ice on lake Champlain, a little south of Burlington, in February, 1810, they discovered a living seal in a wild state, which had found its way through a crack and was crawling upon the ice. They took off their skates, with which they attacked and killed it, and then drew it to the shore. It is said to have been 4½ feet long. It must have reached our lake by way of the St. Lawrence and Richelieu; but it was not ascertained whether the poor (fat) wanderer had lost his way, or having taken *a miff* at society, was seeking voluntary retirement from the world—*of seals.*

ORDER RODENTIA.—*Cuvier.*

This is the same as the order Glires of Linnæus, and embraces those animals, whose teeth are fitted for gnawing. They have two large incisors in each jaw, separated from the grinders by a vacant space. No canine teeth. The grinders in some of the genera have flat or ridged crowns, and in others blunt tubercles. Under jaw articulated by a longitudinal condyle; stomach simple; intestines long; cæcum large; mammæ variable in number. They feed generally on vegetables, but the species with tuberculated grinders are nearly omnivorous.

GENUS CASTOR.—*Linnæus.*

Generic Characters.—Teeth, 20—incisors $\frac{2}{2}$, no canines, grinders, $\frac{4 \cdot 4}{4 \cdot 4}$. Incisors, very strong, smooth on the outside, and angular within; grinders have a fold on the internal edge, and three similar folds on the outer edge of the upper teeth, which are inverted in the lower ones. Eyes, small; ears, short and round; feet, five toed; fore feet short; hind feet longer and palmated; tail, large, flat, and scaly; a pouch near the root of the tail in the male filled with an unctuous, odoriferous secretion.

THE BEAVER.
Castor fiber.—LINNÆUS.

DESCRIPTION.—Fur dense, consisting of two sorts, one coarse, long, and of a chestnut, or reddish brown color, the other shorter, very fine and of smoky or silvery gray; head flattened; nose short and thick; eyes small; ears short, thick, rounded and covered with short fur; neck short; body thick; back arched; tail flat and broad horizontally, oval and covered with oval angular scales; fore legs very short and small; and the fore feet are used as hands for conveying food to the mouth; hind feet with long, hard and callous soles, and long toes connected by a web. The usual length of the beaver from the nose to the origin of the tail, is from 30 to 40 inches, and the tail about 11 inches long and 6 broad at the widest part. The usual weight of a full grown Beaver is stated by Dr. Richardson to be about twenty-four pounds.

HISTORY.—The beaver, though formerly a very common animal in Vermont, is probably now nearly or quite exterminated, none of them having been killed within the state, to my knowledge, for several years. The last, of which I have any account, was killed, in Essex county, 12 years ago.* The vestiges of its labors are, however, still found in "the beaver meadows" in all parts of the country. The peculiarities in the form of the beaver, and especially the remarkable instinct, which guides him in the construction of his dwelling, have always rendered him an object of admiration, and many accounts of him have been published, most of which abounded in exaggeration and fable. The following account by Hearne, who studied the habits of this animal for 20 years, in the fur countries around Hudson's Bay, is pronounced by Dr. Richardson,* who, himself, had the best opportunity for ascertaining its truth, to be the most correct and free from exaggeration, which has ever been published.

"Where beavers are numerous, they construct their habitations upon the banks of lakes, ponds, rivers, and small streams; but when they are at liberty to choose, they always select places where there is sufficient current to facilitate the transportation of wood and other necessaries to their dwellings, and where the water is so deep as not to be frozen to the bottom during the winter. The beavers that build their houses in small rivers and creeks, in which water is liable to be drained off, when the back supplies are dried up by the frost, are wonderfully taught by instinct, to provide against that evil, by making a dam quite across the stream, at a convenient distance from their houses. The beaver dams differ in shape, according to the nature of the place in which they are built. If the water in the stream have but little motion, the dam is almost straight; but when the current is more rapid, it is always made with a considerable curve convex towards the stream. The materials made use of, are drift-wood, green willows, birch and poplars, if they can be got; also mud and stones, intermixed in such a manner, as must evidently contribute to the strength of the dam; but there is no order or method observed in the dams except that of the work being carried on with a regular sweep, and all the parts being made of equal strength. In places which have been long frequented by beavers undisturbed, their dams, by frequent repairing, become a solid bank, capable of resisting a great force both of water and ice; and as the willow, poplar and birch, generally take root and shoot up, they by degrees form a kind of regular planted hedge, which I have seen in some places so tall that birds have built their nests among the branches.

"The beaver-houses are built of the same materials as their dams, and are always proportioned in size to the number of inhabitants, which seldom exceeds four old and six or eight young ones; though, by chance, I have seen above double that number. Instead of order or regulation being observed in rearing their houses, they are of much ruder structure than their dams; for, notwithstanding the sagacity of these animals, it has never been observed that they aim at any other convenience in their houses, than to have a dry place to lie on; and there they usually eat their victuals, which they occasionally take out of the water. It frequently happens that some of the large houses are found to have one or more partitions, if they deserve the appellation; but it is no more than a part of the main building, left by the sagacity of the beaver to support the roof. On such occasions, it is common for those different apartments, as some are pleased to call them, to have no communication with each other but by water; so that, in fact, they may be called double or treble houses, rather than different apartments of the same house. I have seen a beaver-house built in a small island, that had near a dozen different apartments under one roof; and, two or three of these only excepted, none of them had any communication with each other but by water. As there were beavers enough to inhabit each apartment, it is more than probable that each family knew their own, and al-

---

* Letter of the Hon. J. Parker, of Orleans, to the Author, Sept. 27, 1841.

† Fauna Boreali Americana, Part 1. page 108.

ways entered at their own doors, without any further connection with their neighbors than a friendly intercourse, and to join their united labors in erecting their separate habitations, and building their dams where required. Travellers, who assert that beavers have two doors to their houses, one on the landside, and the other next the water, seem to be less acquainted with these animals than others, who assign them an elegant suite of apartments. Such a construction would render their houses of no use, either to protect them from the attacks of their enemies, or guard them against extreme cold weather.

"So far are beavers from driving stakes into the ground, when building their houses, that they lay most of the wood crosswise, and nearly horizontal, and without any other order than that of leaving a hollow, or cavity in the middle; when any unnecessary branches project inward, they cut them off with their teeth, and throw them in among the rest, to prevent the mud from falling through the roof. It is a mistaken notion, that the wood work is first completed and then plastered; for the whole of their houses as well as their dams, are, from the foundation, one mass of mud and wood, mixed with stones, if they can be procured. The mud is always taken from the edge of the bank, or the bottom of the creek or pond, near the door of the house; and, though their fore paws are small, yet it is held so close up between them under their throat, that they carry both mud and stones, while they always drag the wood with their teeth. All their work is executed in the night; and they are so expeditious, that in the course of one night I have known them to have collected as much mud as amounted to some thousands of their little handfuls. It is the great policy in these animals to cover the outside of their houses every fall with fresh mud, and as late as possible in the autumn, even when the frosts become pretty severe, as by this means it soon freezes as hard as a stone, and prevents their common enemy, the wolverene, from disturbing them during the winter. And as they are frequently seen to walk over their work, and sometimes to give a flap with their tail, particularly when plunging into the water, this without doubt, has given rise to the vulgar opinion that they use their tails as a trowel, with which they plaster their houses; whereas that flapping of the tail is no more than a custom, which they always preserve, even when they become tame and domestic, and more particularly so when they are startled."

Judge Parker, who has devoted considerable attention to the habits of our native quadrupeds, after confirming the above statement of Hearne, in relation to the structure of the dams and houses of the beaver, observes: "I have thought the correct judgment exercised by the beaver in the selection of the place for his dam, to be the most remarkable part of his character. The choice seems to be made with reference to the plenty of timber suitable for his food, and the proportion, which the space to be overflowed bears to the length of the dam; and with regard to these, they seem to judge as correctly as man. So far as they have fallen under my own observation, I have always found them at the very best places, which could be selected on the whole stream. One chief object of their pond seems to be, to float timber, which is to serve them for food, to their dwellings; and where the water does not prove deep enough for that purpose, they deepen it by digging a trench along the bottom, and cutting off the logs which lie in their way, with their teeth. I have seen logs 20 inches in diameter, which had been thus cut off and removed."*

Their food during the winter consists principally of the root of the pond lily, *Nuphar luteum*, which they find in the water beneath the ice. They also feed upon the bark of the poplar, birch and willow, which they cut down in the fall and drag into the water opposite the doors of their houses, as a part of their supply for the winter. In the summer they rove about, feeding upon different kinds of herbage and berries, and do not return to repair their houses and lay in their winter stock of wood till towards fall. When they are to erect a new habitation, they fell the timber for it in the spring, but do not begin to build till August, and never complete it till cold weather sets in.

The beaver is a cleanly animal, never allowing any excrement or filth within its lodge. They are said to pair in February and bring forth their young in the latter part of May, producing from four to eight at a litter. Beavers seldom cut down trees which exceed 5 or 6 inches in diameter, and they always leave the top of the stump in the form of a cone. They gnaw all round the tree, but direct its fall by cutting one side higher than the other. The weight of a full grown beaver does not often exceed 30 pounds, though, according to Dr. Williams, they have taken in Vermont weighing from 40 to 60 pounds.†

---

* Letter to the Author.
† His. of Vermont, Vol. I. p. 121.

GENUS FIBER.—*Cuvier.*

Generic Characters.—Teeth, 16—Incisors $\frac{2}{2}$, no canines, grinders $\frac{3-3}{3-3}$. Lower incisors sharp pointed and convex in front; grinders with flat crowns, furnished with scaly, transverse zigzag laminæ; four toes, with the rudiments of a fifth, on the fore feet; five toes on the hind feet, having the edges furnished with stiff hairs, used in swimming, like the membrane of palmated feet; tail long, compressed laterally; both sexes secrete an odoriferous, musky unguent.

THE MUSK RAT.
Fiber zibethicus.—DESM.

DESCRIPTION.—General color, yellowish, or reddish brown, lighter beneath; body thick and flattish, with a short head and indistinct neck; incisory teeth very large; lips covered with coarse hair; nose short; eyes small and lateral, and partly concealed by the hair; ears low, oblong, covered with hair and inconspicuous; tail nearly as long as the body, flattened laterally, and covered with small brown scales, interspersed with short black hairs; legs and feet covered with short, brown shining hair; toes 5 on each foot; thumbs very small; claws strong and sharp; a brown spot beneath the tip of the under jaw. Length of the specimen before me, from the nose to the origin of the tail 13 inches; tail 9½ inches; weight 3½ pounds.

HISTORY.—Musk Rats, or Musquashes, as they are often called, have a strong smell of musk, particularly the males. Their fur is used in the manufacture of hats, and great numbers of their skins are shipped to Europe. Dr. Richardson imforms us that from four to five hundred thousand are annually imported from North America into Great Britain. Musk Rats were very numerous in Vermont when the country was new, and their skins afforded to the early settlers an important article of export. Although now much diminished, they are still found in considerable numbers, inhabiting the banks of our larger streams.

In its aquatic and nocturnal habits, as well as in its appearance and the mode of constructing its dwelling, the Musk-rat is closely allied to the beaver. Like the beaver he is an excellent swimmer, dives well and remains for a considerable time under water. It is only in low swampy situations that the Musk-rat resorts to the construction of habitations above ground.

These are made principally of mud mixed with grass, and in the form of a dome, with a warm bed of leaves and grass within. The only place of entrance is from beneath, and from this there are usually several subterranean passages leading in different directions. When ice forms over the surface of the swamp, they make breathing holes through it, which they sometimes protect from frost by a covering of mud. When disturbed in their dwellings, the Musk-rats retreat through their subterranean passages. They feed principally upon the roots and bark of aquatic plants, but do not, like the beaver, lay in a store of provisions for the winter.

During the winter several families of Musk-rats usually reside together. But when warm weather approaches, they desert their house, and during the summer live in pairs and rear their young, of which they have from three to six at a litter. They are very watchful and shy, seldom venturing abroad during the day time, and hence they are very seldom seen, even in nighborhoods where they are known to abound. They run badly upon the land, but swim with facility and dive instantly on perceiving the flash of a gun, usually giving a smart blow upon the water, with the tail, in the act of diving. They are usually taken in steel-traps. The skins are of little value, seldom bringing more than 17 cts. and often less than 10 cents.

GENUS ARVICOLA.

Generic Characters.—Teeth 16—Incisors $\frac{2}{2}$, no canines, grinders $\frac{3-3}{3-3}$. The grinders are flat on the crowns, and marked with zigzag lines of enamel. Four toes and the rudiments of a fifth on the fore feet; on the hind feet five toes; toes furnished with weak nails, but neither palmated nor furnished with hairs on their borders; ears large; tail round, hairy, and nearly as long as the body.

THE MEADOW MOUSE.
Arvicola riparius.—ORD.

DESCRIPTION.—General color above grayish brown, resulting from the fur, being plumbeous at the base, and tipped with gray and reddish brown; beneath light yellowish lead color; head rather large; ears broad, short, and slightly covered with hair on both sides towards the margin, opening large and apparent; eyes moderately large, black and unconcealed; whiskers few and blackish; tail short and sparsely covered with short stiff hairs; legs and feet slender; toes, four, with a rudiment of a fifth on the fore feet, the second toe longest and the outer shortest; five toes behind, the three middle ones nearly equal. Length of the specimen from which the above description was made, 5 inches; tail 2 inches.

HISTORY.—We have doubtless as many as two or three species belonging to this genus, but they have not been sufficiently examined to enable me to speak with confidence respecting them. Meadow mice are quite common in most parts of the state, and at times they become so greatly multiplied as to do much injury to the meadows and to the stacks of hay and grain. They have their burrows in the banks of streams, and under old stumps, logs and fences; and in neighborhoods where they are plenty, numerous furrows may be seen along the roots of the grass, forming lanes in which they may travel in various directions from their burrows. Their nests are sometimes constructed in their burrows, and are also found at the season of hay harvest, in great numbers, among the vegetation upon the surface of the ground. They are built of coarse straw, lined with fine soft leaves, somewhat in the manner of a bird's nest, with this difference, that they are covered at the top, and the passage into them is from beneath. These nests frequently contain 6 or 8 young ones. The meadow mice, though very prolific, have many enemies which serve in a measure to check their undue multiplication. Large numbers of them are destroyed by owls, hawks, foxes, cats, &c., and the country people, when at labor in the field, are vigilant in putting them to death.

GENUS MUS.—*Linnæus*.

Generic Characters.—Teeth 16—Incisors $\frac{2}{2}$, no canines, grinders $\frac{3-3}{3-3}$. The grinders are furnished with blunt tubercles. Destitute of cheek pouches; fore feet with four toes, and a wart in the place of a thumb, covered with an obtuse nail; hind feet with five toes; nails long, sharp, and incurved; tail long, tapering, naked, and scaly; some part of the hair of the body longer and stiffer than the rest; ears oblong, or round.

THE NORWAY RAT.
Mus decumanus.—PALL.

DESCRIPTION.—General color, light reddish brown intermingled with ash, lighter and grayish beneath; feet pale flesh colored, and nearly naked; tail nearly as long as the body, covered with small dusky scales, with short stiff hairs thinly scattered among them; four toes and a small tubercle in place of a thumb before, five behind; nails small, light horn color, and slightly curved; whiskers of unequal length, partly black and partly white. Total length of the specimen before me, which is a female, from the snout to the tip of the tail, 16 inches; head 1.8; body 7.5; tail 6.7. Six pectoral and six ventral mammæ.

HISTORY.—This rat, which is at present the common rat of the United States, is supposed to have been originally a native of Persia, or India, and was first known in Europe in the early part of the 18th century. It was carried to England, about the year 1750, in the timber ships from Norway, and from this circumstance it received the name of *Norway Rat.* From Europe it was brought over to America, about the commencement of the American Revolution, and is now diffused over the greater part of the continent. The Norway, or, as often called, the Brown rat is very prolific, bringing forth from 10 to 16 at a litter, and but for its numerous enemies, and its own rapacious disposition, it would soon become an intolerable pest. Happily, however, for man, they are not only destroyed by weasels, cats, and dogs, but they are very destructive enemies to one another, both in the young and adult state. They are sometimes caught in traps, but on account of their caution and cunning it requires much art. The surest way of destroying them is by poison, and arsenic is commonly used for that purpose, but so many fatal accidents occur from having this poison about our buildings, that its use is not to be recommended. If poison is to be used for the destruction of rats, the powder of *nux vomica*, mixed with meal and scented with oil of rhodium, should be employed, and it is found very effectual for that purpose. The brown rat is a deadly enemy to the black rat, and destroys it, or drives it from the neighborhood. It also destroys mice. But it does not confine itself to the destruction of noxious animals. It often devours eggs, chickens, and the young of other domestic fowls. It however becomes the greatest nuisance and does most mischief by the destruction of grain, fruit, roots &c. in our granaries and cellars. The graphic character given it by Dr. Godman will not be disputed by any who are acquainted with its habits. "It must be confessed," says the Doctor, "that this rat is one of the veriest scoundrels in the brute creation, though it is a misfortune in him rather than a fault, since he acts solely in obedience to the impulses of nature, is guided by no other law than his own will, and submits to no restraints, but such as are imposed by force. He is, therefore, by no means as bad as the scoundrels of a higher order of beings, who, endowed with superior powers of intelligence, and enjoying the advantages of education, do still act as if they possessed all the villainous qualities of the rat, without being able to offer a similar apology for their conduct. Among quadrupeds this rat may be considered as occupying the same rank as the crow does among birds. He is one of the most impudent, troublesome, mischievous, wicked wretches that ever infested the habitations of man. To the most wily cunning he adds a fierceness and malignity of disposition that frequently renders him a dangerous enemy, and a destroyer of every living creature he can master. He is a pure thief, stealing not only articles of food, for which his hunger would be a sufficient justification, but substances which can be of no possible utility to him. When he gains access to a library he does not hesitate to *translate* and appropriate to his own use the works of the most learned authors, and is not so readily detected as some of his brother pirates of the human kind, since he does not carry off his prize entire, but cuts it into pieces before he conveys it to his den. He is, in short, possessed of no one quality to save him from being universally despised, and his character inspires no stronger feeling than contempt, even in those who are under the necessity of putting him to death."*

THE BLACK RAT.
Mus rattus.—LINN.

DESCRIPTION.—Head elongated; snout pointed; lower jaw very short; eyes large and projecting; ears naked, large, broad and nearly ovate; whiskers long; five flat toes on the hind feet, and on the fore feet four, with a nail representing a thumb; lateral nails, both behind and before, very short; tail nearly naked, and furnished with scales disposed in rings, amounting in some cases to 250; color cinerous black, lighter beneath; whiskers black; top of the feet covered with small white hairs; mammæ 12. Length of the head and body 7 inches, tail 7.5 inches.

HISTORY.—It seems to be a matter of some doubt whether this Rat is indigenous in this country or was introduced from Europe. But whethe introduced, or indigneous, it is certain that they were very numerous here before the introduction of the preceding species. It is stated by Dr. Williams† that neither the Norway rat, nor the Black rat, was known in Vermont till some time after the settlement of the state was commenced, but that, when he wrote, they had become quite common. The Norway or Brown rat is now the common rat in all the older parts of the state; and yet it is but a few years since it was said that none of these rats had ever been seen in the county of Orleans.

THE COMMON MOUSE.
Mus musculus.—LINN.

DESCRIPTION.—Color, dusky gray above and ash gray beneath; forehead, reddish; whiskers, slender, numerous and black; feet, white; nails, reddish with white points; tail, round, sparsely covered with very short hairs, and tapering from the insertion to the extremity; ears large. Total length about seven inches, of which the tail constitutes one half. A variety of this mouse which is wholly white is frequently met with in the neighborhood of lake Champlain, on both sides of the lake, and another variety, less common, is white spotted with black.

HISTORY.—This mischievous little creature, like the preceding, did not exist in North America at the time of the discovery of this continent by the Europeans, but finding its way over in ships, in bales of merchandize, &c., by its great fecundity it filled the country with a rapidity equal to the advancement of the new settlement, and is now very common throughout all the settled parts of the continent. This mouse takes up his residence chiefly in houses, barns and granaries, where he is often exceedingly troublesome, and does much mischief. He is very apt to find his way into cellars and pantries, often by gnawing holes through boards, and he is sure to nibble every kind of eatable that falls in his way. On this account, and on account of the peculiar odor which he communicates to the places which he frequents, the mouse, though a beautiful and sprightly creature, is every where regarded with disgust. The mouse builds its nest very much like that of a bird, lining the inside with wool, cotton or other soft materials. It brings forth young several times during the year, and has from 6 to 10 at a litter, so that its multiplication, when unchecked, is exceedingly rapid. Aristotle, in his history of animals, mentions that a pregnant female of this species was shut up in a chest of grain, and in a short time 120 individuals were counted, from which it would appear that the mouse was as much distinguished on account of its fecundity 2000 years ago as it is at present.

---
* Natural History Vol. 2.—page 78.
† History of Vermont, Vol. 1, p. 113.

GENUS GERBILLUS.—*Desmarest.*

Generic Characters.—Teeth, 16—Incisors $\frac{2}{2}$, no canines, grinders $\frac{3-3}{3-3}$. The grinders are tuberculous: the first with three, the second with two and the third with one tubercle. Head elongated; ears moderately long, rounded at the extremity; fore feet short with four toes and a rudimentary thumb; hind feet long, having five toes with nails; each foot with a proper metatarsal bone; tail long, and more or less hairy.

THE JUMPING MOUSE.
Gerbillus canadensis.—DESM.

DESCRIPTION.—General color, yellowish brown above, grayish yellow on the sides, and yellowish white on the belly; tail tapering, longer than the body, sparsely covered with very short hair, and the tuft at the end very small; head small, narrow and pointed; fore legs very short; hind legs very long; nails slender and sharp; ears moderate and covered on both sides with short hair; upper incisors grooved on the outside. Length of the specimen before me, from the nose to the insertion of the tail 4 inches, head 1 inch, body 3 inches, tail 5 inches, hind leg 2 inches, fore leg $\frac{3}{4}$ of an inch.

HISTORY.—This timid and active little animal is frequently met with in the grain fields and meadows in all parts of the state. When not in motion it might be mistaken for a common field mouse; but its usual method of progression is very different. It sometimes runs on all its feet, but it more commonly moves by leaps on its hind legs, particularly when pursued. It will often clear five or six feet at a leap, and its leaps are made in such quick succession that it is not easily caught. On examination, it is found to differ considerably in form from the mouse, particularly in the great disproportion between the fore and hind legs, the latter being more than twice the length of the former. In this respect it resembles the kangaroo of Australasia, and the jerboa of the eastern continent. They pass the winter in a torpid state and are not usually out in the spring before June.

GENUS ARCTOMYS.—*Geoffroy.*

Generic Characters.—Teeth 22—Incisors $\frac{2}{2}$, no canines, grinders $\frac{5-5}{4-4}$. The incisors are very strong with the anterior surface rounded; grinders furnished with ridges and tubercles. Body thick and heavy; head and eyes large; ears short; paws strong; fore feet with four toes and a rudimentary thumb; hind feet with five toes; nails strong and compressed; tail generally short, hairy.

THE WOODCHUCK.
Arctomys monax.—GMELIN.

DESCRIPTION.—General color, grayish ferruginous brown, paler beneath and approaching to red between the legs; top of the head and nose brown; feet and nails black; whiskers black and stiff, standing in three clusters on each side; tail covered with long reddish brown hair. Length of the specimen before me from the nose to the insertion of the tail $16\frac{1}{2}$ inches; head $3\frac{1}{2}$ inches, body 13 inches, trunk of the tail 5 inches, with the hair extending $1\frac{1}{2}$ inch beyond, fore legs 4 inches, feet $2\frac{1}{2}$ inches; longest nail .6 inch; hind legs $4\frac{1}{2}$ inches; feet 3 inches; largest nail .4 inch. Weight 5 lbs. This though an adult is not one of the largest size.

HISTORY.—The Woodchuck is a common and well known animal in all parts of the state. They are found both in the woods and open fields, where they reside in pairs or families, in holes which they dig in the ground. These holes are usually made beneath a large rock, or stump, or in the side of some dry bank, and are sometimes very extensive, consisting of several apartments with several openings. In these recesses they form their nests of dry leaves and grass in which they spend much of their time in sleep. Their food is entirely vegetable, of which they eat various kinds. They are particularly fond of clover and beans, and are occasionally injurious to the farmers by the extent of their depredations. When feeding they frequently rise upon their haunches to reconoitre, raising their fore feet like hands. In this position, when the weather is fine, they will sometimes sit for hours at the entrance of the holes, but they seldom venture far abroad in the day time. On the approach of cold weather they confine themselves to their holes by closing the passage between themselves and the surface of the ground and spend the winter, like bears, in a torpid state.

The Woodchuck is a cleanly animal, is capable of being tamed, in which state it becomes playful and fond of attention. It is a low-set, clumsy animal, and when the retreat to his hole is cut off, he will boldly face a dog in battle, and is fully a match for one of his own size. His bite, with his long and projecting incisors, is very severe. The female produces from four to six at a litter. The weight of a Woodchuck of the largest size in Vermont when fat is 10 or 11 pounds. Its flesh is sometimes eaten, but is not much esteemed. Sometimes called Ground Hog.

GENUS SCIURUS.—*Linnæus.*

Generic Characters.—Teeth 22—Incisors $\frac{2}{2}$, no canines, grinders $\frac{5\cdot 5}{4\cdot 4}$. The upper incisors are flat in front and wedge-shape at the extremity, the lower are pointed and compressed laterally. The grinders are tubercular. Body small and elongated: head small; ears erect; eyes large; fore feet with four toes and a tubercle instead of a thumb; hind feet with five long toes, all furnished with long hooked nails; tail long and frequently shaggy; two pectoral and six ventral mammæ.

THE GRAY SQUIRREL.
Sciurus cinereus.—GMELIN.

DESCRIPTION.—General color, gray above and white beneath; sides of the head and body, and the exterior of the legs, reddish fawn mixed with gray; inside of the legs and thighs bluish white; tail large and bushy, composed of hairs marked with zones alternately fawn and black, and tipped with white; ears without pencils, rounded and covered with very short hair; whiskers black, 2½ inches long. Length of the specimen before me, from the nose to the insertion of the tail, 10 inches; tail, (trunk 9½, tuft 2,) 11½ inches. Weight 1¼ pound.

HISTORY.—According to Dr. Williams, the Gray Squirrel was formerly the most common squirrel in Vermont. It is still found in considerable numbers but less plentifully at present than some of the smaller species. This as well as some of the other species, in some years, becomes exceedingly multiplied, and then, perhaps, for several years very few of them will be seen. This sudden increase and diminution of their numbers, seems to depend upon two causes, the supply of food and the severity of the winters. Their great multiplication generally follows a mild winter, which was preceded by a productive summer. I believe it to be generally true that when one species becomes very plentiful, the others become so too. The Gray Squirrel prefers woods, which abound in oak, walnut, butternut and chestnut, because these furnish him with such food as he prefers. During the fall they collect a supply of food for the winter, which they carefully deposit in hollow trees or obscure recesses. Their nests which are built with sticks and lined with leaves, are usually placed in the forks of large and lofty trees, or in the hollows of old trees, and in these they spend most of their time during the winter, leaving them only to visit their depositories of food for the purpose of obtaining a supply. This is one of the most active and beautiful of our squirrels. It is easily tamed, and, in captivity, is remarkably playful, but rather disposed to be mischievous, often using its teeth to the injury of the furniture. About a century ago these squirrels were so troublesome in Pennsylvania that government granted a premium of 3*d* a head for their destruction, which in 1749, amounted to £8,000 sterling; from which it would appear the number killed in one year was about 1,280,000.

THE BLACK SQUIRREL.
Sciurus niger.—LINNÆUS.

DESCRIPTION.—Top of the head, back, tail and extremities of the feet, covered with hair of a deep black color; throat, breast and belly brownish black, lighter on the flanks; ears short, black, and not pencilled; smaller and the tail proportionally shorter, and the fur softer than in the preceeding species. Length of the head and body about 8 inches.

HISTORY.—The Black Squirrel is much less common in Vermont than the gray squirrel, particularly in the western parts, and is perhaps, frequently confounded with a blackish variety of the gray squirrel. Having obtained no specimen of this squirrel, I have copied, above, the description contained in Dr. Harlan's Fauna Americana. According to Dr. Williams our largest black squirrels weigh but 2½ lbs., while our largest gray squirrels weigh 3½ lbs.

THE RED SQUIRREL.
Sciurus Hudsonius.—GMEL.

DESCRIPTION.—Color, reddish gray above, and whitish beneath, with a dark line extending along each side, separating the color above from that below; eyes black; whiskers long and black; hairs of the tail cinerous at their base and then black, tipped with red on the upper side, and with yellow on the under. Length of the specimen before me, from the nose to the insertion of the tail, 7½ inches; tail, (trunk 5, hair 1,) 6 inches.

HISTORY.—This animal is every where known in Vermont by the name of Red Squirrel. They are much more common than either of the preceding species, and in some seasons they have multiplied so exceedingly as to be a great annoyance to the farmer, and do considerable damage by their depredations. They spend most of their time in the tops of trees, feeding upon nuts of various kinds, and upon the seeds contained in the burs of spruce and hemlock. Their nests are usually in the hollow of some old tree, and here they lay up for winter their store of provisions, often amounting to several gallons, and consisting of butternuts, beechnuts, acorns, and different kinds of grain. Their food in summer consists of grain, sweet apples, and different kinds of berries, as well as nuts. In the fall and early part of winter they often come around our barns, and purloin their subsistence from our granaries. This squirrel is often called the *Chickaree*, probably from its noisy chatter when alarmed. It is also called the *Hudson*, or *Hudson Bay Squirrel*.

THE STRIPED SQUIRREL.
Sciurus Striatus.—KLEIN.

DESCRIPTION.—Top of the head dark reddish gray; eye-lids whitish; neck gray; back striped, having a black stripe along the spine, then on each side a broad reddish gray stripe, then another black stripe, succeeded by a white stripe, and, lastly, a reddish brown stripe; the throat, belly, and inner surface of the legs, white; head tapering from the ears to the nose; forehead slightly convex; nose covered with short hairs, with a black spot near the extremity; ears short, rounded, and covered with very fine hair, which is reddish brown within; tail less bushy than in the preceding species, blackish above, and red beneath, bordered with gray. Length of the specimen before me, from the nose to the insertion of the tail, 6 inches; tail (trunk 3¼ in., tuft ¾ in.) 4 inches.

HISTORY.—The Striped Squirrel is more common in Vermont than either of the preceding species, and differs from them in being furnished with cheek pouches, in which it carries the food it collects, to its store-house. It also differs from the preceding in having its chief residence in the ground, while the others inhabit hollow trees, and hence it has received the name of *Ground Squirrel*. It is likewise frequently called the *Chipmuck*, or *Chipping Squirrel*, from its note; and it is also called in many places the *Hackee*.

This squirrel is generally seen running along upon the lower rail of fences, or sitting upon stone walls or logs. When frightened they immediately retreat to their holes, which they enter with a peculiarly shrill *chit-te-rie*, indicative of safety, which is as much as to say, "catch me now if you can." When their retreat to their hole is cut off, they become much alarmed, and, in such cases, will sometimes ascend trees, but they betray much timidity, and will seldom go up more than 20 or 30 feet. Their burrows are by the side of stone walls, fences, or the roots of trees, and in places where their food is easily obtained. These burrows are often extensive, with two openings, at considerable distance from each other, and what is remarkable, is that the dirt which has been removed in making the excavation, is no where to be found. This squirrel retires to its burrow on the approach of cold weather, where it spends the winter, subsisting upon its stores of nuts and seeds, which it had carefully provided, and being seldom seen after the beginning of November, before the first of April.

GENUS PTEROMYS.—*Cuvier.*

Generic Characters.—Teeth 22—Incisors, $\frac{2}{2}$, no canines, grinders, $\frac{5-5}{4-4}$. Head round; ears short and rounded; eyes large; fore feet with four elongated toes, furnished with sharp nails and a rudimentary thumb, having an obtuse nail; hind feet with five long toes, much divided, and adapted for seizing; tail long, villose; skin of the sides extending from the anterior to the posterior extremities forming a kind of parachute.

THE FLYING SQUIRREL.
Pteromys volucella.—Desmarest.

DESCRIPTION.—General color, reddish gray above, yellowish white beneath; head large; nose rounded; eyes large, black, prominent, and far apart, and surrounded by a blackish ash color, with a white spot over each; ears broad, rounded, and nearly naked; whiskers black, two inches long; tail long, thickly covered with fine long fur, brown above, lighter beneath, and flattened; a bony appendage, about an inch long, proceeding from the wrist, and used in stretching the flying membrane. Length of the specimen before me, from the nose to the insertion of the tail, 6 inches; tail 5½ inches; spread of the membrane, measured across the breast, 6½ inches.

HISTORY.—This interesting little animal is frequently met with, living in families, in all parts of the State, but is never so greatly multiplied as some of the preceding species of squirrels. They usually inhabit the hollows of trees, and feed upon nuts, grains, seeds and buds. Their wings are not calculated for rising in the air and flying in the manner of bats and birds. Consisting only of an extension of the skin of the flanks, they form only a kind of parachute, by which they are supported for a while in the air, and are thus enabled to sail from one tree to another at a distance of several rods. In proceeding through the forests, they first ascend high upon a tree, and, leaping off in the direction of another tree, and at the same time spreading their wings, they are enabled to sail, while descending, to a considerable distance, and to alight on the tree designated, near the ground. This they ascend, and proceed in like manner to another tree, thus passing to a considerable distance without coming to the ground. Their habits are nocturnal, and, unless disturbed, they seldom leave their nests in the day time. When this animal sleeps, it rolls itself up, and so wraps its large flat tail over its head and limbs as completely to conceal them, and give it the appearance of a simple ball of fur. The flying squirrel is often tamed as a pet, but is more admired on account of its singular form, soft fur, and gentle disposition, than for its sprightliness and activity.

Genus Hystrix.—*Linnæus.*

Generic Characters.—Teeth 20—Incisors, $\frac{2}{2}$, no canines, grinders $\frac{4-4}{4-4}$. The grinders have flat tops, but are furnished with ridges of enamel. Head strong and convex; muzzle thick and turned; ears short and rounded; tongue furnished with spiny scales; fore feet, with four toes, and the rudiment of a thumb; hind feet with five toes; nails strong on all the feet; body covered with spines, intermixed with strong hair; tail more or less long, and sometimes prehensile.

THE HEDGE HOG.
Hystrix dorsata.—Gmelin.

DESCRIPTION.—General color, brownish black; hair rather long, thick, and interspersed with spines or quills, which vary from 1 to 4 inches in length; quills black at the tip, below brownish, and white towards their base. Ears small, and covered by the hair; snout short and thick. Legs and feet covered with hair, the latter armed with long curved nails. Tail thick, flattened, and not prehensile. Length 26 inches; tail 8; height of the back 14.

HISTORY.—The Hedge Hog was originally very common in Vermont, but is now confined principally to the mountainous and woody parts, where it is still found in considerable numbers. This animal is remarkable, principally, on account of the quills or spines, which are intermingled with the hair, on nearly all parts of its body; and as he runs very badly, and is moderate and awkward in all his movements, he relies mostly upon his quills for defence and safety. When his enemy approaches, if allowed sufficient time, he will generally retreat to a fissure among the rocks, or take refuge in the top of a tree, which he ascends with facility; but, if overtaken, he places his head between his fore legs, draws his body into a globular form, and erects his barbed spines, which now project in all directions. In this condition they defy the attack of all enemies but man. The fox, the wolf and the dog attempt to seize him only to be severely wounded in the nose and mouth by the sharp projecting quills. These quills, being barbed at the extremity, and adhering in the wound, are detached from the owner, and by their rankling, and by penetrating deeper and deeper, not only discourage the attack of the assailant, but very often occasion his death. The vulgar notion that this animal has the power of projecting or shooting his quills at his assailant, is without a shadow of foundation.

The quills of the Hedge Hog are highly prized by the aborigines on all parts of the continent, and are used by them in various ways as ornaments of their dresses, pipes and war instruments. For this purpose they are dyed of several rich and permanent colors, cut into short pieces, strung upon threads or sinews, and then wrought into various forms and figures upon their belts, buffalo robes, moccasins, &c., and in these operations they manifest considerable ingenuity and a great deal of patient perseverance.

The Hedge Hog is a solitary, sluggish animal, seldom venturing to much distance from his retreat among the rocks. Their food consists of fruits of different kinds, roots, herbs, and the bark and buds of trees. Their flesh is sometimes eaten, and is esteemed by the Indians as the greatest luxury. They have three or four young at a litter, and their period of gestation is said to be 40 days. The Hedge Hog or American Porcupine, when full grown and fat, weighs about 16 pounds.

GENUS LEPUS.—*Linnæus.*

Generic Characters.—Teeth, 28—Incisors $\frac{4}{2}$, no canines, grinders $\frac{5-5}{6-6}$. The upper incisors are placed in pairs, two wedge-shaped with a longitudinal furrow in front, and two smaller ones intermediately behind; the under incisors square, grinders with flat crowns and transverse laminæ of enamel. Head rather large; ears long; eyes large, projecting laterally; fore feet with five toes; hind feet with four very long toes; all the toes armed with moderate sized nails, which are slightly arched; bottoms of the feet hairy; tail short, hairy and elevated; mammæ from 6 to 10.

THE AMERICAN RABBIT.
Lepus americanus.

DESCRIPTION.—Color, above grayish fawn, varied with blackish brown and reddish; more red about the shoulders than elsewhere; a whitish spot before the eyes and another behind the cheeks; breast and belly white; feet reddish before with the point of the foot fawn color; upper part of the tail the color of the back, beneath white, fur on the body white in winter, but the ears and tail are of the same gray color summmer and winter. Length 14 inches, head $3\frac{1}{2}$ ears $2\frac{1}{4}$, tail 2 inches.

HISTORY.—This animal though strictly a *Hare* has acquired very generally in this country the name of Rabbit. Indeed the name of Rabbit is not only applied to this species, but also to the following, and this is distinguished by the appellation of *Gray* rabbit, on account of its not becoming so white in the winter as the other. This is the most common species of hare throughout the United States, and is also one of the most prolific species. It produces its young three or four times in the course of the year and has from from five to seven at a birth. This animal has been supposed to form burrows in the earth like the European Rabbit, but this is probably a mistake. It is true they are sometimes found in burrows, but it is believed to be only in cases in which they have taken refuge in the holes of foxes or woodchucks.

THE VARYING HARE.
Lepus virginianus.—HARLAN.

DESCRIPTION.—General color, in its *summer dress*, reddish brown, darkest along the back, lighter about the shoulders, and passing into white on the belly. Hairs on the upper parts bluish at their base, then light reddish yellow, and tipped with black. Chin and ears bluish white mixed with reddish brown, the latter margined exteriorly, towards the tip, with black, and slightly edged with white; orbits surrounded by reddish fawn; flanks tinged with orange; sides of the feet whitish; soles covered with long hair of a tawny yellow color. Ears and head of equal length; tail very short; nails long, slightly arched, compressed at the base, and entirely covered by the hair. Incisors above and below nearly equal, the former slightly arched and marked by a longitudinal groove. Length of the specimen before me, which was taken in September, from the nose to the root of the tail, 16 inches; tail, including the fur, 1½; ears 3½; hind foot, 5½. Color, in its *winter dress*, white, or nearly so, resulting from the hairs being bluish at their base, then yellowish fawn, tipped with white.

HISTORY.—This hare is quite common in Vermont, and, in the winter season, is usually called the white rabbit. It is less prolific than the preceding species, producing its young only once or twice a year, and having from 4 to 6 at a time. The young are able to see at birth, and are covered with hair. They are able to provide for themselves in a very few days, after which they receive but little aid from their mothers. The hares feed in summer upon grass, juicy herbs, and the leaves and buds of shrubs, but in winter, when the snow is deep, they gain a precarious subsistence from the buds and bark of bushes and small trees. The bark of the willow, birch, poplar, and the buds of the pine, are with them favorite articles of food. The hares are the most timid and defenceless of all quadrupeds, and no animals have more numerous or formidable enemies. They are pursued and destroyed in great numbers, by men and dogs, by eagles, hawks, and owls, and by all the carnivorous beasts of the forests; and yet, notwithstanding this destruction, nature has sufficiently provided, in their great fecundity, for the preservation of the several species. When pursued, the American rabbit soon becomes wearied, and to avoid being overtaken, takes shelter in some hole in the earth, in a heap of logs, or stones, or in a hollow log, but this species is so fleet as to be in no fear of being overtaken by its pursuers, and, therefore, does not seek concealment. It has been ascertained by measurement that it can leap 21 feet at a bound, and its body is so light in comparison with its broad furry feet that it is enabled to skim easily along the surface of deep snows, while the wearied hounds plunge in at every bound, and soon give up the hopeless pursuit. The skin of the hare is of no value, but the flesh is considered nourishing food.

ORDER RUMINANTIA.

Animals of this order have three kinds of teeth. They have no incisors in the upper jaw, but have usually eight in the lower, which are opposed to a callosity on the upper gums. In some species there are canines only in the upper jaw, and others have them in both. The grinders are twelve in each jaw, marked with two double crescents of enamel on their crowns, of which the convexity is outwards in the lower, and internal in the upper jaw; articulations of the jaw adapted for a triturating motion. The limbs are disposed for walking; the feet with two hoofed toes; the two bones of the metacarpus and metatarsus, consolidated into one; organs of digestion calculated for ruminating, consisting of four stomachs; intestines long; two or four inguinal mammæ. The males have horns, and the females, too, in some species; food always vegetable. The most remarkable faculty of these animals is that of rumination, or of returning the food into the mouth for the purpose of chewing it a second time, called *chewing the cud*, and hence the name of the order, *Ruminantia*.

GENUS CERVUS.—*Linnæus*.

Generic Characters. Teeth 32, or 34— Incisors $\frac{0}{8}$—canines $\frac{0-0}{0-0}$ or $\frac{1-1}{0-0}$ grinders $\frac{6-6}{6-6}$. The canines, where they exist, are bent back and compressed. Head long, terminated by a muzzle; eyes large, pupils elongated transversely; most of the species have a lachrymal sinus; ears long and pointed; tongue soft; horns solid, deciduous, palmated, branched, or simple, in the males; females destitute of horns, except in one species; four inguinal mammæ.

THE MOOSE.
Cervus alces.—LINNÆUS.

DESCRIPTION.—Head long, narrow before the eyes and enlarged towards the mouth, which has some analogy to that of the horse; upper lip exceedingly developed and very thick; nostrils, a lateral slit, more open anteriorly than behind; eyes small, near the base of the horns;

lachrymal pits small; neck short; ears very large and thick; horns, consisting of a very large flattened expansion, furnished with numerous prongs on the external border, with a large isolated branch of the principal stock. Tail excessively short. A tuft of long hair, like beard, beneath the throat, in both sexes, and a protuberance in the same place in the male. Legs long; feet long, and placed obliquely on the soil. Hair coarse and friable. General color fawn-brown. Dimensions, as given by Dr. Harlan: length from the nose to the base of the tail, 6 ft. 10 in.; height before, 5 ft. $2\frac{1}{2}$ in.—behind, 5 ft. $4\frac{3}{4}$ in.; length of the head, 23 in.; ears, 10 in.; horns, 37 in.; neck, 18 in.; tail, $1\frac{1}{2}$ inch. Weight of the horns sometimes 60 pounds.

HISTORY.—Moose were formerly very plentiful in Vermont, and in many places the early settlers depended upon their flesh for no inconsiderable part of the subsistence of their families. They are now exterminated from all portions of the state excepting the county of Essex, in the northeastern part. There they are still found, and several were killed there during the two last winters. The head and horns of one of these, obtained by Judge Parker, of Orleans, and now in his possession, weighed 95 pounds, of which the horns are supposed to constitute one half. The hide and quarters of this Moose, when dressed, weighed a little more than 800 lbs. The height of its horns exceeded 3 feet, and the distance between their tips was more than 5 feet, and larger than this are not often found at the present day. But it would appear from the statement of Dr. Williams that larger individuals were taken in early times. He says that one of these animals in Vermont was found by measure to be 7 feet high, and that the largest Moose were estimated by the hunters to weigh from 1300 to 1400 pounds. The food of the Moose consists of grass, shrubs, the boughs and bark of trees, especially the beech, which they seem to prefer above all others, and a species of maple, *Acer pennsylvanicum*, which is called *Moosewood*. In summer they keep pretty much in families. In winter they herd together, sometimes to the number of 20 or 30 in a company. They seem to prefer cold places; and when the snow is deep they tread it down for a space of several acres, forming what is called a *yard*. Within this space they range, and subsist upon the twigs and bark of the trees, while the snow remains deep upon the ground. In order to eat from the ground, they are obliged to kneel or spread their fore legs, on account of the shortness of their neck. They move with a long shambling trot, and with a rattling of their hoofs, which may be heard at a considerable distance. Their course is swift and straight, and they leap over the highest fences with ease. The males only have horns, which are shed and reproduced annually. The rutting season is in September, and the young are produced about the first of June, usually two at a birth. The female is smaller than the male.* This animal was called *Monsall* by the Algonquin Indians, *Orignal* by the French inhabitants of Canada, and *Moose*, or *Moose Deer*, by the English.†

Since the above was written, I have had an opportunity of examining a living Moose in Burlington. It was a female, two years old, and had then been in captivity about two months, having been taken in Canada, near the north line of this state, in March, 1842. The height at the shoulder was about 6 feet, and it agreed fully with my description, so far as it is applicable to the female, that sex being without horns. It had become so tame as to be led by a halter without difficulty.

THE ELK.
Cervus canadensis.—GMEL.

DESCRIPTION.—Head well formed, tapering to a narrow point; ears large and rapidly moveable; eyes full and dark; horns lofty, graceful, with numerous pointed cylindrical branches, which curve forward. The hair is of a bluish gray color in autumn; dark gray during the winter, and at the approach of spring assumes a reddish, or bright brown color, which it retains during the summer. The croup of a pale yellowish white or clay color. Colors nearly the same in the two sexes; but the females are without horns. Height at the withers, according to Dr. Harlan, 4 feet, the horns 3 feet, first antler 1 foot, second 10 inches, length of the tail 2 inches.

HISTORY.—The horns of the elk have been often found in Vermont, which may be regarded as sufficient proof of the former existence of that animal within the state; and if the animal was found here after the settlement of the state was commenced, it is doubtless now completely exterminated. Elks live in families. Their rutting season is in September, and the young, one and sometimes two in number, are produced in July. Their horns are generally shed in March. This species is said to be still found in numbers

---
* Williams' History, Vol. I, p. 99.
† Harlan, Fauna Americana, p. 232.

in the western states. A specimen of this species, preserved in the Philadelphia Museum, measures seven feet and seven inches from the tip of the nose to the base of the tail, and the horns measure three feet and ten inches. The animal was 13 years old.

THE COMMON DEER.
Cervus virginianus.—GMEL.

DESCRIPTION.—Form light and slender; color reddish fawn in summer, and grayish in winter; horns moderate, with an antler placed high on the inside of each shaft, and two or three others on the posterior side, turned backwards, but varying with the age of the animal; lachrymal pits formed by a fold in the skin; muzzle partially developed; tail proportionally longer than in the preceding species, and thin; no canine teeth. Length 5 feet 5 inches, tail 10 inches, height 3 feet, length of the head 12 inches, of the horns, following the curvature, 22 inches. Weight from 90 to 130 pounds.

HISTORY.—When the country was new this deer was one of the most common and valuable quadrupeds found in our forests, and upon its flesh were the first settlers of the state, to a very considerable extent, dependent for food. Indeed so eagerly was it hunted, and still so anxious were the people for its preservation, that a law for its protection from the 10th of December to the 10th of June was one of the earliest acts of our legislature. But notwithstanding all that has been done for their preservation, their numbers have been constantly diminishing within the state, till they have become exceedingly scarce, except in a few of the most unsettled and woody sections. The range of this species is very extensive, reaching from Canada to the Oronoco in South America. In its form this deer is slender and delicate; and its neck and tail proportionally longer than in most other species; but at the same time it possesses great muscular power, and runs with surprising speed. It is a very timid and shy animal, and, possessing a keen sense of hearing and smelling, it is found to be very difficult to approach within gun shot of him without his taking alarm. In the fall the deer are in good condition, and the venison valuable. In the winter they herd together, and, when the snow is deep, they form what are called "*yards,*" where they tread down the snow and gain a scanty subsistence by browsing the trees and bushes. During this period they become very lean, and neither the skin nor the flesh is of much value. They produce their young in the early part of summer, and have two, and sometimes three, at a birth. The fawns are at first reddish, spotted with white. They lose their spots in autumn and become *gray* in winter. This coat is shed about the first of June and in summer they are nearly *red,* which color continues till August and then changes to *blue.* The skin is said to be thinnest in the *gray,* toughest in the *red* and thickest in the *blue;* the skin and the flesh being most valuable in the blue. The horns of the male are shed in January. The deer is said to manifest great enmity to the Rattle-snake. When it discovers one of these reptiles, it leaps into the air above it and alights upon it with all four of its feet brought together in the form of a square, and this operation is repeated till the hated reptile is destroyed.

DOMESTIC QUADRUPEDS.

Thus far we have confined ourselves to an account of the Quadrupeds which have been found in Vermont in a wild state. In addition to these we have several quadrupeds which have been introduced and are kept in a domesticated state. The following is a list of such as may be regarded as permanent residents.

ORDER CARNIVORA.
Canis familiaris,   The Dog.
Felis catus,   The Cat.

ORDER PACHYDERMATA.
Equus caballus,   The Horse.
Equus asinus,   The Ass.
Sus scrofa,   The Hog.

ORDER RUMINANTIA.
Bos taurus,   The Ox.
Ovis aries,   The Sheep.

There are a few other Quadrupeds, which are sometimes kept as a matter of curiosity, such as the Goat, the English Rabbit, the Guinea Pig, &c.

THE DOG.
Canis familiaris.—Linn.

The Dog has been in a domesticated state from time immemorial; and from him has sprung so great a number of varieties, that it is perhaps impossible to determine which now approaches nearest to the original stock. The dog is mentioned as being a familiar animal nearly two thousand years before the Christian era, but the allusions to him in the Bible seem to imply that he was formerly more sanguinary and savage in his disposition than at present. The dog is the only quadruped which has been the companion of man in every state of society, and in every region and climate of the earth, and no other animal manifests so great and so faithful an attachment to his master as this; and this attachment seems to arise from the purest gratitude, and truest friendship. In works on natural history we have no less than sixty permanent varieties of the dog named and described.* In Vermont, each family in the country usually finds it convenient to keep one dog, and very few have more than one. In our villages a few dogs are kept, (better if fewer,) but as a person's standing in society is not here, as in some countries, indicated by the number of his dogs, the dog mania has never prevailed to any considerable extent, and consequently little pains have been taken to procure rare and popular varieties. As the expense of keeping a dog is generally much more than the profit, and as direful consequences are to be apprehended when dogs are numerous, from the occurrence of hydrophobia among them, we should by no means regret the reduction of the dogs in this state to a moiety of their present number.

THE DOMESTIC CAT.
Felis catus.—Linn.

Our domestic Cat is said by Cuvier to have been originally from the forests of Europe, where it is still found in a wild state. The color of the wild animal is grayish brown on the back and sides, with dark transverse undulations, while below it is lighter colored, and the inside of the thighs and feet are yellowish. There are three bands upon the tail, the inferior third of which is blackish. In the domesticated state this animal varies, as is well known, in the length and fineness of its hair, but infinitely less so than the dog, and is also much less submissive and affectionate. The Cat renders essential service by the destruction of vermin, and most families consider it to their advantage to keep one at least upon their premises. Cats were formerly held in so high estimation on account of their mousing qualities, that in the 10th century laws were passed in England regulating the price of them. It was also enacted, that "whoever stole or killed the cat that guarded the granary of the prince, should forfeit an ewe, with her fleece and lamb, or as much wheat as, when poured upon a cat, suspended by its tail, (the head touching the floor,) would form a heap high enough to cover the creature to the tip of its tail."

Order PACHYDERMATA.

This order is named from the thickness of the skin of the animals which compose it. They have two and sometimes the three kinds of teeth. The four extremities are furnished with toes, variable in number, and terminated with strong nails or hoofs. They have no clavicles; and the organs of digestion are not formed for ruminating. We have no animal of this order existing in Vermont in a wild state, and only three, the Horse, the Ass, and the Hog, which have been introduced.

Genus Equus, Linnæus.
Generic Characters.—Teeth 40—Incisors $\frac{6}{6}$, canines $\frac{1-1}{1-1}$, grinders $\frac{6-6}{6-6}$. Grinders furrowed on each side with flat crowns, and several ridges of enamel; between the canines and grinders a vacant space. Upper lip capable of considerable motion; eyes large; ears rather large, pointed and erect; feet with a single visible toe, covered with a strong hoof; tail with long hair, or in some species with a tuft at the extremity; two inguinal teats; stomach simple and membranous; intestines and cæcum large.

THE HORSE.
Equus caballus.—Linnæus.

This generous and noble spirited animal, next to the sheep and the ox, has probably been the most useful servant of man. At what period he became domesticated we have at present no means of knowing. It must, however, have been soon after the deluge, if not before that event, as there is mention of the horse and his rider in the book of Genesis nearly 2000 years before the Christian era. The horse is the associate and assistant of man in war, in the chase, and in the works of agriculture, of the arts and of commerce. Although wild horses exist at the present day in several parts of the world, yet it is believed that there are now no wild horses, which have descended in a wild state from the original stock.

---
* Brown's Zoological Text Book, Vol. 1, p. 75.

The wild horses in Asia and America are all descended from such as had been formerly domesticated, and had been set at liberty. These wild horses are said to be very numerous, going in troops upon the prairies at the southwest, and that the Indians supply themselves with horses, by catching and taming them. The period of gestation in the horse is 11 months and in the domesticated state the colt is allowed to suck 5 or 6 months. At the age of two years the sexes are separated; at three they are handled and at four are broke to the saddle and harness, and are capable of service and of propagating without injury to themselves. The life of the horse is from 25 to 30 years, but they are not of much value after they reach 20 years. The age of a horse may be pretty nearly ascertained by his teeth. According to Cuvier the milk teeth appear about 15 days after the colt is foaled; at $2\frac{1}{2}$ years the middle ones are replaced; at $3\frac{1}{2}$ the two following ones; and at $4\frac{1}{2}$ the outermost ones or corners. All these teeth have at first indented crowns, which are gradually worn down by use and entirely effaced at 7 years old. The lower canine teeth appear at 3 years old, and the upper ones at 4. They remain pointed till 6, and begin to peel off at 10.

Vermont produces excellent horses and considerable pains have been taken to introduce the best varieties. The greatest part of the labor upon the farms, and nearly the whole of the travel and transportation in this state is performed by horses, and large numbers of fine horses are annually sent to market out of the state. The whole number of horses in Vermont, (including the mules, which are very few,) according to the returns of 1840, was as follows:

Addison,	5,425	Orange,	6,674
Bennington,	3,397	Orleans,	3,462
Caledonia,	5,852	Rutland,	6,200
Chittenden,	4,231	Washington,	4,360
Essex,	1,207	Windham,	4,969
Franklin,	4,427	Windsor,	8,440
Grand Isle,	1,161		
Lamoille,	2,597	Total number,	62,402.

THE ASS.

Equus asinus,—LINNÆUS.

The Ass is distinguished by his long ears, by the tuft which terminates his tail, and by the black cross on his shoulders. His usual color is a brownish gray. He was originally from the great deserts of central Asia, where these animals are still found in a wild state, and where they range in immense herds from north to south, according to the season. The Ass in the domesticated state, is a patient, submissive and serviceable animal, and in many parts of the world is almost the only one employed as a beast of burden. It is much more sure-footed than the horse, and on that account is much used in rough mountainous countries. The hoarseness of the bray of the Ass is well known, and it is produced by two small, peculiar cavities, situated at the bottom of the larynx. The Ass is not kept in Vermont for its labor, but a very few are kept for the production of Mules from the mare.

THE MULE.—The Mule is an unprolific hybrid, produced betwixt the horse and the ass. When the sire was a horse and the dam a she-ass, the offspring was termed *Hinnus* by the ancients, but when the sire was a jack ass and the dam a mare, it was then called *Mulus*. At some periods a considerable number of Mules have been produced in Vermont, but they have always been reared for exportation, none of them being kept within the state for their labor.

GENUS SUS.—*Linnæus.*
Generic Characters.—Teeth 42 or 46—incisors, $\frac{4}{6}$ or $\frac{6}{6}$, canines, $\frac{1-1}{1-1}$, grinders, $\frac{7-7}{7-7}$. Lower incisors directed obliquely forward, the upper ones conical; the canines protruded and bent upwards; grinders simple and tuberculous Body covered with bristles; nose elongated, cartilaginous and furnished with a particular bone to the snout; feet with four toes, the two middle ones only touching the ground, furnished with strong hoofs.

THE COMMON HOG.

Sus scrofa.—LINNÆUS.

The color of the Hog, in a wild state, is blackish brown mixed with gray. Its tusks strong, prismatic, curved outwards and slightly upwards; its body short and thick; its ears erect, and the young are striped with black and white. In the domestic state it is subject to very great variety, both in form and color. Pork or the flesh of the Hog, has always been to the people of Vermont one of the most important articles of food. When the country was new, the first settlers of the state depended, to a very considerable extent, upon the spontaneous productions of the forests for the means of fattening their hogs. Hogs are extremely fond of acorns, beech nuts, and other nuts, and with these the forests abounded. When, on the occurrence of frosts in autumn, these nuts began to fall from the trees, it was the practice of the early settlers to turn their hogs into the woods and let them run till the setting in of winter and the fall of deep snows, when they were usually found in good condition to be butchered. But on account of the great number of bears, wolves and catamounts, which embraced every opportunity to destroy them, the fattening of hogs in this way was, at best, a precarious business. In some places, where a considerable number of hogs were turned into the woods together, a person was kept with them to protect them during the day, and collect them into a place of safety for the night, and often has our blood chilled in our veins as we have heard our fathers narrate, with quivering lips, their bloody struggles with bruin for the possession of a favorite hog. Almost every family in the state fattens one hog, or more than one, for their own use, and by most of our farmers, more or less are fattened for market. Hogs are usually butchered in this state when about 20 months old, and their weight when dressed is from 150, to 400 pounds, according to kind and condition. Considerable pains have been taken within a few years to improve our breed of hogs, and several new varieties have been introduced, one of the latest and most approved of which is called the Berkshire Hog. The Hog is a prolific animal, producing young twice a year, and often having 14 pigs at a litter. The period of gestation is 4 months. The hog increases in size for about 5 or 6 years, and sometimes lives 20 years. The number of hogs in the several counties in Vermont, according to the returns of 1840, was as follows:

Addison,	14,305	Lamoille,	7,287
Bennington,	9,906	Orange,	22,516
Caledonia,	18,991	Orleans,	9,750
Chittenden,	25,310	Rutland,	15,563
Essex,	3,639	Washington,	12,150
Franklin,	8,935	Windham,	29,435
Grand Isle,	3,179	Windsor,	22,834

GENUS BOS.—*Linnæus.*

Generic Characters.—Teeth 32 or 30—Incisors $\frac{0}{8}$ or $\frac{0}{6}$, canines $\frac{0}{0}$, grinders $\frac{6}{6} \cdot \frac{6}{6}$. Head large; forehead straight; muzzle square; horns occupying the crest of the forehead; eyes large; ears funnel shaped; dewlaps on the neck; female with an udder, having four teats; tail long and tufted; horns simple, conical, round with various inflections, sometimes directed laterally.

THE OX.

Bos taurus.—LINN.

We here use the term *ox* in a general sense, to denote *neat cattle*, the male of which is called bull, and the female cow, although it is ordinarily applied to the male in an altered working state. Neither the native country of the ox, nor the time when he was reclaimed from a wild state, is now certainly known. It must, however, have been domesticated at a very early period, as the keeping of cattle is mentioned as an occupation before the flood.* After that event the keeping of cattle and sheep afforded the means of subsistence and constituted the principal part of the wealth of a large proportion of the human race; and has continued to do so down to the present time. We read that when Abraham was in Egypt, 180 years before there is any mention of the horse, he was possessed of sheep and oxen;† and this account of the early domestication and acknowledged value of the ox is confirmed by the records of profane history. This animal was held in so high estimation as to be an object of worship in Egypt, and among the Hindoos was highly venerated and believed to be the first animal created. The traditions of the Celtic nations also enrol the cow among the earliest productions, and represent her as a kind of divinity.

Cattle, like most other domesticated animals, have run into a very considerable number of varieties, and it is now, perhaps, impossible to ascertain which approaches nearest to the original stock. The cattle which were first introduced into this country by the early settlers, were such as were the common cattle of Great Britain 150 or 200 years ago, and from these the present stocks have generally descended, and, till within a few years past, very little pains have been taken for their improvement. These, coming from different parts of England, Scotland and Ireland, consisted of many varieties, which here became amalgamated, and which have here formed what may be called the *American stock*, retaining, like our American people, many both of the good and bad qualities of the races from which it is descended. For many years past much pains have been taken to improve the breeds of cattle, particularly in England, and within a few years some of these improved breeds have been introduced into this country. The most approved of these are the Ayrshire and Durham, and these are doubtless in many respects superior to our native cattle. Still, it is the opinion of many, that the proper method of improving stocks of cattle is not by the introduction of foreign materials, but by selecting, for breeders, from our native stocks, the best varieties, and, from these, those individuals which possess the properties desired in the highest perfection. In this way we shall be sure to have a race of cattle which is adapted to our country and climate, and

---

* Genesis IV—20. † Genesis XII—16.

but a few years would elapse in the pursuance of this policy, before we should be as proud to compare the American stock of cattle with the cattle of foreign countries as we now are to compare the American with foreign nations.

Upon lands which are uneven and rough, the farming operations are carried on to better advantage by oxen than by horses, and on this account large numbers of oxen are kept for labor in Vermont, particularly in the central and eastern parts; but cattle are here raised chiefly for the dairy and for market. No part of our country affords better grazing, and for the production of good beef cattle and good butter and cheese, Vermont may challenge comparison with almost any part of the world. According to the grand list of the state in 1841, there were 31,130 oxen, and 154,669 cows. The number of cattle of every description according to the returns of 1840, was as follows:

Addison,	39,718	Orange,	36,855
Bennington,	16,879	Orleans,	18,293
Caledonia,	32,668	Rutland,	40,029
Chittenden,	24,142	Washington,	25,415
Essex,	6,837	Windham,	42,661
Franklin,	26,965	Windsor,	51,863
Grand Isle,	5,463		
Lamoille,	16,555	Toal number,	384,341

Genus Ovis.—*Linnæus.*

Generic Characters.—Teeth 32—Incisors $\frac{0}{8}$, canines $\frac{0}{0}$, grinders $\frac{6-6}{6-6}$. Horns common to both sexes, often wanting, particularly in the female; thick, angular, wrinkled transversely, pale colored, turning laterally and spirally; ears small; legs slender; hair of two kinds; tail more or less short; two inguinal mammæ.

THE SHEEP.
Ovis aries.—Linn.

In the 4th chapter of the book of Genesis we read that Abel was a keeper of sheep; from which it appears that this animal has existed in a state of domestication from the very beginning of our race. And we learn from history that man has, in almost all ages of the world, depended upon the sheep for a very considerable share of his food and clothing. In the Scriptures the sheep is frequently mentioned, and the lamb, which is the young of this animal, on account of its gentleness and meekness, was employed under the Mosaic dispensation to prefigure the meek and lowly Jesus—" the Lamb of God which taketh away the sin of the world."*

The sheep first introduced into this country by the European settlers, were of a large, hardy, coarse woolled variety, and before the commencement of the present century very little pains had been taken to improve their quality or increase their numbers. The first fine woolled sheep introduced were the Merinos, from Spain, in 1802. In that year Chancellor Livingston imported a buck and two ewes into New York, and Col. D. Humphreys imported 200 sheep of this breed, and placed them on his farm near New Haven, Ct. But these sheep attracted very little attention till the embargo of 1808 and the non-intercourse which followed it had cut off the accustomed supply of woollen goods from England. In 1809 and 1810 nearly 400 Merinos were shipped to this country by the Hon. Wm. Jarvis, then American consul at Lisbon, and these, together with about 2,500 imported by others, were distributed over the greater part of the United States. A considerable number of the Merinos introduced into this country by Consul Jarvis were brought by him to Vermont, and placed upon his unrivalled farm in Weathersfield; and from the importations above mentioned nearly all the Merino sheep in the United States have been derived.

History informs us that Merino sheep existed in Spain as early as the days of Augustus Cæsar, and as the name signifies *beyond sea*, they were probably imported thither from some other country. In 1765, 100 Merino bucks and 200 ewes were transported from Spain into Saxony, and subsequently many more. In these Saxony Merinos the wool became much improved, and from this improved race importations have taken place into the United States, under the name of *Saxony sheep.* The first, consisting of only two or three bucks, were imported in 1823, by Col. James Shepherd, of Northampton, Mass. The two following years a considerable number of Saxony sheep were imported by the Messrs. Searles, of Boston, and the year 1826 witnessed the introduction of no less than 2,500. From these and subsequent importations the Saxony sheep are now scattered into various parts of the country, and in many places crossed with the Merino and the coarse wooled sheep. In Vermont they have been introduced into many towns, but are not very generally diffused over the state.

There are, probably, few countries in the world better adapted to the rearing of sheep than New England, and the soil and climate of the hills of Vermont seem to be peculiarly suited to that purpose. Experience has likewise shown that while the Merino and Saxony sheep thrive here in a remarkable manner, their wool suf-

* John 1: 29.

fers no deterioration in quality, but with suitable attention is rather improved. Sheep require an airy location, both in summer and winter. In summer they thrive much better in elevated, dry pastures than on low, moist lands. In winter they should be yarded from the last of November till the latter part of April, but should never be crammed, in large numbers, into small or tight enclosures. They should be salted weekly both in summer and winter, and at all seasons have free access to pure water. The best season for lambing is thought to be from the 1st to the 10th of May. The daily allowance of food per head for sheep in winter should be 3 lbs. of hay, or 2 lbs. of hay and half a pint of oat meal, or other food equivalent.

Sheep are subject to several diseases, the most common and fatal of which are the *foot-rot* and *scab*. The most approved remedy for the former consists of 3 parts of blue vitriol and 1 of verdigris pulverized as fine as Indian meal and mixed with a sufficient quantity of sharp vinegar to make it as thick as milk. The vinegar should be nearly as hot as boiling water when poured upon the other ingredients, and the mixture should be stirred briskly while hot. This mixture may be put on with a paint-brush, being careful to apply it thoroughly to those parts of the feet which are most inflamed. *For the scab* the best remedy is to immerse the sheep, excepting the head, in a strong decoction of tobacco, scrubbing thoroughly the parts affected. The best time for doing this is immediately after shearing; but it may be done any time during the season. For lambs the decoction should be weaker. *For the bloat* in sheep a great spoonful of castor oil mixed with a teaspoonful of pulverized rhubarb may be given in about a gill of hot water. It may be poured down the sheep's throat with a great spoon.

From 1830 to 1837 wool met with a ready sale, and commanded a high price, in consequence of which the farmers of Vermont, during that period, devoted their chief attention to the production of wool, and the flocks of sheep, in most parts of the state, were increased many fold. The whole number of sheep in the several counties, in 1840, was as follows:

Addison,	261,010	Orange,	156,053
Bennington,	104,721	Orleans,	46,669
Caledonia,	100,886	Rutland,	271,727
Chittenden,	110,774	Washington,	110,672
Essex,	14,188	Windham,	114,336
Franklin,	87,385	Windsor,	234,826
Grand Isle,	27,451		
Lamoille,	40,920	Total number,	1,681,818

CHAPTER III.

BIRDS OF VERMONT.

Preliminary Observations.

Birds are organized for flight; have a double respiratory and circulating system, and produce their young by eggs. They are distinguished from all other vertebrated animals by being clothed with feathers. Their whole structure is adapted for flying. Their bones are hard and hollow, which give them at the same time lightness and strength. Their lungs are attached to their ribs, and are composed of membranes penetrated by orifices, which permit a free passage of the air into almost all parts of the body. Birds have long necks, and bills composed of horny substance, but they are always destitute of teeth. Their organ of smell is situated at the base of the bill, and is generally hid by the feathers. Their tongue is principally cartilaginous, and their taste probably imperfect. Their eyes are so constructed that their sight is very acute, whether the object be near or distant. In addition to the eye-lids, they have a membranous curtain to cover and protect the eye. Birds which fly by day have no external ear, but owls, or such as fly by night, have one, but it is not so much developed as in quadrupeds. The brain of birds is remarkably large. Their wind-pipe consists of entire rings, and, at the lower end, where it branches off to the lungs, it is furnished with a glottis This is called the lower larynx, and with this the voice of birds is produced, which has great compass, owing to the large volume of air contained in the air vessels.

Most birds undergo two moults annually. In some species the winter plumage differs considerably from that of the summer; and the male and female also vary in color in many species. The digestion of birds is rapid in proportion to the activity of their life and the force of their respiration. Their stomach is composed of three parts; namely, a crop, a membranous stomach, and a gizzard. The gizzard is armed with two strong muscles, and, by the assistance of small stones, which the fowl swallows, grinds up the food, and thus performs the office of mastication.

The velocity with which birds travel through the air exceeds that of any terrestrial animal. Eagles, and many other birds, fly at the rate of 60 miles an hour. Most birds are migratory, very few comparatively spending the whole year in the same neighborhood. The crow, the partridge, and a few species of woodpeckers, owls, hawks, and water fowl, are all which are known to reside permanently in Vermont. Several species are seen here in winter which are never seen in summer, and many are seen to pass northerly in the spring and return to the south in the fall, which make scarcely any stop with us.

The characters by which birds are distinguished into orders and genera are derived principally from the formation of the bill and feet. We have adopted the classification of Temminck, which is followed by Mr. Nuttall, in his valuable Manual of Ornithology. The following are the Orders.

I. *Rapaces*—birds of prey.
II. *Omnivores*—living on all kinds of food.
III. *Insectivores*—feeding on insects.
IV. *Granivores*—feeding on grain
V. *Zygodactyli*—with the toes disposed in opposite pairs.
VI. *Tenuirostres*—birds with slender bills.
VII. *Alcyones*—with three toes before, united, and one behind; the tarsi being very short.
VIII. *Chelidones*—with three toes before, divided, or only united at the base by a short membrane; the back toe often reversible.
IX. *Columbœ*—with toes before entirely divided, and one behind.
X. *Gallinœ*—with three toes before, united by a membrane; the back toe joined to the tarsus above the joint of the other toes.
XI. *Grallatores*—with long slender legs, naked above the knee; three toes before and one behind, all nearly on the same level.

XII. *Pinnatipedes*—with the tarsi slender and compressed; three toes before and one behind, with a rudimentary membrane along the toes, the posterior one joined interiorly to the tarsus.
XIII. *Palmipedes*—with short feet, more or less drawn up to the abdomen; anterior toes partly or wholly connected by a membrane.

The following table contains a list of the Birds of Vermont, arranged in the order in which they are described in the subsequent pages.

BIRDS OF VERMONT.

ORDER RAPACES—*Birds of Prey.*
Falco leucocephalus, Bald Eagle.
" *chrysaetos,* Golden Eagle.
" *haliætus,* Fish Hawk.
" *lineatus,* Red-should'd Hawk.
" *pennsylvanicus* Broad winged Hawk.
" *fuscus,* Slate colored Hawk.
" *peregrinus,* Large footed Hawk.
" *palumbarius,* Gos-Hawk.
" *Cooperi,* Cooper's Hawk.
" *cyaneus,* Marsh Hawk.
" *borealis,* Red-tailed Hawk.
" *columbarius,* Pigeon Hawk.
Strix asio, Screech Owl.
" *funerea,* Hawk Owl.
" *nyctea,* Snowy Owl.
" *virginiana,* Great-horned Owl.
" *cinerea,* Cinereous Owl.
" *brachyotus,* Short-eared Owl.
" *nebulosa,* Barred Owl.
" *acadica,* Saw-Whet.
" *americana,* Barn Owl.

ORDER OMNIVORES—*Food of all kinds.*
Sturnus ludovicianus Meadow Lark.
Icterus baltimore, Baltimore Oriole.
" *phænicus,* Red Winged Black Bird.
" *pecoris,* Cow Black Bird.
" *agripennis.* Bob-o-link.
Quiscalus versicolor, Crow Black Bird.
" *ferrugineus,* Rusty Black Bird.
Corvus americanus, Common Crow.
" *corax,* Raven.
" *cristatus,* Blue Jay.
" *canadensis,* Canada Jay.
Parus atricapillus, Chicadee.
" *hudsonicus,* Hudson Bay Titmouse.
Bombycilla carolinensis, Cedar Bird.

ORDER INSECTIVORES—*Living on Insects.*
Lanius borealis, Butcher Bird.
Muscicapa tyrannis, King Bird.
" *fusca,* Phœbee.
" *virens,* Wood Pewee.
" *acadica,* Small Pewee.
" *canadensis.* Spotted Flycatcher.
Vireo flavifrons, Yellow throated Vireo.
" *noveboracensis* White eyed Vireo.
" *olivaceus,* Red eyed Vireo.
" *solitarius,* Solitary Vireo.

Turdus rufus,	Brown Thrush.	ORDER ALCYONES.—*Halcyons.*	
" *felivox,*	Cat Bird.	*Alcedo alcyon,*	Belted King Fisher.
" *migratorius,*	Robin.	ORDER CHELIDONES—*The Swallow Tribe.*	
" *Wilsonii,*	Wilson's Thrush,	*Hirundo purpurea,*	Purple Martin.
" *noveboracensis*	New York Thrush.	" *rufa,*	Barn Swallow.
" *aurocapillus,*	Golden crowned do.	" *fulva,*	Cliff Swallow.
" *solitarius,*	Hermit Thrush,	" *bicolor,*	White bellied Swal.
Sylvia coronata,	Yellow crowned Warbler	" *riparia,*	Bank Swallow.
" *petechia,*	Yellow red poll do.	*Cypselus pelasgius,*	Chimney Swallow.
" *æstiva,*	Summer Warbler.	*Caprimulgus vociferus,*	Whip-poor-Will.
" *maculosa,*	Spotted Warbler.	" *virginianus,*	Night Hawk.
" *rubricapilla,*	Nashville Warbler.	ORDER COLUMBÆ—*The Pigeon Tribe.*	
" *virens,*	Black throated Green do.	*Columba migratoria,*	Passenger Pigeon.
" *pinus,*	Pine Creeping do.	" *carolinensis,*	Carolina Dove.
" *cærulea,*	Cœrulean Warbler,	ORDER GALLINÆ—*Gallinaceous Birds.*	
" *Blackburniæ,*	Blackburn's Warbler	*Meleagris gallopavo*	Wild Turkey.
" *icterocephala,*	Chestnut sided do.	*Perdix virginianus,*	Quail.
" *canadensis,*	Black throated do.	*Tetrao umbellus,*	Partridge.
" *trichas,*	Maryland yellow throat.	" *canadensis,*	Spruce Partridge.
" *vermivora,*	Worm eating Warb'r	ORDER GRALLATORES—*Wading Birds.*	
" *varia,*	Black & White Creeper.	*Caledris arenaria,*	Sanderling Plover.
Regulus calendulus,	Ruby crowned Wren	*Fulica americana,*	Common Coot.
" *tricolor,*	Fiery crowned Wren	*Grus americana,*	Whooping Crane.
Troglodytes ædon,	House Wren.	*Ardea nycticorax,*	Night Heron.
" *hyemalis,*	Winter Wren.	" *Herodias,*	Great Blue Heron.
" *americanus,*	Wood Wren.	" *virescens,*	Green Heron.
Sialia Wilsonii,	Blue Bird.	*Totanus Bartramius*	Upland Plover.
Anthos spinoletta,	Brown Lark.	" *chloropigius,*	Solitary Tatler.
ORDER GRANIVORES—*Living on Seeds.*		" *macularius,*	Spotted Tatler.
Emberiza nivalis,	Snow Bunting.	*Scolopax Wilsonii,*	Common Snipe.
" *graminea,*	Bay winged Bunting	*Rusticola minor,*	Woodcock.
" *savanna,*	Savannah Bunting.	ORDER PINNATIPEDES—*Lobe-footed Birds.*	
Fringilla melodia,	Song Sparrow.	*Podiceps carolinensis,*	Pied-bill Dobchick.
" *hyemalis,*	Snow Bird.	ORDER PALMIPEDES—*Web-footed Birds.*	
" *canadensis,*	Tree Sparrow,	*Larus Bonapartii,*	Bonapartian Gull.
" *socialis,*	Chipping Sparrow.	" *atricilla,*	Black headed Gull.
" *juncorum,*	Field Sparrow.	*Anser canadensis,*	Canadian Goose.
" *palustris,*	Swamp Sparrow.	*Anas sponsa,*	Wood Duck.
" *tristis,*	Gold Finch.	" *boschas,*	Mallard.
" *linaria,*	Pine Linnet.	" *obscura,*	Dusky Duck.
" *iliaca,*	Ferruginous Finch.	" *discors,*	Blue winged Teal.
" *pennsylvanica*	White throat. Finch.	*Mergus merganser,*	Goosander.
" *leucophrys,*	White crown. Finch.	*Colymbus glacialis,*	Loon.
" *arctica,*	Arctic ground Finch.		
" *erythrophthalma*	Towhe-ground Finch		
" *purpurea,*	Purple Linnet.		
Pyrrhula enucleator,	Pine Grosbeak.		
Loxia curvirostra,	Common Cross bill.		
" *leucoptera.*	White Winged do.		

BIRDS OF PREY.

ORDER ZYGODACTYLI—*The toes in pairs.*

Coccyrus americanus Yellow bill Cuckoo.
" *dominicus.* Black billed Cuckoo.
Picus auratus, Gold wing. Woodpecker.
" *erythrocephalus,* Red headed do.
" *varius,* Yellow bellied do.
" *villosus,* Hairy Woodpecker.
" *pubescens,* Downy Woodpecker
" *arcticus,* Arctic three toed do.

ORDER TENUIROSTRES—*Slender bill Birds.*

Sitta carolinensis, White breast. Nuthatch.
" *canadensis.* Red bellied Nuthatch
Certhia familiaris, Brown Creeper.
Trochilus colubris, Ruby throat Hum'g Bird.

Birds of this order are distinguished by their hooked bills and powerful claws. They pursue and destroy other birds and small quadrupeds; and they are among birds what the carnivora are among quadrupeds.

GENUS FALCO.—*Linn. and Tem.*

Generic Character.—The head covered with feathers; the bill hooked, commonly curved from the base; cere colored and more or less hairy at the base; the lower mandible obliquely rounded, and both sometimes notched; the nostrils lateral, rounded, or ovoid, situated in the cere and open; tarsus clothed with feathers or scaly; the toes, three before and one behind—the exterior toe commonly united to the adjacent one by a membrane; nails sharp, strongly hooked, movable and retractile; tail feathers, twelve.

This Genus embraces the Eagles, Falcons, Hawks, Kites and Buzzards, and is divided by modern Ornithologists into no less than ten genera; but we deem it unnecessary to give the distinctive characters of these genera in this work.

THE BALD EAGLE.
Falco leucocephalus.—LINNÆUS.

DESCRIPTION.—Color of the body and wings deep lively brown or chocolate; head, upper part of the neck, tail and tail coverts clear white; bill, cere and feet yellow, with the soles of the feet rough and warty; iris light yellow. Length of the female 3 feet, spread of the wings 7 feet; male 2 or 3 inches shorter. The white of the head and tail is not clear till the third year, being previously blended with grayish brown.

HISTORY.—The Bald Eagle is found in the northern parts of both continents, but is much more common on the western than on the eastern continent. It is found in all parts of the United States, and is frequently seen in Vermont, but is not known to breed within the state. This Eagle is the adopted emblem of our country, but we should hesitate to acknowledge him to be the true representative of our national character. He has the reputation of being a free-booter, living by robbing the fish hawk of his honest gains. For this purpose he takes his stand upon some lofty tree growing near the shore, and when he sees the fish hawk rise from the water with his prey, he commences the pursuit, and the fish hawk, in order to effect his own escape, is compelled to abandon the fruit of his labor, which is immediately secured by the eagle and borne away to his nest. When this eagle cannot procure a sufficient supply of fish, which is its favorite food, it preys upon other birds, and small quadrupeds and reptiles. The nest of the Bald Eagle is built in the top of some lofty tree. It is constructed of sticks lined with coarse grass. The eggs, according to Audubon, are from two to four, and are of a dull white color. They are usually hatched in May, and require the aid of the parents in procuring food till September.

THE GOLDEN EAGLE.
Falco chrysaëtos.—LINN.

DESCRIPTION.—Bill bluish gray at the base, black at the tip; cere yellow; eyebrows light blue; iris chestnut; fore part of the head, cheeks, throat and under parts, deep brown; hind head, posterior and lateral parts of the neck light brownish yellow, the shafts and concealed parts of the feathers deep brown. The back deep brown, glossy, with purplish reflections; wing coverts lighter; primary quills brownish black; the secondaries, with their coverts brown, those next the body more or less mottled with brownish white, excepting at the ends; edges of the wings at the flexure pale yellowish brown. Tail dark brown, lighter towards the base, with a few irregular whitish markings; tail long, slightly rounded. Wings long; 4th quill longest, and the 6 first abruptly cut out on the inner webs. Length 38 inches, spread of the wings 7 feet; bill along the back 2⅜ inches; edge of lower mandible 2½; tarsus 4½; middle toe and claw 4½; hind claw 2¾. Extremities of the folded wings 1 inch short of that of the tail.—*Audubon.*

HISTORY.—The Golden Eagle, though rare, is occasionally seen in Vermont and has sometimes been known to build its nest and rear its young within the state. The nest is placed upon the inaccessible shelf of some rugged precipice, and consists of a few sticks and weeds barely sufficient to keep the eggs from rolling down the rocks. The eggs are two or three in number, 3½ inches long, of a dull white color with undefined patches of brown. These eagles feed upon young fawns, hares, raccoons, wild turkies, partridges and other quadrupeds and birds, but will feed on putrid flesh, only when severely pressed by hunger.

The following description is drawn from a specimen preserved in the museum of the College of Natural History of the University of Vermont.

DESCRIPTION.—General color grayish chocolate brown resulting from the feathers being dark chocolate edged with brownish ash; feathers white at the base, which makes it appear spotted with white when the feathers are disturbed; tail with irregular whitish marks towards the base. Bill clear blue-black; upper mandible obtusely toothed; tarsus roundish, two thirds feathered; feet strong, toes rasp-like on the underside. Length from the point of the bill to the end of the tail 3 feet 7 inches, folded wing 26 inches: tail beyond the folded wings 6.5 inches; from the tip of the upper mandible along the curve to the cere 2.5, width of the cere .9, under mandible 2.9, depth of the upper bill 1.2, middle toe without the nail 2.5 inches.

This eagle was killed several years ago near Burlington. It was discovered sitting upon the beach apparently asleep, and in that condition it was approached and killed with an oar. It would appear from the partially feathered tarsus to belong to the family of sea eagles, and 1 was at first disposed to consider it the young of the Bald Eagle, but by measuring I found it to be larger than the adult of that species. Though it differs somewhat in color, it resembles Audubon's figure of the Washington Eagle more nearly than any other.

whole coast of the United States and is also seen along the lakes and rivers in the interior. It usually arrives in New England about the first of April and departs to the south again in the fall. According to Audubon some of them winter about New-Orleans. This hawk subsists, as its name would imply, principally upon fish, which it takes by hovering over the water and plunging upon them as they rise near the surface and then bears them off in its talons. They sometimes catch fishes in this way weighing four or five pounds. They breed all along the coast of the middle states. Their nest is usually placed in the top of a large tree near the shore and is of great size, sometimes measuring four feet in diameter and the same in height. It is composed of sticks intermingled and lined with sea weed and grass. The eggs are 3 or 4 in number, of an oval form, yellowish white color and spotted with reddish brown. The arrival of the Fish Hawk along the sea coast in the spring is hailed with joy by the fishermen, who regard it as the harbinger of the arrival of shoals of fishes.

THE FISH HAWK.
Falco haliætus.—SAVIG.

DESCRIPTION.—General color of the upper parts dusky brown, tail barred with pale brown. The upper part of the head and neck white, the middle part of the crown dark brown. A broad band of brown from the bill down each side of the neck; upper parts of the neck streaked with brown; under parts whitish; anterior tarsal feathers tinged with brown. Bill brownish black, blue at the base and margin; cere light blue; iris yellow; feet pale greenish blue tinged with brown; claws black. Length 23 inches; spread of the wings 54; bill, along the back, 2; tarsus 2¼; middle toe 3.—*Audubon.*

HISTORY.—The Fish Hawk is quite common during the summer along the

THE RED-SHOULDERED HAWK.
Falco Lineatus.—GMEL.

DESCRIPTION.—Color of the head, neck and back, yellowish brown, resulting from the feathers being dark brown, edged with ferruginous; wings, and wing coverts spotted and tipped with white; tail dark brown, tipped with white, crossed by four narrow grayish white bars. Breast and belly bright ferruginous, with a black line along the shafts of the feathers, and spots of yellowish white. Vent, femorals, and under tail coverts, of a light ochrey tint, with some of the feathers spotted with brown, and the outer femorals long and barred with ferruginous. Legs and feet bright yellow; bill and claws dark horn color. Length of the specimen before me, 19 inches; folded wing 13, reaching beyond the third white bar on the tail; tail 8, reaching 2½ beyond the folded wings.

HISTORY.—In Vermont this hawk passes, with several other species, under the general name of Hen Hawk, but is sometimes distinguished as the Red Hen Hawk. It confines itself more to the woods than several other species, where it may be seen flying among the trees, or sitting upon a limb watching for the appearance of a squirrel, or some other small animal, upon which he may make a repast. This hawk breeds in Vermont. Its nest is about the size of the crow's nest. It is placed in the forked branch of a high tree, made of sticks, lined with moss. Its eggs, usually four or five in number, are laid in April. They are of a broad, oval form, granular on the outside, and of a light blue color, spotted towards the small end with reddish brown. Whenever their nests are approached, they manifest much uneasiness, and their *Keé-oó* becomes very loud and angry.

THE BROAD-WINGED HAWK.
Falco pennsylvanicus.—WILSON.

DESCRIPTION.—General color of the head, back and wings above brown, tinged with buff on the neck; wings very faintly barred with black; tail short with three brownish white bars, and narrowly terminated with the same. Breast brownish buff spotted with white; belly, sides and femorals, white with the feathers thickly marked with large hastate spots of yellowish brown; vent and under tail coverts white with a few spots. A brown stripe from the mouth towards the throat; bill bluish black, nostrils oval, head large and flattened above; cere and legs yellow; legs short and strong; tarsus shielded with parallel scales; anterior outer toes slightly connected; space between the nostril and eye bristly; wings broad, the fourth quill longest; the three first abruptly notched on their inner webs. Length of the specimen before me, which is a female, 15 inches; spread of the wings 33½.

HISTORY.—This hawk bears a considerable resemblance to the preceding; it is, however, though smaller, proportionally more thick and robust, less ferruginous, has a shorter tail, and is without the white marking on the exterior of the wings. The Broad-winged Hawk breeds in Vermont, and the specimen from which the above description was made, was shot, while building her nest, in Burlington, in April, 1840. Within her were found five eggs in different stages of enlargement, one of which appeared to be fully grown with shell quite hard and in a condition to be deposited in the nest. Its color was light sky-blue finely specked with brown towards one end, with a smooth surface. The nest of this hawk is about the size of the crow's, built in the top of a tree with sticks, and lined with grass, roots and moss.

THE SLATE-COLORED HAWK.
Falco fuscus.—GMELIN.

DESCRIPTION.—Form slender; general color above reddish slate, the feathers being brown slate slightly edged with rufous; scapulars and upper tail coverts with large concealed white spots; wings obscurely barred with dark and light brown; tail with alternate bars of blackish brown and dark ash, five of each, the terminal bar being ash edged with white; chin, throat and belly yellowish white, with a line or brown stripe along the shafts of the feathers on the chin and throat, and large tear shaped reddish brown spots on the belly; thighs reddish, lighter on the outside, with large hastate spots on the outside, making them appear barred; under tail coverts pure white; bars on the under side of the wings and tail distinct; legs and feet yellow; claws black; bill bluish black; cere greenish yellow; iris bright yellow. Length of each of two specimens before me 13.4 inches, spread of the wings 24 inches, folded wing 8, tail 6.2, reaching 3.5 beyond the folded wings, tarsus 2.5, bill along the ridge .6; along the gap .8.

HISTORY.—This hawk is very common in Vermont, and generally passes under the name of Pigeon Hawk. It is usually seen in our fields and pastures, flying very swiftly near the surface of the ground in search of its prey, which consists of small birds, mice and reptiles. It sometimes approaches our dwellings and carries off young chickens. This species is very widely diffused over our country, being found, according to Audubon, as far south as Texas, and according to Richardson as far north as lat. 51°. The nest of this hawk is built sometimes in rocky cliffs and sometimes on trees. The eggs are usually four or five in number, rounded at both ends, of a livid white color, blotched with chocolate. This is the Sharp-shinned Hawk, figured and described by Audubon in his Birds of America, I—100, plate 25.

THE LARGE-FOOTED HAWK.
Falco peregrinus.—GMEL.

DESCRIPTION.—Head and hind neck grayish black, tinged with blue; the rest of the upper parts dark bluish gray, indistinctly barred with deep brown. Quills blackish brown, with elliptical reddish white spots on their inner webs. Tail grayish brown, marked with about twelve bars. Throat and fore neck white; a broad band of blackish blue from the angle of the mouth downwards; sides, breast and thighs reddish white, transversely marked with dark brown spots in a longitudinal series; under wing feathers whitish, transversely barred. Bill blackish blue at the tip, pale green at the base; cere oil green; bare orbital space orange; iris hazel; feet lemon yellow; claws brownish black. Length 16½ in.; spread of the wings 30 inches.—*Audubon.*

HISTORY.—This hawk is common to both the eastern and western continents. It is found in most parts of the United States, and, according to Audubon, has, within a few years, become much more common than formerly. I am not sure that any of this species have been taken in Vermont, but, from their being common in neighboring states, the probability of their existence here is so strong that I have thought it best to place it in my list. According to Nuttall it builds its nest in the most inaccessible clefts of rocks, and lays 3 or 4 eggs, which are of a reddish yellow color, spotted with brown.

THE GOS HAWK.
Falco palumbarius.—LINN.

DESCRIPTION.—Adult male, dark bluish gray above; the tail with four broad bands of blackish brown; the upper part of the head grayish black; a white band, with black lines, over the eyes; lower parts white, narrowly barred with gray, and longitudinally streaked with dark brown. *Young,* brown above; the feathers edged with reddish white; the head and hind neck pale red, streaked with blackish brown; the lower parts yellowish white, with oblong longitudinal dark brown spots. Length 24 inches; spread of the wings 47.—*Audubon.*

HISTORY.—This hawk is rare in Vermont, but is sometimes met with in the northern part of the state. The Gos-Hawk in Europe is sometimes trained for falconry. Its disposition is very savage, and it is withal so much of a cannibal as sometimes to devour its own young. Their ordinary food consists of young hares, squirrels, young geese, partridges, pigeons, and other smaller birds and quadrupeds. It builds its nest in the manner of the crow, in the central part of the top of a high tree. Its eggs, usually 3 or 4, are of a bluish white, marked and spotted with brown.

COOPER'S HAWK.
Falco Cooperi.—BONAP.

DESCRIPTION.—Tail rounded; tarsi moderately stout. Adult *male,* dull bluish gray above; the tail with four broad bands of blackish brown, and tipped with white; upper part of the head grayish black; lower parts transversely barred with light red and white; the throat white, longitudinally streaked *Female* similar, with the bands on the breast broader. *Young,* umber brown above, more or less spotted with white; the tail with four blackish brown bars; lower parts white; each feather with a longitudinal, narrow, oblong brown spot. Length, male 20 in., female 22,—spread 36, 38.—*Aud.* Legs and feet yellow; cere greenish yellow; iris bright yellow. Tail reaches 5 inches beyond the folded wing.—*Nuttall.*

HISTORY.—This is quite a common hawk in Vermont, and, with several others, passes under the general name of Hen Hawk. Nor is the name in this case inappropriate, since this hawk, more frequently perhaps than any other, bears off hens and chickens from the farm yard. This hawk breeds in this state, and its nest, according to Audubon, is usually placed in the forks of the branch of an oak, towards the top, and resembles that of the crow, being composed of crooked sticks, lined with grass and a few feathers. But that they do not build upon trees exclusively appears from the fact that a nest of this hawk, containing two eggs, was found, a few years ago, by George H. Peck, Esq., built upon the ground, in Burlington. The eggs are usually 3 or 4, almost globular, large for the size of the bird, of a dull, white color, strongly granulated and rough.

THE MARSH HAWK.
Falco cyaneus.—LINN.

DESCRIPTION.—Color of the *male* bluish gray; quill feathers white at their origin, and black towards the extremities; internal base of the wings, rump, belly, sides, thighs, and beneath the tail, white, without spots; upper part of the tail cinereous gray, with ends of the feathers whitish. Iris and feet yellow. *Female*, dirty brown above, with the feathers bordered with rusty ; beneath rusty yellow, with large longitudinal brown spots ; quills banded exteriorly with dark brown and black ; interiorly with black and white ; rump white, with rusty spots; two middle tail feathers banded with blackish and dark gray ; lateral feathers banded with yellowish red and blackish. Length 22 inches. Male 1 or 2 inches less. *Young* very similar to the female.—*Nuttall.*

HISTORY.—This very common species of hawk is also known by the name of Hen Hawk and Hen Harrier. It is very widely diffused, being found in Europe, Africa, North and South America, and the West Indies. This hawk builds its nest upon the ground in swampy woods, or in marshes covered with sedge or reeds. It selects a spot a little elevated above the surrounding marsh, and the nest is compactly built of dry reeds and grass. The eggs are usually four, bluish white, and sometimes sprinkled and marked with pale reddish brown. This hawk feeds upon partridges, plovers, and smaller birds, and also upon lizards, frogs, and snakes.

THE RED-TAILED HAWK.
Falco borealis.—GMEL.

DESCRIPTION.—General color dusky brown tinged with ferruginous above, beneath whitish with dark hastate spots ; wings dusky, barred with blackish ; tail rounded, extending 2 inches beyond the wings, of a bright brown or brick color, with a single band of black near the end and tipped with brownish white. Chin white, bill grayish black ; iris, cere, sides of the mouth and legs yellow, breast somewhat rust colored ; vent and femorals pale ochreous, the latter with a few heart shaped spots of brown. Length 20 to 22 inches, spread of the wings 45 inches.—*Nuttall.*

HISTORY.—The Red Tailed Hawk, according to Audubon, is a constant resident in all parts of the United States. This hawk feeds upon young hares and other small quadrupeds and birds. He is so strong and powerful as to be able to overcome and bear off doves, goslings and dunghill fowls, and his depredations upon the farmer's poultry yard are by no means of rare occurrence. And yet he is so shy and wary, that it is extremely difficult to approach near enough to shoot him with a gun, of the use of which he, like the crow, seems to have an intuitive knowledge. The best method of getting a shot at these wary birds in open land is to approach them on horseback. The Red-Tailed Hawk breeds in Vermont. Its nest is built in the fork of a lofty tree, and is composed of sticks, twigs, coarse grass and moss. The eggs are 4 or 5, of a dull white color, blotched with brown and black.

THE PIGEON HAWK.
Falco columbarius, LINN.

DESCRIPTION.—Whole upper parts of a deep dusky brown except the tail which is crossed by five narrow whitish bars ; beneath yellowish or reddish white, spotted and streaked with brown. The bill is of a light bluish gray, tipped with black ; cere and skin round the eye greenish ; iris deep hazel ; legs yellow ; claws black ; feathers on the thighs remarkably long. *Female* with the cere and legs greenish yellow ; upper parts dark grayish brown ; the lower pale and spotted as in the male. *Young* with the head reddish brown, streaked with dusky, in other respects resembling the female. Length of the male 11 inches, spread of the wings 23.—*Nutt. Aud.*

HISTORY.—The Pigeon Hawk is much less common than several other of the smaller species of hawk. Audubon informs us that this hawk breeds in Nova Scotia, New Brunswick and Labrador. The nests are usually placed upon the top of small firs with which those countries abound, at the height of 10 or 12 feet from the ground. They are built of sticks slightly lined with moss and feathers. The eggs are usually five, and are an inch and three quarters in length. Their ground color is a dull yellowish brown, thickly clouded with irregular blotches of dull dark reddish brown. This hawk is shy and watchful, seldom being seen out of the forests. It feeds upon small birds, mice and reptiles.

GENUS STRIX.

Generic Characters.—Beak compressed, bent from its origin ; base surrounded by a cere, covered wholly, or in part, by stiff erect hairs ; head large, much feathered ; nostrils lateral, rounded, open, pierced in the anterior margin of the cere, concealed by hairs directed forwards ; eyes very large ; orbits surrounded by feathers ; legs and feet feathered, frequently to the very claws ; feet with three toes before and one behind, separate ; the exterior reversible ; first quills dentated on their anterior border, the third longest.

This Genus embraces the Owl Family, and is now divided by naturalists into no less than six genera. The owls are called nocturnal birds of prey, because they seek their prey chiefly by night. The pupil of the Owl's eye is so large

THE SCREECH OWL.
Strix asio.—LINN.
Bubo asio.—Aud. Birds Am. I—147, pl. 40.

DESCRIPTION.—Upper parts pale brown, spotted and dotted with brownish black; a pale gray line from the base of the upper mandible over each eye; quills light brownish gray, barred with brownish black; their coverts dark brown; secondary coverts with the tips white; throat yellowish gray, lower parts light gray, patched and sprinkled with brownish black; tail feathers tinged with red. *Young*, with upper parts light brownish red; each feather with a central blackish brown line; tail and quills barred with dull brown; a line over the eye and the tips of the secondary coverts reddish white; breast and sides light yellowish gray, spotted and lined with brownish black and bright reddish brown; the rest of the lower parts yellowish gray; the tarsal feathers pale yellowish red. Length 10 inches; spread 23.—*Aud.*

HISTORY.—This little owl is found in nearly all parts of the United States, but is much more common in northern than in southern sections. The Screech Owl is by no means rare in Vermont, and many a Green Mountain lad, as he has been passing through a wood in a dark night has felt his hair rise, his heart leap, and himself flying as upon wings of the wind, at the terrific scream of this bird, perched in a tree just over his head. Although more common in the fall and fore part of winter, many of them spend the summer and rear their young in this state. Their nest, which is made of grass and feathers, is placed at the bottom of a hollow tree or stub, often not more than 6 or 8 feet from the ground. The eggs are white, of a globular form, and usually 4 or 5 in number. Only one brood is raised in a season. The young become fully feathered in August, when they appear as described above. This owl is often designated as the Little Screech Owl, and is also called the Mottled Owl.

THE HAWK OWL.
Strix funerea.—GMELIN.
Surnia funerea—Aud. Am. Birds, I—112, pl. 27.

DESCRIPTION.—Tail long, much rounded, the lateral feathers two inches shorter than the middle. Upper part of the head brownish-black, closely spotted with white; hind neck black, with two broad longitudinal bands of white spots; the rest of the upper part dark brown, spotted with white; tail with eight transverse bars of white, the feathers tipped with the same; facial disks grayish white, margined with black; lower parts transversely barred with brown and dull white.—*Aud.* Bill yellow; feet thickly feathered; nails horn-color.—*Nutt.* Length of the male 16 inches; spread of the wings 32; female larger.

HISTORY.—This species forms the connecting link between the hawks and the owls, having, in several respects, a considerable resemblance to both, and hence its name, *Hawk-Owl.* We are informed by Dr. Richardson that this owl is common throughout the fur countries from Hudson's bay to the Pacific ocean, and that it is more frequently shot than any other. It must, however, be a rare bird in the United States, generally, since the indefatigable Audubon confesses that he has never seen it alive. But it is because he has not visited the north part of our own state that he has been denied this pleasure; for he is assured by no less authority than Dr. Thomas M. Brewer, of Boston, that the Hawk-Owl is so common about Memphremagog lake in Vermont, that a dozen of them may be procured by a good gunner in a day, and that their nests, which are in hollow trees, are frequently met with. Its eggs, according to Richardson, are white, and usually two in number.

THE SNOWY OWL.
Strix nyctea.—LINNÆUS.
Surnia nyctea.—Aud. Am. Birds, I—113, pl. 28.

DESCRIPTION.—General color white, more or less spotted and barred with brown; the tail rounded and extending a little beyond the folded wings; the second and fourth quills equal, the third longest; bill bluish black, curved from the base; upper mandible thickly studded with stiff, bristly white feathers; throat and legs covered with soft, pure white down, which becomes hairy upon the feet, and nearly conceals his long, black, and sharp claws. Length of the specimen before me 27 inches; spread of the wings 56 inches; longest quill 15 inches.

HISTORY.—The principal residence of this species of owls is in the northernmost parts of both the eastern and western continents. It is very common in Lapland, Iceland, and in the countries around Hudson's Bay, and its large size and thick downy plumage are well fitted to resist the climate of those icy regions. "In those dreary wilds, surrounded by almost perpetual winter, he dwells, breeds and obtains his subsistence. His white robe renders him scarcely discernible from the overwhelming snows where he reigns like the boreal spirit of the storm. His loud, hollow, barking growl, *'whowh 'whowh, 'whowh, hah, hah, hah*, and other more dismal cries, sound like the unearthly ban of the infernal Cerberus, and heard amidst a region of cheerless solitude, his lonely and terrific voice augments rather than relieves the horrors of the scene."* The Snowy Owl seeks his food by day as well as by night, and in the midst of winter many of them are compelled to proceed to the southward to procure the means of subsistence. At such times they are seen, usually in pairs, in various parts of the U. States. They do not make their appearance in Vermont until winter is fully set in, and leave us with the earliest indications of spring. They breed in the regions far to the north, and are said to make their nest upon steep rocks, or old pine trees, and to lay two eggs, which are of a pure white. They feed upon other birds, mice, rats, and other small quadrupeds.

THE GREAT HORNED OWL.
Strix virginiana.—GMEL.

Bubo virginianus.—AUD. Am. Birds, I—143, pl. 39.

DESCRIPTION.—Bill black; iris bright yellow. Above whitish and ferruginous, thickly mottled with dusky; face ferruginous, bounded by a band of black. A whitish space between the bill and the eyes. Beneath marked with numerous transverse dusky bars on a yellow and white ground; vent paler. Feet covered with hair-like pale brown feathers; tail rounded and broad, reaching an inch beyond the wings, mottled with brown and tawny and crossed with 6 or 7 narrow bars of brown; chin whitish. Horns broad, 3 inches long, formed of 12 or 14 feathers, with black webs and edged with

* Nuttall.

brownish yellow. Length of the male 21 inches, female 2 inches longer.—*Nutt.*

HISTORY.—This is one of the largest species of American Owls, and is found through all the regions from the gulf of Mexico to Hudson's bay. It breeds in this state and in some of the unsettled woody parts is quite common. Its nest, which is large, is built of dry sticks and lined with leaves and some feathers. The eggs are from three to six in number, about the size of those of the common hen, but rounder and of a yellowish white color. This owl is often called the *Cat Owl*, from the resemblance of its face to that of the cat. It confines itself mostly to the retired and dark thickets of the forests, and particularly to thickets of spruce and other evergreens, and, in many places during the summer these owls may be heard responding to one another their *waugh ho! waugh ho! waugh hoo*—during the whole night. Their food consists of various kinds of birds, hares, squirrels and other quadrupeds, and they sometimes come around our barns, and carry off our domestic fowls. These owls are said sometimes to have pounced upon cats, mistaking them perhaps for rabbits, but finding themselves to have caught a Tartar, they are generally very willing to relinquish their grasp.

THE CINEREOUS OWL.
Strix cinerea.—GMEL.

Syrnium cinereum.—Aud. Am. Birds, I—130, pl. 35.

DESCRIPTION.—Upper parts grayish brown, variegated with grayish white in irregular undulated markings; the feathers on the upper part of the head with two transverse white spots on each web; the smaller wing-coverts of a darker brown, and less mottled than the back; the outer scapulars with more white on their outer webs; primaries blackish-brown toward the end, in the rest of their extent marked with a few broad light-gray oblique bands, dotted and undulated with darker; tail similarly barred; ruff-feathers white towards the end, dark brown in the centre; disks on their inner sides gray, with black tips, in the rest of their extent grayish-white with 6 bars of blackish-brown irregularly disposed in a concentric manner; lower parts grayish-brown, variegated with grayish and yellowish white; feet barred with the same. Length $30\frac{1}{2}$ inches; spread, 48.—*Aud.*

HISTORY.—This is the largest species of owl known in this country. It is only occasionally met with in the northern parts of the United States, but further north it is by no means a rare bird, being according to Dr. Richardson common in the woody districts between Hudson's Bay and the Pacific ocean, as far north as the 68° of latitude. Dr. R. found a nest of one of these owls on the 22d of May, containing three young. It was built of sticks on the top of a balsam poplar, and was lined with feathers. The eggs are said to be spotted. This owl is rarely seen in this state, but occasionally makes his appearance here in the depth of winter.

THE SHORT-EARED OWL.
Strix brachyota.—LATHAM.
Otus brachyotus--Aud. Am. Birds I—140, pl. 38.

DESCRIPTION.—Ear-like tufts inconspicuous, consisting of 2 or 3 short feathers ; general color ochreous spotted with blackish-brown ; face round, the eyes blackish ; tail ochreous with about 5 brown bands, not extending beyond the wings, and tipped with white ; beneath yellow with longitudinal spots of blackish-brown ; iris bright yellow ; bill black ; feet and toes feathered. *Female* with the general tints paler. Length from 13 to 15 inches.—*Nutt.*

HISTORY.—This species migrate to the south in the fall, and during the winter are so numerous in Florida that Audubon says that he has shot no less than seven of them in a single morning. They proceed to the north on the approach of spring for the purpose of rearing their young, but some of them are known to spend the summer, and, occasionally, to breed as far south as Pennsylvania. This owl is found in Vermont, and I am assured by Dr. Brewer that it breeds in the northeastern part of the state. It builds its nest upon the ground, and its eggs, which are about four, are of a dull bluish white color. The short-eared owl is attracted by nocturnal fires, and will sometimes approach so near as to be knocked down with a stick.

THE BARRED OWL.
Strix nebulosa.—LINNÆUS.

DESCRIPTION.—General color umber-brown, spotted and barred with white and yellowish white above ; beneath whitish, barred transversely on the breast and longitudinally on the belly with umber brown, and having large sagittate spots of the same on the feathers towards the tail; tail long, reaching 4 inches beyond the folded wings, rounded, tipped with white, convex above, and crossed by six broad bars of umber brown, separated by narrow bars of yellowish white ; plumage in front of the eye ends in long black hairs ; bill yellow ; legs covered with feathers, extremities of the toes covered with scales ; nails long, sharp, and of a dark horn color. Length 20 inches.

HISTORY.—The Barred Owl inhabits both the eastern and western continents. It is found in all parts of the United States, and is one of the most common owls found in Vermont. It does not confine itself to the woods, but comes around our dwellings and is often seen among our shade trees and orchards in the midst of our villages. I have before me two specimens, both of which were shot in the village of Burlington. Their food consists of young hares, squirrels, mice, grous and other birds, and also of frogs and other reptiles. They sometimes destroy chickens. This owl, according to Audubon, does not build a nest, but lays its eggs, in the latter part of March, upon the soft rotten wood in a hollow tree, and sometimes in the old nest of a crow or red-tail hawk. The eggs are of a globular form, pure white, with a smooth shell and from 4 to 6 in number.

THE SAW-WHET.
Strix acadica.—GMEL.
Ulula acadica.--Aud. Am. Birds; I—123, pl. 33.

DESCRIPTION.—General color above olivaceous brown, scapulars and some of the wing-coverts spotted with white ; the first six primary quills obliquely barred with white ; tail darker, with two narrow white bars ; upper part of the head streaked with grayish-white ; ruff white, spotted with dusky. Lower parts whitish ; the sides and breast marked with broad elongated patches of brownish-red. Length of the male 7½ inches, spread 17. Female 8½, 18.—*Audubon.*

HISTORY.—This little owl is not uncommon in Vermont, and it is generally known by the name of *Saw-Whet;* and this name is derived from the sound of its peculiar note, which resembles that of the filing of the teeth of a large saw. People, who are unacquainted with this bird, travelling in the forest, are often deceived by its note, supposing themselves to be approaching a saw-mill, while far remote from any settlement. Audubon relates that he himself was several times deceived in this way. This bird is sometimes called the *Little Owl,* or 'Little Acadian Owl.' It is retired and solitary in its habits, confining itself during the day to evergreen and other thickets of the forest. For rearing its young, the Saw-Whet takes possession of the old nest of a crow, or some other large bird, or of a hollow cavity of an old tree. The eggs are of a form approaching to globular, are of a glossy-white color, and are from three to six in number. This owl feeds upon mice, beetles, moths and grasshoppers.

THE BARN OWL.
Strix americana.—AUDUBON.

DESCRIPTION.—Bill pale grayish yellow; claws and scales brownish yellow. General color of the upper parts grayish brown, with light yellowish-red interspersed, produced by very minute mottling, each feather having towards the end a central streak of deep brown terminated by a small oblong grayish-white spot; wings similarly colored; secondary coverts and outer edges of primary coverts with a large proportion of light brownish-red, fading anteriorly into white, each feather having a small dark brown spot at the tip. Length and spread, male 17, 42; female 18, 46.—*Audubon.*

HISTORY.—This owl, though very common in the southern states, is so rare at the north-east, that Audubon says that he has never seen it to the eastward of Pennsylvania, and yet I am assured by Dr. Brewer that it is not only found in Vermont, but breeds here. This owl is entirely nocturnal in its habits, and when disturbed in the day time flies about in a irregular, bewildered manner. Audubon supposes its food to consist entirely of small quadrupeds. This owl is said to bear a close resemblance to the *Strix flammea*, or White Barn Owl.

OMNIVOROUS BIRDS.

These have the bill robust, medium-sized, and sharp on the edges; upper mandible more or less convex, and notched at the point; feet with four toes, three before and one behind; wings of medium length; quill feathers terminating in a point. They live, for the most part, in companies or flocks and are monogamous. The greater part of them build their nests on trees, but some of the species occupy the crannies of old walls, and some build upon the ground. Their principal food consists of insects, worms and carrion, to which they often add grain and fruit.

GENUS STURNUS.—*Linnæus.*

Generic Characters.—The bill in the form of a lengthened cone, depressed and somewhat blunt, with the edges vertical; above somewhat rounded. Nostrils partly closed by an arched membrane. The tongue narrowed, sharp, and cleft at the point; the hind nail longest and largest; the first quill short, the second and third longest.

THE MEADOW LARK.
Sturnus ludovicianus.—LINNÆUS.

DESCRIPTION.—The color above is variegated with black, bright bay and ochreous; beneath and a line over the eye bright yellow; a black crescent on the breast; tail wedge-form, feathers pointed, and the four outer ones nearly all white; bill brown above, bluish white beneath, conical with deep rounded sinuses at the base; legs and feet large, reddish white. The sexes differ but little in color, but in the young the yellow is much fainter. Length of the specimen before me 10 inches; folded wing, 5.

HISTORY.—The Meadow Lark is a harmless bird, and is common in all parts of the United States, and particularly so in Vermont, where it breeds in large numbers. Their residence is chiefly in meadows and old fields. They build their nest in some thick tuft of dry grass. It is usually constructed of the coarse grass, lined with finer blades of the same, and approached by the bird through a concealed covered way, and hence they are not readily found. The eggs are large and white, with a bluish tint, and marked with brownish spots. They are usually 4 or 5 in number. The food of the Meadow Lark consists of the larvæ of various kinds of insects, worms, beetles and grass seeds; but it does not meddle with fruits and berries. It is of a shy, timid and retiring disposition, usually spending the whole summer in the moist meadows, and only retiring from them on the approach of winter.

GENUS ICTERUS.—*Brisson.*

Generic Characters.—Bill in the form of an elongated sharp pointed cone, somewhat compressed, rounded above, and rarely somewhat curved; with the margins inflected. Nostrils oval, covered by a membrane. Tongue sharp and cleft at the tip. Tarsus longer than the middle toe; inner toe but little shorter than the outer, and nearly equal to the hind one; middle toe longest; hind nail twice as large as the others. Wings sharp; first and second primary, but little shorter than the third and fourth, which are longest. The female very different from the male, and the young resemble the female.

THE BALTIMORE ORIOLE.
Icterus Baltimore.—BONAPARTE.

DESCRIPTION.—Color of the shoulders, rump, lateral tail feathers, breast and belly bright orange; head, back, wings, middle tail feathers and chin black; wing feathers and coverts slightly edged with white on their outer webs; bill bluish horn color; legs, feet and nails brownish; iris hazel. In the female and young the orange is pale, and the parts which are black in the male are grayish; tail even; hind toe and nail strongest; bill very acute; 2d and 3d primaries equal and longest. Length of the specimen before me 7 inches; folded wing, 4¾.

HISTORY.—The Baltimore Oriole, or *Golden Robin*, as he is here more commonly called, is one of our most gay and lively birds. It arrives in Vermont in the early part of May, and about the beginning of June may be seen busily engaged in the construction of its nest. For this purpose they usually select a flexible branch of a tree standing on the side of a gentle declivity. The nest is suspended from this by strings or threads in the form of a pendulous cylindrical pouch 5 or 6 inches in depth. The exterior is formed of strings, strips of bark and other fibrous substances, and the interior lined with grass, moss, wool, hair or downy substances. The eggs are usually 4 or 5 in number. They are white with a faint tinge of blue, and are usually marked at the large end with irregular brownish lines and spots. The period of incubation, according to Audubon, is 14 days, and the same pair frequently rear two broods in a season. Though shy and suspicious, they seem to prefer building their nests upon the high trees in the open land by the side of roads and about farm-houses. They feed their young principally with soft caterpillars, and the male and female both unite in this labor. The food of the old birds consists mostly of caterpillars and insects of different kinds. They are also fond of cherries, currants and strawberries, but do not often commit depredations upon these fruits in our gardens. They are thought to possess an extraordinary relish for green peas, as they sometimes attack those growing in our gardens. They split open the pod without detaching it from the vine, and, as is generally supposed, for the purpose of obtaining the young and tender peas. But Mr. Peabody informs us that it has been ascertained by Dr. Harris, that the Oriole opens the pods not for the sake of the peas, but for the grub of the pea-bug; and that instead of mischief, he is performing a service, for which he is more deserving of gratitude than reproach. Although we have several birds which occasionally do a little mischief in our fields and gardens, it is at least doubtful whether we have any which would not be found to be beneficial rather than otherwise, were their history fully known. From its manner of building, this bird is often called he *Hang Bird*, or *Hang Nest*.

THE RED-WINGED BLACK-BIRD.
Icterus phœniceus.—DAUD.

DESCRIPTION.—Color of the *male* rich glossy black, with the exception of the lesser wing coverts, in which the lower row of feathers is of a buff orange color tipped with white, and the rest of a bright scarlet; legs, feet and bill glossy black, the latter an elongated, straight, sharp-pointed cone, slightly flattened in front; iris hazel; tail rounded, reaching 2 inches beyond the folded wings. Length of the specimen before me 9 inches, the folded wing 5 inches, spread of the wings 13½ inches. The *female* is considerably smaller than the male, and her general color dull reddish brown. The lesser wing coverts usually exhibit something of the reddish and orange hue, but seldom, if ever, is the bright scarlet observed in the female.

HISTORY.—This singularly marked bird usually arrives in Vermont early in April, and takes up its residence in flocks in the marshes and swamps. Here they commence building their nests about the mid-

THE COW BLACK-BIRD.

dle of May. These are usually constructed in a thicket of alders, or other bushes, at the height only of a few feet from the ground, and are made of the leaves of flags, swamp-grass, &c., something in the form of that of the Golden Robin. The eggs, varying from 3 to 5 in number, are bluish white, with irregular faint purple markings on the larger end. About the beginning of September they begin to collect in flocks, and sometimes do considerable damage to the unripe corn. But it is believed that the advantage derived from these birds in the destruction of larvæ and insects in the spring of the year vastly more than compensates for all the damage they do. It is stated by Kalm, that after a great destruction of these and the common Black-Birds for the legal reward of 3d. per dozen, in 1748, the worms and grubs multiplied so exceedingly as to destroy a great part of the grass in New England.*

THE COW BLACK-BIRD.
Icterus pecoris.—Tem.

DESCRIPTION.—Color glossy black with violet reflections from the back and breast; head and neck above and below dusky cinamon brown; bill robust, conical,acute, slightly compressed towards the end, and of a glossy black color; upper mandible rounded and encroaching a little upon the forehead, sides of the lower mandable inflected; nostrils basal and partly covered; neck short, body robust; tarsus compressed, acute behind and covered anteriorly with seven longish scutella; toes free, lateral ones nearly equal; legs, feet, and claws brownish black. Tail rather short and slightly forked. Wings longish, curved, slightly rounded and the 2d and 3d quills longest. Length of the specimen before me 7 inches; folded wing 4½, spread of the wings 12, tail reaches 1 inch beyond the folded wing. Female less than the male, and of a dusky color.

HISTORY.—The Cow Black-Bird derives its name from its habit of being much among the cattle as they are feeding in the pastures. Its food consists almost entirely of insects, and it might be regarded as a public benefactor were it not for certain habits which render it detestable and prevent its receiving the credit to which its good qualities would otherwise entitle it. Being strangers to the joys which spring from conjugal fidelity and having a strong aversion to domestic cares, this bird contrives to escape them by laying its eggs in the nests of other birds. This it does in the absence of the owners of the nest, and when the owners return they usually manifest much uneasiness and make strong efforts to throw out the intruded egg. When they do not succeed in this, they often build a flooring over the strange egg and elevate the sides so as to form a new nest within the old. But in many cases circumstances will not allow them time for this labor, and then they are obliged patiently to submit to the imposition. The egg of the Cow-Bird is always hatched first, and the young by its superior size often smothers the lawful heirs. The proprietors of the nest, however, feed the foundling and treat it with the same kindness as if it were their own offspring.

A case of this intrusion of the Cow Black-Bird occurred in Burlington in 1840, in the garden of my friend R. G. Cole, Esq. Cashier of the Burlington Bank. He had noticed a pair of common yellow birds, *Fringilla tristis*, busily engaged for several days in building a nest upon one of his trees. A day or two after he had supposed it complete, he noticed that it had suddenly undergone a very considerable enlargement, so much so that his curiosity was excited, and upon examining it he found that it consisted of two nests, one within the other, and that the lower nest contained an egg of the Cow Black-Bird. The upper nest was entirely of cotton, and upon the circumstance being known, it was found that my friend Mr.S.E. Howard,whose yard is adjacent to the garden containing the nest, had observed two birds eagerly searching his premises for building materials, and that he had, with his accustomed liberality, purposely thrown out several handfuls of cotton, all of which disappeared in the course of a few hours, and were found neatly wrought into the nest above-mentioned.

The egg of the Cow Black-Bird is a little larger than that of the Blue bird, oval, whitish tinged with green and spotted with brown. Its notes are affected and unpleasant.

* Travels in North America, 1—372.

THE BOB-O-LINK.
Icterus agripennis.—BONAP.

DESCRIPTION.—The spring dress of the *male*:—the top of the head, wings, tail, sides of the neck, and whole under plumage, black, with the feathers frequently skirted with brownish yellow ; back of the head yellowish white ; scapulars, rump, and tail coverts white, tinged with ash ; extremities of the tail feathers similar to those of the woodpeckers ; bill bluish black ; legs dark brown. Color of the *female*, the *young*, and the male, in autumn and winter, varied with brownish black and brownish yellow above, dull yellow beneath. Length of the specimen before me 7 inches ; spread of the wings 11½ inches.

HISTORY.—This is a common bird in the summer throughout the United States. In many parts it is called the *Rice* Bird, or Rice Bunting, from the circumstance of its feeding much upon wild rice. It is also sometimes called the *Skunk Black Bird*, from the resemblance of its black and white markings to those of the skunk. But *Bob-o-link* is its most common designation. This bird does not usually make its appearance in Vermont till the latter part of May, and the males are generally seen a few days earlier than the females. They take up their residence in the low meadows, and upon these and the neighboring ploughed fields they destroy vast numbers of insects and larvæ; and this kind of food being abundant, they seldom leave it for the purpose of doing injury by feeding upon grain or fruits. Hence they are rather regarded as benefactors, and being of an animated, jovial turn, though somewhat boisterous, they are received on their return in the spring with a hearty welcome. The Bob-o-link builds its nest on the ground, among the grass. It is placed in a slight depression and constructed of grass, coarse on the outside and lined with that which is finer. The female lays from 4 to 6 eggs, which are of a dull yellowish white color, spotted with brown. About the last of July the males put off their black and white nuptial dress, and assume the gray, unostentatious garb of the female and the young, and by the middle of August they begin to collect in flocks in the swamps and wet meadows, and soon after leave for a more southern climate.

GENUS QUISCALUS.—*Vieillot.*

Generic Characters.—Bill bare, compressed from the base, entire, with sharp edges bent inwards ; upper mandible forming an acute angle with the feathers of the head, curved from the middle, projecting beyond the lower, and provided with a long heel within. Nostrils oval, half closed by a membrane. Tongue cartilaginous, flattened, torn at the sides and cleft at the point. Tarsus a little longer than the middle toe; inner toe free, outer one united at the base to the middle one. Wings moderate in length ; 1st primary equal to the 5th, and but little shorter than the 2d, 3d, and 4th, which are longest. Tail of 12 feathers, more or less rounded.

COMMON CROW BLACK-BIRD.
Quiscalus versicolor.—VIEILLOT.

DESCRIPTION.—*Color* of the head, neck, and breast, deep violet, with greenish and purplish reflections; back, belly, and scapulars dark bronze color; wings and tail reflecting various shades of purple, with green blue and coppery tints. Bill and legs black. Upper mandible longer, but not so stout as the lower, and the keel within large. Feet and claws strong. Iris bright gamboge yellow. Tail of 12 feathers, rounded or wedge form, and reaching 3 inches beyond the folded wings. Length of the specimen before me 12 inches ; tail 5¼; folded wings 5.7 ; bill above 1.2, to the angle of the mouth 1.4. Length of the female usually 11 inches.

HISTORY.—The Crow Black Bird is an active and sociable bird, which warns us by his loud, clanking note, late in the spring, that he is once more in our fields and gardens, apparently unconscious that there can be any objection. He is one of those creatures concerning which it is difficult to say whether they are friends or foes ; sometimes they are the one and

THE RUSTY BLACK-BIRD.

sometimes the other, and it is only by striking a balance between the service and injury, that we can determine how to regard them. That he pulls up corn for the sake of the seed is undeniable; but it is also true that he devours immense numbers of insects, grubs and caterpillars. Perhaps it may be possible to secure his services and prevent his depredations. Some attempts to effect this object have already been made, by soaking the seed in some solution, which shall make it less palatable to the bird.* Crow Black Birds build their nests in communities, sometimes on bushes and sometimes on lofty trees, and several nests are frequently seen upon the same tree. The nest is composed outwardly of mud and coarse grass, and is lined inwardly with fine grass, hair, &c. The eggs, usually 5 or 6, are greenish, spotted with dark olive. Only one brood is usually reared in a season. About the time the leaves fall in autumn the old and young collect in very large flocks and commence their migration to the south, laying the whole country under contribution as they advance.

THE RUSTY BLACK-BIRD.
Quiscalus ferrugineus.—LATH.

DESCRIPTION.—General color of the *male* deep black, with greenish and bluish reflections; bill and feet black; iris pale yellow. Wings long; second quill longest; tail long, slightly rounded; plumage soft, blended, and glossy. Bill straight, tapering, and compressed from the base; nostrils, basal, oval, half closed above by a membrane. Body rather slender; feet strong; tarsus covered anteriorly with a few long scutella. Length 9¼ inches; spread 14½, in males. General color of the *female* brownish black; the sides of the head over the eyes, and a broad band beneath it, light yellowish brown; the feathers of the lower parts more or less margined with brownish. Bill, iris, and feet as in the male.—*Audubon.*

HISTORY.—The Rusty Black Bird, called also the *Rusty Grakle*, passes through this state in its spring and fall migrations, and is sometimes seen here in considerable flocks, particularly in the fall. Some of them probably breed in the north part of the state. They resemble the Red-winged Black Birds in their habits and in the construction of their nests, which are built upon low bushes in moist meadows. The eggs are 4 or 5, of a light blue color, streaked and dashed with lines of brown and black.

* Peabody.

GENUS CORVUS—*Linnæus.*

Generic Characters.—Bill thick, straight at its base, slightly bent towards the point; nostrils basal, open and hidden by reflected bristly feathers; feet with three toes before and one behind, divided; the tarsus longer than the middle toe; wings pointed; first quill short, third and fourth longest. The tail consists of 12 feathers.

THE CROW.
Corvus americanus.—AUDUBON.

DESCRIPTION.—Color black and glossy, with violet reflections from the wings, tail and shoulder feathers; tail rounded, and extending an inch and a half beyond the folded wings; bill, legs, feet and claws black; bristly feathers incumbent upon each side of the bill covering the nostrils; the fourth quill feather longest; usual length 19 inches.

HISTORY.—The Crow is found in all parts of the world, and is one of the few large birds which pass the whole winter in Vermont. During the winter the Crows reside in flocks, but on the approach of spring they separate into pairs, and retire into the forests for the purpose of rearing their young. During this period they are vigilant, suspicious, and upon any real or supposed intrusion upon their purpose they become very noisy. They build their nests upon lofty trees, and usually select for that purpose such as have thick tops, in which the nests can be more effectually concealed. On this account the pine and other evergreens are often chosen. The nest is constructed exteriorly of sticks, plastered with earth, and lined with moss, wool, or other soft substances. Their eggs, from 4 to 6 in number, are of a pale green color, marked with streaks and blotches of brown. The Crow is omnivorous, devouring insects, worms, carrion, fish, grain, fruits, snakes, frogs and other reptiles, and also the eggs of other birds. In the spring of the year he does the agriculturist considerable damage by pulling up the young Indian corn for the sake of the kernel, on which account a bounty of 10 cents a head for his destruction was, for a time, authorized by legislative enactment. To prevent his depredations upon the corn fields various kinds of scare-crows have been devised, but that which is most commonly resorted to at present, consists in stretching threads of cotton yarn across the field in various directions. To compensate for the mischief which they do, it must be acknowledged that crows do the farmer some service by the destruction of grubs and insects, besides acting as general scavengers in removing the carcases of dead animals.* It is said they know how to break open nuts and shellfish, in order to eat what is within, by letting them fall from a great height upon the rocks below; and there is a story that, as a certain ancient philosopher was walking along the sea-shore gathering shells, one of these unlucky birds, mistaking his bald head for a stone, dropped a shell-fish upon it, and thus killed at once a philosopher and an oyster.*

The crow is easily tamed, and soon learns to distinguish those who have the care of him, but is of a thievish propensity, and often carries off valuable articles and hides them by thrusting them into holes and crevices.

THE RAVEN.
Corvus corax.—LINNÆUS.

DESCRIPTION.—Color of the plumage deep black, glossed with blue and purplish blue, the lower parts with green; feathers of the foreneck lanceolate and elongated; tail much rounded, reaching 2 inches beyond the wings; nasal feathers half the length of the bill; bill and feet black; iris dark chestnut brown. Length 26 inches, spread 50.—*Aud. Rich.*

HISTORY.—The Raven is a well known bird, being found in almost all parts of the world. Dr. Richardson says that it abounds in the fur countries, and extends its migrations northward even to the polar seas. It has for several years been less frequently seen in Vermont than formerly, and it was always a rare bird here compared with the crow. It feeds principally upon the carcasses and offals of the larger animals which are slain by hunters or wolves, or that die by disease. The Raven does not, like the crow, build its nest upon a tree, but in the inaccessible clefts of lofty precipices. The Raven is easily tamed, and manifests much attachment to its keeper. It may be taught to imitate the human voice and to articulate many words very distinctly.

* Nuttall.

THE BLUE JAY.
Corvus cristatus.—LINNÆUS.

DESCRIPTION.—General color light blue above, grayish white beneath; a stripe of black passes over the head and down on each side of the neck, forming a collar under the throat; a black spot before each eye connected by a black line over the base of the bill; crest pale blue in front, approaching to black on the back part; outer webs of the primaries, and both webs of the secondaries and wing coverts bright blue, the two latter barred with black and tipped with white; tail of 12 feathers, wedge-form, bright blue, barred with black excepting the two outer feathers, and tipped with white excepting the two inner ones; mouth, bill, legs, feet and claws black. Length of the specimen before me 11 inches.

HISTORY.—The Blue Jay is one of our most elegant and lively birds. It is common in every part of the United States, and is found as far north as the 56th° of latitude. It breeds in Vermont as well as in almost or quite every other state in the Union. They are somewhat migratory, most of them proceeding to the south in the fall. Audubon says they are very numerous in the southern states during the winter. They are most plentiful in Vermont in autumn, when they commit depredations upon fields of corn and oats. The greater part of them proceed to the south before winter sets in, but some remain with us after the snows fall, and purloin a scanty subsistence from our corn cribs and granaries. These birds are truly omnivorous, feeding upon almost any thing which falls in their way. In the summer season it destroys the eggs and young of other birds. When confined in a cage with several other birds, it has been known to kill and devour them all. The Blue Jay is a very active, noisy bird, and is capable of imitating the voice of the sparrow-hawk so nearly as to frighten all the small birds in the neighborhood. Its nest, which is composed of twigs and fibrous roots, is built in trees. The eggs are 4 or 5, of a dull white color, spotted with brown.

THE CANADA JAY.
Corvus canadensis.—LINN.

DESCRIPTION.—General color dark leaden gray; hind head black; forehead, collar beneath, and tip of the tail brownish white; interior veins of the wings brown and partly tipped with white; bill and legs black; iris dark hazel; plumage of the head loose and prominent; tail long and wedge-shaped. Sexes alike in color. Length 11 inches; spread, 15.—*Nuttall.*

HISTORY.—This jay, which is called in some places the *Whiskey Jack*, and in others the *Carrion Bird*, inhabits principally between the 44th and 65th parallels of north latitude. It is found in the state of Maine, and in the north parts of New Hampshire, Vermont and New York, but is seldom seen further to the southward. It breeds in each of the states above named. The nest is usually placed in the thick top of a spruce or fir, at the height of 6 or 8 feet from the ground. It is placed near the trunk of the tree, and is made of twigs and fibrous roots, lined with moss and grass. The eggs are from 4 to 6, of a light gray color, faintly marked with brown. They feed, during the summer, upon worms and insects, and, during the winter, they are driven by necessity to feed upon the buds and leaves of spruce and fir.

GENUS PARUS.—*Linnæus.*

Generic Characters.—Bill short, straight, conic, compressed, entire, edged and pointed, having bristles at the base; the upper mandible longer, rounded above and slightly curved; nostrils at the base of the bill, rounded and concealed by the advancing feathers; tongue blunt and cleft or entire, and acute; feet rather large, toes almost wholly divided; the nail of the hind toe strongest, and most curved; fourth and fifth primaries longest. The *female* and *young* differ but little from the adult male. Moult, annual; plumage, long and slender.

THE CHICADEE.
Parus atricapillus.—LINN.

DESCRIPTION.—The whole upper part of the head, nape, chin and throat, velvet black; a white line from the nostril passing beneath the eye, spreads out upon the side of the neck; back ash color; quill and tail feathers brownish black, edged with grayish white; belly brownish white, deepening into brownish yellow upon the sides and beneath the tail; bill black; legs and feet bluish; fifth quill feather longest; fourth and sixth nearly as long; tail long and rounded. Length 5½ inches, tail 2¾; folded wing 2.7, spread of the wings 6¾.

HISTORY.—The Chicadee, or Blackcap- Titmouse, seems to be common through the whole continent, from Mexico to the 65th degree of north latitude. They rear their young in all parts of the United States. For that purpose they take possession of the hollow of a decayed tree or of the deserted holes of the woodpecker, or where these are not to be had they excavate a cavity for themselves in some rotten stub of a tree. The materials of which the nest is composed, according to Audubon, vary in different districts, but are generally the hair of quadrupeds in considerable quantities, and disposed in the shape of a loose bag or purse lining the inside of the excavation, while others have said that without constructing any nest, they lay their eggs, usually 6 or 8, upon the dry rotten wood at the bottom of the cavity. The eggs are white, with specks of brownish red. This industrious little bird resembles the wood-peckers in many of its habits, running round upon the trunks and limbs of the trees with the greatest ease, frequently with its back downward, while searching for its food. Late in the fall, they may be seen in considerable numbers about our orchards and shade trees, and they doubtless render essential service by destroying the eggs and larvæ of insects which have been deposited in the crevices of the bark, to be hatched the next spring

THE HUDSON BAY TITMOUSE.
Parus hudsonicus.—LATH.

DESCRIPTION.—General color dull leaden, tinged with a light brown; head umber brown; throat and fore neck black, with a band of white under each eye; breast and belly grayish white, sides light yellowish brown. Bill black, short, straight, slightly convex and acutely pointed; iris dark brown; feet lead color. Length 5 inches, spread 7. *Female* resembles the male, but the colors are duller.—*Audubon.*

HISTORY.—This species is much less common in Vermont than the preceding, and is not often seen farther to the southward than the north part of this state. It breeds in the state of Maine, and some of them very probably rear their young in the northeastern part of this state. Its nest, like that of the preceding, is in the hollow cavity of an old tree, and one, which Audubon found in Labrador, was completely lined with fur.

GENUS BOMBYCILLA.—*Brisson.*

Generic Characters.—Bill short, straight and elevated; upper mandible slightly curved towards the tip, and provided with a strongly marked tooth; nostrils at the base of the bill, oval, open, hidden by stiff hairs directed forward; tongue cartilaginous, broad at the tip and lacerated; feet with three toes directed forward, and one backward, the exterior united to the middle toe. Wings moderate, 1st and 2d primaries longest; the spurious feathers very short. Sexes alike in appearance and both crested.

THE CEDAR, OR CHERRY BIRD.
Bombycilla carolinensis.—BRISSON.

DESCRIPTION.——Head, neck, breast, back and wing coverts yellowish brown, brighest on the front of the crest and darkest on the back; frontlet black, with a black line over the eye extending backward under the crest; chin blackish, a white line along the margin of the under jaw; belly yellow; vent white; wings dusky; rump and tail coverts dark ash; tail of the same color deepening into dusky and broadly tipped with bright yellow; more or less of the secondaries of the wings sometimes ornamented with small vermillion colored appendages, resembling sealing wax. The bill, legs and claws are black; iris red. In the female the tints are duller. Length 7½ inches.

HISTORY.—This species inhabits all parts of the United States. It is most common in the southern states during the winter and in the northern during the summer. These birds are very social in their habits, usually living in small flocks, even during the period in which they are rearing their young; and hence we usually find several of their nests in the same neighborhood, and often within a few rods of each other. The nest is usually placed in the top of a spruce or hemlock, at the height of 15 or 20 feet from the ground, and is constructed with sticks, roots and grass, lined with lint, down and other soft substances. The eggs, usually 4 or 5 in number, are of a pale clay-white, spotted with umber at the large end. These birds, which mostly migrate to the south in the fall, return to Vermont in April, and are found here during the summer in large numbers. During the early part of summer they feed upon worms and insects, and render an essential service by the destruction of these and the catterpillars, which infest our orchards; but this service is soon forgotten, and when the little bird claims for his reward, a few of the cherries, which he has protected, he is only answered by the gun of the ungrateful and cruel gardener. Although they feed upon fruits and berries of various kinds, they seem to be more fond of cherries and the berries of red cedar than any others, and hence their name *Cherry Bird*, or *Cedar Bird.*

INSECTIVEROUS BIRDS.

In birds of this order the bill is either short or of moderate length. It is straight, rounded or awl-shaped. The upper mandible is curved and notched towards the point, most commonly provided at the base with stiff hairs directed forward. The feet have three toes before and one behind, all on the same level. The outer toe is united to the middle one as far as the first articulation. Their food is insects in the summer, but principally berries during the colder part of the year. Their voices are, for the most part, melodious.

GENUS LANIUS.—*Linnæus.*

Generic Characters.—Bill of medium size, strong, straight from the base, considerably compressed; upper mandible much bent, toothed and hooked towards the tip, which is acute; base of the bill without a cere, furnished with strong bristles directed forward; nostrils close to the base, lateral, nearly round, half closed by a vaulted membrane, and nearly concealed by the bristles; tarsus longer than the middle toe; feet with three toes before and one behind, free; the third and fourth quills longest.

THE BUTCHER BIRD.

Lanius borealis.—VIEILLOT.

DESCRIPTION.—Color above pale cinereous, becoming nearly white towards the tail; wings and tail brownish black, with a black bar extending from the nostril through the eye to the neck; beneath white, beautifully waved with pale brown; outer feathers of the tail partly white and a whitish spot on the wings just below their coverts; legs and feet black; bill and claws bluish black. Tail rounded, extending 3 inches beyond the folded wings; third primary longest. Length of the specimen before me 10 inches, spread 13.

HISTORY.—The Butcher-Bird, or, as he is, perhaps, more generally called, the *Great Northern Shrike*, though frequently seen in Vermont, is not very common. The specimen from which the above description and figure were made, was shot in Burlington in May, 1842. Dr. Richardson says that this bird is common in the woody districts of the fur countries as far north as the 60th parallel of latitude. Many of them migrate to the south in the fall, but some remain in the fur countries through the winter. Its nest is built in the fork of a tree, of grass and moss, and lined with feathers. The eggs, 5 or 6 in number, are of a pale bluish gray, spotted at the large end with dark yellowish brown. Like the king bird it attacks eagles, hawks and crows, and drives them from the neighborhood of its nest.

GENUS MUSCICAPA.—*Linnæus.*

Generic Characters.—Bill medium sized, rather stout, angular, considerably widened and flattened towards the base, which is guarded by longish bristles; upper mandible notched towards the end and bent at the tip; nostrils basal, lateral and ovoid, partly hid by hairs; tarsus the same length as the middle toe or a little longer; inner toe free, or scarcely united at the base; hind nail more curved than the rest, and larger than that of the middle toe; wings long and somewhat sharp; first quill very short, the second shorter than the third and fourth, which are longest.

THE KING BIRD.

Muscicapa tyrannus.—BRISSON.

DESCRIPTION.—Color of the head when the feathers are smooth, shining velvet black, but when the feathers are ruffled a spot of bright ochrey yellow appears on the crown; back brownish black; wings very dark, hair brown, the secondaries and wing coverts edged with gray; tail even, pitch black, tipped with white, and extending far beyond the wings; breast light ash; belly white; bill, legs and feet black; bill wide at the base gradually narrowing to the tip; upper mandible with convex sides, meeting in an obtuse ridge and hooked at the point; short, stiff bristles at the angle of the mouth; second quill longest. Length 8 inches, spread of the wings 14.

HISTORY.—The King Bird, or Tyrant Fly-catcher, as he is sometimes called, spends the winter at the south, beyond the limits of the United States. Early in the spring he proceeds to the north and during the summer is found rearing its young in all parts of the United States, and, according to Richardson, as far north as the 57th parallel of latitude. It arrives in Vermont in the early part of May, and in the summer is common in all parts of the state. Its nest is built in the tops of orchard and forest trees, at various heights from the ground, and is composed of coarse dry grass, weeds and loose pieces of bark, compactly connected and bedded with down, tow and woolly substances, and lined with fine fibrous roots, grass, and hair. The eggs are from 3 to 5, of a bluish white color, marked with spots of deep bright brown. The same pair frequently rears two broods in a season. The food of the king bird consists almost entirely of insects, such as beetles, crickets, grasshoppers and various kinds of flies and catterpillars, and the only harm, which he is accused of doing, is that of catching a few honey bees as they are gathering honey from the flowers, which is very trifling compared with the services which he renders the farmer and gardener. The king bird manifests no fear of the larger birds, but whenever, during their breeding season, a hawk or crow comes near his nest, he boldly attacks him, pounces upon his back, and persecutes him till he is glad to abandon the neighborhood.

THE PHŒBE.
Muscicapa fusca.—BONAP.

DESCRIPTION.—General color above brown with an olive tinge, darker on the head; wings and tail blackish brown, the feathers having the appearance of being faded and worn, and the color of their shafts dark umber; an indistinct grayish circle around the eye, the pupil of which is bluish black and the iris dark hazel; belly yellowish white; tail slightly forked. Bill broad, hooked at the point, and wholly black; legs and feet black with sharp claws. Length of the specimen before me 6½ inches; folded wing 3.4; tail 2.7 and reaching 1.4 beyond the folded wings. The 3d quill longest, 2d and 4th equal.

HISTORY.—This well known and familiar bird arrives from the south about the beginning of April and retires again in October. During the summer it is found in all parts of the state. It seems to prefer building its nest beneath bridges, in sheds and under the eaves of barns. The nest is usually constructed of mud and moss, and lined with grass, hair and other fibrous substances, and is sometimes built upon the top of beams, and at others stuck upon the sides. The eggs are 4 or 5, and are white and unspotted. These birds become very much attached to places where they have reared their young, and the same pair will resort to a particular locality for that purpose, many years in succession. In illustration of this statement I will mention one, of several cases which have fallen under my own observation. About the year 1826 two of these birds built a nest upon a shelf in my wood-shed, and for two years in succession raised broods of young-ones in the same place. The third year when the young were about half grown the female bird disappeared. The male bird remained about the nest, but, not feeding the young ones, they died. The male staid till fall and then left, but returned alone in the spring; and for three successive summers that bird sung his solitary and sad lament for her to whom his young heart and early vows had been plighted, around the place which had been the scene of mutual joys. The name of this bird is derived from the sound of its note. It is also called the *Pewit Flycatcher.*

THE WOOD PEWEE.
Muscicapa virens.—LINN.

DESCRIPTION.—Color dusky brownish olive; head brownish black, slightly crested; below pale yellowish, inclining to white. Tail forked; 2d primary longest; 1st much shorter than the 3d, and longer than the 6th. Length 6 inches; spread 10. The female a little smaller.—*Nutt.*

HISTORY. This species bears considerable resemblance to the preceding, but differs from it in its habits and notes. It arrives later in the spring, and confines itself principally to the thickets and forests. Its nest is usually attached to the horizontal branch of a tree, and is very curiously constructed of grass, fine roots, lichens and cobwebs, held together by a glutinous cement, and is so thin as to appear almost transparent. The eggs are 4 or 5, of a light yellowish hue, spotted with reddish brown towards the large end.

THE SMALL PEWEE.
Muscicapa acadica.—GMEL.

DESCRIPTION.—Color above dusky olive green; yellowish white beneath, inclining to ash on the breast; wings dusky brown, crossed with two bars of dull white; outer edge of the 1st primary, edges of the secondaries, and ring around the eye, whitish; under wing coverts pale yellow; 2d, 3d, and 4th primaries nearly equal and longest. Tail pale dusky brown, notched; legs and feet black. Sexes nearly alike. Length 5½ inches; spread 9.—*Nuttall.*

HISTORY.—This species is common during the summer in all the northern parts of the United States and Canada, but none of them were seen by Audubon or his party in Labrador. It breeds in this state, and usually fixes its nest in the upright forks of a small tree, at a height of from 8 to 30 feet from the ground. The eggs, from 4 to 6 in number, are white and unspotted. It feeds, like the other species of this genus, upon bees, flies and moths.

THE SPOTTED FLY-CATCHER.
Muscicapa canadensis.—LINN.

DESCRIPTION.—Male with the upper parts ash-gray; the feathers of the wings and tail brown, edged with gray; the head spotted with black; loral space, a band beneath the eye proceeding down the side of the neck, and a belt of triangular spots across the lower part of the fore neck, black; lower parts, and a bar from the nostril over the eye pure yellow; lower wing and tail coverts white; the third quill longest, the second and fourth but little shorter; tail rounded. *Female* similar to the male, but the colors fainter. *Young* with the neck unspotted. Length 5, spread 9.—*Audubon.*

HISTORY.—This bird, according to Audubon, gives a decided preference to mountainous districts, and particularly to such as are covered with a thick growth of underwood and shrubbery. We are informed by the same high authority that its nest is placed in the fork of a bush, made of moss and lined with grass—that the eggs, usually 5, are white, with a few spots of bright red towards the large end. It probably breeds in Vermont, but I have no positive proof of the fact.

GENUS VIREO.

Generic Characters.—Bill rather short, a little compressed, and furnished with bristles at its base; upper mandible curved at the extremity and strongly notched; the lower shorter and recurved at the tip; nostrils basal, rounded; tongue cartilaginous and cleft at the point; tarsus longer than the middle toe; wings rather acute, the 2d or 3d primary longest. Female resembles the male, and both sexes more or less tinged with olive green.

THE YELLOW-THROATED VIREO.
Vireo flavifrons.—VIEILLOT.

DESCRIPTION.—Color yellow-olive above, belly white; throat, breast, frontlet and line round the eye yellow; lesser wing-coverts, lower part of the back and rump, ash; wings nearly black with two white bars; tail blackish, a little forked; primaries edged with pale ash, secondaries with white; exterior tail feathers edged with white; legs, feet and bill grayish-blue; iris hazel. The yellow of the female and young duller. Length 5½, spread 9.—*Nuttall.*

HISTORY.—This species rears its young in the south part of the state. Its nest is suspended upon the limb of a tree, and is constructed of strips of bark and fibrous substances, which are cemented together with saliva. The eggs are about 4 in number, are white and spotted towards the larger end with blackish.

THE WHITE-EYED VIREO.
Vireo noveboracensis.—BONAPARTE.

DESCRIPTION.—Yellow olive above, white beneath; sides, line round the eye and spot near the nostrils yellow; wings dusky, with two yellow bands; tail dusky brown, forked; bill, legs and feet light bluish-gray; iris white. Length 5¼; spread 7.—*Nutt.*

HISTORY.—This species constructs its nest very much in the manner of the preceding, but usually builds nearer the ground. It lays 4 or 5 eggs, which are white, spotted towards the large end with brown.

THE RED-EYED VIREO.
Vireo olivaceous.—BONAP.

DESCRIPTION.—General color above yellow olive; crown dark ash; a light gray line from the upper mandible passes over the eye and widens behind it, with a dark line above and another below, extending from the eye to the rictus; all beneath whitish, tinged with light yellow under the wings and on the sides; wing and tail feathers brownish black, with their outer margins yellow olive; 2d and 3d primaries longest; bill brown above, lighter beneath, straight, abruptly bent and notched at the point; nostrils roundish, basal; a few weak bristles at the angle of the mouth; iris bright brick red; legs bluish gray; tail slightly forked. Length 6 inches; tail 2.4; folded wings 3.3; bill above .5, to the angle of the mouth .75; tarsus .7.

HISTORY.—This is probably the most common species of Vireo found in Vermont. They arrive early in May, and take up their residence in the forests and the lofty trees around our fields and gardens. Their song is loud, lively, and energetic. They feed principally upon insects and catterpillars. Their nest is constructed of strings, strips of bark, and fibrous substances, agglutinated together into the form of a pouch. The eggs are 3 or 4, white, with a few blackish brown spots towards the large end. The cow black-bird lays its egg in the nest of this bird more frequently than in any other. The specimen from which the foregoing description was made, was shot in Burlington.

THE SOLITARY VIREO.
Vireo solitarius.—VIEILLOT.

DESCRIPTION.—Dusky olive above ; belly white ; head bluish gray ; breast pale cinereus, inclining to reddish gray on the throat ; flanks and sides of the breast yellow ; wings dusky brown, with two white bands ; tail emarginate and nearly black ; primaries and tail feathers bordered with light green ; a line of white from the nostril to the eye, which it encircles ; bill short, broad ; upper mandible black, lower pale bluish gray ; iris hazel. Female with the head dusky olive and the throat greenish. Length 5 in. ; spread 8.—*Nut.*

HISTORY.—This is a rare bird in this state ; but is said to resemble the preceding species in its habits. It suspends its nest from the forked twigs of bushes, and lays 4 or 5 eggs, which are light flesh color, with brownish red spots towards the large end.

GENUS TURDUS.

Generic Characters.—Bill of moderate dimensions, with cutting edges, compressed and curved towards the point ; the upper mandible generally notched towards the extremity, the lower roundish ; a few scattered bristles at the angle of the mouth ; nostrils basal, lateral, rounded, and half closed by a naked membrane ; tongue notched at the tip ; feet rather stout ; tarsus longer than the middle toe, which is attached at the base to the outer one ; wings rather short ; the third, fourth and fifth quill longest. The female and young differ little from the male, excepting the young are more spotted. They moult annually.

THE BROWN THRUSH.
Turdus rufus.—LINNÆUS.

DESCRIPTION.—All the upper parts, and the under side of the tail, bright reddish brown ; breast and belly yellowish white, marked with long pointed dusky spots ; wings crossed by two whitish bars, relieved with black ; tail long, reaching near 4 inches beyond the wings, and rounded ; bill long, slightly arched, black above, and whitish below near the base ; nostrils naked ; short, stiff, black bristles over the angle of the mouth ; legs, feet and claws dusky brown ; tarsus scutilated in front ; middle toe much the longest ; iris bright orange. Length 11 in. ; spread of the wings 13 inches.

HISTORY.—This bird is known in many places by the name of French Mocking Birk, and surely no bird, if we except the Mocking bird (*Turdus polyglottus*), excels it in the variety and sweetness of its song. It arrives here from the south the latter part of April, and commences building its nest early in May The nest is commonly built upon the ground, or but little elevated above it, in some little thicket, and is constructed with sticks and lined with fine fibrous roots. The eggs are 4 or 5 in number, of a greenish white color, and sprinkled all over with reddish brown spots. During the period of incubation the male will often sit and sing for hours upon the top of a neighboring tree. His music is original, but varied, full, and charming. The food of the Brown Thrush consists of insects, worms, berries, and fruits of various kinds. This bird is known in many places by the name of *Thrasher*, or *Red Thrasher.*

THE CAT-BIRD.
Turdus felivox.—VIEILLOT.

DESCRIPTION.—General color dark slate, lighter beneath ; top of the head, bill, and inside of the mouth, black ; under tail coverts reddish chestnut ; bill a little hooked at the point ; legs and feet brown ; first quill very short, the 4th and 5th longest ; quill feathers lighter on the outer edges ; tail long and rounded. Length 8½ inches ; spread of the wings 11¼ in.

HISTORY.—The Cat Bird is very common in all parts of Vermont, where it arrives from the south in the early part of May. This bird, like most others of the family, is an excellent songster, and may be heard in almost every neighborhood during the early part of summer, ushering in the dawn with his cheerful strains. When this bird is disturbed while rearing its young, its note is harsh and unpleasant, somewhat resembling the mewing of a cat, and from this circumstance it undoubtedly received the name of Cat Bird. The Cat Bird builds its nest in a thicket of bushes, at the height of 5 or 6 feet from the ground. It is constructed with sticks and briars, and lined with fine thread-like roots, which are of a dark color. The eggs are 4 or 5, of a bluish green color, and without spots. Like the Mocking Bird, the Cat Bird is often known to imitate the notes of other birds, and sounds of various kinds. The food of the Cat Bird is similar to that of the preceding species, being made up of worms, beetles, cherries, and various other insects, fruits and berries.

THE AMERICAN ROBIN.
Turdus migratorius.—LINNÆUS.

DESCRIPTION.—Color of the head, back of the neck and tail brownish black ; the back and rump dark ash ; breast dark reddish orange ; belly and vent white ; chin white, spotted with brownish black; wings blackish brown ; the exterior edges of the feathers faded and grayish; exterior tail feathers white at their inner tip ; three white spots margin the eye. The bill is lemon yellow, with a brownish tip ; legs and feet dark brown. The young, during the first season, spotted with white and dusky on the breast. Length 9 inches.

HISTORY.—This universal favorite is found, during the summer, throughout nearly the whole of North America, They retire to the south late in autumn. where they pass the colder part of the winter; but, returning early to the north, reach Vermont usually about the 20th of March ;* and their arrival is always hailed with joy, as the unerring harbinger of approaching spring. While the snow continues upon the ground, the Robin subsists principally upon the berries which remain upon the sumach, mountain ash and red cedar. The Robin, as is well known, is a very familiar bird, and seems to seek to place its nest where it shall be under man's protection. And hence we find its nest most frequently in gardens and orchards. The nest is sometimes built upon a fence, a wall, or a stump, but more commonly in the fork of an apple-tree or other small tree. It is constructed with grass and mud firmly bedded together, and lined with fine straw

* See page 13.

and blades of grass. The eggs, usually 5, are of a bluish green color and unspotted. During the summer their food consists of worms, insects, and various kinds of berries. The Robin is easily tamed, and in the domesticated state may be taught to imitate not only the notes of other birds, but various strains of music.

WILSON'S THRUSH.
Turdus Wilsonii.—BONAPARTE.

DESCRIPTION.--Upper parts uniform light reddish-brown, a little deeper on the head; quill and tail-coverts light olive brown, the outer webs of the former like the back ; lower parts grayish-white, the sides and lower part of the neck, and a small portion of the breast tinged with pale yellowish brown, and marked with small, faint and undecided triangular brown spots ; wings with the 3d quill longest ; the 4th scarcely shorter, and slightly exceeding the second. Length 7; spread 12.—*Audubon.*

HISTORY.—This species arrives from the south in the early part of May, and immediately commence the construction of their nests. These are built in low, thick bushes, in the dark parts of the forests, sometimes upon the ground, but more commonly from 1 to 3 feet above it. The eggs, 4 or 5 in number, are of an emerald green without spots, and differ very little from those of the Cat Bird, with the exception of being a little smaller. They usually raise two broods in a season.

THE NEW YORK THRUSH.
Turdus noveboracensis.—NUTTALL.

DESCRIPTION.—Color of the whole upper plumage a uniform deep hair brown ; stripe over the eye and whole under surface pale primrose yellow, marked with pencil-shaped spots of the color of the upper plumage ; inner wing coverts yellowish gray, spotted with brown near the edge of the wing; bill dark umber brown above, paler beneath ; legs brownish flesh color. The three first quills nearly equal and longest; tail nearly even ; lateral toes nearly equal ; nails small and of the color of the bill. Length 5¾ inches; tail 2½ ; folded wing 3 ; bill from the angle of the mouth ¾ inch.

HISTORY.—The Aquatic Thrush is quite a common bird in Vermont, but is of retiring habits and therefore seldom seen except in the thickest parts of the forests. Its nest is built upon the ground and is constructed of leaves and moss, and lined with fine roots and sometimes with hair. The eggs are 4 or 5, of a yellowish white color and pretty thickly sprinkled towards the large end with two shades of reddish brown. The specimen from which the above description was made was obtained, with its nest and eggs, in Burlington, in June, 1840. This bird from its preference to neighborhoods of water is sometimes called the *Aquatic Thrush.*

THE GOLDEN-CROWNED THRUSH.
Turdus aurocapillus.—WILSON.

DESCRIPTION.—Color above rich yellow-olive; the tips of the wings and inner vanes of the quills dusky brown; the 3 first primaries nearly equal; a dusky line from the nostril to the hind head; crown brownish orange; beneath white; the breast covered with deep brown pencil-shaped spots; legs pale flesh-color; bill dusky above, below whitish. Crown of the female paler. Length 6, spread 9. *Nuttall.*

HISTORY.—This bird is pretty common in nearly all parts of the United States, but is shy and retiring, and found only in the thickets of the forests. Its oven shaped nest is placed in the side of a dry and mossy bank and is constructed with great neatness. It is formed of grass and covered with leaves and sticks, having the place of entrance upon the side. The eggs are 4 or 5, whitish, irregularly spotted with reddish brown. The food of this bird consists wholly of insects and their larvæ.

THE HERMIT THRUSH.
Turdus solitarius.—WILSON.

DESCRIPTION.—Color above plain deep olive-brown, below dull white; upper part of the breast and throat cream color; the dusky brown pencillated spots carried over the breast and under the wings where the sides are pale olive; tail and coverts as well as the wings strongly tinged with rufous; legs pale flesh color; bill short black above, flesh-colored below; iris large and nearly black; tail short and emarginate; 3d primary longest. The *female* darker, with the spots on the breast larger and more dusky. Length 7½; spread 10½.—*Nutt.*

HISTORY.—The Hermit Thrush is said to inhabit every part of the United States. It is a solitary bird living wholly in the woods, and is said by Nuttall to be scarcely inferior to the Nightingale in its powers of song. Its nest according to Audubon is placed upon the limbs of trees a few feet from the ground, and is composed of dry weeds and leaves, and neatly lined within with fine grass. The eggs, from 4 to 6, are of a light blue color, sprinkled with blotches towards the large end.

GENUS SYLVIA.—*Latham.*

Generic Characters.—Bill straight, slender, awl-shaped, higher than wide at the base, and usually furnished with scattered bristles; lower mandible straight, upper sometimes notched; nostrils lateral, oval, situated at the base of the bill, and partly covered by a membrane; tarsus longer than the middle toe; inner toe free; hind nail shorter than the toe; wings short.

THE YELLOW-CROWNED WARBLER.
Sylvia coronata.—LATHAM.

DESCRIPTION.—Back dark ash, spotted or striped with black; crown, sides of the breast and rump bright yellow; wings and tail black, with the outer vanes of the feathers margined with white or light ash; wing coverts tipped with white, forming two white bars across each wing; outer tail feathers on each side with a large white spot on their inner vane; breast white, spotted with black; belly and vent white; bill black, straight, slightly bent at the point and rounded above and below; legs and feet black; tail forked; the 2d, 3d and 4th primaries nearly equal; 1st but little shorter. Winter dress and that of the young paler, and of an olivacious hue. Length of the specimen before me 5½ inches; spread of the wings 7½ inches.

HISTORY.—The Yellow-crowned Warbler, or *Myrtle Bird,* as it is sometimes called, is common in Vermont, and I am informed by Dr. Brewer that they breed in the north part of the state. The nest, according to Audubon, is placed upon the horizontal branch of a fir or other evergreen. It is compactly built of sticks and strips of bark, and lined with hair, feathers and down. The eggs are of a rosy tint, thinly spotted with reddish brown towards the large end. Their food is insects and caterpillars in summer and they feed upon seeds, and myrtle and other berries during the winter.

THE YELLOW RED-POLL WARBLER.
Sylvia petechia.—LATH.

DESCRIPTION.—Male with the crown deep brownish red; upper parts yellow olive streaked with brown; rump greenish yellow without streaks; wings and tail dusky brown with the feathers edged with whitish or yellowish; a bright yellow streak from the nostril over the eye; lower parts yellow; the sides of the neck, its lower part, and the sides of the body streaked with deep red; the three outer quills nearly equal; tail emarginate. Colors of the *female* duller. The *young* dull light greenish brown, tinged with gray. Length 5½, spread 8½.—*Aud.*

HISTORY.—Very little is yet known of the history of this bird. During the winter it is found in large numbers in the southern states, and early in the spring passes through New England, to rear its young at the north and returns again in the fall. Audubon found them plentiful in Labrador and Newfoundland, in August, feeding their young, but did not succeed in discovering any of their nests.

THE SUMMER WARBLER.
Sylvia æstiva.—LATH.

DESCRIPTION.—Greenish yellow above; crown and beneath bright golden yellow; breast and sides with long spots of reddish orange; wings and tail brown, edged with yellow; tail emarginate; bill grayish blue; legs pale. *Female* with the colors duller, and the breast unspotted. *Young* greenish olive above, with the throat yellowish white. Length 5, spread 7.

HISTORY.—This is one of our most beautiful and musical Warblers. It arrives in Vermont in the early part of May, and the female is soon engaged in the construction of her nest, while the male is spending the most of his time in cheering her and the neighborhood with his song. The Summer Warbler seems to delight in building its nest and rearing its young in our orchards and on the trees around our dwellings, as if conscious of its ability to afford us pleasure by its music. Several pairs of these birds are now (June 24, 1842,) rearing their young and warbling in the heart of our village, and two have their nest on a tree in my garden. It is built of a few coarse straws, shreds of bark, and woolly lint, lined with horsehairs and bristles. The eggs are 4, of a yellowish white color, sprinkled with specks of pale brown towards the large end. It is said that the Cow-Black Bird often deposits its eggs in the nests of these birds, and that they are in the habit of incarcerating them in the manner described on page 69; and, as I have learned since that article was printed, that the nest there described was built about the beginning of June, much earlier than the *Fringilla tristis* usually builds; it is probable that the yellow bird there mentioned, was the *Sylvia æstiva*, or Summer Yellow Bird, as this is often called.

THE SPOTTED WARBLER.
Sylvia maculosa.—LATH.

DESCRIPTION.—Crown ash; back blackish; tail coverts, tail and wings black, the latter crossed by two bars of white; rump and beneath bright yellow; breast spotted with black; vent white; legs brown; bill, front, lores and behind the ear black. *Female* with the breast whitish, and the colors duller. Length 5, spread 7½.—*Nutt.*

HISTORY.—This beautiful species is only occasionally seen in its passage towards the north in the spring. It is said to build its nest around Hudson's Bay, upon the willows. It is considered one of the most musical and most beautiful of the American Warblers.

THE NASHVILLE WARBLER.
Sylvia rubricapilla.—WILSON.

DESCRIPTION.—Yellowish green, or olive above; breast, chin and under tail coverts yellow; belly whitish; head and neck dark ash, inclining to olive; crown deep chestnut; wings and tail hair brown; feathers more or less edged with yellow on the outer vanes; tail slightly forked; bill brownish, straight and very sharp; legs and feet brownish yellow. The *female* is said to be paler beneath, grayish and without the chestnut on the crown. Length of the specimen before me, which is a male, 4¼ inches, spread of the wings 6½ in.; the 2d and 3d primaries longest; the 1st and 4th nearly equal.

HISTORY.—This species was discovered by Wilson near Nashville, Tennessee, and is represented by ornithologists as being a very rare bird. Audubon says he has never seen more than three or four of them. The specimen from which the above description was made, was shot in Burlington, in the spring of 1840, and is the only one I have seen.

BLACK-THROATED GREEN WARBLER.
Sylvia virens.—LATH.

DESCRIPTION.—Color yellowish green above; beneath whitish; front, cheeks, sides of the neck, and line over the eye, yellow; chin and throat to the breast black; wings and tail dusky, the former with two white bars, and the latter with the three lateral feathers, marked with white on their inner webs; bill black; legs and feet brownish. *Female* with the chin yellow, and the throat blackish, tinged with yellow. Length 5, spread 7¾.—*Nutt.*

HISTORY.—This species, though rare, probably breeds in this state. Mr. Nuttall found one of their nests in Massachusetts, in June, 1830. It was in a low, thick and stunted Virginia juniper, and was made of fibrous bark, and lined with feathers, grass, and a few hairs. The eggs were 4, whitish, sprinkled towards the large end with brown and blackish.

PINE CREEPING WARBLER.
Sylvia pinus.—LATH.

Sylvicola pinus.—Aud. Am. Birds, II.—37, pl. 82.

DESCRIPTION.—Male with the upper parts yellowish green, inclining to olive, the rump brighter; streak over the eye; eye-lids, throat, breast and sides bright yellow, with a greenish tinge; the rest of the lower parts white; wings and tail blackish brown; secondary coverts and first row of small coverts tipped with dull white; primaries edged with whitish, secondaries with brownish gray; outer two tail feathers with a patch of white on their inner web near the end. Wings moderate, first three quills nearly equal; tail emarginate. *Female* and *young* brownish above, other colors duller. Length 5, spread 8.—*Aud.*

HISTORY.—This is one of the most common species of Warblers in the United States, being met with from Louisiana to Maine, but more abundantly at the south than at the north. It resembles the Creepers in running upon the trunks of trees. Its nest is placed high upon the limbs of trees, and is composed of dry grass and roots, lined with hair. The eggs, from 4 to 6, have a light sea-green tint, and are sprinkled with reddish brown dots, thickest towards the large end.

THE CŒRULEAN WARBLER.
Sylvia cærulea —WILS.

DESCRIPTION.—Wings long, 3 outer quills nearly equal, 1st and 2d longest; upper parts fine light blue, brighter on the head; the back marked with longitudinal streaks of blackish; a narrow band of black from the forehead along the lore to behind the eye; two white bands on the wings; quills black, margined with pale blue; tail slightly emarginate; feathers black, edged with blue, with a white patch on the inner web of each toward the end; lower parts white, with a band of dark bluish gray across the foreneck, and oblong spots of the same along the sides. *Female* with the upper parts light bluish green, the lower yellowish; *young* like the female. Length 4½, spread 8.—*Audubon.*

HISTORY.—This species is not very common in the northern part of the United States. Its nest, according to Audubon, is built upon bushes, constructed with stalks and fibres of vines, and lined with moss. The eggs are 4 or 5, white, spotted at the large end with reddish.

BLACKBURN'S WARBLER.
Sylvia Blackburniæ.—LATH.

DESCRIPTION.—The head striped with black and orange; back black, skirted with ash; wings black, with a large lateral patch of white; throat and breast reddish-orange, bounded by streaks and spots of black; belly dull yellow, streaked with black; vent white; tail a little forked, 3 lateral feathers white on the inner web; cheeks black; bill and legs brown. *Female* yellow, without orange, and black spots fewer. Length 4½, spread 7.—*Nutt.*

HISTORY.—This is a rare bird in the United States. But few of them are seen in Vermont, and yet it is said that some of them rear their young here. The nest is placed in the fork of a small tree but a few feet from the ground, and is lined with hair and feathers. The eggs are white, sprinkled with red towards the large end.

THE CHESTNUT-SIDED WARBLER.
Sylvia icterocephala.—LATH.

DESCRIPTION.—Crown yellow; feathers of the back and rump black, edged with greenish white; wings dusky, the primaries edged with white and the secondaries with greenish yellow; the first and second row of coverts broadly tipped with light yellow, forming two bars on each wing; a triangular black spot beneath the eye; chin and belly white; sides, from the black beneath the eye to the thighs, and across the breast, bright chestnut; tail forked, dusky above, white beneath; legs, feet and bill dusky; iris hazel. Length 5, spread 7.

HISTORY.—This beautiful warbler is represented by Audubon as being extremely rare in all parts of the United States. The specimen, from which the above description was drawn, was killed in Burlington, on the 11th of June, 1842, and it is thought to be rather a common bird here, and I have but little doubt that it breeds in this state, although I have never seen its nest. Audubon professes himself ignorant of their breeding places; but Nuttall and Peabody assure us that several of their nests have been found in Massachusetts.

THE BLACK-THROATED WARBLER.
Sylvia canadensis.—LATH.

DESCRIPTION.—Light blue slate above; beneath white; wings and tail dusky black, the latter wedge-shaped, edged with blue, feathers pointed, external ones with a large white spot; throat, cheeks, upper part of the breast and sides under the wings, deep black; legs and feet dusky yellow; bill black; a white spot on the wings. The black in the *female* dusky ash, or wanting. Length 5, spread 7½.—*Nutt.*

HISTORY.—This species is rare and very little known. Its nest, according to Audubon, is placed on the horizontal branch of a fir, 6 or 8 feet from the ground. The eggs, 4 or 5 in number, are of a rosy tint, sprinkled with reddish-brown at the large end.

THE MARYLAND YELLOW-THROAT.
Sylvia trichas.—LATH.

DESCRIPTION.—Yellow-olive above, inclining to cinereous on the crown; front and wide patch through the eye black; throat, breast and vent yellow, fainter on the belly; wings, and unspotted wedge-shaped tail, dusky brown; quills of both edged with yellow-olive; bill black above, pale beneath; legs pale flesh-color; iris dark hazel. *Female* without black on the face, and beneath dull yellow. Length 5, spread 7.—*Nutt.*

HISTORY.—This is quite a common bird. It arrives from the south in the early part of May. Its nest, according to Peabody, is constructed on or near the ground, among dry leaves, brush or withered grass. The eggs, 4 or 5, are white, with blotches and lines of brown chiefly towards the large end.

THE WORM-EATING WARBLER.
Sylvia vermivora.—LATH.

DESCRIPTION.—Dusky olive above except the wings and tail, which are umber brown. Head buff, marked with 4 longitudinal stripes of umber brown; breast orange buff, mixed with dusky; vent waved with dusky olive; bill blackish above, below flesh colored; legs pale flesh color; iris hazel; bill stout. Length 5¼, spread 8.—*Nuttall.*

HISTORY.—This active and industrious little bird is said to arrive late from the south and retire early, and resembles somewhat the Chicadee in its manners and notes. Its nest, according to Audubon, is made of dry mosses, hickory and chestnut blossoms, and the eggs are 4 or 5, cream colored, with a few dark red spots near the large end. The nest is usually placed between two twigs, 8 or 9 feet from the ground.

BLACK AND WHITE CREEPER.
Sylvia varia.—LATH.

DESCRIPTION.—The crown white, bordered on each side by a band of black, which is again bounded by a line of white passing over each eye; ear feathers black, as well as the chin and throat; wings the same, with 2 white bars; breast back, sides, and rump spotted with black and white; tail and primaries edged with light gray, the coverts black, bordered with white; belly white; legs and feet dusky yellow; bill rather long, black above, paler below. *Female* with the crown wholly black, and without the black ear-feathers. Length 5, spread 7½.—*Nuttall.*

HISTORY.—This bird is found in most parts of the United States, and in many of its habits is closely allied to the Creepers and Nuthatches. It seldom perches upon the branches of trees, but creeps spirally round upon the trunk and large limbs, searching for insects and their eggs in the crevices of the bark. Dr. Brewer informs us that this bird builds its nest upon the ground. It is composed externally of coarse straw, and lined with hair. The eggs, about 4 in number, are white, with a few brownish red spots, chiefly towards the large end.

GENUS REGULUS.—*Cuvier.*

Generic Characters.—Bill short, straight, very slender, subulate, compressed from the base, and narrowed in the middle, furnished with bristles at the base, and with the edges somewhat bent in; the upper mandible is slenderly notched, and a little curved at the tip. Nostrils basal, oval, half closed by a membrane, and additionally covered also with two small projecting, rigid, decompound feathers. Tongue bristly at the tip. Feet slender; tarsus longer than the middle toe; lateral toes nearly equal; the inner one free; hind toe stoutest. Wings short, rather acute; 3d and 4th primaries longest; tail notched.

THE RUBY-CROWNED WREN.
Regulus calendulus.—STEPHENS.

DESCRIPTION.—Color above olivaceous, yellowish on the rump and grayish on the head, with a bright vermillion colored spot on the hind head, which is partly concealed by the dark feathers; wings and tail brownish black, with the outer edges yellow; wing coverts terminated with white, forming a whitish bar upon the wings; a yellowish white line around the eye; beneath, brownish white on the neck, changing into yellowish white on the belly; upper mandible slightly curved near the tip; legs, toes and nails long, slender, and of a smoky brown color. Length 4; spread 5½.

HISTORY.—The history of this little songster is very imperfectly known. It is found during the winter, in considerable numbers, in the southern states, and, in the northern states, is frequently seen in its migrations to the north and south, in spring and fall. Audubon has no doubt but that it breeds in Labrador, but neither he nor any other of our ornithologists has succeeded in finding its nest. The beautiful specimen from which the above description was made, was killed in Burlington on the 26th of April, 1842.

THE FIERY-CROWNED WREN.
Regulus tricolor.—NUTT.

Regulus satrapa.—Aud. Am. Birds, II—165, pl. 132.

DESCRIPTION.—Color above ash gray on the neck, and the back yellowish olive; cheeks grayish white; crown flame colored, bordered with yellow and black; beneath whitish, tinged with olive gray; bill slender and rather short; bristles at its base; plumage loose and tufty; 4th primary longest; the first very short; legs rather long, tarsus slender. Length 4; spread 7.—*Audubon.*

HISTORY.—This is an active little bird, and is often seen in company with the creepers and titmice, searching for flies and insects. It is put down by Dr. Brewer as breeding in this state. Audubon found it rearing its young in Labrador.

GENUS TROGLODYTES.—*Cuvier.*

Generic Characters.—Bill slender, subulate, somewhat arched and elongated, also acute, compressed, and without notch; mandibles equal. Nostrils basal, oval, half closed by a membrane. Tongue slender, the tip divided into 2 or 3 small bristles. Feet slender; tarsus longer than the middle toe; inner toe free; posterior with a larger nail than the rest. Wings short, concave and rounded; 3d, 4th, and 5th primaries longest.

THE HOUSE WREN.
Troglodytes ædon.—VIEILLOT.

DESCRIPTION.—Color above reddish-brown, darkest on the head and neck, lighter towards the rump, feathers mostly barred with dusky; beneath dull pale gray, nearly white on the belly; sides and under tail coverts barred with brown; a yellowish line from the upper mandible over the eye; cheeks yellowish gray, spotted with brownish red; bill dark brown above, lighter beneath; iris hazel; feet flesh color; wings short, 3d and 4th quills longest; tail rather long. Length 4½, spread 5½.

HISTORY.—This familiar and interesting little bird is common in all parts of the United States, from April until the beginning of October, when it retires to the south: but the place where it winters seems yet to remain unknown. The House Wren is sprightly, active and diligent, and has received its name in consequence of its delighting to make its residence in our orchards, gardens, and about our houses. Its nest is formed with coarse sticks, shreds of bark, hair, &c., in some natural or artificial cavity, such as a hollow stump, or post, or the vacant space at the foot of a brace in the frame of a building, or a box provided for it by the gardener. And whatever the cavity selected, it seems to be its object to fill it with sticks and other articles, leaving room only for itself and young. The eggs, from 6 to 8, are of a reddish flesh-color, sprinkled with reddish-brown. Audubon has represented this wren as feeding its young in a nest constructed in an old hat. The Wren manifests great antipathy to the cat, and will scold her till she is out of sight.

THE WINTER WREN.
Troglodytes hyemalis.—VIEILLOT.

DESCRIPTION.—Dark brown above, crossed with transverse dusky touches, except on the head and neck, which are plain; the black spots on the back terminate in minute points of dull white; the same colored points are seen on the first row of wing-coverts; the primaries are crossed by alternate rows of cream color; throat, line over the eye, sides of the neck and breast dirty white, with minute transverse touches of drab; belly and vent mottled with sooty black, deep brown, and white, in bars; tail very short; legs and feet pale clay-color; bill straight, half an inch long, dark brown above, whitish beneath; iris hazel. Length 3½, spread 5.—*Nutt.*

HISTORY.—This sprightly and musical little bird bears a very strong resemblance to the preceding, and might easily be mistaken for it. It may, however, be distinguished by its shorter tail, more slender bill, and by having the under parts more distinctly barred. The nest of this wren is built upon, or very near the ground, at the foot of a tree, or by the side of a rock. It is formed of moss and leaves, and lined with hair, and has its entrance on the side. This bird is said to lay from 10 to 18 eggs, but the nests, discovered by Audubon, contained no more than 6. Their color is light blue, spotted with reddish brown. The song of this wren is very agreeable and loud for the size of the bird.

THE WOOD WREN.

Troglodytes americanus.—AUD.

DESCRIPTION.—Bill of moderate length, nearly straight, slender, acute; neck short; body rather full; plumage soft, blended, slightly glossed; wings short, broad; 4th and 5th quills longest; tail rather long, graduated; general color above dark reddish brown, duller and tinged with gray on the head, indistinctly barred with dark brown; wings and tail waved with dark brown, edges of the outer primaries lighter; under parts pale brownish gray, barred more or less distinctly. Length 4⅞, spread 6¼.—*Aud.*

HISTORY.—This new species was discovered by Audubon in the summer of 1832, in the state of Maine, where it breeds in hollow logs in the woods, seldom if ever making its appearance in cleared land. The color of the egg of the Wood Wren is dull yellowish white, with blotches and streaks of purplish-red and blackish-brown. This wren breeds in Vermont, and Audubon describes an egg procured in this state by Dr. Brewer. Late in the fall of 1840, I saw a pair of these wrens in a little wood in Burlington, and watched them for some time. They were silent except a low chirp occasionally, and were intently and diligently searching for spiders and insects upon the sides and beneath the logs.

GENUS SIALIA.—*Swainson.*

Generic Characters.—Bill of ordinary length, nearly straight, about as broad as high at the base; upper mandible rounded carinated towards the base, notched and curved at the tip; tongue cartilaginous, shortly lacerate at the base, and emarginate at the point; nostrils basal, open, partly obstructed by an internal tubercle, the nasal fosse extensive and depressed; tarsus rather robust, a little shorter than the middle toe; inner toe free; the hind one stoutest, longer than the nail; wings rather long and acute; 1st and 2d primaries longest, the 3d scarcely shorter.

THE BLUE BIRD.

Sialia Wilsonii.—SWAINSON.

DESCRIPTION.—Color sky-blue above; ferruginous, passing into brownish white, beneath; vent white; wings full and broad; inner vanes of the quills and their shafts dusky, outer vanes blue; bill and legs black; inside of the mouth yellow. Colors of the *female* duller than in the male. Length 6½, spread 11¼.

HISTORY.—This well known and familiar bird is found in all parts of the United States and of the British North American provinces. It is every where a great favorite, and its return in the spring is hailed with hardly less joy than that of the Robin. It seems to delight in being around our dwellings, and rears its young in hollow stumps and posts and in little boxes made for that purpose and placed on upright poles. The nest consists of a slight lining of the cavity with a few straws and feathers. The eggs are usually 5, of a pale blue color and without spots. They often raise two or three broods in a season. Their food consists almost entirely of insects, such as beetles, spiders and grasshoppers, and, on account of their destruction of these, they are, like most others, real benefactors of the farmer, and richly deserve his protection. Birds seem to be specially designed by Providence to prevent the undue increase of noxious insects, and so useful are they that, in general, whoever destroys a bird, destroys a friend. Blue Birds are very common in all parts of Vermont, and their agreeable warble is heard from March till October.

Genus ANTHUS.—*Linnæus.*

Generic Characters.—Bill straight, slender, cylindric, and subulate towards the point, with edges somewhat inflected towards the middle, and at the base destitute of bristles; the base of the upper mandible carinated, with the point slightly notched and declining. Nostrils basal, lateral, half closed by a membrane. Feet slender; tarsus longer than the middle toe; inner toe free; hind toe shortest, with the nail generally long and nearly straight; wings moderate; three first primaries longest; secondaries notched at the tip; two of the scapulars nearly equal to the longest primaries; tail rather long and emarginate.

THE BROWN LARK.
Anthus spinoletta.—BONAP.

DESCRIPTION.—Grayish brown above, with a darker shade in the centre of each feather; beneath and line over the eye, white; breast and flanks spotted with grayish brown, or blackish; tail feathers nearly black, the outer one half white, upon the 2d and often upon the 3d, a conic white spot; lower mandible straight and livid, the upper blackish; legs chestnut; iris hazel. *Female* more spotted below. *Young* dark brown, inclining to olive; strongly spotted on the beast.—*Nutt.*

HISTORY.—The Brown Lark is met with in every part of the United States as a bird of passage. It feeds upon insects and seeds, and may often be seen running along the margin of ponds and streams, and in old fields in pursuit of these. It was found by Audubon breeding abundantly on the coast of Labrador, and Dr. Brewer obtained its eggs from Coventry, (now Orleans), in this state. The nest is placed at the foot of a wall or rock, curiously formed of bent grass, and partly buried in dark mould. The eggs are usually 6. Their ground color is a deep reddish chestnut, darkened by numerous dots, and various lines of reddish brown. This bird is also called the American Petit, or *Titlark.*

GRANIVOROUS BIRDS.

The Birds of this order have a strong, short, thick, and more or less conic bill, which extends back upon the forehead. The ridge of the upper mandible is usually somewhat flattened, and both portions of the bill are generally without the toothed notch. The feet are arranged with 3 toes before and 1 behind. The wings are of moderate dimensions. These birds spend the summer in pairs, but assemble together in the fall and migrate in large flocks.

Genus EMBERIZA.—*Linnæus.*

Generic Characters.—Bill short, robust, conic, somewhat compressed, and without notch; the margins contracted inward, a little angular towards the base; the upper mandible rounded above, acute, smaller and narrower than the lower; the palate with a longitudinal bony tubercle; the lower mandible rounded beneath, and very acute. Nostrils basal, small, partly covered by the feathers of the forehead. Tarsus about equal to the middle toe; the lateral toes equal; outer united at the base to the middle toe. Wings with the 1st primary almost equal to the 2d and 3d, which are longest. Tail even or emarginate.

THE SNOW BUNTING.
Emberiza nivalis.—LINNÆUS.
Plectrophanes nivalis.—Aud. Am. Birds, III—55 pl. 155.

DESCRIPTION.—*Male,* in winter, with the head, neck, lower parts, a great proportion of the wings, including the smaller coverts, secondary coverts, several secondary quills, the bases of the primaries and their coverts, and the greater part of the outer tail feathers on each side, white; the head and hind neck more or less tinged with brownish red; the upper parts reddish gray, or yellowish red mottled with black, the concealed part of the plumage being of the latter color; the bill brownish yellow. *Female,* in the winter, with the white less extended. *Young,* at this season, like the female, but browner. *Male,* in summer, with the back, scapulars, inner secondaries, terminal portion of the primaries, and 4 middle tail feathers deep black; all the other parts pure white; bill black. *Female* with the black parts tinged with brown, and more or less reddish brown on the head and rump. Length 7; spread 13.—*Audubon.*

HISTORY.—The Snow Buntings spend the great part of the year in high northern latitudes. They breed, according to Dr. Richardson, in the most northerly part of the continent, and on the islands of the arctic ocean. The nest is made of dry grass in the crevices of rocks, and lined with deer's hair and feathers. The eggs are greenish white, spotted and blotched with umber. They usually make their appearance in Vermont in December, in the midst of storms of snow. They arrive in flocks, frequently in company with the Tree Sparrow and Blue Snow Bird, and, in descending upon our gardens and fields, to collect their scanty pittance of seeds from the dry weeds which rise above the snow, they always come down in a spiral direction, passing several times around the spot on which they are to alight. They are much more plentiful in some winters than in others, and are generally known by the name of *White Snow Bird.*

THE BAY-WINGED BUNTING.
Emberiza graminea.—GMEL.

DESCRIPTION.—General color of the upper parts light brown, streaked and mottled with darker; lesser wing-coverts reddish-brown; first quills margined externally with white; outer tail feathers marked with an oblique band of white; a narrow circle of white round the eye; throat and breast yellowish white; the latter and fore part of the cheeks streaked with dark brown; sides and belly yellowish brown, fading into white towards the tail, and sparsely streaked with dark brown; wings with the 3d and 4th quills longest; plumage compact; tail rather long; tarsus, toes, and claws flesh color. Length 5¾, spread 10.—*Aud.*

HISTORY.—The Bay-Winged Bunting, or Finch, is found in all the northeastern portion of the United States. I learn from Dr. Brewer that it breeds in Vermont as well as other parts of New England, and that its nest is placed upon the ground without concealment, but that it uses much art in decoying enemies from the neighborhood of it.

THE SAVANNAH BUNTING.
Emberiza savanna.—WILS.

DESCRIPTION.—General color above pale reddish brown, spotted with brownish black; the edges of the feathers being of the former color; lower parts white, the breast spotted and the sides streaked with deep brown; cheeks and space over the eye light citron yellow; bill dusky above, pale brown beneath; wings and tail short, the latter emarginate; head rather large; neck short. Length 5½, spread 8½.—*Aud.*

HISTORY.—The Savannah Bunting, or Savannah Finch, as he is also called, is, according to Audubon, one of the most abundant and hardy species in the United States. It breeds in this state, and constructs its nest very much in the manner of the Song Sparrow, at the foot of a tuft of grass, or in a low bush. The eggs, from 4 to 6, are of a pale bluish color, softly mottled with purplish brown.

Genus Fringilla.—LINNÆUS.

Generic Characters.—Bill short, robust, conic on all sides and generally without a notch; upper mandible wider than the lower, somewhat turgid and a little bent at the tip, without keel, depressed at the upper part, and often prolonged into an angle entering the feathers of the forehead; nostrils basal, round, covered by the feathers; tongue thick, acute compressed and bifid at the tip; tarsus shorter than the middle toe; toes disconnected at the base; hind nail largest. Wings short; 1st and 2nd primaries but little shorter than the 3d and 4th, which are longest.

THE SONG SPARROW.
Fringilla melodia.—WILS.

DESCRIPTION.—Crown brownish chestnut, divided longitudinally by a grayish line; line over the eye light ash, becoming white towards the bill; mottled above and below with brown, chestnut and ash; much lightest on the belly, each feather being marked with brown along the middle, surrounded by chestnut and edged with ash, giving the bird a striped appearance, particularly on the back and lower part of the breast; wings and tail chestnut brown; bill dark horn color, lighter below; legs light flesh-colored; feet and nails dusky. Length 6¼ inches; spread of the wings 8½ inches. Tail wedge-form, 2 inches longer than the folded wings; 1st primary short, 3d and 4th longest.

HISTORY.—This is one of our most common and familiar sparrows. It arrives early from the south, and in company with the Blue Bird and Robin, ushers in the spring with its cheerful notes, while the snows are yet lingering upon the ground.* This sparrow breeds in all parts of the United States and Canada. The nest is usually placed upon the ground but is sometimes a little elevated above it in a low bush. It is usually formed of dry grass and lined with hair. The eggs, usually 5, are of a bluish gray color, thickly spotted with different shades of brown. They are very prolific, frequently raising three broods in a year. The Song Sparrow is common in our gardens, orchards and meadows, preferring the open fields and low bushes to the woods. They feed upon worms, insects, larvæ and seeds.

---

* For the time of their appearance see Part I—13.

THE BLUE SNOW-BIRD.
Fringilla hyemalis.—LINNÆUS.

DESCRIPTION.—General color dark brownish ash, or bluish slate above and on the breast; belly white; feathers on the back slightly tinged with ferruginous; wings and central tail feathers dark slate; outer tail feather on each side pure white, and the next white wholly or in part; tail forked, the lateral feathers curving outward towards the tip; bill short, acute; bill, legs and feet brownish in summer, pale flesh-color in winter; claws slender and compressed. *Female* and *young* tinged with brown. Length 6 inches, spread of the wings 9 inches.

HISTORY.—This is one of our most common and numerous species, and in the spring and autumn they are met with in every part of the state. Late in the fall they mostly migrate to the south, and in the early part of summer they mostly retire from the low lands either beyond the limits of the state to the north, or to the central mountainous districts for the purpose of rearing their young. They breed in large numbers in all the mountain towns, through the whole length of the state. The nest is built upon the ground by the side of a rock, stump, tuft of grass, or in the side of a dry bank, and is composed of small sticks and withered grass. The eggs, from 3 to 5, are of a pale green, brushed and spotted with darker. They breed in small numbers in the low lands in this state. I found one of their nests in Burlington, near Winooski river, on the 27th of July, containing 3 young nearly fledged. The most common note of this bird is a sharp *chip*, and hence it is often called the *Chipping Bird*, or Blue Chipping Bird.

THE TREE SPARROW.
Fringilla canadensis.—LATHAM.

DESCRIPTION.—Crown of the head bright bay, slightly mottled with ash color; a stripe over the eye, white at its commencement near the bill, and backwards fading into pale ash; sides of the neck, chin and breast pale ash; on the centre of the breast an obscure dark spot; from the lower angle of the bill and behind the eye proceeds a small stripe of chestnut; back varied with black, bay, brown and drab; wings marked with two white bars; outer feathers edged with white, inner with pale brown; bill black, yellowish beneath; tail forked, feathers black, edged with white; vent white; legs slender, dusky brown; feet black. Length of specimen before me 6 inches; spread 9 inches.

HISTORY.—This beautiful little sparrow is a winter resident in Vermont. It arrives in flocks from the north about the first of November, and proceeds again northerly about the first of April. During the winter these sparrows are often seen in flocks by themselves or in company with the snow buntings, gathering their scanty pittance of seeds from the weeds which rise above the snow in our fields and gardens. They are sometimes seen seeking shelter, in the midst of woods, from the winds and storms. Some of them rear their young in Vermont, but the greater part breed farther north, in the neighborhood of Hudson's Bay. They build their nest among the herbage, with mud and dry grass, and line it with hair or down. They lay 4 or 5 eggs at a litter, which are of a pale brown, spotted with darker color.

THE CHIPPING SPARROW.
Fringilla socialis.—WILS.

DESCRIPTION.—Frontlet nearly black; crown bright chestnut; back varied with brownish-black, ash and bay; wings and tail dark chestnut brown; line over the eye, chin and vent white; breast and sides of the neck pale ash; rump dark ash; bill blackish above, dark flesh-color below; legs and feet slender, pale flesh-color; hind nail a little shorter than the toe; first four primaries nearly equal; tail forked, reaching $1\frac{1}{4}$ inch beyond the folded wings. Length 5 inches, spread of the wings $7\frac{1}{2}$ inches.

HISTORY.—Of all our sparrows this is the most familiar and most common. It breeds abundantly in every part of the state, and seems to take much pains to place its nest as near as possible to our dwellings, or close by the side of the most frequented walks in our yards and gardens. Sometimes it is placed upon a lilach or other shrub so near to a window as to be easily reached with the hand. The female will sit upon her nest with apparent unconcern while people are almost constantly passing and repassing within 2 or 3 feet of her. The nest is rather slight, and always composed, internally, of hair, and hence it is often called the *Hair Bird*. The eggs, 4 or 5, are bright greenish blue, with a few spots of brown of different shades. They usually raise two or three broods in a season.

THE FIELD, OR RUSH SPARROW.
Fringilla juncorum.—NUTT.

DESCRIPTION.—Above varied with bay, drab and dusky; crown chestnut; cheeks throat and breast pale brownish drab; belly and vent white; tail dusky, forked and edged with whitish; bill and legs reddish cinnamon color; hind nail as long as the toe; the 3d primary longest, the 1st shorter than the 6th. Length 5¾ in.—*Nutt.*

HISTORY.—This species very much resembles the Chipping Sparrow, but the bay above is brighter, and the tail proportionably longer. It builds its nest of dried grass, upon the ground, in the shelter of a low bush or grassy tuft. The eggs are so thickly sprinkled with ferruginous as to appear almost wholly of that color.

THE SWAMP SPARROW.
Fringilla palustris.—WILS.

DESCRIPTION.—Blackish brown above, belly white; crown bright bay, undivided, bordered with blackish; line over the eye, sides of the neck, and breast ash color; wings and tail dusky, the primaries edged with brownish white, the secondaries with bay; bill dusky; iris hazel; legs stout and long, and with the feet pale brownish horn color. *Young* spotted with black and olive brown. Length 6; spread 8.—*Nuttall.*

HISTORY.—This species is aquatic in its habits, and resides principally in low wet lands and swamps, and hence its name, *Swamp Sparrow.* It arrives from the south in April, and builds its nest in a tuft of rank grass in the midst of a marsh. The eggs are 4 or 5, of a dirty white color, spotted with reddish brown.

YELLOW BIRD, OR AMERICAN GOLD FINCH.
Fringilla tristis.—LINNÆUS.

DESCRIPTION.—-General color of the male, in summer, rich gamboge yellow, fading into white towards the tail; crown and frontlet black; wings and tail black, varied with white; smaller wing feathers and coverts tipped and edged with white; tail sharply forked, with the feathers acutely pointed, and shaded off into white on their inner webs towards the tips; bill conical, acute, brownish yellow, and the gap straight; legs, feet and claws slender, and of a yellowish brown color. *Female, young,* and *male,* in autumn, brownish olive above, yellowish white beneath. Length 5 in.; spread 8. Four first primaries nearly equal.

HISTORY.—The Yellow Bird, or American Gold Finch, is common in summer from tropical America to the 50th parallel of north latitude. It arrives in Vermont later than several of the other sparrows, and is later in rearing its young. It seldom builds its nest till some time in July, and is less disposed to build in the immediate vicinity of our dwellings than several others of the family. The nest is usually placed in the top of a young forest tree, from 15 to 30 feet from the ground, and is composed of the dry bark of herbaceous plants, thickly bedded with cotton-like down of the Canada thistle. The eggs, 4 or 5, are white and without spots. This bird seems to be extremely fond of the seeds of the thistle, and of other compound flowers; and it often visits our gardens for the purpose of feeding upon lettuce and flower seeds. They soon become reconciled to the cage, and their song is nearly as sonorous and animated as that of the Canary Bird.

THE PINE LINNET.
Fringilla pinus.—WILSON.

DESCRIPTION.—Color dark flaxen, spotted with blackish; wings black, with two yellowish white bars; quill shafts and lateral tail feathers on the lower half yellow; rump, breast and sides spotted and streaked with blackish brown; bill dull horn color; legs purplish brown; iris hazel. Length 4¾; spread 8¼.

HISTORY -The Pine Linnet passes most of the year to the northward of the United States; but, in the depth of winter, often makes its appearance here and in states still further south. Of its history we know very little.

THE LESSER RED-POLL.
Fringilla linaria.—LINNÆUS.

DESCRIPTION.—-General color of the upper plumage yellowish gray, darkly streaked with blackish brown; wings and tail feathers blackish, slightly edged with white, with two narrow yellowish white bars on each wing; crown bright deep crimson, with a crimson tinge on the rump and sides of the throat; a brownish black band around the base of the bill, and reaching down upon the throat; belly bluish white, spotted and striped with brown upon the sides and beneath the tail; feathers on the thighs yellowish brown. Bill slender, straight, acutely pointed, yellowish on the sides, and brown above and below towards the tip; wings long, the three first quills longest, and nearly equal; tail sharply forked; legs, feet and claws black; claws slender, curved, acute, the hind one much the longest. Length of the specimen before me 5¼ inches; tail 2¼; folded wing 3.

HISTORY.—This elegant species is seldom seen among us, excepting in the winter, when they often appear in large flocks. They breed, according to Audubon, in Maine, Nova Scotia, and Labrador, and a few probably rear their young in this state. Dr. Richardson says that it is a permanent resident of the fur countries, where it may be seen in the coldest weather. Its nest resembles that of the Yellow Bird. The eggs, usually 5, are bluish green, spotted with reddish brown towards the large end.

THE FERRUGINOUS FINCH.
Fringilla iliaca.—MERREM.

DESCRIPTION.—Above varied with reddish brown and gray; beneath white, largely spotted with bright bay and dusky; head and neck cinereous, the feathers margined with ferruginous; wings and tail rust color, inclined to reddish brown; 1st and 2d row of wing-coverts tipped with white; bill stout, dusky above; iris hazel. Length 6, spread 9½.—*Nutt.*

HISTORY.—Most of this species spend the summer to the northward of the United States, and appear among us only during their spring and fall migrations. Some few of them, however, breed in the northern states, and I am informed by Dr. Brewer that they rear their young in the north part of this state. They build their nest upon the ground, and their eggs, 4 or 5, are of a dull greenish hue, irregularly blotched with brown.

WHITE-THROATED FINCH.
Fringilla pennsylvanica.—LATH.

DESCRIPTION.—The head striped with dusky and white; a yellow line from the nostril to the eye; upper parts varied with dusky, bay and light brown; shoulder of the wing edged with greenish yellow; cheeks and breast cinereous; throat and belly white; legs pale flesh-color; bill bluish horn-color; iris hazel. *Female* below, and stripes on the head, light drab. Length 7, spread 9½.—*Nutt.*

HISTORY.—This large and handsome Finch, or Sparrow, spends the winter, in large numbers, in the southern states, but, on the approach of spring, proceed to the north and rear their young throughout the whole region, from New England to the Fur Countries about Hudson's Bay. A few of them breed in the north part of Vermont. Their nest is built upon the ground, made of grass, and lined with hair and feathers. The eggs are pale green, marbled with reddish brown.

WHITE-CROWNED FINCH.
Fringilla leucophrys.—TEMM.

DESCRIPTION.—Crown white, line surrounding it and through each eye black; back streaked with dark rusty brown and pale bluish white; wings dusky, with two white bands; tertials black; rump and tail coverts drab; chin and belly whitish; vent pale ochreous; tail long, rounded, dusky, broadly edged with drab; bill, legs and feet cinnamon brown. *Female* with the colors duller. Length 7½, spread 10.—*Nutt.*

HISTORY.—This species is seen here only during its spring and fall migrations. Audubon informs us that it breeds in Newfoundland, Labrador and still further north. Their nest is built upon the ground, made of moss and lined with hair. The eggs, usually 5, are of a sea-green color, mottled and blotched with different shades of brown.

ARCTIC GROUND FINCH.
Fringilla arctica.—SWAIN.

DESCRIPTION.—The head, neck above and below, scapulars, all the wing coverts and tail pitch black; some of the breast feathers fringed with white; back scapulars, and wing coverts striped or tipped with white; quills hair brown; middle of the breast and belly pure white; sides, flanks and under tail coverts deep and bright ferruginous; bill black; legs pale brown. *Female* with upper plumage ferruginous-brown. Length 8¾, tail 4.—*Nutt.*

HISTORY.—This species is migratory, spending the summer and rearing its young in the Fur Countries, and retiring in the winter to warmer regions. Dr. Brewer informs me that it breeds also about Coventry, (now Orleans,) in this state. The nest is made of grass and leaves upon the ground, and the eggs, 4 or 5, are white, spotted with reddish chocolate.

TOWHE-GROUND FINCH.
Fringilla erythrophthalma.—LINN.

DESCRIPTION.—Upper parts black; belly white; flanks and vent bay; tail rounded, 4 outer feathers partly white; a white spot on the wing below the coverts and an interrupted white margin on the primaries; bill black. *Female* olive brown where the male is black, the head and throat inclining to chestnut; 3 only of the lateral tail feathers marked with white. Length 8, spread 11.—*Nutt.*

HISTORY.—This common bird derives

its name *Tow-he* from the sound of its note, when calling to its mate. It is found in all parts of the United States and Canada, but retires to the southern states to pass the winter. This bird breeds in Vermont. Its nest is built upon the ground, and the eggs, from 4 to 6, are white, tinged with flesh-color, and spotted with reddish brown.

THE PURPLE LINNET.
Fringilla purpurea.—GMELIN.

DESCRIPTION.—Head, breast and rump deep rich lake, approaching to crimson, and fading into rose color on the belly; feathers on the back brownish lake fringed with ash, producing a spotted appearance; vent and under tail coverts white; wings and tail dusky, edged with reddish white; bill grayish, dark horn color, having a fringe of cream-colored feathers at the base; tail forked; legs and claws brown; head and neck rather large; outline of each mandible a little convex; nostrils nearly concealed by the feathers. *Female* and young brownish above, and yellowish white beneath, without the crimson. Second and third primaries longest: 1st and 4th a little shorter. Length 6 inches, spread of the wings 9 inches.

HISTORY.—This beautiful and cheerful little songster arrives from the south about the beginning of April, and continues till October. Although the greater part of them proceed still further north to spend the summer, considerable numbers of them are known to rear their young in this state. Their nest is usually built upon a cedar, a fir or other evergreen, and is described by Dr. Brewer as being rudely made of grass and weeds, and lined with roots. The eggs are bright emerald green. These birds are often tamed and kept in cages, where they sing very pleasantly.

GENUS PYRRHULA.—*Brisson.*

Generic Characters.—Bill short, robust, thick, convex-conic, turgid at the sides, compressed at the point, the upper mandible acute, and obviously curved, as well as the inferior more or less; palate smooth and scooped; nostrils basal, lateral, rounded and most commonly concealed by the feathers; tongue thick and somewhat fleshy; tarsus shorter than the middle toe, which is united at the base to the outer; wings rather short; the 3 first primaries graduated, the 4th longest; tail square or slightly rounded. *Female* differs considerably from the male. They moult generally twice in a year.

THE PINE GROSBEAK.
Pyrrhula enucleator.—TEMM.

DESCRIPTION.—General color red; wings and tail dark cinereous, wing coverts forming two white bands; quills, lesser coverts and tail-feathers tinged with crimson; under plumage more red than the upper, except the middle of the belly, vent and tail coverts, which are bluish-gray; bill blackish brown; legs black. Tail broad and forked; 1st quill slightly shorter than the 2d, which hardly exceeds the 3d. Length $11\frac{1}{4}$, tail $4\frac{1}{4}$, wing $4\frac{2}{3}$. —*Richardson.* Length given by Audubon, $8\frac{1}{2}$; by Nuttall, 9.

HISTORY.—The Pine Grosbeak, or Bull Finch, inhabits the northern parts of both continents, and, according to Audubon, is a constant resident in the state of Maine, and to the northward to Hudson's Bay, where it builds its nest upon small trees, and feeds upon the seeds of the white spruce and other trees. They are seen in most parts of the United States only in the winter.

GENUS LOXIA.—*Brisson.*

Generic Characters.—Bill robust and convex, with the mandibles crossing each other, and compressed towards the points, which are extended in the form of crescents. Nostrils basal, lateral, rounded, hidden by the advancing hairs of the front. Tongue cartilaginous, short, entire and pointed. Tarsus nearly equal to the middle toe; toes divided to the base; hind nail largest, much curved. Wings moderate, 1st and 2d primaries longest. Tail notched. Female and young differ from the adult male.

THE COMMON CROSS-BILL.
Loxia curvirostra.—LINN.

DESCRIPTION.—General color dull light red inclining to vermilion, darker on the wings, with quills and tail feathers brownish black; lower parts paler, nearly white on the belly; plumage blended, but firm; tail short, small, emarginate. *Female* with the upper parts grayish-brown tinged with green, the rump dull grayish yellow. *Young* with the colors duller and more inclining to yellowish green. Length 7, spread 10.—*Aud.*

HISTORY.—This species is quite common in this state and to the northward of it, but further south is seldom seen, except in the winter. It feeds principally upon the seeds of the different kinds of pines and spruces, and its crossed mandibles are peculiarly fitted for extracting them from the cones. This bird breeds in Vermont, and its egg was obtained by

Dr. Brewer from Coventry (now Orleans,) in this state. Its color is greenish white, thickly covered, more especially towards the large end, with very brown spots. They are said to breed in winter, and to have their nests in pines, spruces and firs.

WHITE WINGED CROSS-BILL.
Loxia leucoptera —GMEL.

DESCRIPTION.—General color of the male rich carmine, inclining to crimson, dusky on the middle of the back; scapulars, wings, tail and upper tail coverts, black; two broad bands of white on the wings; sides brownish streaked with dusky; wings pointed, 3 outer primaries longest; tail emarginate. *Female* with the upper parts dusky, the feathers margined with grayish-yellow; rump, breast and lower parts yellow, streaked with dusky. Length 6¼, spread 10⅔.—*Aud.*

HISTORY.—The White Winged Cross-Bill resides mostly to the northward of the United States, and comes hither in flocks during the winter. They are, however, according to Audubon, not uncommon in New Jersey and Pennsylvania, where a few of them breed. Mr. Hutchins says that this migratory species reaches Hudson's bay in March, where it breeds, making its nest of grass, mud and feathers, in pine trees, and laying 5 white eggs marked with yellowish spots.

YOKED-TOED BIRDS.

In this order the form of the bill is various, but in general more or less arched and hooked. The toes are always in pairs directed two backward and two forward, and hence they received the name *Zygodactyli*, or yoked-toed. The hind exterior toe is, however, often reversible.

GENUS COCCYZUS.—*Vieillot.*

Generic Characters.—Bill strong, compressed with a distinct ridge and slightly bent from its base; under mandible straight, sloping at the tip; nostrils basal half covered by a naked membrane; tongue short, narrow and acute; tarsus naked, longer, or about the length of the longest toe; two anterior toes united at the base; nails short and but little curved; wings rather short; 3d and 4th primaries longest.

YELLOW BILLED CUCKOO.
Coccyzus americanus.—BONAPARTE.

DESCRIPTION.—Color above dark grayish-brown, with greenish and yellowish silky reflections; tail long, the two middle feathers the color of the back; the others dusky gradually shortening to the outer ones, with large white tips, the two outer scarcely half the length of the middle ones; below white; feathers of the thighs large and hiding the knees as in the hawks; legs and feet pale greenish-blue; iris hazel; lower mandible and lower part of the upper mandible yellow. *Female*, with the 4 middle tail-feathers without white spots. Length 12, spread 16.—*Nutt.*

HISTORY.—The Yellow-billed Cuckoo, returns from the south about the first of May and is much oftener heard than seen, as it keeps itself for the most part concealed in the thick tops of trees and bushes. It breeds in the southern part of the state. Its nest is placed on the horizontal branch of a small tree, and is very slovenly put together. The eggs, from 2 to 4, are of a pale bluish green color. This cuckoo destroys many catterpillars, beetles and other insects, but he gets a share of his living less creditably by sucking the eggs of other small birds. His note is coarse and unpleasant. The cry of this bird has been thought to presage rain, and hence it is sometimes called the *Rain-Crow.*

THE BLACK-BILLED CUCKOO.
Coccyzus dominicus.—NUTTALL.

DESCRIPTION.—General color above light hair brown with glossy bronze reflections; beneath white approaching to brownish ash on the throat, breast and towards the tail; tail feathers, excepting the two middle ones, tipped with white; a naked space of a bright brick red color around the eye; bill as long as the head, compressed laterally, arched and acute; upper mandible brownish black; lower, bluish; tarsus and feet bluish and scutilated; nostrils basal, lateral and partly closed by a membrane; legs rather short; body slender; tail long, graduated, consisting of 10 feathers. Length of the specimen before me 11½ inches; folded wing 5¼; tail 6, and reaching 3¾ beyond the folded wing; gape 1.2, bill above .9.

HISTORY.—This species is believed to be more common in Vermont than the preceding, but resembles it in appearance and mode of living. It, however, arrives later and passes the breeding season more in the woods. Their nests are made of twigs and lined with moss, but are very flat and shallow. The eggs, from 3 to 5, are of a bluish green color, and smaller than those of the preceding species.

GOLDEN-WINGED WOODPECKER. RED-HEADED WOODPECKER.

Genus Picus.—*Linnæus.*

Generic Characters.—Bill large or moderate, usually straight, pyramidal, compressed, cuneate, and edged like scissors towards the point; nostrils basal, oval, open, partly concealed by bristly feathers at the base of the bill; tongue long, extensile and vermiform; legs strong; feet robust, suited for climbing; two toes before, united at the base, and usually two behind, divided; 1st primary very short, 3d and 4th longest; tail cuneiform, with 12 feathers, the lateral ones being very short.

The Woodpeckers resemble one another in their habits and manner of life. Their nests are in excavations in old trees, and the young of most of the species emit a rank disagreeable odor. They do some injury by pecking holes in the bark of our fruit trees, in the pursuit of their favorite food; but it is trifling compared with the service which they render by the destruction of eggs, larvæ and insects.

GOLDEN-WINGED WOODPECKER.
Picus auratus.—Linn.

Description.—Upper plumage umber brown barred transversely with black; upper part of the head cinereous with a crimson red crescent behind; cheeks and throat bright cinnamon color; from the lower mandible descends a stripe of black to the throat; a black crescent on the breast; under plumage generally yellowish white, beautifully spotted with black, the spots circular on the breast, and hastate or heart-form towards the tail; under side of the wings and tail and the shafts of most of the larger feathers, saffron yellow; rump white; tail coverts white, notched and banded with black; tail black above with some of the feathers slightly edged and tipped with yellowish white; bill bluish black; legs grayish blue; iris dark hazel. Length 11½ inches; spread 19; length of the bill 1½.

History.—This is our largest, and one of our most common Woodpeckers. It is known by several names, such as Flicker, Yellow Hammer, and Partridge Woodpecker. This Woodpecker spends the winter in the southern states, and returns some time in April. Their nest is made by excavating a cavity in an old tree with their bill, and they have been known in this way to make a winding borough in solid oak, 15 inches in length. The eggs, usually 6, are pure white.

RED-HEADED WOODPECKER.
Picus erythrocephalus.—Linn.

Description.—Color of the head, neck and throat rich crimson; fore part of the back, scapulars and wing coverts bluish black; greater quills, anterior border of the wings, and tail pitch black; secondaries, rump and all the under parts of the body white; tail forked, several feathers tipped, and the two outer ones edged with white; shafts of the secondaries black; bill greenish blue, darker towards the tip, stout and slightly arched; iris yellowish brown. Colors of the *female* dull. Head and neck of the *young* grayish. Length 9, spread 16; 3d primary longest.

History.—The Red-Headed Woodpecker, although at present by no means rare in Vermont, is much less common than formerly. They pass the winter in the southern states, and return in the early part of May. Their migrations, according to Audubon, are performed in the night. They are remarkably fond of sweet apples, and are often seen in orchards. Their nest is excavated in the trunk or large limb of an old dead tree. The eggs are about 6, white and marked with reddish spots at the large end.

YELLOW BELLIED WOODPECKER.
Picus varius.—Wilson.

Description.—Color varied with black, white, yellow and crimson; fore part of the head and throat crimson; back mottled with black, white and pale yellow; wings black, with most of the feathers spotted and tipped with white; tail mostly black, with the two central feathers white, spotted with black on their inner webs, and some of the outer ones tipped with yellow; breast and belly light yellow; sides under the wings dusky yellow, spotted longitudinally with black; legs and feet dusky blue, inclining to green; feet four toed; bill blackish horn color, long and stout. Female, with the throat and back of the head whitish; young with a broad white band across the wings, and nearly without yellow on the back. Length 8; spread 14.

HISTORY.—This species is common throughout the continent, from the tropic to the 53d degree of north latitude. During the summer they confine themselves principally to the forests, where they rear their young in cavities excavated in old trees. Their eggs are white, and usually 4 or more. The cavity in which they rear their young is often excavated to the depth of from 15 to 24 inches in the solid wood.

THE HAIRY WOODPECKER.
Picus villosus.—LINNÆUS.

DESCRIPTION.—Color varied with black and white above; wholly white beneath; back clothed with long, loose, downy feathers; wings brownish black, thickly spotted with white; tail pointed, forked, outer feathers white, with an umber tinge at the extremity, second feather on each side black at the lower part, central and longest feathers pitch black; the crown, a stripe down the back of the neck, and a spot on each side of the head back of the eye, black; occipital band red in the male and black in the female; bill and claws bluish horn color; bill covered at the base with yellowish white hairy feathers, black at their extremity. Length 9; spread 15.

HISTORY.—This species is spread very extensively over the country, and in this state is much more common than the preceding, being often seen in the open fields and upon our orchard and shade trees. Its nest is constructed in the manner of the preceding species, and it lays about 5 white eggs.

THE DOWNY WOODPECKER.
Picus pubescens.—LINNÆUS.

DESCRIPTION.—Color of the top and sides of the head, wings and middle tail feathers, black; the chin, two stripes along the sides of the head, a stripe down the back, and numerous roundish spots on the wings, pure white; under plumage pale ash gray; outer tail feathers yellowish white, barred with black; feathers long, loose and downy on the back; head of the male crossed by a scarlet band, which is black in the female; nasal feathers tawny white; bill and claws bluish black; legs greenish; four toes on each foot. Total length of the specimen before me 6.2 inches; spread 11 inches; folded wings 4 inches.

HISTORY.—This is our smallest and, by far, our most numerous species of Woodpecker. In color it has a very close resemblance to the preceding, but differs from it very considerably in size. It is a permanent resident in this state, but as it rears its young for the most part in the forests, it is not much seen during the summer, but on the approach of autumn it makes its appearance upon our orchard and shade trees in considerable numbers. This is one of the most diligent of the feathered tribe, and may be recommended as a pattern of industry and perseverance. So intent is it in searching for eggs, larvæ and insects, that it scarcely heeds what is doing around it, and may often be approached so near as almost to be taken into the hand before it will abandon its business.

ARCTIC THREE TOED WOODPECKER.
Picus arcticus.—SWAINSON.

DESCRIPTION.—Back velvet black, with bluish and greenish reflections; crown saffron yellow; 5 rows of white spots on the quills; sides of the neck and under plumage white, thickly barred with black; two middle tail feathers brownish black; outer ones barred with black and tipped with white; bill bluish gray above, whitish beneath; legs lead colored. Length 10½ inches; wings 5.—*Richardson.*

HISTORY.—This large species of Woodpecker is very rare in comparison with the preceding. It is marked in a list kindly furnished me by Dr. Brewer, as breeding in this state, in the vicinity of Burlington. It has usually been confounded by ornithologists with the *Picus tridactylus,* or Common Three-Toed Woodpecker; The hind toe is completely versatile, and may be placed forward perfectly on a level with the others.

SLENDER BILLED BIRDS.

Birds of this order have the bill long, or moderately extended, partly arched and awl-shaped; it is also entire and acute or sometimes wedge-shaped at the extremity. The feet have three toes before and one behind, the outer united at the base to the middle one; hind toe generally long; the nails extended and curved. In their habits and method of running upon the trunks and branches of trees, they bear considerable resemblance to the woodpeckers.

GENUS SITTA.—*Linnæus.*

Generic Characters.—Bill straight, moderate sized conic-awl-shaped, round and sharp edged towards the point; lower mandible usually recurved from the tip; nostrils basal, orbicular, open, half closed by a membrane, and partly hid by the advancing bristly feathers of the face;

tongue short, wide at the qase, notched and hard at the tip; feet robust, hind toe stout and long; wings short; tail rather short consisting of 12 feathers. Sexes similar in color.

WHITE-BREASTED NUTHATCH.
Sitta carolinensis.—BRISSON.

DESCRIPTION.—General color dark lead above, grayish white beneath; head and neck black above, white on the sides and beneath; central part of the wing feathers and wing coverts black, edged with lead color or white; ferruginous tinge about the vent; bill bluish black, lighter beneath towards the base, long and straight; upper mandible longest; feet and legs dusky; hind toe stout and long with a large nail; claws all hooked and sharp; 2d 3d and 4th primaries longest and nearly equal. Length 5½ inches; spread 11.

HISTORY.—The White breasted Nuthatch is a permanent resident throughout nearly the whole of North America, and is very common in this state. During the fall and winter they come into our orchards and yards, where their rough *quank*, two or three times repeated, may be often heard as they run around like the Woodpecker upon the trunks of the trees. Early in the spring they retire to the forests, where they rear their young in the hollow of a tree or large limbs. The eggs, usually 5, are of a dull white color, spotted with brown at the large end.

THE RED-BELLIED NUTHATCH.
Sitta canadensis.—LINNÆUS.

DESCRIPTION.—Lead color above, reddish, or rust-color on the belly; head and neck above and line through the eye, black; a white stripe above and below the eye and on the margin of each wing; lateral tail feathers black and white, central ones lead color; feet and legs dusky; hind toe stout and long; bill black, large, long and straight; 3d primary longest, 2nd and 4th nearly as long. Length 4½ inches, spread of the wings 8 inches.

HISTORY.—This species resembles the preceding in general appearance and habits, but is said to have a predilection to pine forests, feeding much upon the oily seeds of evergreens. The flight of the Nuthatches is short, seldom extending farther than from one tree to another; and yet they have great powers of flight, since Audubon saw one come on board his vessel 300 miles from the shore. The specimens from which both preceding descriptions were made were obtained in Burlington.

GENUS CERTHIA.—*Linnæus.*

Generic Characters.—Bill long, or middling, more or less arched, entire three-sided, compressed, slender and acute; nostrils basal, naked, pierced in grooves, and half closed by a small membrane; tongue acute; feet slender; inner toe free, somewhat shorter than the outer; hind toe longer and more robust; nails much curved, that of the hind toe largest; wings rather short, spurious feathers small; tail of 12 feathers, elastic, ridged, and acuminate. The sexes and young nearly alike.

THE BROWN CREEPER.
Certhia familiaris.—LINNÆUS.

DESCRIPTION —Color varied with dusky brown, ferruginous, and white above, white beneath; rump bright rust color; tail rusty brown, as long as the body, with the extremity of each feather attenuated to a sharp rigid point, as in woodpeckers; under tail coverts tinged with rusty; 3d and 4th primaries longest, and all the primaries, excepting the two first, with a yellowish white spot near the middle; legs and feet brownish. Length 5¼ inches; spread 7 inches.

HISTORY.—This industrious little bird is seldom seen in the summer, on account of its passing that season in the depth of the forests, but on the approach of winter he may be seen upon the trees in more open places, diligently seeking for its food. It very much resembles the smaller Woodpeckers and Nuthatches in its habits, hopping about upon the trunk of the tree, searching every nook and crevice in the bark for spiders, insects, eggs and larvæ. The Brown Creeper breeds in this state, and for this purpose it takes possession of the deserted hole of a squirrel or woodpecker. The nest, according to Audubon, is loosely formed of grasses and lichens, and lined with feathers. The eggs, from 6 to 8, are yellowish white, irregularly marked with red and purplish spots. Nuttall found one of their nests in Roxbury, Ms., upon the ground by the side of a rock, containing 4 young.

GENUS TROCHILUS.—*Linnæus.*

Generic Characters.—Bill long, straight, or curved, tubular, very slender, with the base depressed and acuminated; upper mandible nearly enveloping the under one; tongue long, extensible, bifid and tubular; nostrils basal, linear, and covered by a membrane; legs very short; tarsus shorter than the middle toe; fore toes almost wholly divided; wings long and acute; first quill longest.

THE COMMON HUMMING-BIRD.
Trochilus colubris.—LINNÆUS.

DESCRIPTION.—The whole upper plumage shining golden green; wings glossed brownish black; tail broad, dusky, outer feathers tipped with white, or rusty white; throat and breast of the male with changeable ruby-colored, greenish and orange reflections; bill black and a little arched; legs and feet dusky black; nails very sharp and hooked. Female and young yellowish white beneath. Length 3½ inches, spread of the wings 4¼ inches; length of the bill along the gape 1 inch, nearly.

HISTORY.—Of American Humming-Birds there are said to be upwards of 100 species, but of the very few species which venture beyond the tropics, this is the only one which visits Vermont. It arrives in May, and during the summer is seen in all parts of the state collecting its food, which consists of insects and nectar from the various flowers. While many of them extend their migrations still further north, and rear their young on the very confines of the arctic circle, considerable numbers of them stop by the way, and not a few of them breed in this state. The puny nest, constructed of lichens and down, cemented together with saliva, is placed upon a large branch of an orchard or forest tree, at heights varying from 4 to 40 feet from the ground. The eggs, 2 in number, are white, and the period of incubation 10 days. While rearing its young the Humming-Bird bravely attacks the King Bird and the Martin, and drives them from the neighborhood of its nest.

HALCYONS.

In this order the bill is long, sharp-pointed, almost quadrangular and straight, or slightly curved; feet very short; the tarsus articulated; the middle toe united with the outer, commonly to the second joint, and with the inner toe to the first articulation. The female and young differ but little in color from the adult male.

GENUS ALCEDO.—*Linnæus.*

Generic Characters.—Bill long, straight, quadrangular, compressed, and sometimes curved at the point; nostril basal, lateral, oblique, and nearly closed by a naked membrane; tongue short and fleshy; legs and feet short; tarsus shorter than the middle toe; hind nail smallest; wings rather short.

THE BELTED KING FISHER.
Alcedo alcyon.—LINNÆUS.

DESCRIPTION.—General color bluish slate; the primaries, the central parts of the secondaries and of the feathers forming the crest, and the shafts generally of the dorsal plumage, pitch black; a small spot before and another under the eye, spots on the wing and tail feathers and their tips, and all the under plumage, white, except the band around the neck, which is bluish slate; bill straight; claws brownish black; legs small and short. Length 12 inches; spread 20 inches. Female shorter, with some parts ferruginous and more white on the wings.

HISTORY.—The King Fisher is found along the borders of streams and ponds, in all parts of the United States, and is quite common in all parts of this state. It feeds principally upon small fishes, which it takes by darting upon them as they are gliding near the surface of the water. The note of the King Fisher is a rough grating *crackle.* Its nest is formed by perforating horizontally the side of a steep bank, in the manner of the Bank Swallow. These perforations sometimes extend 5 or 6 feet into the bank, with an enlargement at the extremity for the reception of the nest, which consists only of a few twigs, grass and feathers. The eggs are white, and usually 6 in number. Their period of incubation is 16 days.

THE SWALLOW TRIBE.

The birds belonging to this order have a very short bill, which is much depressed and very wide at the base; upper mandible curved at the point; legs short; three toes before, and one behind which is frequently reversible; nails hooked; wings very long and acute. The sexes and young are nearly alike. They feed on insects, which they catch flying. They migrate to tropical countries to spend the winter.

Genus Hirundo.—*Linnæus.*

Generic Characters.—Bill short, triangular, depressed, wide at the base, and cleft nearly to the eyes; upper mandible notched and a little hooked at the point; nostrils basal, oblong, partly closed by a membrane and covered by the advancing feathers of the frontlet; tongue short, bifid; tarsus short; toes and claws long and slender, three before and one behind; the exterior united as far as the first joint of the intermediate one; wings long; the first quill longest; tail of 12 feathers, and forked.

THE PURPLE MARTIN.
Hirundo purpurea.—Linnæus.

Description.—Color of the head, whole body and scapulars black, with a rich glossy shade of bluish purple; wings and tail pitch black, with little gloss; bill, legs and claws black; margins of both mandibles inflexed in the middle; nostrils basal and oval. *Female* brownish black above, with very little of the purple gloss; belly brownish white with hair brown spots; breast brownish gray. Length 8 inches; spread of the wings 16 inches.

History.—The Purple Martin is the largest of our swallows, and is more intimate with man than any other undomesticated bird. It returns from the south about the last of April, and formerly reared its young in the hollows and excavations in old trees; but since the country has become settled, habitations have been provided for this general favorite in almost every neighborhood, by the erection of martin boxes. Its nest is made of leaves, straw and feathers; and the eggs, from 4 to 6, are pure white and without spots. The Martins have sometimes arrived so early in the spring as to become chilled to death in their houses during a cold storm. This was the case a few years ago in the vicinity of Burlington. The flight of the Martin is very rapid, and, like the redoubtable King Bird, it pursues and boldly attacks eagles, hawks and crows, and drives them from the neighborhood of its dwelling. There is said to be a tradion that the Martin was not seen in New England till about the time of the revolution. It is, however, mentioned by Kalm as being common in New Jersey in 1749. They usually depart to the south about the middle of August.

THE BARN SWALLOW.
Hirundo rufa.—Gmelin.

Description.—Color above and band on the breast steel-blue; front and beneath chestnut brown, paler on the belly; tail forked, with a white spot on the lateral feathers, the outer ones narrow and an inch and a half longer than the next; legs dark purple; iris hazel. *Female* with belly and vent rufous-white. Length 7, spread 13.

History.—This swallow is, perhaps, more generally diffused over the state and better known than either of the other species; but it would seem that their numbers have rather been diminishing for several years past in this state, while those of the Cliff Swallow have been vastly multiplied. This swallow arrives in Vermont about the 28th of April. (See *page* 13.) They generally build their nest against a rafter or beam in the barn. It is formed principally of mud, and lined with fine grass and a few feathers. The eggs, usually 5, are white, spotted with reddish brown.

FULVOUS, OR CLIFF SWALLOW.
Hirundo fulva.—Vieill.

Description.—Top of the head, back, upper side of the tail and wings brownish black, with violet reflections from the head, back and wing coverts; forehead marked with a crescent of yellowish white; chin, throat and sides of the neck brownish red; rump yellowish red; belly white tinged with reddish brown; bill black, short, depressed, and very broad at the base. Wings long, slender; first quill longest, second nearly as long; tail even, extending as far as the folded wings. Length of the specimen before me 5½ inches; folded wing 4¼.

History.—This swallow seems to have been hardly known to ornithologists till about the year 1815, when they were noticed near the Ohio river in Ohio and

Kentucky. In 1817 they made their appearance at Whitehall, at the south end of lake Champlain, and shortly after at Randolph, Richmond, and some other places in this state. In unsettled places they build their nests upon the sides of rocky cliffs, but here they are usually placed beneath the eaves of barns and other buildings. They are constructed principally of clay or mud, in the form of a retort or gourd, and are lined with dry grass. The eggs, usually 4, are white, spotted with brown. These swallows always build their nests in companies, and are so remarkably gregarious, that from 50 to 100 of their nests may often be counted at the same time beneath the eaves of a single building.

WHITE-BELLIED SWALLOW.
Hirundo bicolor.—VIEIL.

DESCRIPTION.—Color above light glossy greenish blue; wings and tail brownish black; belly white; the closed wings extend a little beyond the tail, which is forked; tarsus naked. *Female* like the male, but less glossy. Length 5½ inches; spread 10.

HISTORY.—This Swallow is much less common in Vermont than the other species. Their nests are made of grass and lined with feathers, and are placed in various situations, such as beneath the eaves of old buildings, or in hollow trees, and they not unfrequently take possession of Blue bird and Martin boxes. The eggs, 4 or 5, are pure white.

THE BANK SWALLOW.
Hirundo riparia.—LINN.

DESCRIPTION.—Color above, and band on the breast, cinereous brown; beneath white; wings brownish black; tail forked, with the outer feathers edged with white; tarsus naked, excepting a few tufts of downy feathers behind; chin slightly fulvous. Length 5¼ in.; folded wing 4 in., and reaching nearly to the extremity of the tail.

HISTORY.-The Bank Swallow, or Sand Martin, is gregarious, like the Cliff Swallow, and may be found in companies in all parts of the state which afford suitable places for its habitation. These are usually sandy cliffs on banks of rivers. They commence 2 or 3 feet below the surface of the bank, and perforate the ground in a horizontal direction to the distance of from 2 to 4 feet, and at the further extremity they place their nest, which is composed of a little dry grass and a few feathers. The eggs, usually 5, are pure white. Often from 30 to 60 or more of these Swallow holes may be counted in a bank, in the space of one or two rods. The voice of this swallow is a low mutter.

GENUS CYPSELUS.—*Illiger.*

Generic Characters.—Bill very short, triangular, cleft to the eyes, depressed; the upper mandible slightly notched and curved at the point; nostrils lateral, contiguous, large, partly covered by a membrane; tongue, short, wide and bifid at the tip; feet very short; toes divided, hind toe shortest, reversible, generally directed forward; nails retractile, channeled beneath; wings very long. Sexes and young nearly alike in plumage.

THE CHIMNEY SWALLOW.
Cypselus pelasgius.—TEMMINCK.

DESCRIPTION.—General color sooty brown, approaching to black, lightish about the throat and over the eye; legs and feet bluish, muscular, with exceeding sharp claws; the folded wings very narrow and long, extending 1½ inch beyond the tail, which is short and rounded, with the shafts of the feathers reaching beyond the vanes into sharp, strong, and very elastic points; 2d quill of the wings longest. Length from the end of the bill to the extremity of the tail, 4½ inches; to the extremity of the folded wings 6 in.; spread of the wings 12 inches.

HISTORY.—The Chimney Swallow is one of our most singular birds. It arrives from the south, where it has spent the winter, about the beginning of May. On their arrival here before the country was much settled, they took up their residence in large flocks in particular hollow trees, which, in consequence, received the name of Swallow Trees. Three of these trees, all large hollow elms, are mentioned by Dr. Williams (Hist. I—140) as being particularly noted in this state soon after the settlement was commenced. One of these was in Middlebury, one in Bridport, and the other in Hubbardton. About the beginning of May the Swallows were observed to issue from these trees early in the morning in immense numbers, and to return into them again just before dark in the evening. The same phenomena were also observed in the latter part of summer, before the entire disappearance of the swallows and as their departure to the south was not observed, they were generally believed to spend the winter in these trees in a torpid state. Before this country was much settled, Chimney Swallows built their nests on the interior surface of large hollow trees, but they now take advantage of unoccupied chimneys for that purpose, and for roosting places. The nest is formed of slender twigs, interlocked and cemented together, and to the chimney or tree, by an adhesive mucilage secreted by the stomach of the architect. The eggs are white, and usually 4. This Swallow is often called the *Chimney Swift*.

GENUS CAPRIMULGUS.—*Linnæus.*

Generic Characters.—Bill extremely short, feeble and cleft beyond the eyes; upper mandible usually surrounded with spreading bristles, sometimes hooked at the tip, the margin turned outward; nostrils basal, wide, partly covered by a feathered membrane; tongue small, acute and entire; tarsus partly feathered; anterior toes united by a membrane to the first joint; hind toe reversible, nails short; wings long; tail of 10 feathers; the sexes distinguishable by their plumage; the young similar to the adults.

along the streams and low lands in various parts of the state, even up to the northern boundary. For a nest this bird makes a slight excavation upon the surface of the dry ground, in the forest, usually by the side of a rock, a log, or a pile of bushes; and, in this, about the 1st of June, the female lays two eggs, which are of a bluish white color, thickly blotched with dark olive. The young, like chickens, are able to run about and hide themselves as soon as they are hatched; and being without a nest, and very nearly the color of the ground, they very easily escape notice.

THE NIGHT HAWK.

Caprimulgus virginianus.—BRISSON.

DESCRIPTION.—General color dark liver brown, often with a greenish gloss; the head, neck, back, scapulars and wing coverts spotted with white, and yellowish brown; quills of the wings brownish black, with a broad bar of white across the middle, above and below; a broad sagittate spot of pure white on the throat, and white across the tail in the male; under plumage and inner wing coverts marked with alternate bars of dark liver brown and yellowish white; wings swallow-like, reaching a little beyond the tail; 1st quill longest, 2d nearly as long; bill blackish without bristles; legs short, pale brown. Length 9½ inches; spread 23 in. *Female* 9 inches long, and color ochrey about the head and throat.

HISTORY.—The Night Hawk arrives in Vermont in May, and is very common, during the summer, in all parts of the state. They rear their young in meadows and old fields. The eggs, which are only two, are laid upon a bare spot of ground, without any manner of nest. They are of a muddy white color, thickly freckled all over with reddish brown. During the period of incubation the males are often sporting upon the wing, and emitting their sharp squeak, high in the air, towards the close of the day, occasionally precipitating themselves towards the earth, emitting at the same time their peculiar *pŏō-ō-ō*, and then rising quickly to their former height. This sport is usual-

THE WHIP-POOR-WILL.

Caprimulgus vociferus.—WILSON.

DESCRIPTION.—Variegated above with black, brownish white and rust color, with fine streaks and sprinkles; upper part of the head brownish gray, marked with a longitudinal stripe of black; tail of 10 feathers rounded, the 3 outer feathers white at their extremities; the 4 middle ones without white at the ends, but with herring-bone figures of black, and pale ochre; cheeks and sides of the head brick color; chin black with small brown spots; a semi-circle of white across the throat; breast and belly mottled and streaked with black and ochre; bristles on the cheeks much longer than the bill; middle claw pectinated; female less than the male. Length 9½, spread 19.—*Nutt.*

HISTORY.—The Whip-poor-will arrives in Vermont early in May, and his plaintive note is soon heard in the groves,

ly continued till nearly dark, and hence this bird, probably, received the name of Night Hawk, or *Night Jar.*

GENUS COLUMBA.—*Linnæus.*

Generic Characters.—The bill, in this Genus, is of moderate size, compressed, vaulted, turgid towards the tip, which is more or less curved. The base of the upper mandible is covered with soft skin, protuberant at its base, in which the nostrils are situated. Nostrils medial, longitudinal. Tongue acute, entire; feet short, robust; tarsi reticulated; toes divided; wings moderate; tail of 12 or 14 feathers.

THE PASSENGER PIGEON.
Columba migratoria.—LINN.

DESCRIPTION.—General color of the upper plumage and breast light umber brown; rump bluish, belly and under tail coverts dirty white; nearly all the feathers above and on the breast tipped with yellowish white, forming little crescent-shaped bars; outer webs of the primaries edged with buff or rufous; tail of 12 feathers, with middle pair dark brown, and longest, the others with a basal spot of rufous and a central black spot or band on the inner web, outer feathers shortest, and white, excepting the spots, much longer than the folded wings; bill black; legs and feet dull red; breast of the male with a reddish tinge. Length 15 inches; spread 23 inches.. 1st and 2d primaries equal and longest.

HISTORY.—The American Wild Pigeon is met with in greater or less numbers throughout the whole region from Mexico to Hudson's Bay. These birds are remarkably gregarious in their habits, almost always flying, roosting and breeding in large flocks. When the country was new there were many of their roosts and breeding places in this state. Richard Hazen, who run the line between this state and Massachusetts, in 1741, stated that to the westward of Connecticut river, he found pigeons' nests so thick upon the beech trees that 500 could be counted at one time. At Clarendon, according to Dr. Williams, (Hist. vol. I—137,) the pigeons bred in immense numbers. The trees were loaded with nests for hundreds of acres; 25 nests being frequently seen upon one tree, and the ground beneath was covered with their dung to the depth of two inches. These accounts are far exceeded by what is told of their roosting and breeding places at the west, where they often covered thousands of acres, and all the trees and under growth were killed in consequence. From 90 to 100 nests have frequently been counted on a single tree. The nests are made of twigs, the eggs are 2 and white. Pigeons are much less abundant in Vermont than formerly, but they now, in some years, appear in large numbers.

THE CAROLINA DOVE.
Columba carolinensis.—LINNÆUS.

DESCRIPTION.—General color above pale yellowish brown; below brownish yellow; crown and upper part of the neck greenish-blue; forehead and breast vinaceous; black spot under the ear; bill blackish, purplish-red at the base; tail of 14 feathers, with the 4 lateral ones black near the extremity, and white at the tip. Length 12, spread 17.—*Nutt.*

HISTORY.—The Carolina Dove, called also the *Turtle Dove*, is not very common in Vermont. Dr. Brewer saw a flock of them near Woodstock in August, 1839; and they have been occasionally seen in other parts. From its plaintive *àgh-còo-còo-còo*, it is sometimes called the *Mourning Dove*. They are by no means shy, are said to be easily tamed, and their flesh is pronounced equal to that of the Woodcock.

GALLINACEOUS BIRDS.

Birds of this order have the bill short and convex; the upper mandible vaulted, curved from the base or only at the point; nostrils basal, partly covered by an arched rigid membrane; feet stout, tarsus long; toes usually three before and one behind, the latter articulated higher than the rest, scarcely touching the ground at the tip, sometimes wanting; wings generally short and concave; tail consisting of from 10 to 18 feathers. Colors of the female less brilliant than those of the male. Our domestic land fowls, as hens, turkies and peacocks, belong to this order.

GENUS MELEAGRIS.—*Linnæus.*

Generic Characters.—Bill entire, and at the base covered by a membrane which is prolonged into a pendulous, fleshy, conic, erectile, hairy carbuncle; nostrils oblique; tongue fleshy and entire; feet rather long; tarsus naked, provided with a blunt spur in the male; middle toe longest; nails wide and blunt, flat beneath; wings short; 1st primary smallest, 4th and 5th largest; tail of 14 to 18 wide feathers, and capable of a vertical expansion; head small, naked and warty; a pendulous tuft on the lower part of the neck. *Female* smaller; colors duller and more obscure.

THE WILD TURKEY.
Meleagris gallopavo.—LINNÆUS.

DESCRIPTION.—Upper part of the back and wings yellowish-brown of a metallic lustre, changing to deep purple, the tips of the feathers broadly edged with velvet black; primaries dusky, banded with white; tail of 18 feathers, ferruginous thickly waved with black, and with a black band near the extremity; lower part of the back and tail coverts deep chestnut, banded with green and black; legs and feet purplish-red; iris hazel; beneath duller. *Female* and *young* with the colors less brilliant. Length 48, spread 68.—*Nutt.*

HISTORY.—The Wild Turkey, which was formerly common throughout our whole country, has every where diminished with the advancement of the settlements, and is now become exceedingly rare in all parts of New England, and indeed in all the eastern parts of the United States. A few of them, however, continue still to visit and breed upon the mountains in the southern part of the state. The Domestic Turkey sprung from this species, and was sent from Mexico to Spain in the 16th century. It was introduced into England in 1524, and into France and other parts of Europe about the same time.

GENUS PERDIX.—*Latham.*

Generic Characters.—Bill entire and bare; upper mandible vaulted and strongly curved towards the point; nostrils basal, lateral, half closed by a vaulted naked membrane; feet naked, fore toes united by a membrane to the first articulation; hind toe less than half the length of the inner; nails incurved, acute; head wholly feathered, often with a naked space around the eye; tail short, rounded, and deflected, consisting of from 12 to 18 close feathers. Female and young scarcely differ in plumage from the male.

THE QUAIL.
Perdix virginiana.—LATH.

DESCRIPTION.—Cinnamon brown above, varied with black and whitish; crown, neck and upper part of the breast reddish brown; line over the eye and throat pure white, the latter bounded with a black crescent; wings dusky, coverts edged with yellowish white; belly yellowish white, varied with wide arrow heads of black; tail ash colored, finely spotted with reddish brown; bill black; iris hazel; legs and feet light lead color. Length 9, spread 14.—*Nutt.*

HISTORY.—This bird, generally known as the Quail in New England, is in other places more commonly called the *American Partridge.* It is not found in this state at present very plentifully, but is more common in the southwestern parts than elsewhere. They generally go in small flocks, spending most of the time on the ground, and in autumn are often seen gleaning in fields from which corn and grain have been harvested. The Quail is very prolific, laying from 10 to 18 eggs, which are white, in a nest formed partly in the ground, under the shelter of a tuft of grass. Frequent attempts have been made to domesticate the Quail, but with very little success.

GENUS TETRAO.—*Linnæus.*

Generic Characters.—Bill short, robust, arcuated above, convex and bent towards the tip, naked at the base; nostrils basal, half closed by an arched membrane, and hidden by small feathers; tongue short, fleshy, and pointed; tarsus feathered and spurless in both sexes; three toes before united to the first joint; hind toe half as long as the inner, and roughened.

THE PARTRIDGE.
Tetrao umbellus.—LINN.

DESCRIPTION.—General color above and beneath black, pale chestnut, and yellowish white, marbled, and disposed in spots, bars and lines. Ruff brownish black with greenish or cinnamon colored reflections. Quills liver brown, their outer webs barred near the base and mottled towards the tip with cream yellow; 4th quill longest. Tail with alternate undulating bars of brownish black, gray and faint chestnut, the subterminal bar being brownish black and broad; a light stripe from the nostril to the eye. Bill dark horn color, short, arched, and covered at the base by feathers; head and neck small; body bulky; tarsus feathered half way down before and some lower behind. Wings short and broad. Tail large, fan like, of 18 feathers. Length 18, spread 24.

History.—This bird, which is usually known as the Partridge in New England, is called the *Pheasant* in most other parts of the United States, and by ornithological writers is more commonly distinguished as the *Ruffed Grouse*. It is quite common and a permanent resident in all parts of Vermont. The nest of the Partridge is upon the ground by the side of a bush or log, and is very simple, consisting only of a few leaves. The eggs, usually about 12, are of a yellowish white color, and the young run about, like chickens, after their clucking mother, as soon as they are hatched. They are exceeding wild and difficult to tame, and it is amusing to see how quick they will hide themselves under leaves and logs whenever they are approached. The male of this species is distinguished for his peculiar *drumming*, which is performed, standing upon a log in a thick part of the woods, and rapidly beating his sides for about half a minute at a time, with his wings. This operation is repeated about once in 8 or 10 minutes, and the sound produced, somewhat resembling distant thunder, is often heard at the distance of half a mile. Their flesh is much esteemed for food.

THE SPRUCE PARTRIDGE.
Tetrao canadensis.—Linn.

Description.—Upper parts marked with semi-circular bars of black and yellowish brown, the paler color always forming the terminal bar; outer edge of the wings, primary coverts and quills clove brown; tail black tipped with orange; breast and belly with feathers blackish tipped with white; cheeks and throat barred and mottled with white; bill and nails black; fringed comb over the eye bright red; toes pectinated. Length 17, wing 7½.—*Rich.*

History.—This Grouse, which is called, at different places, the Spruce, the Wood or the Swamp Partridge, from its favorite places of resort, is seldom seen in Vermont excepting in the most northerly parts, and there it is scarce, compared with the preceding species. Its food in winter is said to consist principally of the leaves of the white spruce, and its flesh has then a strong, disagreeable flavor. In summer it is better, but still inferior to the preceding. Its nest is upon the ground, and the eggs, which are usually not more than 5 or 6, are said to be varied with yellow, white and black. It is known to breed in several towns in Orleans county.

WADING BIRDS.

In this order the bill varies in form, but is usually straight, and carried out into a lengthened and compressed cone, though rarely it is depressed, or flat. The legs are long and usually naked some distance above the knees; toes usually long and slender, three before and one behind, the latter on a level, or a little more elevated than the rest. Most of the Waders are more or less nocturnal in their habits. The sexes differ but little in external appearance. They live along the borders of seas, lakes and rivers, and feed upon fish, reptiles and insects.

Genus Calidris.—*Illi. Temm.*

Generic Characters.—Bill of moderate size, slender, straight, rather soft, flexible in every part, compressed from its base, with the point depressed, flattened and wider than the middle. Nasal groove elongated nearly to the point of the bill; nostrils lateral. Feet slender, the 3 toes all directed forward and almost entirely divided to their base. Wings of moderate size; the first quill longest.

THE SANDERLING PLOVER.
Calidris arenaria.—Illiger.

Description.—Color above mottled with black, white and yellowish; wings brownish black, with the shafts and tips of the quills, and a broad band extending across the whole wing, with the exception of the first 4 primaries, white. All the under plumage white, excepting a broad collar round the lower part of the neck, which is grayish; bill, legs, feet and nails black; iris hazel; two middle tail feathers longest, brownish, and edged with yellowish white. Folded wings a little longer than the tail; thighs feathered more than half way down; nails short; upper mandible longest, and curved a little at the point. *Winter* plumage nearly white. Length of the specimen before me 7½; folded wing 5; spread 14; bill, along the ridge, 1.

History.—This beautiful species, according to Dr. Richardson, breeds on the coast of Hudson's Bay. Its nest is rudely made of grass in marshes, and the eggs are 4, dusky, spotted with black. This plover is only occasionally met with in Vermont, along the shores of our lakes and ponds. The specimen from which the above description and figure were drawn was shot in Burlington, in September, 1841.

Genus Grus.—*Pallas.*

Generic Characters.—Bill a little longer than the head, strong, straight, compressed, attenuated, and obstuse at the point; ridge of the bill elevated; mandibles with a wide furrow on each side of the base; nostrils in a furrow in the middle of the bill, pervious, posteriorly closed by a membrane; feet long and robust, naked for a large space above the knee, middle toe united to the outer one by rudimental membrane, hind toe articulated high on the tarsus; wings moderate 2d, 3d, and 4th primaries longest, secondaries broader than the primaries, tail short, of 12 feathers.

THE WHOOPING CRANE.
Grus americana.—Temm.

Description.—The forehead, crown and cheeks covered with orange colored warty skin, with a few black hairs; hind head ash-color; the rest of the plumage pure white, except the primaries, which are brownish black; bill and iris yellow, legs and naked part of the thighs black. From the base of each wing arise numerous large flowing feathers, which project over the tail and tips of the wings, some of them being loose and webbed like those of the Ostrich; length 48, bill 6, height 60.—*Nuttall.*

History.—This bird is one of the largest of the feathered tribes in the United States, and is known in Vermont only by being occasionally seen during its migrations. It is common in summer in the fur countries where it breeds. Its two eggs are bluish white and as large as those of the swan. When wounded, says Dr. Richardson, he has been known to put the fowler to flight and fairly drive him from the field.

Genus Ardea.—*Linn. Tem.*

Generic Characters.—Bill long, robust, straight, pointed, compressed to an edge, the ridge rounded; upper mandible slightly furrowed; nostrils lateral, basal, situated in the furrow, and half closed by a membrane; orbits and lores naked; legs long, slender, lower part of the thighs without feathers; middle toe united to the outer one by a short membrane; hind toe on the same level with the other three; wings of moderate dimensions, obtuse; 1st primary nearly equal to the 2d and 3d, which are longest: tail short, rounded, containing 10 or 12 feathers.

THE NIGHT HERON.
Ardea nycticorax.—Wilson.

Description.—General color nearly white; front, occipital feathers and line over the eye pure white; crown, back and scapulars greenish; tail coverts, wings and tail pale ash; lower parts yellowish cream-color; legs yellowish green; bill black, 4½ inches along the gap. Without crest in autumn. *Young* brown streaked with rufous white. Length 28, spread 48.—*Nutt.*

History.—Vermont is about the limit of the northern migration of this Heron, and here it is rare. It is usually called the *Qua* Bird. It breeds all along the Atlantic coast to the southward of New England. They build their nests in trees in the retired parts of swamps, and frequently there are two or three nests on the same tree. The eggs, about 4, are of a pale greenish-blue color, and as large as those of the common hen.

THE GREAT HERON.
Ardea Herodias.—Linnæus.

Description.—General color grayish ash; crest brownish, the middle of the feathers striped with whitish; back of the neck ash; small feathers on the wings edged with ferruginous; feathers on the neck and breast white in the centre, edged with brown, giving a striped appearance· thighs naked some distance above the knees; feathers on the upper part of the thighs buff; legs brownish, tinged with yellow; chin, cheeks and sides of the head whitish; quills slate color; tail a little longer than the folded wings; generally two tapering feathers in the crest

5 or 6 inches long. Length of the specimen from which the above description is drawn, from the point of the bill to the extremity of the tail, 46 inches; height, when standing, 40 inches; length of the bill, from the angle of the mouth, 7 in.; folded wing 19; tarsus 7½; longest toe 5.

HISTORY.—The Great Blue Heron is frequently seen in the neighborhood of lake Champlain. The specimen from which the above description was drawn was shot near Burlington, and is now in the Museum of the College of Natural History of the University. They are said to rear their young in companies, making their nests with sticks in the tops of tall trees. The eggs, usually 4, are larger than those of the hen, light green, and unspotted.

THE GREEN HERON.
Ardea virescens.—LINN.

DESCRIPTION.—Color of the back, tail, crown and wings dark glossy green, approaching to black; wing feathers mostly tipped with white; wing coverts and scapulars tipped and edged with white and ferruginous; neck above and on the sides dark wine color; chin and line under the angle of the mouth, white; throat and under side of the neck, with the feathers, white, tipped or margined with brownish; belly brownish white; lore and iris bright yellow; bill black, lighter beneath and yellowish towards the base; legs and feet greenish yellow; feathers on the back of the head and neck long; tail short, consisting of 12 feathers; the 1st and 4th primaries a little shorter than the 2d and 3d, which are longest. Length 17 inches; spread 23; folded wing 7½; bill from the angle of the mouth 3; along the ridge 24 inches.

HISTORY.—The Green Heron, better known by a more disgusting name, is very common in many parts of the state. It seems to prefer the solitude of swamps and marshes, where it feeds upon fishes and reptiles, and also upon dragon flies and other insects. It builds its nest upon trees, and lays 4 blue eggs. They come from the south about the first of May, and return in October.

GENUS TOTANUS.—*Bech. Temm.*

Generic Characters.—Bill of moderate length, straight, or a little recurved, flexible at the base, hard and acuminate at the point; both mandibles furrowed on each side to the middle; nostrils in the furrow, basal, linear and pervious; legs long and slender; feet with three anterior toes, the exterior united to the middle one, sometimes to the second joint; wings of medium length; tail of 12 feathers, generally short.

THE UPLAND PLOVER.
Totanus Bartramius.—TEMMINCK.

DESCRIPTION.—General color above blackish, the feathers edged with tawny rufous; lower part of the back and upper tail coverts pitch black; wings brownish black above, shaft of the first primary white, and most of the primaries with concealed white spots or bars on their inner webs; chin and belly white; under tail coverts tinged with rufous; brownish sagittate spots on the breast and sides; under sides of the wings barred and waved with brown and white; tertials long; bill blackish above and at the point, yellowish below; tongue sagittate; 1st primary longest; length 12 inches; spread of the wings 22 inches; bill from the angle of the mouth 1½ inch.

HISTORY.—This species was first described by Wilson, who named it *Bartramius* in honor of his friend Bartram. It is quite common in the western parts of this state during the summer, and resides principally in meadows, feeding upon grasshoppers and other insects. Its nest is made upon the ground usually in a little clump of bushes. They are a shy bird and quite plain in appearance when seen at a distance, but closely viewed their colors appear beautifully variegated, especially beneath. They live for the most part, in pairs or families.

THE SOLITARY TATTLER.
Totanus chloropygius.—VIEILLOT.

DESCRIPTION.—The whole upper plumage dark hair brown, interspersed with small, irregular, marginal spots of white, and usually slightly glossed with green reflections; the lateral tail feathers with their coverts regularly barred with black and white, the bars being broadest on the former; middle tail feathers dark brown, with small white spots on the edges; primaries, their shafts and coverts brownish black, unspotted, the shaft of the 1st primary a little lightest; a short stripe over the eye, the chin, belly and under tail coverts white; neck and breast spotted or striped with brownish; under side of the wings next the base and axiliaries finely barred or waved with brown and white; bill brown, with the nasal groove two thirds its length; legs and feet dusky olive. Length 8¼ inches, tail 2¼, folded wing 5, bill 1¼, tarsus 1.3.

HISTORY.—This bird is often seen along the shores of our streams and ponds, and, as it spends the whole summer with us, it doubtless breeds here; but I have not known of its nest being found. According to Dr. Richardson it breeds in most of the intermediate districts between Pennsylvania and the northern extremity of the continent, depositing its eggs upon the beach, without forming any kind of nest. It is generally seen running along upon the shore, frequently stopping, and often nodding, or balancing its head and tail, and hence its vulgar appellation is *Tip-up*.

THE SPOTTED TATTLER.
Totanus macularius.—TEMMINCK.

DESCRIPTION.—Color glossy olive brown, waved with dusky; one or more of the outer tail feathers white, barred with black; quills dusky brown, the two outer plain, the next marked with an oval white spot on their inner webs; secondaries white on their inner webs and tipped with white; below white, tinged with gray at the sides of the neck, with roundish dusky spots; bill yellow below, black at the tip; legs waxyellow; iris hazel. Length 7½. *Young* white below, without spots.—*Nutt.*

HISTORY.—This bird is often called the *Peet-Weet*, from its shrill and peculiar note. It resembles the preceding species in general appearance, and in most of its habits, particularly in that of balancing or wagging its tail, and it bears the same vulgar name of *Tip-up*, the two kinds not being distinguished from each other by ordinary observers. This species is much the most numerous of the two, and breeds in this state in considerable numbers. The nest is made in a tuft of grass, with a thin lining of hay. The eggs, usually 4, are of a dull cream color, spotted with brown, most thickly towards the large end. The female, when alarmed, practices much art for the safety of her young.

GENUS SCOLOPAX.—*Linn.*

Generic Characters.—Bill long, straight, slender, compressed, soft and flexible; the point depressed, dilated, tumid and obtuse, minutely tuberculated or dotted, projecting over the lower mandible; both mandibles furrowed to the middle. Nostrils in the furrow of the bill, basal, lateral, linear, pervious and covered by a membrane. Feet and legs moderate, slender, 4 toed, naked space above the knee small; toes entirely divided. Wings moderate, the 1st and 2d primaries longest and nearly equal. Tail short, rounded, consisting of 12 or more feathers.

THE COMMON SNIPE.
Scolopax Wilsonii.—TEMMINCK.

DESCRIPTION.—Tail rounded, of 16 feathers, with a bright ferruginous, subterminal bar; back and scapulars black, with bronzy reflections; rump dusky, faintly mottled and barred with pale yellowish brown; crown black, divided by an irregular line of pale brown, and another of the same tint passes over each eye; neck and upper part of the breast pale brown, with small, dusky, longitudinal spots; chin white tinged with brown; bill brown, blackish at the tip. Length 11 to 11½, spread 17, bill 2⅛ to 2¾.—*Nutt.*

HISTORY.—This species, which is nearly related to the European Snipe, is found throughout the whole of America from Hudson's bay to the equator. This bird arrives from the south early in the spring, and spends the summer in low, moist grounds, breeding in swamps, where it lays its eggs in a hollow loosely lined with a little grass. The eggs are 4, of a yellow-olive color, speckled with different shades of brown. The young leave the nest as soon as they are hatched. The flesh of the Snipe is in high estimation on account of its exquisite flavor, on which account it is eagerly sought by the sportsman. They are frequently seen striking their bill into the black marshy soil. Their food consists principally of worms, leeches and aquatic insects.

GENUS RUSTICOLA.—*Vieill..*

Generic Characters.—Bill similar to that of the Snipe, but more robust, with the extremity attenuated and not depressed; the under mandible is also deeply grooved beneath. Eyes placed far back in the head. Legs short, robust and wholly feathered to the knees: tarsus shorter than the middle toe; toes cleft from the base, and the hind nail truncated. The 1st or 4th primary longest. Tail of 12 feathers.

THE WOODCOCK.
Rusticola minor.—NUTTALL.

DESCRIPTION.—Back darkly marbled with black ferruginous and ash; chin white; throat grayish; belly yellowish white; thighs and posterior parts beneath bright ferruginous; crown black, crossed with three light ferruginous bands, the middle one broadest. A black stripe from the eye to the angle of the mouth, and another from the bill up the frontlet; front part of the head grayish; marbling on the wings lighter and finer than on the back; legs and feet light flesh color; bill dusky horn color, nearly black at the tip; nails brownish black, small. First 4 primaries nearly equal, 3 first narrow. Length of the specimen before me 11 inches, folded wing $5\frac{1}{2}$, bill 2.9.

HISTORY.—The Woodcock is quite common in Vermont, although very seldom seen, on account of its nocturnal habits. It feeds and moves from place to place almost exclusively in the night. This bird returns from the south early, and selects a breeding place in the woods. The nest is made upon the ground, of grass and leaves. The eggs, usually 4, are of a yellowish clay color blotched with purple and brown. The young leave the nest as soon as hatched, but are unable to fly for 3 or 4 weeks. During the period of incubation the peculiar note of the male may often be heard morning and evening, while he rises spirally into the air and then descends again to the neighborhood of the nest. The flesh of the Woodcock, like that of the Snipe, is highly esteemed and eagerly sought, on account of its delicious flavor.

GENUS RALLUS.—*Linn.*

Generic Characters.—Bill varying in length, thick at the base, and generally straight and compressed; upper mandible furrowed on each side; somewhat arched and curved at the extremity, with its base extending upwards between the feathers of the forehead; nostrils situated in the furrow of the bill above its base, oblong, pervious and covered at the base by a membrane; tongue narrow, acute and fibrous at the tip; forehead feathered; legs small, with a naked space above the knee; toes wholly divided; wings moderate, rounded; tail of 12 feathers, not extending beyond their coverts. Plumage of the sexes, in general, nearly similar.

THE VIRGINIA RAIL.
Rallus virginianus.—LINNÆUS.

DESCRIPTION.—Upper part black, the feathers edged with olive brown; cheek and stripe over the eye ash; over the lores, the under eye-lid and chin white; wing coverts chestnut; quills deep dusky; throat, breast and belly reddish brown; sides and vent black, with white bars; legs and feet dusky reddish brown. Length 10, spread 14. The *female* a little less, and paler.—*Nutt.*

HISTORY.—This bird is sometimes called the Clapper Rail, but more commonly the *Small Mud Hen.* It is met with in fresh water marshes in most parts of the United States, during the summer, but migrates to the south on the approach of winter. With its neck stretched out and its short tail erected, it runs with great speed: but, when closely pursued, frequently rises upon the wing, yet seldom flies far at a time. It breeds in this state, making its nest in the wettest part of the marsh, of rushes and withered grass. The eggs, from 6 to 10, are of a pale cream color, sprinkled with brownish-red and purple. The female is so much attached to her eggs that she will sometimes suffer herself to be taken in the hands sooner than abandon them.

LOBE-FOOTED BIRDS.

This order takes its name from the circumstance of the toes of the different species being, in most cases, margined with a membrane. They are aquatic in their habits, and swim and dive with facility. They live in small flocks along the sea coast, and along the shores of lakes and ponds, feeding upon fish, reptiles, worms and vegetables. The sexes are nearly alike in plumage.

GENUS FULICA.—*Briss. Linn.*

Generic Characters.—Bill shorter than the head, stout, nearly straight, conical, compressed, higher than broad at base, acute at tip; mandibles equal, furrowed each side at the base, the upper covering the margins of the lower, and spreading out into a naked membrane over the forehead;

lower, boat-like; nostrils in a furrow,medial lateral, concave, oblong, pervious, half closed by a turgid membrane; feet moderate, far back; naked space above the knee small; tarsus compressed, almost edged behind; anterior toes very long, nearly divided to the base, margined on each side by a broad scolloped membrane; hind toe bearing on the ground, edged on the inner side by an entire membrane; wings moderate, rounded, 2d and 3d primaries longest; tail short, narrow, of 12 or 14 feathers; sexes and young nearly alike in plumage.

THE COMMON COOT.
Fulica americana.—GMEL.

DESCRIPTION.—Head and neck velvet black; fore part of the back, scapulars and wing-coverts blackish gray; tertiaries, tips of the scapulars, rump and tail-coverts clove brown, with a greenish tinge; quills, tail and vent pitch black; under tail coverts and tips of the secondaries white; bill pale horn color, with a chestnut ring near its tip; under plumage lead-gray; legs and toes bluish green, the scolloped membrane mostly lead color. Length 16 inches.—*Rich.*

HISTORY.—The American Coot is found throughout nearly the whole continent, and seems almost indifferent to climate, regulating its migrations principally by the scarcity or abundance of food, which consists of seeds, grasses, worms, snails, insects, and small fishes. It is nocturnal in its habits, and is said to perform its migrations by night.

GENUS PODICEPS.—*Lath.*

Generic Characters.—Bill moderate, robust, hard, straight, and compressed, conically elongated and acute; upper mandible deeply and broadly furrowed on each side at the base, somewhat curved at tip; the lower boat-shaped; nostrils in the furrow, basal, lateral, concave, oblong, pervious, posteriorly half closed by a membrane; feet turned outward, situated far back; the thigh almost hidden in the belly; tarsus much compressed; anterior toes greatly depressed, connected at the base by a membrane, forming a broad lobe round each toe; nails wide and flattened; wings short and narrow; tail, none. *Female* similar to the male in plumage.

THE PIED-BILL DOBCHICK.
Podiceps carolinensis.—LATH.

DESCRIPTION.—Upper plumage dusky brown; secondaries obliquely tipped with white; a roundish black spot under the chin; throat and cheeks below brownish gray; patch on the breast dotted or clouded with brownish white and black; belly almost white, mottled under the wings and on the flanks; rump dusky; bill with a broad black band around its middle, including the nostrils; legs black; iris hazel. Length 14 in.—*Nuttall.*

HISTORY.—These birds make some stop in our waters during their fall migration, but are not known to breed in this state. They feed upon fishes and water-insects. When alarmed they conceal themselves by sinking in the water, with only the end of the bill, by which they are enabled to breathe, elevated above the surface, and this is not easily seen. From this and other singular habits they have received the name of *Water-Witches.*

WEB-FOOTED BIRDS.

In this order, which consists wholly of Water Birds, the bill is much varied in form; the legs short, generally placed far back; the anterior toes wholly or partially connected by webs, and, in some families, all the toes are united by one membrane; the hind toe articulated; interiorly upon the tarsus, or wholly wanting.

GENUS L   's.—*Linnæus*

Generic Characters.—Bill moderate, strong, hard, compressed, with the edges sharp and curved inward, a little bent at the tip; nostrils lateral, longitudinal, linear, open and pervious; feet rather slender; tarsus nearly equal to the middle toe; web entire to the tips of the toes; hind toe very small and high on the tarsus; wings long and acute; tail even, of 12 feathers. Female smaller than the male; otherwise alike.

THE BONAPARTIAN GULL.
Larus Bonapartii.—SWA. & RICH.

DESCRIPTION.—Head bluish black; back and upper part of the wings light lead color, or pearly gray; neck, tail and whole under plumage pure white; the outer edge of the first primary and the extremities of the others, black, in some cases slightly tipped with white; in some cases the outer edge of the 2d primary is edged with a line of black; bill shining black, nearly straight, a little turgid and notched near the tip; inside of the mouth legs and feet light bright red; folded wings 2 inches longer than the tail which is slightly rounded. Length 15 inches, the folded wing 10; bill along the gape, 1½.

HISTORY.—This beautiful Gull is often seen in small flocks in Lake Champlain, but is most plentiful in autumn, when those which have been rearing their young at the north are proceeding southward to spend the winter. Numbers of them are however said to breed upon the islands in lake Champlain, particularly upon those called the Four Brothers. They feed principally upon insects and are distinguished by a peculiarly shrill and plaintive cry. Their flesh is esteemed good food. The specimen from which our description was made, was shot, with several others belonging to the same flock, in Shelburne Bay.

THE HERRING GULL.
Larus argentatus.—BRUNN.

DESCRIPTION.—*Winter plumage.* Top of the head, region of the eyes, occiput, nape and sides of the neck white, each feather with a longitudinal pale brown streak; front, throat, all the lower parts, back and tail white; top of the back, scapulars, and the whole wing bluish ash; primaries blackish towards the end terminating in white; bill ochre yellow; orbits and iris yellow, the latter pale; feet reddish flesh-color. *Summer plumage*, with the head and neck pure white. *Young* blackish ash, mottled with yellowish rusty. Length about 24 inches.—*Nuttall.*

HISTORY.—The Herring Gull derives its vulgar name from the circumstance of its feeding much upon Herrings, which it catches by following the shoals. They are common to the milder parts of both continents, and are not uncommon in lake Champlain, where numbers of them breed upon the small, uninhabited islands. The Rev. G. G. Ingersoll has procured the eggs of this Gull from one of the islands called the Four Brothers, situated five or six miles from Burlington. Their ground color is light olive, irregularly spotted with dull reddish-brown and dirty ash. The nest is usually made of sticks upon the ground or a rock, but Audubon found them at the Bay of Fundy, breeding upon low fir trees.

GENUS ANSER.—*Brisson.*

Generic Characters.—Bill moderate, stout, at the base higher than broad, somewhat conic, cylindrical, depressed towards the point, and narrowed and rounded at the extremity; upper mandible not covering the margins of the lower; the ridge of the bill broad and elevated; the nail somewhat orbicular, curved and obtuse; marginal teeth short, conic and acute; nostrils medial, lateral, longitudinal, elliptic, large, open and pervious, covered by a membrane; tongue thick, fleshy and fringed on the sides; feet central, stout, webs entire; wings moderate, acute; quills strong; tail rounded. Sexes similar in plumage.

THE CANADA GOOSE.
Anser canadensis.—BONAPARTE.

DESCRIPTION.—Head, two thirds of the neck, greater quills, rump and tail pitch black; back and wings broccoli-brown, edged with wood-brown; base of the neck before and the under plumage yellowish gray, with paler edges; flanks and base of the plumage generally brownish-gray. A few feathers about the eye, a large kidney-shaped patch on the throat, the sides of the rump, and tail coverts, pure white; bill and feet black; neck long. Length 41, tail 9, wing 19½.—*Rich.*

HISTORY.—The Wild Goose is well known in all parts of the United States as a bird of passage. In Vermont they are seen in large numbers during their spring and fall migrations, and it is not uncommon for them to alight in our lakes and ponds to feed and rest themselves, where they are frequently shot, but they are not known to breed within the state. Their principal breeding places are further north between the 50th and 67th parallels of latitude. They lay 6 or 7 greenish-white eggs in a nest rudely made upon the ground. The residents about Hudson's bay depend much upon geese for their supply of winter provisions, 3 or 4,000 of them being killed annually, and barrelled up for use. In their migrations, Wild Geese usually fly in large flocks, arranged in the form of the letter ▷, with the vertex of the angle forward. Sometimes they alight in fields and meadows, and, not unfrequently, they are compelled to alight in consequence of being bewildered and lost in thick fogs and severe storms.

Under such circumstances numbers of them are frequently shot.

Genus ANAS.—*Linn.*

Generic Characters.—Bill broader than high at the base, widening more or less at the extremity, somewhat flattened, obtuse and much depressed towards the point; marginal teeth lamelliform, weak; upper mandible convex, curved and furnished with a slender nail at the end; the lower narrower, flat, and entirely covered by the margins of the upper; nostrils basal, approaching together, oval, open, pervious, and partly closed by a membrane; tongue stout and obtuse, fringed at the sides; neck about the length of the body; feet central, small, weak, web entire; wings moderate acute; quills long, 1st and 2d longest; tail of from 14 to 16 feathers. Plumage of the sexes different.

SUMMER, OR WOOD DUCK.
Anas sponsa.—LINNÆUS.

DESCRIPTION.—Top of the head, crest, and about the eyes, different shades of green, with purple reflections; crest and side of the head marked by two white lines, one terminating behind the eye and the other extending to the bill; a black patch on each side of the neck; chin, back part of the cheek, and ring round the neck white; lower part of the neck and breast bright chestnut-brown, spotted with white; back, scapulars, wings and tail exhibiting a play of green, purple, blue, gray, and velvet black; a hair-like, splendent, reddish purple tuft on each side of the rump; belly whitish; flanks yellowish gray, beautifully waved with black, the tips of the long feathers, and also those on the shoulder, broadly barred with white and black. On most of the plumage is a play of colors with metallic lustre; bill higher than wide at the base, narrowed towards the point, flesh color above, with a black spot between the nostrils and at the tip; black below; tail of 14 wide rounded feathers, longer than the folded wings. *Female* without the tufts on the rump, the fine lines on the flanks, with shorter crest, and less vivid plumage, mostly of a brownish hue. Length of the specimen before me (male) 20 inches; the folded wing 8½.

HISTORY.—The Wood Duck is one of the most beautiful birds seen in this state, and is one of the very few permanent residents here. Their food consists of tadpoles, insects and worms, and also of beechnuts and various kinds of berries. Their flight is rapid and graceful, and they also swim and dive well. Their sense of hearing is very quick, and when alarmed they sometimes conceal themselves in the water, with the bill only above the surface. Their nests are upon trees, usually in the hollow of a broken and decayed trunk, or large limb, and the eggs, from 8 to 14, are yellowish white, and a little smaller than those of the common hen. The young, when hatched, are carried down in the bill of the parent, and then conducted to the water. The flesh of this Duck is esteemed for food.

THE MALLARD.
Anas boschas.—LINN.

DESCRIPTION.—Head and upper part of the neck green, with blue and dark purple reflections; collar around the neck white; feathers of the breast dark reddish chestnut, slightly edged with white; scapulars, back and parts beneath sprinkled and waved with blackish on a white ground, much lighter towards the tail; rump and tail coverts blackish green; sides of the rump partly, and interior of the wings wholly, white; folded wing shorter than the tail; bill yellow; iris reddish brown; legs orange; *Female* and *young* brownish varied with yellowish and blackish. Length of the specimen before me, which is a male, 26 inches; folded wing 11; bill 2.1; tarsus 1.8; longest toe 2.4; width of the bill 1.1.

HISTORY.—This is our common domestic duck in its wild state. It is frequently seen in small flocks in lake Champlain, but is more plentiful at the south and southwest. The specimen from which the above description was made, was shot in the lake near Burlington in May 1842. It is finely preserved and is now in the museum of the College of Natural History of the Vermont University. Their nest is made upon the borders of rivers and lakes at some distance from the water. The eggs, from 10 to 18, are bluish white. The female frequently covers her eggs when she leaves them. The young are led to the water as soon as hatched and are at once able to swim and dive with great expertness. Wild ducks feed upon fish, aquatic insects and plants; and they fly in the form of the letter ▷, with the meeting of the two lines directed forward.

THE DUSKY DUCK.
Anas obscura.—GMEL.

DESCRIPTION.—Upper part of the head deep dusky-brown, with small streaks of drab on the fore part; the rest of the head and greater part of the neck dull yellowish-white, each feather marked down the centre with a line of blackish-brown; inferior part of the neck and whole lower parts dusky, the feathers edged more or less broadly with brownish white; upper parts the same, but deeper; speculum blue, with green and amethyst-red reflections; wings and tail dusky; the tail feathers sharp pointed; bill greenish ash; legs and feet dusky yellow; *female* browner. Length 24, spread 38.—*Nutt.*

HISTORY.—This Duck is said to be found only in North America. It is met with throughout the United States and British provinces, from Florida to Labrador, and is generally but improperly called the Black Duck. It is found alike along the sea coast, in salt marshes, and along the fresh water rivers and lakes. They breed in marshes, making their nests of weeds, and laying from 8 to 12 eggs, which are of a dull ivory white and about the size of those of the common duck. Their voice, or quack, is also similar to that of the common duck.

THE BLUE-WINGED TEAL.
Anas discors.—LINN.

DESCRIPTION.—Upper surface of the head and under tail coverts brownish black; a white crescent from the forehead to the chin bordered with black; sides of the head and neck purple; base of the neck above, back, tertiaries and tail coverts brownish-green; fore parts marked with semi-ovate pale brown bars; lesser wing coverts pure pale blue; speculum dark green; primaries, their coverts and the tail liver brown; sides of the rump and under wing coverts white; under plumage reddish-orange, glossed with chestnut on the breast, with blackish spots; bill bluish-black; feet yellow. *Female* brownish, without the white before the eye and on the rump, and the purple tint on the head and neck. *Young* without the green speculum; in other respects like the female. Length 18 inches.

HISTORY.—The Blue-Winged Teal inhabits, according to the season, all parts of the continent up to the 58th parallel of latitude. It arrives in this state from the south in the latter part of April, and I have before me a specimen which was shot in Winooski river, at Burlington, about the first of May, 1842. They feed upon insects and vegetables, and are said to be particularly fond of wild rice. They usually become very fat, and their flesh is highly esteemed for food.

GENUS MERGUS.—*Linnæus.*

Generic Characters.—Bill long, or moderate, straight, nearly cylindrical, slender, and broad at the base; the edges serrated, and the teeth subulate, sharp, and inclining backwards; the upper mandible hooked and furnished with a nail at the tip; nostrils lateral, open, situated near the middle of the bill; legs short, strong, placed far back; three anterior toes webbed to their points: hind toe articulated high with a broad membrane: wings moderate, acute: 1st and 2d primaries longest: tail short and rounded. Female and young differ considerably from the male.

THE GOOSANDER, OR SHELDRAKE.
Mergus merganser.—LINN.

DESCRIPTION.—Color of the *old male* above nearly black; head and upper part of the neck greenish black, with reflections; belly white, shaded with rose color. Humeral wing coverts blackish; lower part of the back and the tail ash; bill red on the sides, but black above and below; iris reddish; legs vermillion. *Female* and *young* above light slate or grayish ash, shafts of the feathers darker; secondary wing feathers and their coverts white on the posterior part; head, crest and neck reddish brown; chin and upper part of the breast gray; belly yellowish white; wings black, 2d quill longest; bill reddish brown above, red below; legs and feet reddish yellow; webs brownish.—Length of the specimen before, which is a female, 25 inches; folded wing 9½; spread 32; bill, from the angle of the mouth, 2¾; tarsus 2½; longest toe 3 inches.

HISTORY.—The Goosander inhabits the northern parts of Europe, Asia and America, where they breed and spend the greater part of the year. On the approach of cold weather they migrate towards the south, but still many of them spend the winter in high northern latitudes. They are occasionally met with in our lakes and rivers at nearly all seasons, but are not found in Vermont in very large numbers. The specimen from which a part of the above description was made, was shot in Winooski river Sept. 4, 1841. This fowl is very voracious, and feeds principally upon fishes, of which the stomach of the one above described contained the fragments of several, one of which was three inches long. The rough incurved papillæ upon the tongue, and the sharp serratures along the edges of the bill, seem admirably adapted for seizing and retaining its finny prey.

GENUS COLYMBUS.—*Linnæus*.

Generic Characters.—Bill longer than the head, stout, straight, nearly cylindrical, compressed, with the point subulate and acute; the edges bent in, sharp and entire; nostrils basal, concave, and half closed by a membrane; feet large, placed far behind; tibia almost drawn up into the belly; tarsus strong, compressed; the three anterior toes very long, united to their tips by webs; hind toe small, touching the ground merely at the tip, united to the outer toe by a rudimental membrane; wings moderate; 1st and 2d primaries longest; tail short, rounded and composed of 18 or 20 feathers; the sexes alike in plumage.

THE LOON, OR GREAT NORTHERN DIVER.
Colymbus glacialis.—LINN.

DESCRIPTION.—Head and back of the neck glossy black; back grayish black spotted with white, the spots squarish and largest on the middle of the back, roundish forward, and very small towards the rump; beneath white; neck spotted with black, with a black and whitish ring; wings brownish black above, without spots; legs black; bill dark horn color. Length of the specimen before me to the extremity of the tail 35 inches, folded wing 14 inches, bill to the angle of the mouth 4½ inches, foot to the extremity of the longest nail 5½ inches. The first quill longest.

HISTORY.—The Loon, or Great Northern Diver, is found in the northern parts of both the Eastern and Western Continent. In this country it resides principally in the lakes in the interior, spending nearly its whole time in the water. It dives with great facility, and is able to remain for a long time under water. Its legs are situated so far back that it is with the greatest difficulty that it walks at all upon land. The Loon is not uncommon in our lakes and ponds, where numbers of them spend the summer and rear their young. Their nest is upon the ground near the margin of a pond, and somewhat elevated above the surface of the water. The eggs are about the size of those of the domestic goose, of a dark smoky olive color, blotched with umber brown. The flesh of the Loon is tough and unpalatable.

DOMESTIC FOWLS.

The only birds we have in a state of permanent domestication are the Goose, the Turkey, the Duck, the Barn-door fowl, the Peacock the Guinea Hen and the Dove.

THE COMMON GOOSE, *Anas anser*, which has acquired so many colors in our poultry yards, originated from a wild species, which is gray, with a brown mantle undulated with gray, and an orange colored beak. The name of the species in a wild state is *Anser cinereus*. Geese are kept in considerable numbers in this state, principally for their feathers.

THE DOMESTIC TURKEY, *Meleagris galloparo*, in its wild state, has been already described on page 101. In the domesticated state it has acquired a variety of colors and undergone some change in form and size. Turkeys are raised for their flesh which is highly valued.

THE DOMESTIC DUCK, *Anas domestica,* sprang from the common Mallard Duck, *Anas boschas.* See page 109. The change produced in the Duck by domestication is much less than in the two preceding species. Very few of them are raised in this state, and these are kept rather for curiosity than profit.

BARN-DOOR FOWL, *Gallus domesticus.* This species, denominated the Cock and the Hen, varies almost infinitely in colors, and very considerably in size and form. It has been in a domesticated state from time immemorial, and more or less of them are kept by almost every family in the state. Their flesh and eggs form almost indispensable articles of food ; and with suitable attention and precaution against mischief, the keeping of hens for their eggs is not unprofitable.

THE PEACOCK, *Pavo cristatus.* The Peacock is said to have been originally from the north of India, and to have been introduced into Europe by Alexander the Great. It is celebrated only for the magnificence and beauty of its plumage.

THE GUINEA HEN, *Numida meleagris.* The Guinea Hen was originally from Africa. Its slate colored plumage is everywhere springled with small round white spots. In its wild state it lives in flocks, in marshes.

THE DOVE. Our common Dove is said to be descended from the Rock Dove, *Columba livia.*

The three last are kept only in small numbers, as a matter of curiosity.

The following table contains the estimated value of Poultry in the several counties in this state, according to the returns of the census of 1840.

Addison Co.,	$8,637	Orange,	$14,395
Bennington,	9,414	Orleans,	4,269
Caledonia,	10,029	Rutland,	13,092
Chittenden,	8,014	Washington,	15,840
Essex,	1,744	Windham,	13,854
Franklin,	5,912	Windsor,	20,313
Grand Isle,	1,873		
Lamoille,	4,192	Total value,	$131,578

CHAPTER IV.

REPTILES OF VERMONT.

Preliminary Observations.

Reptiles are usually regarded as disagreeable and loathesome objects, though many of them, on account of their singular structure and habits, are highly interesting. These animals have cold red blood, with a dry skin, which is naked or covered with scales, and, in many species, periodically renewed. Their temperature usually corresponds with that of the medium in which they are situated. When the temperature is down to freezing they become torpid. They are found largest and most numerous in the hottest portions of the earth.

The bones of reptiles are in general softer than those of quadrupeds and birds, and vary much in their connection and number in the different genera. Frogs and toads have no ribs ; serpents have them detached without a sternum ; tortoises have them all united together ; and lizards have them like birds. Some of these animals have four feet, others two, and others none. Some are fitted for leaping, others for crawling, and others for swimming, and several for all these modes of progression. Their circulation is imperfect, their sensations obtuse, and they are in general sluggish in their habits.

Reptiles all produce their young by means of eggs; these are not, however, hatched by the parent, but deposited in situations favorable for their developement. In some genera the young are produced perfect, while in others they are of a widely different form, being shaped like, and having the habits of a fish, and like insects undergoing a transformation before arriving at perfection, of which the tadpole and frog afford a familiar example

In his classification of Reptiles, Cuvier adops the arrangement of Brongniart, who takes the characters of his orders from the principal organs, in conjunction with the animal functions. In this arrangement they are divided into the four following orders.

I. *Chelonia,* or Tortoises. Body covered with a shield, or plate.

II. *Sauria,* or Lizards. Body covered with scales.

III. *Ophidia,* or Serpents. Destitute of feet.

IV. *Batrachia,* or Frogs, &c. Body covered with a naked and loose skin.

The following is a list of the Reptiles found in Vermont, arranged in the order in which they are described in the subsequent pages.

ORDERS OF REPTILES.

ORDER CHELONIA—*Tortoises.*
Emys picta,   Painted Tortoise.
" *insculpta,*   Sculptured Tortoise.
Emysaurus serpentina, Snapping Tortoise.
ORDER SAURIA.—*Lizards.*
There are none of this order found in the state.
ORDER OPHIDIA—*Serpents.*
Coluber sirtalis,   Striped Snake.
" *saurita,*   Ribband Snake.
" *ordinatus,*   Brown Snake.
" *occipito maculatus,* Spotted-neck Snake.
" *punctatus,*   Ringed Snake.
" *vernalis,*   Green Snake.
" *constrictor,*   Black Snake.
" *eximius,*   Chicken Snake.
" *sipedon,*   Water Snake.
Crotalus durissus,   Rattle Snake.
ORDER BATRACHIA.—*Batrachians.*
Rana pipiens,   Bull Frog.
" *fontinalis,*   Spring Frog.
" *halecina,*   Leopard Frog.
" *palustris,*   Pickerel Frog.
" *sylvatica,*   Woods Frog.
" *horiconensis,* Horicon Frog.
" *melanota,*   Black Frog.
Hylodes Pickeringii, Pickering's Hylodes.
Hyla versicolor,   Tree Toad.
" *squirella,*   Peeping Tree Frog.
Bufo americanus,   Common Toad.
Salamandra symmetrica Symmetrical Salamander
" *dorsalis,*   Many Spotted do.
" *salmonea,*   Salmon colored do.
" *tigrina,*   Tiger Salamander.
" *venenosa,*   Violet colored do.
" *erythronota,*   Red-backed do.
" *glutinosa,*   Glutinous do.
" *bislineata,*   Two lined do.
Menobranchus maculatus, Proteus.

ORDER I—CHELONIA.
TORTOISES.

Animals of this order have four feet, a heart with two auricles, and the body enveloped in two plates, or shields, formed of the vertebræ and ribs above and sternum beneath. Tortoises have no teeth, but their jaws are invested with a bony substance which serves as a substitute for teeth. The sexes may in general be distinguished by the cavity in the sternum of the male. They possess great tenacity of life, moving for a long time after their heads are cut off. They require little nourishment, and can pass months, and even years, without eating.

GENUS EMYS.—*Brongniart.*

Generic Characters.—Shell depressed, solid; sternum broad, solid, immoveable, firmly joined to the shell, consisting of twelve plates, and four supplemental ones; extremities palmated, anterior with five nails and posterior with four; head of ordinary size; tail long.

THE PAINTED TORTOISE.
Emys picta.—SCHNEIDER.

DESCRIPTION.—Shell oblong, oval, rather depressed, smooth, and of a dusky brown color; all the dorsal and lateral plates margined with yellow; a reddish yellow line along the middle of the back; first vertebral plate quadrangular, wider on the fore part and slightly elongated behind, the second six sided, the third quadrangular, the fourth six sided, narrow behind, the fifth seven sided; the first lateral plate four sided, upper edge narrow, the lower rounded; the second and third nearly square. The intermediate marginal plate is narrow, with a notch on each side; all the rest are either oblong or square, each having a red spot in the centre, surrounded by irregular concentric red lines; marginal plates mostly red beneath; sternum reddish yellow, serrated before; pectoral plates narrow; caudal plates triangular, rounded behind; head and skin generally dark brown; an oblong yellow spot behind each eye, and another upon the back part of the head; cheeks and chin striped with yellow, becoming red on the neck; legs striped and spotted with red; tail with two yellow stripes above and two red ones on the sides, which unite beneath in one; eyes small, pupil black; iris golden, with a broad black stripe through the middle. Length of the shell of the specimen before me 5 inches; width $4\frac{1}{4}$; height $2\frac{1}{2}$.
Plates D. 5, L. 8, M. 25, S. 12.*

HISTORY.—This is our most common species of tortoise, and exists in large numbers in the coves along the margin of lake Champlain and in the stagnant waters about the mouths of our rivers. It is very aquatic in its habits, and is seldom seen more than a few feet from the water. In the spring of the year, when the marshes are inundated, hundreds of these animals may be seen at a time, sitting upon the rocks and logs which lie partly above the water, and basking in the sun. On approaching them they immediately plunge into the water and disappear. When the

---

* D—dorsal, L—lateral, M—marginal, S—sternal.

painted Tortoise is first hatched it is very thin and nearly circular, and the color of the sternum deep red. As it grows the back becomes more elevated and the sides compressed, and the red of the sternum usually assumes a yellowish hue, and in some cases the red entirely disappears, leaving the sternum wholly yellow. It feeds upon shell-fish, insects and reptiles.

THE SCULPTURED TORTOISE.
Emys insculpta.—Le Conte.

Description.—Shell oval, slightly carinated and emarginate behind; all the plates with yellowish radiating lines and striæ, cut by other concentric striæ ; first vertebral plate pentagonal, the 2d, 3d and 4th subhexagonal, the 5th octagonal; six of its faces anterior ; 1st and 4th lateral plates pentagonal, 2d and 3d subheptagonal ; intermediate marginal plate very narrow ; the first pentagonal projecting a little beyond the next ; the rest mostly quadrangular ; the three plates on each side of the caudal plates slightly revolute ; sternum notched behind, yellow and striated, all the plates being marked with a large black spot on their posterior part ; plates under the throat triangular ; all the rest quadrangular ; skin granulated or scaly, reddish black above, dull red beneath ; head, nails and tail black ; jaws dark horn color, marked with yellow. Length of the shell of the specimen before me 6½ inches ; width 5⅓; width of the head 1 inch ; length of the tail beyond the shell 1½ inch ; height 3 inches.

Plates D. 5, L. 8, M. 25, S. 12.

History.—This species, when fully grown, is a little larger than the preceding. It is not so aquatic in its habits, it being frequently found at a considerable distance from the water, and being often met with in the woods, it is sometimes called the Wood Tortoise. The Sculptured Tortoise not only resorts to coves, and the deep, still waters of rivers, but is frequently found taking shelter in the deep, narrow rills in our pastures and meadows. The lateral plates seem in this species to be subject to some variation. In one of my full grown specimens the lateral plates are only three, instead of four, upon each side. Food of this species the same as of the preceding.

Genus Emysaurus.—*Dumeril.*

Generic Characters.—Head large, covered with small plates; snout short; jaws hooked; two warts beneath the chin; sternum immoveable, cruciform, composed of ten plates ; three sterno-costal plates ; fore feet with five claws, hind feet with four ; tail long, surmounted with a scaly crest.

THE SNAPPING TORTOISE.
Emysaurus serpentina.—Linnæus.

Description.—General color dark greenish brown above, lighter and yellowish beneath ; upper shell oval, depressed and notched behind ; vertebral plates scabrous ; lateral marked near the base with concentric striæ; marginal oblong, the six posterior ones forming six obtuse teeth, projecting backwards ; sternum narrow, lozenge-shaped, pointed and entire at both ends ; head, neck and limbs very large and strong ; jaws sharp, hooked ; skin of the neck and legs granular above and warty beneath ; two prominent warts under the chin ; fore legs with rows of broad sharp scales ; hind legs with several broad scales beneath ; claws strong, five before and four behind ; tail straight, about two thirds the length of the shell, tapering, and crested with large bony prominences, which gradually diminish towards the end ; sides and under part of the tail covered with smaller scales. Length of the shell, of the specimen before me, 11 inches ; width 9 inches ; tail 8 in. ; head 3¼ in. long, 2½ wide.

Plates D. 5, L. 8, M. 25, S. 11.

History.—This is the largest species of Tortoise found in Vermont, often weighing from 15 to 18 or 20 lbs. It is much more disposed to bite than the preceding species. It will seize upon a stick held towards it, and suffer itself to be raised by it from the ground sooner than relinquish its hold ; and hence it is usually called in New England the Snapping Turtle, or Tortoise. At the south it is called the Alligator Tortoise, from the resemblance of its crested tail to that of the Alligator. This species is often found at a considerable distance from water, and will live a long time without water. It feeds upon fishes, reptiles, and young birds, and is said sometimes to catch chickens.

ORDER II—SAURIA.
LIZARDS.

These have elongated bodies, covered with scales, usually four feet; some with claws and some without; an elongated tail; mouth furnished with teeth. No species of this order has been observed in Vermont. The reptiles usually called Lizards here all belong to the Salamander family.

ORDER III—OPHIDIA.
SERPENTS.

Serpents have a heart with two auricles, an elongated, cylindrical body, destitute of feet, and for the most part covered with scales. They move by means of the folds and flexure of their bodies. They are sometimes divided into *venomous* and *non-venomous*. The Rattle Snake is the only venomous or poisonous serpent found in Vermont.

GENUS COLUBER.—*Linnæus.*

Generic Characters.—Body long, cylindrical and tapering, head oblong, covered above with smooth polygonal plates; above covered with rhomboidal scales, imbricate, reticulated, carinated, or smooth; abdomen with transverse plates; beneath the tail with double plates; anus transverse, simple; jaws furnished with sharp teeth; without poisonous fangs. Some species are oviparous, and others ovo-viviparous.

THE STRIPED SNAKE.
Coluber sirtalis.—LINNÆUS.

DESCRIPTION—Upper part of the body dark brown, with a narrow yellow line extending from the head along the back to the tail, and a broader parallel stripe of the same color on each side joining the abdominal plates; belly greenish yellow; abdominal plates marked on each side with two black spots; scales oblong, carinated, small on the back and increasing in size towards the abdomen; head flattened, covered with ten plates, one at the nose, two pair back of this, three between the eyes, and behind these two larger ones; pupil of the eye black, iris reddish; small sharp teeth in the jaws and palate Of three specimens before me, the first, 22 inches long, has 154 abdominal plates, and 75 pair of subcaudal scales, the second, 21 inches long, has 146 plates, and 62 pair of scales, and the third 27 inches long, of which the tail measures 6, 141 plates and 60 pair of scales.

HISTORY.—This is the most common and generally diffused species of snake in Vermont, and is universally known by the name of *Striped Snake.* It is perfectly harmless, excepting sometimes to catch a chicken, gosling, or young turkey or duck, and rob birds' nests of their eggs, or young. They also feed upon toads and frogs. Serpents do not chew their food like quadrupeds, but whatever they eat they swallow whole. Their jaws are so constructed as to be separable at the joint, which enables them to swallow animals much larger than themselves; and instances of their swallowing such animals fall under the observation of every field laborer. Often does a large sluggish snake lie in his way, with a portion of his body distended to near the size of his fist. On killing and opening him, a large frog, toad, or other animal is found, which the gormandizer had caught, lubricated and swallowed alive; and for the digestion of which all the energies of the animal were now employed. Often have we ourselves been startled by the piercing and mournful cry of a poor frog, which had been caught by one of these animals; and how indignant have we been, on going to the spot, to see the horror-stricken sufferer, with his hind quarters ingulfed in the throat of a huge snake, vainly struggling with his fore feet to extricate himself, and at the same time uttering a most piteous moan. Under such circumstances it has afforded us real satisfaction to destroy the cruel aggressor and liberate his wretched victim. For the purpose of robbing birds' nests this snake will climb fences and bushes several feet from the ground. The usual length of this snake is about two feet, of which the tail constitutes one fourth. He sometimes attains the length of about three feet.

THE RIBBAND SNAKE.
Coluber saurita.—LINN.

DESCRIPTION.—Form more slender and graceful than that of the striped snake, which it resembles in the arrangement of its stripes. A bright yellowish white line begins between the posterior plates on the head and extends along the back to the extremity of the tail. On each side of this, commencing at the orbit of the eye, is a shining black line which fades into brown towards the posterior extremity. Then comes a narrow yellow line on each side, commencing half an inch back of the angle of the mouth, which also fades into umber brown towards the tail. Below these, on each side, is a broad, well-defined stripe of umber brown, slightly bronzed, embracing a row of large scales, whose keels form a distinct lateral line, and extending down upon the abdominal plates and subcaudal scales. The margin of the upper jaw, the under jaw and belly are white; all the colors fainter and blended towards the tail. The upper jaw margined by 15 and the under by 21 marginal plates; two rows of teeth in the upper and one in the lower jaw, all small and sharp. Length of the specimen before me 29 inches; to the vent 20, tail 9. Head covered with 10 plates, the posterior largest. Abdominal plates 165, subcaudal scales 110 pair.

HISTORY.—I forwarded a specimen of this snake to my friend Dr. Storer, of Boston, who, in acknowledging its reception, says that it "is without any question the *sirtalis*." After so decided an opinion from such high authority, it may be thought presumption in me to introduce it as a different species; but knowing it, from my own observations, to differ very considerably from the common *C. sirtalis*, both in appearance and habits, and finding it to agree *as* nearly with the descriptions which I find of the *C. saurita*, I have ventured to describe it under that name, that the differences between it and the *sirtalis* may be seen. Besides differing in form and color, and in the much greater number of subcaudal scales, it is far more lively and quicker in all its motions, and so far as my own observation extends is always found in low grounds, and at no great distance from water. Among hundreds of the *C. sirtalis* which I have seen upon the high lands and mountains in this state, I have never met with an individual answering to the description here given. Shaw calls the color of the stripes of both these species bluish-green, from which it is probable that his descriptions were made from specimens preserved in spirits, since the yellow stripes in these serpents, under such circumstances, assume that hue.

THE BROWN SNAKE.
Coluber ordinatus.—LINNÆUS.

DESCRIPTION.—Brownish ash or clay color above, lighter beneath; a light stripe along the back from the head to the tail, on each side of which is a row of black spots, and two rows of similar spots, but much smaller, along the extremities of the abdominal plates on each side, the spots becoming obsolete towards the tail; scales carinated, small on the back but increasing in size towards the belly; head small, covered with ten plates of an olive brown color, the two posterior, and the middle one between the eyes, largest. The upper jaw is margined by 14 scales, and the lower by 12, besides the tip; an oblique black band crosses the angle of the mouth, and another a little back of it on the upper part of the neck; teeth in both jaws, and two rows of hooking teeth in the palate; eyes small; iris bright hazel. Length of the specimen before me about 15 inches; abdominal plates 130; a small part of the tail broken off.

HISTORY.—This plain and harmless little snake is frequently met with, but is less common than several other species. I have met with only two or three individuals in Burlington. It feeds upon insects.

THE SPOTTED-NECK SNAKE.
Coluber occipito-maculatus.—STORER.

DESCRIPTION.—Color above varying in the specimens before me, six in number, from light ash gray and reddish brown to nearly black; belly from a light brick red to a very dark copper color; three fulvous spots on the neck, one at the occiput above, and one below, on each side; in some of the specimens a row of blackish scales, usually slightly marked with white on each side of the dorsal line, and another row at the commencement of the abdominal plates; in others the color above is uniform; 12 plates margin the upper jaw besides the one at the snout; snout and under jaw yellowish white, and a white spot at the angle of the mouth; throat grayish, gradually passing into red on the abdomen; width of the head equal to that of the body; neck small, body gradually enlarges from the neck to near the vent, where it is largest; tail short and sharply pointed, contained $4\frac{1}{2}$ times in the total length; iris reddish hazel. Length of the longest specimen 9.9 inches, tail 2.2, with 119 abdominal plates and 45 pairs of subcaudal scales; another about the same length had 122 plates and 46 pair of scales; the shortest 3.7 in., tail .8, plates 119, scales 42 pair; the others not counted.

HISTORY.—This mild and inoffensive little snake, though very common in and about Burlington, is seldom seen in the early part of summer. They begin to make their appearance abroad about the beginning of September, and during that month, and the greater part of October, they are in some years met with in large numbers, varying in length from 3 to 10 or 11 inches, which is about the extent to which they grow. The shade of color above seems to be as various as the individuals. In the whole number which I have examined I have not found two alike; but in all, the contrast between the color above and that of the belly is very marked, and the spots on the neck and at the angle of the mouth have been constant, and in most cases very plain.

THE RINGED SNAKE.
Coluber punctatus.—LINNÆUS.

DESCRIPTION.—Color above uniform bluish brown, approaching to black in some specimens; beneath yellow; margin of the upper jaw, lower jaw and band round the neck, yellowish white; a row of small black spots along each side of the abdomen at the meeting of the dark color above with the light color below; usually a similar row of spots along the middle of the abdomen from the chin to the vent, but this is wanting in the specimen before me. Head flattish, about the width of the body, neck but little smaller than the body. Length 13 inches, tail 3, plates 164, scales 60 pair.

HISTORY.—This snake is of a timid disposition, being seldom seen abroad, but is often met with in different parts of the state, concealed under stones, logs, and the bark of old, decayed trees. Its food consists principally of insects.

THE GREEN SNAKE.
Coluber vernalis.—DE KAY.

DESCRIPTION.—Color above beautiful grass green; beneath greenish, or yellowish white; margin of the upper jaw yellowish; pupil black, upper edge of the iris yellow, below grayish brown. Scales not keeled, smooth, rhomboidal, with the acute angles truncated, giving them the appearance of unequal sided hexagons. Head flattened and covered with 10 plates, one at the snout, two pair behind these, then 3 plates between the eyes, 2 larger ones behind these upon the occiput, upper jaw bordered by 15 scales, including the one at the snout; nostril circular, and near the end of the snout. Length of the specimen before me $18\frac{1}{4}$ inches, head $\frac{1}{6}$ in, from the snout to the vent $11\frac{1}{2}$, tail 6., width of the head .3. Tail terminated in a sharp, horn-colored spine. Abdominal plates 131, sub-caudal 170 in the two rows.

HISTORY.—This beautiful and lively little snake is very common in the western parts of the state, and particularly in the neighborhood of lake Champlain. It is perfectly harmless, and feeds principally upon insects. On the east side of the Green Mountains in this state, it is quite rare, if found at all.

THE BLACK SNAKE.
Coluber constrictor.—LINNÆUS.

DESCRIPTION.—Color above almost black; beneath, slate-color; neck, margin of the jaws, and snout, yellow. Plates on the top of the head very large; that at the snout convex, projecting, yellow bordered with black at the upper and lateral margins; first pair of plates nearly quadrangular; the second, pentagonal; middle plate between the eyes hexagonal and largest of the three; 16 plates border the upper jaw; eyes large; nostrils large, vertical, situated between the 2d and 3d plates back of the snout; three pair of elongated plates on the throat just back of the chin; back of these two pair of smaller ones; back covered with large rhomboidal smooth scales. Length 51 inches, tail 11. Abdominal plates 184, scales 85.—*Storer.*

HISTORY.—This snake is met with only in the south and southwestern parts of the state, and even there it is not very common. It sometimes grows to the length of 6 feet, and runs with great speed, on which account it is sometimes called the Racer. It is perfectly harmless, and feeds upon toads, frogs, meadow mice and small birds, swallowing them whole. It was formerly very generally believed to possess the power of fascination, and Dr. Williams adduces (Hist. I —485,) the testimony of several persons in support of the opinion, but the notion is now very generally exploded.

THE CHICKEN SNAKE.
Coluber eximius.—DE KAY.

DESCRIPTION.—Color light ash, with numerous large ocellated wood brown spots surrounded with black, which cover more than half of the upper surface. A row of these spots, which are very large, passes from the head along the back to the extremity of the tail; another row of similar but smaller spots passes along each side, the spots lying intermediate between those on the back; belly light flesh color, with quadrangular brownish spots; iris reddish orange. Body elongated; size nearly uniform from the head to the vent, and covered above with rhomboidal scales, each having two punctures, or indentations, near the posterior extremity. Head covered with 10 plates, the central one between the eyes triangular, and the two posterior ones very large; upper jaw margined by 14 and the lower by 18 scales, besides the one at the tip; tail terminated in a blunt horny spine. Length of the specimen before me 32 inches, tail 4½, head 1, width ½ the length. Abdominal plates 206, subcaudal scales 46 pair.

HISTORY.—This snake is occasionally met with in all parts of the state, but is not very common. It is called the *Chicken Snake* on account of its occasionally destroying young chickens. It is also called the *House Snake*, because it is often met with in and about old houses; and the *Milk Snake* from its supposed fondness for milk. In some places it is known by the name of the *Chequered Adder*, or *Thunder-and-Lightning Snake*. This snake sometimes exceeds five feet in length, with a circumference in the largest part of more than 4 inches. They feed principally upon toads, frogs and salamanders, and are supposed also to catch mice. The opinion seems to be prevalent that this snake is poisonous, but we have seen no evidence adduced in its support. It is very sluggish in its habits and movements, and may be often seen stretched along in the side of a stone wall, basking in the sun.

THE WATER SNAKE.
Coluber sipedon.—LINNÆUS.

DESCRIPTION.—Color above dark brown with large club-shaped spots upon the sides of light yellowish brown surrounded by blackish, which join the light color of the belly, and usually run to a point on the back, sometimes meeting, but more commonly alternating with the spots on the opposite side; belly mottled with blackish, yellowish-brown and yellowish-white, the latter mostly triangular, and in longitudinal rows; darker beneath the tail. Body thick in proportion to the length, and nearly uniform in size from the neck to near the vent, after which it tapers rapidly to a point; scales strongly carinated, especially on the posterior part of the body. Length of the specimen before me 28½ inches, tail 7½, plates 140, scales 72 pair.

HISTORY.—This Snake is never seen at much distance from the water, but is quite common in the marshes and grassy coves along the margin of lake Champlain, and about the mouths of our large rivers. It sometimes grows to the size of a man's wrist, and is generally avoided as venomous. It feeds upon frogs and salamanders.

GENUS CROTALUS.—*Linnæus.*

Generic Characters.—Head large, triangular, rounded in front, covered with plates anteriorly; vertex and occiput with scales; a deep pit between the eye and nostril, upper jaw armed with poisonous fangs; body elongated, thick; tail short and thick, terminating in a rattle, which is a corneous production of the epidermis; plates on the abdomen and under the tail.

THE BANDED RATTLE SNAKE.
Crotalus durissus.—KALM.

DESCRIPTION.—Upper parts yellowish-brown, with rhomboidal black spots along the back, margined with bright yellow; upon the sides of these rhombs a black band is continued to the sides of the body, where it terminates in an irregular quadrate black spot; tail black; under parts yellow, with fuliginous dots and blotches; scales on the back elongated, carinated, larger and less carinated on the sides; top of the head flattened, scales upon the top small, on the sides large, pentagonal—on the edges of the jaws quadrangular; snout terminated by one plate; a quadrangular plate on each side of this; directly back of these a smaller one in which are the circular nostrils, situated obliquely, pointing forwards; above the two lateral plates, two others are situated; the first meeting the snout anteriorly, and the second extending some distance beyond the nostrils behind; a large plate at the anterior angle of the eye, separated from the nostrils by two quite small ones, at the anterior inferior angle of which is the aperture for the poison; a large plate over the eye; two still larger upon the throat. Length 37 inches, head 1½, width of the head one inch. Rattles, 6; abdominal plates 170, caudal 24.—*Storer.*

HISTORY.—This is the only poisonous reptile known to exist in Vermont; and although Rattle Snakes were formerly found here in considerable numbers, they were mostly confined to a very few localities, from which they have now nearly disappeared, but still the remembrance of these localities is, in most cases, preserved in the name of "Rattle Snake Hill," or "Rattle Snake Mountain." The Rattle Snake feeds upon young birds, mice, and reptiles. Its poisonous fangs are situated in the upper jaw, and used only as weapons of defence; and as it always gives warning with its rattles before it strikes, cases of persons being bitten by it in this state have been extremely rare, and in no case, within my own knowledge, fatal. The rattles consist of horny portions of the tail loosely attached to one another, and it has generally been supposed that a rattle is added every year, and that the number of rattles indicates the age of the animal. But this is a mistake. In some cases several new rattles are added in a year, and in others none at all. The Rattle Snake has also been supposed to possess the power of fascination, by which it charmed birds and squirrels, causing them to leap into its mouth, but the opinion is totally erroneous. The motions of this serpent are moderate, and its body thick and clumsy, in which respect, as well as in the form of the rattles, which are not spiral, our figure is erroneous, being much too slender

ORDER IV.—BATRACHIA.
FROGS AND SALAMANDERS.

In animals of this order the heart has but one auricle, and the body is covered with a naked skin. In their mature state they are provided with lungs; but before their transformation they breathe by branchiæ or gills. This order may be divided into two families. The Frog Family and the Salamander Family, or the tailless and the tailed batrachians.

I.—FROG FAMILY.

This family embraces the Frogs, Tree Frogs and Toad. Their common mode of progression is by hops or leaps.

GENUS RANA.—*Linnæus.*

Generic Characters.—Body covered with a smooth skin; upper jaw furnished with a row of minute teeth; another interrupted row in the middle of the palate; no post-tympanal glands; posterior extremities long, and in general fully palmated; fingers four; toes five in number.

THE BULL FROG.
Rana pipiens.—LINNÆUS.

DESCRIPTION.—Color above yellowish green, approaching to brownish olive towards the posterior parts, and sparsely spotted with pale rusty brown; the posterior extremities with a few brownish bars; head and upper lip green; tympanum elliptical, large, rusty round the margin, greenish in the middle; under lip, chin and throat yellow; other parts beneath yellowish white; nostril mid-way between the eye and the snout, and the distance between the nostrils equal to the distance from the nostril to the snout; eyes prominent, pupil black, iris reticulated with black and yellow; a cuticular fold from the orbit passes over and down behind the tympanum, and, upon the shoulder, meets another fold passing from the mouth along the lower part of the abdomen; skin granulated. Length of the head and body of the specimen before me 5½, posterior extremities 8; hind feet fully webbed; greatest diameter of the tympanum .7.

HISTORY.—This is the largest frog found in Vermont, often growing considerably larger than the specimen above described. It is very common in various parts of the state, particularly in the neighborhood of lake Champlain. It is very aquatic in its habits, being seldom seen at a distance of more than a few feet from the water. It feeds upon worms, water insects and small molluscous animals. The stomach of the specimen from which the above figure and description were made, contained the elytra of large coleopterous insects.

THE SPRING FROG.
Rana fontinalis.—Le Conte.

Description.—Head and anterior portion of the body above green, irregularly spotted with brown; posterior parts brownish or greenish ash, spotted with black; snout yellowish; chin yellowish white; posterior margins of the jaws black, or spotted with black; belly white and skin very smooth; skin above and on the posterior parts of the thighs granulated; eyes very prominent, pupil black, surrounded by a golden line; iris finely mottled with black and golden, and surrounded by a golden line; tympanum yellowish brown; a dark colored band along the posterior of the fore leg; hind legs darker, irregularly barred and blotched with black; nostril nearer the eye than the snout; a cuticular fold from the orbit along the side of the back, from which a fold passes down behind the tympanum. Anterior toes 4 in., posterior 5. Length $3\frac{1}{2}$, posterior extremities $5\frac{1}{2}$.

History.—This frog is found more generally diffused over the state than any other. It is common in most of the small streams, and especially about springs, and hence its name, Spring Frog.

THE LEOPARD FROG.
Rana halecina —Kalm.

Description.—Upper part of the body brownish bronze, marked with large, distinct, circular, oblong and irregular spots, of a dark green or brown color, and usually surrounded by a delicate light, or yellowish green border; usually two irregular rows of spots along the back, and one, two, or three still more irregular along each side; sides separated from the back by an elevated bronze-colored ridge; fore legs with spots, and hind legs with spots and bars, similar to those on the body; a black line along the margin of the upper lip, excepting at the point; tympanum small, bronze-colored, and nearly round; eyes prominent, pupils black, and iris varied with black and bronze, the latter forming a long line over the pupil; throat and belly white and smooth; feet palmated; the fourth toe much larger than the rest, and tubercles beneath the joints of all the fingers and toes. Length of the specimen before me, which is of about the usual size, $3\frac{1}{2}$ inches; length of the hind leg to the end of the longest toe $5\frac{3}{4}$ inches.

History.—This is one of the most common and least aquatic of all our frogs. During the summer, it is met with in fields and moist meadows, at a great distance from any water. It was called by Kalm, who first described it, the *Shad Frog*, from its making its appearance in the Spring at the same time with the Shad, but it is better known by the name of Leopard Frog, on account of its ocellated spots.*

THE PICKEREL FROG.
Rana palustris.—Le Conte.

Description.—Color brownish ash above; throat and belly white; flanks and under sides of the limbs yellow; back, sides, upper sides of the limbs, and the margin of the under jaw spotted, or barred with brownish black. Spots along the back squarish, in two longitudinal rows, with two rows of similar, but smaller spots, on each side below the lateral line, which is distinct, of a bronzy hue, and extends from the eye to the posterior part of the body. There are usually two spots between the eyes and one in front; hind legs barred with brownish black, and a few spots of the same on the fore

---

* Frogs seem to be able to subsist for an unlimited length of time in a torpid state. There have been repeated and well authenticated instances of their being dug up, in this state, from depths and under circumstances which made it nearly certain that they must have lain there for many centuries. Dr. Williams (Hist. I—150, 479) has given the particulars respecting a considerable number of frogs which were dug up in Windsor, Castleton and Burlington, at depths of from 5 to 30 feet below the surface of the ground. A number of those dug up in Burlington were preserved in spirits in the museum of the University, where I frequently saw them, and although they were all lost when the college edifice was burnt, in 1824, I think I can safely say from present recollections, that they were all of the species *Rana halecina*, which is at present our most common species. In 1822 a living frog was dug up in Bridgewater, at the depth of 26 feet from the surface of the ground.

legs: nose pointed; eyes prominent; iris dark golden; tympanum small and nearly the color of back; a brownish line from the snout to the eyes; tubercles on the lower surface of the toes at the joints. Length of the head and body 3 inches.

HISTORY.—This prettily marked frog bears considerable resemblance to the preceding species, and like it varies, in the different specimens, very much in the brilliancy of the colors and the form of the spots. It was named *palustris*, by Le Conte, on account of his finding it about salt marshes, but it is equally common about fresh water streams, ponds and marshes.

THE WOODS FROG.
Rana sylvatica.—LE CONTE.

DESCRIPTION.—Color varying from light drab to reddish brown above and whitish beneath, often with rusty patches in the young; a longitudinal black line commences at the point of the nose, and, widening as it extends backward so as to involve about two thirds of the eye and the whole of the tympanum, terminates at the shoulder; usually a fine black line along the margin of the upper lip, with a yellow line separating it from the vitta passing through the eye; hind legs with broad, obscure, blackish, transverse bands. Length when fully grown about 3 inches.

HISTORY.—This frog is found in all parts of the state, and, though frequently met with in moist meadows, is much more common in woods, and hence its name, Woods Frog. This, like the Leopard Frog, is often seen at a great distance from any water. It varies greatly in the intensity of its general colors, varying from nearly black to light reddish brown or almost white, but is readily distinguished from all the other species by the black vitta or stripe passing through the eye and embracing the tympanum. The young are usually darkest colored and become lighter as they increase in age and size.

THE HORICON FROG.
Rana horiconensis.—HOLBROOK.

DESCRIPTION.—Head large, with snout rather pointed, the whole dusky green above; nostrils lateral, nearer the snout than the orbits, eyes large, prominent, and beautiful, pupil black, iris reticulated, black and golden; tympanum large, bronzed with a light spot in the centre; upper lip light bronze, with dusky bars; above this an indistinct band of bluish white, with black spots, which extends from near the snout under the orbit and tympanum, to the shoulders; lower jaw, chin, and throat white. Body robust, dark olive, interspersed with irregular black spots, with an elevated cuticular fold on each side, of lighter color, from the orbit to the posterior extremities; abdomen silvery white. Anterior extremities dusky above, white below; posterior dark olive above with transverse black bars; posterior part of the thighs granulated and flesh colored, feet dusky, above and below. Length 3½ inches.—*Hol.*

HISTORY.—This frog was found by Dr. Holbrook, at the outlet of lake George, and, if found there, there can be no doubt of its existence in Vermont. I think I have met with it in Burlington, but at the time supposed it to be the Spring Frog.

THE BLACK FROG.
Rana melanota.—RAFINESQUE.

DESCRIPTION.—Back olivaceous black; a yellow streak on the sides of the head; chin, throat, and inside of the legs whitish with black spots; belly white, immaculate: total length, 2½ inches. *Raf.*

HISTORY.—I give this on the authority of Rafinesque, who says that it inhabits lake Champlain and lake George.

GENUS HYLODES.—*Fitzinger.*
Generic Characters.—Mouth furnished with a tongue; teeth in the upper jaw and palate; tympanum visible; extremities slender; tips of the fingers and toes terminating in slightly developed tubercles.

PICKERING'S HYLODES.
Hylodes Pickeringii.

DESCRIPTION.—Color varying from yellowish ash to light olive above, with irregular brown markings and numerous small brown spots; hind legs faintly banded with brown; beneath, whitish yellow and granulated; head rather broad; nose blunt; fore feet with four toes, one disposed like a thumb for clasping; hind feet slightly webbed, with five toes, and two tubercles on the heel; all the toes terminated in small tumefactions or soft tubercles; a considerable cavity between the orbits; a dark marking on each side of the head embracing the tympanum. Total length of the head and body about 1 inch.

HISTORY.—I have two fine specimens of this beautiful little animal, both of which I captured in Burlington. The first measures just 1 inch from the snout to the posterior of the body. I captured it in a dry pine grove, October 6, 1840. Though the weather was cool it was very active, and it was with difficulty that I succeeded in taking it. Its leaps were often from four to six feet. It would bound into the air and cling to the small limbs and bushes 4 or 5 feet from the ground. The other I caught in August, 1840, near what is called the High Bridge. The length of the head and body is .8 in.

GENUS HYLA.—*Laurenti*.

Generic Characters;—Body is generally elongated; upper jaw and palate furnished with teeth; tympanum apparent; no post-tympanal glands; fingers long, and, with the toes, terminating in rounded viscous pellets.

THE COMMON TREE TOAD.
Hyla versicolor.—LE CONTE

DESCRIPTION.—General form like that of the common toad, with the posterior portion more slender. Usual color above, light ash with irregular brownish blotches, frequently cruciform between the shoulders, and commonly two brown bars crossing the thighs and hind legs; belly white and granulated; flanks and under side of the thighs orange; head broad; snout blunt; pupils black; iris golden, reticulated with black; anterior extremities rather small; four toes before and five behind on each foot, all terminated by tumefactions or pellets. Usual length 2 inches.

HISTORY.—The Tree Toad is so called on account of its often being found upon trees, which it climbs by means of the pellets upon its toes. By these it is able to sustain itself upon the smooth surface of a perpendicular pane of window glass. They for the most part remain silent and concealed during the day time, but during warm rainy weather they sometimes become very noisy, and ascend upon logs, fences, and trees, but as they assume very nearly the hue of the object upon which they are situated, they are not readily discovered. They feed and move from place to place mostly by night, but when discovered during the day, they will often suffer themselves to be taken in the hand without making any effort to escape. In their general form they resemble the common toad.

THE PEEPING TREE FROG.
Hyla squirella.—BOSC.

DESCRIPTION.—Form slender; semi-transparent; color brownish red above, with obscure, irregular, brown blotches, bars, and specks on the upper side of the head, body, and legs; chin and throat greenish; belly and under side of the thighs yellowish white, with the flanks and posterior of the thighs light orange, a cuticular fold along each side; eyes small, pupil black, iris golden; a large cavity on the head between the orbits; head broader than long; mouth large, tongue fleshy; minute teeth on the upper jaw and palate; upper jaw margined with whitish; bones of the head very thin and transparent; limbs slender; 4 toes on the anterior and 5 on the posterior feet, all terminated in rose colored pellets; one toe on each fore foot disposed like a thumb for clasping; hind feet palmated. Length of the specimen before me, 1.1 in.; head, .3; thighs, .5; tarsus to the end of the toes, .7; greatest width of the head, .35

HISTORY.—This species, though not so common as the preceding, is met with in different parts of the state, but is much oftener heard than seen. During the warm summer evenings its shrill *peep* is heard to a great distance. It ascends trees and is often found concealed between the loose bark and wood of old decayed trees. This species, in its general form, has a-nearer resemblance to the frogs than to the common toad. The specimen from which my figure and description are made was captured in Burlington.

Genus Bufo.—*Laurenti.*

Generic Characters.—Head short ; jaws without teeth : tympanum visible ; behind the ear is a large glandular tumor, having visible pores ; body short, thick, swollen, covered with warts or papillæ ; posterior extremities but slightly elongated.

THE COMMON TOAD.
Bufo americanus.

Description.—Color of the back and outside of the limbs reddish brown, with brownish blotches edged with black and surrounded by a dull yellowish line, with a light ash colored stripe from the top of the head along the middle of the back to the posterior extremity of the body. Belly dull yellowish white, sprinkled with brown spots. Two very large porous glands back of the eyes. The body above covered with warts or tubercles, the color of the central part of which is usually ferruginous; body beneath granulated. Tympanum small. Eyes brilliant; iris beautifully reticulated with black and golden. Four toes on the anterior feet, five on the posterior, with a hard excrescence forming the rudiment of a sixth toe ; hard tubercles on the under side of the feet and toes. Head rather large. Length 3½ in.

History.—The toad, which has been too long looked upon with disgust, and regarded rather as an enemy than a friend, is beginning to be viewed by horticulturists as a benefactor, and there can be no doubt that it renders an essential service by the destruction of noxious insects, and deserves rather to be cherished than driven from cultivated grounds. During the day the toad usually sits motionless in some retired, obscure place, watching for flies and other insects, and when any one approaches within suitable distance, he suddenly darts out his tongue, to which the insect adheres, and he seldom fails of returning it to his mouth with the prey attached to it. During the night they venture abroad, and are often met with in large numbers in places where few if any are to be found in the day time.

II.—SALAMANDER FAMILY.

Genus Salamandra.—*Brongniart.*

Generic Characters.—Body elongated ; tail long ; extremities four ; fingers four; toes five; no tympanum ; numerous small teeth in the jaws and palate ; tongue as in frogs ; no sternum ; ribs rudimental ; pelvis suspended by ligaments.

This genus comprehends those animals which are generally known by the name of efts and newts.

SYMMETRICAL SALAMANDER.
Salamandra symmetrica.—Harlan.

Description.—Color brownish orange above, bright orange beneath ; on each side of the spine a row of from three to seven ocellated spots of beautiful vermillion color, with the surrounding circle black ; the sides and under parts of the body sprinkled with minute black points, extending from the chin to near the extremity of the tail ; head flattened ; nose blunt ; eyes bright and not very prominent, with two longitudinal ridges between them ; four toes on the fore feet, five on the hind ; skin on the body and legs roughened by minute tubercles. The specimen before me has six ocellated spots on each side of the spine, and measures 3.3 inches. Length of the tail, which is cylindrical, next the body, and flattened vertically towards the extremity, 1.7 inches.

History.—This species of Salamander is frequently met with in different parts of the state, but is less common than several of the following species. It exists throughout the United States, from Maine to Florida. It is found in water, under old logs in moist places, and is sometimes seen crawling abroad on the wet ground after a shower. Its motions are rather moderate. It feeds upon spiders and small insects.

MANY-SPOTTED SALAMANDER.
Salamandra dorsalis.—Harlan.

Description.—General color olive above, with a slight tinge of green, and varying from sulphur yellow to reddish orange beneath ; a row of ocellated vermillion colored spots, with a blackish halo on each side of the dorsal line, which vary in number and size in different individuals; the whole surface of the body, limbs and tail thickly sprinkled with minute black dots. The head is short, rather broad behind, and pointed at the snout, with the nostrils near the extremity; eyes rather prominent, pupils black, iris light yellow; tail roundish at the base, then compressed laterally through its whole length, and very thin at the extremity; fore legs and feet small and delicate, with 4 small toes; hind legs nearly twice as large, with 5 toes. Length of the largest of two specimens before me, 3.7 inches; head and neck .6; body 1.1; tail 2.

HISTORY.—This is one of the most common species of Salamander in Vermont, and is eminently aquatic, spending nearly all the time in the water. When kept in a vessel of water it rises to the surface every few minutes for the purpose of taking in air. It is an animal of considerable activity, and its movements are often very sudden. It is perfectly harmless, and usually manifests much anxiety to conceal itself from view. This salamander seems to be much annoyed by a species of parasitic animals. One of the specimens before me has at least 20 upon it at this moment. They are soft animals, resembling a snail in appearance, but more pointed at the two extremities. They move in the manner of caterpillars, by reaching forward and then bringing up its posterior. They fasten themselves upon the salamander by their mouths, in the manner of the lampreys or bloodsuckers, and adhere with such force as not to be easily separated. The animal upon which they are fastened seems to be in much agony, and frequently struggles, but in vain, to rid himself of them. When fully extended they measure one third of an inch. On being taken from the water, they die as soon as the water which adheres to them is evaporated.

SALMON-COLORED SALAMANDER.
Salamandra salmonea.—STORER.

DESCRIPTION.—Color yellowish brown above, salmon color at the sides, with a bright salmon-colored line from the nostril to the upper part of the orbit; upper jaw pale salmon color, with a few brown spots; lower jaw, and body beneath whitish; light salmon color beneath the tail.

Head large and flat; snout obtuse; nostrils small; a strongly marked cuticular fold upon the neck; eyes remote and very prominent; pupil black; iris copper-colored; body elongated and cylindrical; posterior extremities twice the size of the anterior. Tail longer than the body, rounded at the root, compressed laterally and pointed at the tip. Length 6¼; tail beyond the vent 2¼.—*Storer.*

HISTORY.—This species was first described and named by Dr. Storer, of Boston, from a specimen found by Dr. Binney, in Vermont, and his description, with a figure, was published in Dr. Holbrook's Amer. Herpetology, Vol. III—101. A description is also given in Dr. Storer's Report, p. 248. I have a specimen of this salamander, taken in Bridgewater, but as it is not fully grown I give Dr. Storer's description. It is found upon moist lands.

THE TIGER SALAMANDER.
Salamandra tigrina.—GREEN.

DESCRIPTION.—Color blackish above, marked irregularly and thickly with roundish, oblong and angular yellow spots of different sizes; belly brownish gray; legs the same color as the body, with a few yellow spots on the outside. Head rather large; snout rounded; eyes black and prominent; four toes on the fore feet, 3d the longest; 5 on the hind feet, 3d and 4th longest; hind legs about twice the size of the fore legs; a distinct cuticular fold under the throat; tail longer than the body, roundish at the base, but soon becoming flattened, and edged towards the extremity and terminated in a flattened point. Hind legs midway between the snout and the extremity of the tail.—Length of the specimens before me 3 in., but it grows larger.

HISTORY.—This Salamander is frequently met with in Vermont, living in swamps and marshes. I obtained 3 good specimens of this species from the stomach of a Ribband Snake, *C. saurita*, besides some others which were partly digested. The snake from which they were taken measured about 2 feet, and the salamanders 3 inches. On the 4th of August, 1842, I caught with a scoop-net more than a dozen salamanders, out of a small muddy pool in Burlington, which I suppose to belong to this species. They were about 3 inches in length, of a brownish yellow color, and most of them were in the larva state, having the fin along the back, and the branchiæ remaining, but from several of them these appendages had disappeared. I have kept two of the former and one of the latter, in a vessel of water, up to this time, August 17, 1842. The branchiæ and fins have vanished, their color has become quite dark, and the yellow spots are making their appearance very distinctly.

VIOLET-COLORED SALAMANDER.
Salamandra venenosa.—BARTON.

DESCRIPTION —Color above dark grayish brown, with a row of large roundish bright yellow spots on each side of the dorsal line, which unite into a single row towards the extremity of the tail ; several of these spots on the head and upper sides of the legs ; color lighter beneath, with some minute white spots ; tail roundish at the base, but slightly flattened through the greater part of the length, and terminated in a flattened rounded point; snout bluntly rounded ; eyes not very prominent; hind legs midway between the snout and end of the tail. Length of the specimen before me 6¼ inches; width across the head .6, across the body .5.

HISTORY.—This large species is not very common in Vermont. The specimen from which my description is made was found in a marshy place in Burlington.

RED-BACKED SALAMANDER.
Salamandra erythronota.—GREEN.

S. erythronota. } GREEN.
S. cinerea.

DESCRIPTION.—Sides brownish, and often with minute light specks, fading into steel-gray on the belly, usually a broad brownish red stripe along the back ; belly dark steel gray, lighter and yellowish towards the chin ; head above darker than the body; form slender, cylindrical; tail nearly cylindrical, and longer than the head and body ; vent midway between the snout and the extremity of the tail ; head broader than the body, short in front of the eyes; snout bluntly rounded ; eyes prominent, lively, pupil black, iris golden. A distinct cuticular fold on the throat; legs slender, brownish ; toes short, 4 before and 5 behind. Length of the longest of two specimens before me 3.4 inches ;' from the snout to the fore legs .5 —to the hind legs 1.55; from the hind legs to the point of the tail 1.85 ; width of the head .2.

HISTORY.—This salamander is quite common in Vermont, and is probably the least aquatic of all our salamanders. It is often met with under the rotten logs on dry pine plains ; and also in ledgy places in the hard wood forests, under the loose stones and among the decayed leaves. Its appearance is lively, and its motions often very sudden. Aided by a sudden vibration of the tail, it has the power of leaping several times its length. I have before me two specimens, both found in Burlington, one with a brownish red stripe along its back, and answering to Dr. Green's *S. erythronota*, and the other, which is a little larger, answering to his *S. cinerea*. The stripe on the back seems to be the only difference, and I believe they are now regarded by herpetologists as belonging to the same species.

THE GLUTINOUS SALAMANDER.
Salamandra glutinosa.—GREEN.

DESCRIPTION.—Whole upper part of the body dark brown, sprinkled with distinct light blue spots ; sides light colored from the blue spots becoming confluent ; abdomen lighter, exhibiting the spots more numerous and distinct than the back ; eyes prominent, wide apart, of a deep black color; head flattened above ; nostrils small ; legs color of the body and spotted like it; anterior feet 4 toed, posterior 5 toed and unusually long ; tail, length of the body, much compressed throughout its whole extent, save the extremities, the anterior of which is circular, the posterior pointed. Length 6 inches ; head .75 ; width of the head .5.—*Storer.*

HISTORY.—This species I have not seen in Vermont. I insert it on the authority of Prof. Adams, who informs me that there is a Vermont specimen of it in the Collections of Middlebury College.

THE TWO-LINED SALAMANDER.
Salamandra bis-lineata.—GREEN.

DESCRIPTION.—Tail longer than the body, tapering, compressed, and pointed ; snout oval ; back cinereous, with two and sometimes three dark lines, if three, the middle one broadest near the head, and about the length of the body, the lateral ones extending from behind the eyes to the end of the tail; sides cinereous ; beneath whitish or yellowish; anterior toes 4., posterior 5. Length 3 inches.—*Green.*

HISTORY. This salamander I have not seen in Vermont, but Prof. Adams informs me that he has a Vermont specimen which belongs to this species. According to Dr. Green it inhabits shallow waters, appears early in spring, and is very active.

GENUS MENOBRANCHUS.—*Harlan.*

Generic Characters.—Head large, flattened, truncate, two rows of teeth in the upper jaw, a single row in the lower; teeth small, conical, pointed; gills and tail persistent during life.

THE PROTEUS.
Menobranchus maculatus.—BARNES.

DESCRIPTION.—General color dark cinereous gray, produced by minute yellowish specks on a dark bluish ground, and irregularly interspersed with circular spots about the size of a pea, of a darker hue; the throat and central parts of the abdomen nearly white; a brownish stripe commencing at the nose and extending backwards over the eye; the margin of the tail often of an orange tinge, with blackish blotches near the extremity. The head is large, flattened, and the snout truncated; eyes small and far apart; mouth large; throat contracted with a transverse fold in the cuticle beneath; tongue large and fleshy; teeth small and sharp, two rows in the upper jaw and one in the lower. The gills are external, large, and each consists of three delicately tufted or fringed lobes, which, when vibrating in the water, are of a fine blood-red color; body cylindrical, covered with a smooth mucous skin; tail long, flattened and broad vertically, and rounded at the end like that of an eel; legs four, each foot furnished with four toes resembling fingers, but without nails, although the cuticle at the extremities is dark colored, having much the appearance of nails. The total length of the specimen before me, and from which the above figure and description are made, is 12¼ inches, and this is about the usual length.

HISTORY.—This singular reptile was first described by Schneider, about the year 1799, from a specimen obtained from lake Champlain.* This specimen was probably obtained at Winooski falls, which were, for some time, the only known locality of this animal, and where more or less of them are now taken every spring, upon the hooks suspended on night lines for taking fishes. The fishermen formerly considered them poisonous, and when they found them upon their lines they were glad to rid themselves of them by cutting the lines and letting them go with the hook in their mouths; but they are now found to be perfectly harmless and inoffensive. This animal is seldom seen excepting in the months of April and May, and this is the season for depositing its eggs. In a specimen taken on the 13th of April, 1840, I found about 150 eggs of the size of a small pea and, apparently just ready to be extruded. The food of this reptile consists of various kinds of worms and insects. The stomach of the one above mentioned contained two hemipterous insects, each three fourths of an inch long, the wings and bodies of which were entire, besides numerous fragments of other insects. Of the habits of this animal very little is known. It seems to spend the greater portion of the time about falls, concealed in the inaccessible recesses and crevices of the rocks below the surface of the water, and not to venture much abroad excepting at the season for depositing its eggs. Although it passes nearly the whole time in water, it is truly an amphibious animal, having lungs for breathing in the atmosphere, as well as branchiæ for breathing in water. It does not, however, breathe in water by receiving the water into its mouth and passing it out through the gills, in the manner of fishes, but simply by the vibrations of its branchiæ in the water. When kept in a vessel containing a large quantity of water, or in which the water is frequently renewed, it manifests but little disposition to rise to the surface for atmospheric air. But when the quantity of water is small,

---

* The following is Schneider's description, and our reptile answers to it in almost every particular.
Corpus ultra 8 pollices longum et fere pollicem, crassum, molle, spongiosum, multis poris pervium, in utroque latere tribus macularum rotundarum, nigrarum seriebus variegatum; cauda compressa et anceps, utrinque maculata, inferiore acie recta, superiore curvata, in finem teretiusculum terminatur. Caput latum et planum: oculi parvi, nares anteriores in margine labii superioris, maxillæ superioris geminæ ut inferioris dentes conici, obtusi, satis longi; lingua lata, integra, anterius soluta: apertura oris patit usque ad oculorum lineam verticalem; labia piscium labiis similia; pedes dissiti quatuor, tetradactyli omnes, absque unguiculis; ani rima in longitudinem patet; branchiæ utrinque ternæ extus propendent, appositæ superne totidem arcubus cartilagineis, quorum latus internum tubercula cartilaginea, velut in piscium genere, exasperant. &c.

FISHES OF VERMONT.

PRELIMINARY OBSERVATIONS.

and not often changed, it soon finds the air in the water insufficient for its purpose, in which case it comes to the surface, takes in a mouthful of air, and sinks again with it to the bottom. After retaining the air for a time, probably long enough for the consumption of its oxygen in the lungs, it suffers it to escape through the mouth and gill openings, and it is seen to rise in small bubbles to the surface. This animal is said to be found in several places at the west, particularly in streams falling into lake Ontario, where it is said sometimes to attain the length of two feet. The length of those taken at Winooski Falls varies from 8 to 13 inches. I have never seen one which exceeded 15 inches. The best figure of our animal which I have seen published is in the Annals of N. Y. Lyceum, vol. I, plate 16. The description and figure in Dr. Holbrook's American Herpetology do not answer to our Menobranchus, but as Prof. G. W. Benedict has furnished Dr. H. with an accurate colored figure, drawn from a living specimen by the Rt. Rev. J. H. Hopkins, we hope to see it correctly represented in a future volume of his splendid and valuable work. We are strongly inclined to believe the animal which he describes to be a different species from ours. Notwithstanding what he and others have said in proof of the identity of the *Triton lateralis* of Say, the *Menobranchus lateralis* of Harlan, Holbrook, and others, with the reptile described by Schneider, I am strongly inclined to the opinion that they are different species. I have therefore given the name suggested by Prof. Benedict, and adopted by Barnes, the preference, and have described our animal under the name of *Menobranchus maculatus*, that being descriptive of our reptile, and the other not so.

CHAPTER V.

FISHES OF VERMONT.

Preliminary Observations.

FISHES constitute the Fourth Class of the animal kingdom. They are vertebrated animals, with cold red blood. They respire by means of branchiæ, or gills, and they move in water by means of fins. Their entire structure is as evidently fitted for swimming as that of birds is for flight. The tail is the principal organ of motion, and progression is effected by striking it alternately from right and left against the water. The mean specific gravity of fishes is the same as the fluid in which they live, so that no effort is required to keep them suspended, and a large part of them are furnished with an air bladder, by the compression or dilatation of which they can vary their specific gravity, and thus rise or descend without the aid of their fins.

The *head* of fishes is usually larger in proportion to the size of the body than that of other animals; and although it is subject to great variety of form, it in almost all cases consists of the same number of bones as is found in other oviparous animals. These bones are separate in young fishes, but in older ones become united and consolidated so as to make it difficult to distinguish them. The *nostrils* are simple cavities placed at the front of the snout, and usually double. The cornea of the *eye* is very flat, and has but little aqueous humor, but the crystalline is hard and globular. The *ear* of fishes is very obscure, and, having neither eustachian tubes nor tympanal bones, their sense of hearing must be very imperfect. The head is attached to the body in such manner that its motion is exceedingly limited. The *tongue* varies in different families: in some it is fleshy, but in many cases it is osseous and frequently covered with teeth, so that their sense of taste must be very obtuse. The *body* of fishes is in most cases covered with scales, which cannot allow much sensibility to the touch. This imperfection is, probably, supplied in some cases by the fleshy cirri, with which several species are furnished. The *teeth* of fishes vary almost infinitely in number, form and situation. Besides the jaws, they are often found upon the tongue and palate, and not unfrequently in the throat and at the base of the gills, while some families are entirely destitute of them. The *stomach* is generally simple and the intestines short.

The sexes of fishes are distinguished by

the male having a milt and the female a roe. The roe is composed of a multitude of eggs, which the female deposits in some suitable place. After their extrusion, they are impregnated by the male, and left to hatch, without the further aid or care of the parents.

Fishes are long-lived animals, and their fecundity is very remarkable. We have authenticated accounts of a pike having lived 260 years, and a carp 200; and Leuwenhoek computed the number of eggs in the roe of a cod fish to be 3,686,760, and in that of a flounder to be 1,357,400.

In a country like Vermont, situated so remote from the ocean, and watered only by small fresh water streams and lakes, a very great variety or abundance of fish is hardly to be expected; and yet it is a notorious fact, that when the country was new all our waters swarmed with fishes of various kinds. Salmon and Shad were taken in the greatest plenty and perfection in Connecticut river; and the former together with the salmon trout, were abundant in lake Champlain, and in most of the streams connected with it. In the spring of the year, when these fishes were ascending our streams to their breeding places, they could be taken at the falls and rapids in scoop-nets, or in baskets fastened to poles, in almost any quantities desired. Brook trout, weighing from one to three pounds, were plentiful in nearly all our streams and ponds. But with the clearing and settling of the country these kinds of fishes have diminished till the three former have become extremely rare, and the latter, though still numerous in many parts, are seldom taken exceeding half a pound in weight. For the production of this state of things several other causes have operated besides their diminution by fishing. The salmon and shad have probably been driven from our waters, chiefly by the erection of dams across nearly all our streams, which prevent their ascent to their favorite spawning places. Freshets, also, which have become more sudden and violent since the country has become cleared, have swept out the logs and other obstructions, which formed their places of resort and concealment, and have thus tended not only to diminish the number of our fishes, but to prevent their attaining so great magnitude as formerly. Those fishes of our lakes which do not ascend far up our streams to deposit their spawn, have not been so much affected by these causes. These, however, though still taken in considerable quantities, are not so abundant as formerly.

Cuvier divides fishes into two sub-classes. I. *Osseous Fishes*, or such as have hard, solid bones. II. *Cartilaginous Fishes*, or such as have cartilage in the place of bones. Most of our fishes belong to the first of these divisions. The following is a Catalogue of Vermont Fishes, arranged in the order in which they are described in the subsequent pages.

I. OSSEOUS, OR BONY FISHES.

ORDER I.—ACANTHOPTERYGII.

Family I.—Percidæ.

Perca serrato-granulata, Common Perch.
Lucio-Perca americana, Pike Perch.
Pomotis vulgaris. Common Sun Fish.
" *megalotis*, Big Eared Sun Fish.
Centrarchus æneus, Rock Bass.
" *fasciatus*, Black Bass.
Etheostoma caprodes, Hog Fish.

Family II.—Scienidæ.

Corvina oscula, Sheep's Head.

ORD. II.—MALACOPTERYGII ABDOMINALES

Family I.—Cyprinidæ.

Catastomus cyprinus, Carp Sucker.
" *oblongus*, Lake Mullet.
" *teres*, Sucker.
" *nigricans*, Black Sucker.
" *longirostrum*, Long Nosed Sucker.
Leuciscus pulchellus, Common Dace.
" *crysoleucas*, Shiner.
" *atronasus*, Brook Minnow.
Hydrargyra fusca, Mud Fish.

Family II.—Esocidæ.

Esox estor, Common Pike.
" *reticulatus*, Pickerel.

Family III.—Siluridæ.

Pimelodus vulgaris, Horned Pout.
" *nebulosus*, Bull Pout.
" *cænosus*, Cat Fish.

Family IV.—Salmonidæ.

Salmo salar, Salmon.
" *namaycush*, Namaycush, or Longe.
" *fontinalis*, Brook Trout.
Osmerus eperlanus, Smelt.
Coregonus albus, White Fish.

Family V—Clupidæ.

Alosa vulgaris, Shad.
Hiodon clodalus. Winter Shad.
Lepisosteus oxyurus, Bill Fish.
" *lineatus*, Striped Bill Fish.

ORD. II.-MALACOPTERYGII SUBBRACHIATI

Family, Gadidæ.

Lota maculosa, Ling.
" *compressa*, Eel-pout.

ORDER IV.—MALACOPTERYGII APODES.

ORDERS OF FISHES.

Family, Murenidæ.
Murena vulgaris, Common Eel.
" *bostoniensis,* Black Eel.
" *argentea,* Silver Eel.

II. CARTILAGINOUS FISHES.
Family I.—Sturionidæ.
Acipenser rubicundus, Round Nosed Sturgeon.
" *oxyrhynchus,* Sharp Nosed Sturgeon.
Family II.—Cyclostomidæ.
Petromyzon nigricans, Blue Lamprey.
Ammocœtes concolor, Mud Lamprey.

I. OSSEOUS, OR BONY FISHES.

ORDER I.—ACANTHOPTERYGII.

Spinous rayed Fishes.

Fishes of this order are recognized by the spines which occupy the place of the first rays of the dorsal fin, or the rays of the first dorsal when there are two. Sometimes, instead of a first dorsal, there are only a few free spines.

I. PERCIDÆ, OR PERCH FAMILY.

GENUS PERCA.—*Cuvier.*

Generic Characters.—Two separate dorsal fins; rays of the first spinous; tongue smooth; teeth in both jaws, in front of the vomer, and on the palatine bones; preoperculum notched below and serrated on the posterior edge; operculum bony, ending in a flattened point directed backwards; branchial rays seven; scales rough, hard, and not easily detached.

THE COMMON PERCH.

Perca serrato-granulata.—CUV.
Cuv. et Val. Hist. Nat. des Poiss., II—47.

DESCRIPTION.—Body deep and thick, but becoming slender and nearly cylindrical towards the tail; head rather small, and tapering towards the snout; both jaws and palate covered with small teeth; color greenish, or yellowish brown above; sides yellow, crossed by 7 transverse brownish bands; belly white; lateral line parallel to the curve of the back; tail concave. Preoperculum narrow, and its edge armed with small spinous teeth, those on the lower margin larger, with their points directed forwards; the operculum radiated with granulated rays, terminating posteriorly in a spine, with several spinous denticulations beneath, and grooves extending forward from them. The edges of the inter-operculum and sub-operculum are finely serrated, and the latter is prolonged into a membranous point lying under the spine of the operculum. Humeral bones grooved and usually serrated. Jaws equal; eyes rather large; iris yellowish; dorsal and caudal fins brownish; pectorals orange on the lower part; the others more or less ruddy. The first dorsal more than twice as long as high, with a black spot or clouded with black towards the posterior part, the second two thirds as long as the first. Depth of the body to the total length of the fish as 1 to 4. Length of the specimen before me 12 inches, depth 3, thickness 2. Rays, B. 7, P. 0, V. 1|5, D. 13|[1]|14, A. 2|7, C. 17.*

HISTORY.—The Yellow Perch is one of the most common fishes found in lake Champlain, and in the mouths of the rivers falling into this lake. They are taken both with the seine and hook, but chiefly with the latter. In the winter they are caught by cutting holes in the ice. They vary from 8 to 12 and even 14 inches in length, and are carried round for sale from house to house in the villages along the lake, at all seasons of the year, neatly scaled and dressed ready for cooking. In this condition they are sold at from 10 to 20 cents a dozen, according to the season, and their abundance. The flesh of the Perch is white, firm and agreeable to the palate, but is rather dry and bony.

This fish agrees throughout with Dr. Mitchell's description of his *Bodianus flavescens,* and is undoubtedly the species from which his description was drawn. Cuvier, having obtained specimens of this and another species which very closely resemble it, from the waters of the United States, gave to this species the name of *P. serrato-granulata,* on account of its serrated and granulated gill covers; to the other, distinguished from this by the want of granulations, by its smaller size and greater number of brown bands upon its sides, he gave the name of *P. flavescens.*

GENUS LUCIO-PERCA.—*Cuvier.*

Generic Characters.—In the form of the body and situation of the fins like a Perch; head more like a Pike; edge of the pre-operculum with one simple emargination; some of the maxillary and palatine teeth long and pointed.

---

* The letters indicate the fins, and the figures the number of rays in each; B. Branchial rays; P. Pectoral; V. Ventral; D. Dorsal; A. Anal; and C. Caudal.

AMERICAN PIKE-PERCH.
Lucio-Perca americana.—CUVIER.

Cuv. et Val. Hist. Nat. des Poiss. III. p. 122, pl. 16.
Fauna Boreali Americana, Fishes, p. 10.

DESCRIPTION.—Body tapering and cylindrical towards the tail; color nearly black above, sides brown and orange, belly yellowish or bluish white, tail and fins spotted with black on a yellow ground, but varying much in different individuals; head depressed; eyes large, pupil transparent, iris yellow; lower jaw longer than the upper; two rows of teeth in the upper jaw and one in the lower; teeth hooking inward and many of them long; operculum terminated by a membranous point, preoperculum serrated and spinous at the angle; a bony plate over the pectoral fin; rays of the first dorsal fin spinous.

Rays, Br. 7, P. 13, V. 6, D. 14—21, A. 14, C. 17.

HISTORY.—The usual length of this fish is from fourteen to twenty inches, and its weight from one to four pounds. It is taken very plentifully from the waters of lake Champlain and its tributaries. It is a firm, bony fish, but as the bones are large and easily separated from the flesh, they are much less troublesome than in the Perch, and some other species. Its flesh is well flavored, though not so juicy and rich as that of our White Fish and some few others. In the form of its body and the situation of its fins, it closely resembles the Perches, but its head and teeth are more like the Pikes, and hence its name, *Lucio-Perca*, or *Pike-Perch*. This fish is called by Dr. Williams, in his History of Vermont, the *White Perch*, but is generally known in Vermont simply by the name of *Pike*, while the fish usually called Pike in other places is here called *Pickerel*. This fish, on the contrary, is called Pickerel in Canada. We have another species of this genus, probably the *L. canadensis*, but I am unable to say so positively at present.

GENUS POMOTIS —*Cuvier.*

Generic Characters.—A single dorsal fin; 6 gill rays on each side; teeth small and crowded; body compressed and oval; a membranous prolongation at the angle of the operculum.

SUN FISH, OR POND PERCH.
Pomotis vulgaris.—CUVIER.

Shaw's Zoology, IV—482. Lit. and Phil Trans.N. Y., 1-403 Fauna Boreali Americana, p. 23. Storer's Report, p. 11.

DESCRIPTION.—Color brownish green above; below yellow; sides bluish, spotted with brownish, umber, and dark purple; sides of the head striped longitudinally with undulating deep blue lines, with umber spots; a large black spot, edged with silvery above and below, on the posterior angle of the operculum and its skinny prolongation, terminating backward in bright scarlet; all the fins brownish, portions of the dorsal and caudal spotted finely with black; head between the eyes smooth, dark green, with 3 pores, or pits, the lines connecting which form very nearly an equi-lateral triangle; teeth minute and sharp in both jaws; upper jaw protractile; under jaw longest; mouth small; nostrils double, with a pore, making it appear triple; eyes large and round; back regularly curved from the nape to the posterior of the dorsal fin; lateral line parallel to the curve of the back. Depth of the body to the total length of the fish, as 1 to 3, nearly; commencement of the anal fin equi-distant from the two extremities; usual length about 5 inches.

Rays Br. 6, P. 13, V. 1|5, D. 9|12, A. 3|10, C. 17.

HISTORY.—This is a very common fish in the coves along the margin of lake Champlain, and about the mouths of our rivers. Though extensively known by the name of *Sun Fish*, and *Pond Perch*, it is, perhaps, more generally known by the name of *Pumpkin Seed*. It is also sometimes called *Bream*. This fish, though said in Jardine's Naturalists' Library to be of unobtrusive colors, is one of the highest colored and most beautiful fishes found in our waters—'oftentimes vieing in brilliancy with the tropical fishes.' The Sun Fish, though often taken with other fishes in the seine, is more commonly taken with the hook, at which it bites with avidity. Its flesh is white and palatable, but the fish being small, thin, and bony, is little sought as an article of food.

THE BIG-EARED SUN FISH.
Pomotis megalotis.—RAFINESQUE.

Icthelis megalotis, Ichthyologia Ohiensis, p. 29.

DESCRIPTION.—Color brownish olive above, head darker; sides approaching to chestnut; belly coppery, or ruddy white; sides of the head and body with flexuose greenish, or bluish stripes and spots. Membranous prolongation of the operculum very long and wholly black; eyes dark, the pupils being black, and iris brown. Tail and fins brownish. All the colors less brilliant than in the *Pomotis vulgaris*, its mouth proportionably larger, its tail less forked, and its pectorals broader and less pointed. Depth contained a little more than twice in the total length. Length of the specimen before me 4½ inches, depth 1.9, height of the pectoral 0.9, length of the black portion of the prolongation of the operculum 0.4.

Rays, B., P. 4|, V. 1|5, D. 10|11, A. 3|10, C. 18.

HISTORY.—The specimen from which the above figure and description were drawn, was taken in Connecticut river at Barnet. It bears considerable resemblance to the preceding species, and is there known by the same vulgar names. It may however readily be distinguished by the greater prolongation of the black membranous portion of the gill cover, and the absence of the scarlet termination, as well as by its greater depth in proportion to the length, its nearly even tail, deeper cleft mouth, and its broader and less pointed pectoral fins.

GENUS CENTRARCHUS.—*Cuv. et Val.*

Generic Characters.—Body oval, compressed ; one dorsal fin ; teeth like velvet pile, on the jaws, front of the vomer, palatine bones and the base of the tongue ; preoperculum entire ; angle of the operculum divided into two flat points ; anal spines from 3 to 9.

THE ROCK BASS.
Centrarchus æneus.—CUVIER

Cichla ænea, Le Sueur Jour. Ac. Sc. Phil. II, p 214.
Centrarchus æneus, Cuv. et Val. iii, pl. 48.—Fauna Boreali Americana (Fishes) p. 18.

DESCRIPTION.—Form elliptical ; body deep and thin. Back dark; sides yellowish, approaching to white on the belly ; a quadrangular black spot in the centre of each scale, giving the sides a striped appearance from the gill-opening to the tail. Scales large on the sides, with the exposed part circular, and the concealed part finely grooved and truncated at the base; smaller on the back, belly, cheeks and operculum; lateral line parallel to the curve of the back, containing 42 scales ; opercula scaled, preoperculum serrated at the angle ; the operculum terminates backward in two thin lobes, with an acute notch between, and a dark colored membranous prolongation ; plate above the pectoral smooth. Teeth small and thick like velvet pile in both jaws, on the vomer, and on the edges of the palatine bones. Eyes large and dark. Vent anterior. Ventral fins directly under the pectorals; anal commences under the 8th spinous ray of the dorsal ; dorsal and anal expanded posteriorly. The first ventral, the first twelve dorsal, and the first six anal rays spinous. Length of the specimen before me 7½ inches, from the snout to the vent 3¼ ;—to the posterior margin of the operculum 2¼ ; depth 2¾, and contained near twice and a half in the total length.

Rays, Br. P. 14, V. 1|5, D. 12|10, A. 6|9, C. 17.

HISTORY.—This fish is here known by no other name than Rock Bass. It is quite a common fish in lake Champlain, and its larger tributaries. It is usually taken with the hook along the precipitous rocky banks of the lake and rivers, and from this circumstance it derives its name. It is considered a very good fish for the table, and its weight is usually about half a pound.

THE BLACK BASS.
Centrarchus fasciatus.—LE SUEUR.

Cichla fasciata, Le Sue. Jour. Ac.Sc.Phil.II, p 214.

DESCRIPTION.—Form somewhat eliptical, compressed, a little convex on the sides, and pointed forwards. Color greenish above, lighter and faintly mottled on the sides, and grayish white beneath; sides of the head line, light green ;

THE BLACK BASS.

scales firm, moderate on the sides and operculum, but very small on the cheeks, back of the neck, throat and belly. *Preoperculum* with its upper limb nearly vertical and nearly at right angles with the lower, without spines or serratures; interoperculum and suboperculum scaly up on the upper side, and smooth below; operculum triangular, with a membranous prolongation posteriorly, and the bony part terminating posteriorly in two thin lobes, with a deep notch between them, the lower lobe, which is largest, ending in several short spines; teeth small, sharp and numerous in both jaws, on the lower anterior edges of the palatine bones, and on the vomer with a small cluster near the base of the triangular tongue, all standing like the pile on velvet, but hooking a little inward, those on the jaws largest. *Fins* small, brownish, and their soft parts covered with a rather thick mucous skin; the dorsal rounded behind, low at the junction of the spinous and soft parts, and the spinous rays capable of being reclined, imbricated and concealed in a longitudinal groove along the back; ventrals a little behind the pectorals; the anal under the posterior portion of the dorsal, and extending a little further back; tail slightly emarginate, with the lobes rounded. Vent a trifle nearest the posterior extremity; eyes moderately large; lower jaw a trifle longer than the upper, with several visible pores along its margin. Length of the specimen before me 19 inches; the greatest depth equals one third of the length, exclusive of the tail.

Rays Br. 6, P. 17, V. 1|5, D. 10|15, A. 3|11, C. 17.

HISTORY.—The Black Bass, by which name this fish is here generally known, ranks as one of the best fishes taken from our waters; but, as is apt to be the case with good fishes, it is much less abundant than several other species which are greatly its inferior in point of quality. It is usually taken with the seine, and its weight varies from one to five or six pounds.

GENUS ETHEOSTOMA.—*Rafinesque.*

Generic Characters.—Body nearly cylindrical and scaly; mouth variable with small teeth; gill cover double or triple, unserrate with a spine on the operculum, and without scales; branchial rays six; rays in the ventral six, one of which is spiny, no appendage; dorsal more or less divided into two, with all the rays of the anterior portion spiny; vent nearly medial.

THE HOG FISH.
Etheostoma caprodes.—RAF.

Rafinesque Ich. Ohiens. p. 38. Kirt. Rep. Zool. Oh o, p. 168. Boston Jour. Nat. His. III—346.

DESCRIPTION.—Body lengthened and cylindrical; head elongated, flattened on the forehead, with the snout protruded and rounded like that of the hog; under jaw narrower and shorter than the upper; mouth beneath, small. Color yellowish, darkly spotted and barred with brown above and on the sides; belly yellowish white; 10 brown bars or blotches on the sides, the posterior one at the base of the tail black, with about 20 less distinct bars above and between these passing over the back; caudal and dorsal fins finely spotted or barred with brown; pectoral, ventral and anal transparent, unspotted and yellowish; posterior part of the head above nearly black, but lighter towards the snout; eyes middling size, prominent; pupil black, surrounded by a bright line and a yellowish silvery iris; tail slightly lunated; scales ciliated and rough; operculum terminated posteriorly in a sharp spine; minute teeth in both jaws and on the vomer; lateral line straight; ventrals behind the pectorals and under the anterior part of the second dorsal. Length 3.2 inches; pectoral fin as long as the head.

Rays, Br. 6, P. 14, V. 6, D. 14|14, A. 12, C. 17.

HISTORY.—This fish, though its vulgar name might be thought to imply the contrary, is certainly one of the most symmetrical and beautiful fishes found in our waters. It received the name of *Hog Fish* from a resemblance in the form of its snout and lower jaw to those of that quadruped. It is quite common in the mouths of the streams which fall into lake Champlain, but being a slender fish, and never exceeding 4 or 5 inches in length, no account is made of it as an article of food, and very little is known of its habits. It swims low in the water, and when at rest usually lies at the bottom.

II.—SCIENIDÆ OR SCIENA FAMILY.
GENUS CORVINA.—*Cuvier.*

Generic Characters.—Head gibbous, cavernous, and scaly; stones in the sack of the ear very large; no canine nor palatine teeth; all the teeth small and crowded; preoperculum dentated; branchial rays seven; anal fin short, with the second spine robust and strong.

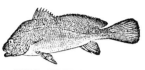

THE SHEEP'S HEAD.
Corvina oscula.—LE SUEUR.
Sciena oscula. Le Su., Jour. A. N. Sci., ii, p. 252.

DESCRIPTION.—Back elevated; body deep, thick through the abdomen, and compressed to an edge along the back, and slender near the tail; head declining; snout short, rounded, with three small openings at the end, and large pores near the tip of the lower jaw; mouth rather small, lips distinct; teeth in both jaws conic and crowded, the outer series largest; eyes large, round, and near the snout; nostrils double, the posterior much the largest, and very near the eye; head and opercula covered with scales; preoperculum coarsely serrated; base of 2d dorsal, pectoral, anal and caudal fins covered with scales; the 9 rays of the first dorsal, 1 ray of the 2d dorsal, the first ventral and two first anal rays, spinous; the 1st dorsal and 1st anal spine very short, the 2d large and stout; scales rough. Color brownish gray above, sides silvery, and pearly white, or cream color, beneath; head with livid purple reflections; dorsal, pectoral, anal and caudal fins brownish; ventrals yellowish; lateral line parallel to the arch of the back, and visible on two-thirds of the length of the tail; tail rounded; height of the second dorsal nearly uniform, the posterior reaching the base of the caudal; depth of the fish contained 3 times in the total length. Length of the specimen before me 17½ inches; greatest depth just behind the pectorals 5½.

Rays Br. 7, P. 16, V. 1|5, D. 9—1|31, A. 2|8, C. 18.

HISTORY.—This fish is quite common in lake Champlain, and is here generally known by the name of Sheep's Head. It is also found in the western lakes and the Ohio river, where it is more commonly called the White Perch. This fish, taken from the Ohio river, is said to be fat, tender, and well flavored; but ours is lean, tough, and bony, and seldom eaten. It received its vulgar name from its resembling in appearance the *Sargus ovis*, which is also called Sheep's Head on account of its 'arched nose and smutty face;' but the resemblance is in appearance only, for while the latter is considered one of the most delicious fishes for the table, the former is seldom carried to the table.

ORD. II—MALACOPTERYGII ABDOMINALES.
Soft rayed abdominal fishes.

The Malacopterygii are distinguished by having nearly or quite all of the fin-rays soft and branching as in the trout, and the order abdominales embraces the soft-rayed fishes, whose ventral fins are situated far back upon the abdomen, as in the trout, sucker and pickerel.

I.—CYPRINIDÆ, OR CARP FAMILY.
GENUS CATASTOMUS.—LE SUEUR.

Generic Characters.—Back with a single dorsal fin; gill membrane three rayed; head and opercula smooth: jaws toothless and retractile; mouth beneath the snout; lips plaited, lobed, or carunculated, suitable for sucking; throat with pectinated teeth. This Genus embraces the Suckers of the United States, of which there are about 20 species.

THE CARP SUCKER.
Catastomus cyprinus.*—LE SUEUR.
Jour. Acad. Sci. Phil., vol. I. p. 91, plate.

DESCRIPTION.—Form gibbous; back arched, thin and sharp; belly thick and flattened between the pectoral and ventral fins. Head small and sloping; snout short; eyes rather small, pupil black, iris golden yellow; nostrils large and double; mouth small and lunated. Color light silvery brown, with golden reflections above, approaching to yellowish white, or cream color below. Scales very large, excepting along the base of the dorsal fin, of a semi-rhomboidal form, and beautifully radiated; the lateral line first bends downward, then nearly straight; 40 scales on the lateral line and 13 in the oblique row, extending from the beginning of the dorsal to the middle of the ventral fin. Fins brownish flesh-color, all the rays coarse; the dorsal commences at the highest part of the back, a little forward of the ventrals, and terminates nearly

---

* This species was removed by Cuvier from the genus *Catastomus*, of Le Sueur, to his own sub-genus *Labeo*, which is distinguished from the Catastomus by the greater length of the dorsal fin.

over the middle of the anal, three or four of the first rays being much elongated, the others short; the anal fin slightly lunated, the caudal forked with pointed lobes. The swimming bladder divided in three sacks, connected by tubes. Length of the specimen before me from the snout to the extremity of the tail 16 inches,—to the tail 13, to the vent 10,—to the middle of the gill opening 3½; greatest depth 5; greatest thickness 2½; height of the front part of the dorsal 4½; length of the dorsal 5, scale on the side .8 by .7.

Rays, Br. 3, P. 16, V. 10, D. 28, A. 9, C. 18.

HISTORY.—This fish, though said to be common further south, is only occasionally taken in our waters, and here varies from 1 to 3 or 4 pounds in weight. It is considered a very good fish for the table, but like the others in this family it is wanting in firmness.

THE LAKE MULLET.

Catastomus oblongus.—MITCHELL.

Cyprinus oblongus—Mitchell. Trans. Lit. and Phil. Soc. of N. Y., 1--459.

DESCRIPTION.—Form gibbous; back arched; body deep and thick; head short and smooth; mouth under, small and toothless; gill openings narrow. Color above dark brown, lighter with bronzy reflections on the sides, and dirty cream-color beneath; scales large with radiating striæ, and arranged in about 13 longitudinal rows on each side; lateral line medial and nearly straight, but not very conspicuous. Dorsal fin brownish, the other fins lighter and usually more or less ruddy; pectorals, situated low and far forward upon the throat; ventrals under the middle of the dorsal; the anal reaching the base of the caudal; tail deeply forked; swimming bladder in three sacks connected by tubes. Length of the specimen before me 25 inches, depth in front of the dorsal 6, thickness 3, height of the dorsal 3.2. Weight 6¼ lbs.

Rays, B. 3, P. 17, V. 9, D. 16, A. 9, C. 18.

HISTORY.—This fish is described by Dr. Mitchell under the name of the *Chub of New York*. It is here very generally known by the name of *Mullet*, under which name several species of lake suckers are confounded, although it belongs to a family of fishes entirely distinct from the real Mullet. This is one of our most common fishes, and in the spring and early part of summer is caught with the seine in large quantities, both in lake Champlain and in the mouths of its larger tributaries. The flesh of this fish is rather soft, and is considerably filled with the knots of fine bones so common to this family, and yet it is regarded as a very good fish for the table. There are various methods of cooking it, but it is generally most highly esteemed when baked. The fish grows to a larger size, and is taken in lake Champlain in larger quantities than any other species of this family. Their usual length is from 15 to 20 inches, and their weight from 2 to 5 pounds. But individuals are often taken which are much larger, weighing, in some cases, 9 or 10 pounds. The usual price, when fresh, is from 3 to 4 cents a pound.

THE SUCKER.

Catastomus teres.—MITCHELL.

Cyprinus teres -Mitchell. Trans. Lit. and Phil. Soc of N. Y., I--459.

DESCRIPTION.—Body lengthened, thick and subcylindrical, the head one-sixth the total length; color blackish brown above, darkest on the head, often tinged with green; sides brownish, often with golden reflections from the scales; belly white, and sometimes yellowish; dorsal and caudal fin brown; the other fins ruddy, or yellowish brown. Head rather small, and with the cheeks and opercula smooth; eyes small, iris golden, but very dark in some specimens; nostrils large, double and very near the eye in front. Scales of middling size, radiated, with 17 in the oblique row extending from the anterior base of the ventral to the posterior ray of the dorsal, the middle scale being crossed by the lateral line which is straight in the middle of the body, and contains 61 scales. Pectoral fins situated very near the gills, the dorsal on the middle of the back, and about as long as high; the ventrals rather small, under the middle of the dorsal; the anal far back, reaching the base of the caudal, and its length contained 2½ times in its height; the tail forked; all the fin rays coarse, particularly those of the anal fin. The swimming bladder in two sacks connected by a tube. Length of the specimen before me 22½ inches, from the snout to the posterior edge of the gill covers 4.4, from the gill to the base of the tail along the lateral line 15. Its greatest depth 5.4, thickness 3, and its weight 5¼ lbs.

THE BLACK AND LONG-NOSED SUCKERS. THE DACE.

Rays, Br. 3, P. 18, V. 10, D. 13, A. 8, C. 18.

HISTORY.—This is generally known on the west side of the Green Mountains by the name of Sucker, or Black Sucker, while another species is known by the same names on the east side of the mountains. This fish is quite common in lake Champlain, and in most of the large streams and ponds connected with it.

THE BLACK SUCKER.
Catastomus nigricans.—LE SUEUR.
Jour. Acad. Nat. Science, 1—102. Storer's Report, Fishes of Mass., p. 86.

DESCRIPTION.—Color of the back black; sides reddish yellow with black blotches; beneath white, with golden reflections; scales moderate in size; head quadrangular, one fifth the length of the fish; top of the head of a deeper black than the body; eyes moderate, oblong; pupils black; irides golden; mouth large; corrugations of the lips very large, particularly those of the lower lip; lateral line, rising back of the operculum on a line opposite the centre of the eye, makes a very slight curve downwards and then pursues nearly a straight course to the tail, and contains 60 scales; back between the head and dorsal fin rounded; pectoral, ventral and anal fins reddish; caudal and dorsal blackish; height of the dorsal equal to two thirds its length; third and fourth rays of the anal reach the base of the caudal. Length of the specimen from which the description is drawn 15 inches.

Rays, D. 13, P. 18, V. 9, A. 8, C. 18.—*Storer.*

HISTORY.—This I suppose to be the common Sucker on the east side of the Green Mountains in this state; but not having obtained any good specimen of it, I have copied above Dr. Storer's description, which was made from a specimen obtained from Walpole. They frequently weigh 3 or 4 pounds, and exceed 20 inches in length.

THE LONG-NOSED SUCKER.
Catastomus longirostrum.—LE SUEUR.
Journal Academy Nat. Sciences, Phil., 1—102.

DESCRIPTION.——Body sub-cylindric, straight, delicate; head flat; eyes large, irides yellowish white; aperture of the mouth greatly arcuated, and large; scales very small and roundish; color of the body above reddish, paler on the sides; abdomen white, with a bluish tint; lateral line curved above the pectoral fin. Dorsal fin deeper than broad, quadrangular; the extremity of the anal fin does not reach the base of the caudal; head horizontal, terminated in a long snout. Length of the individual described 5 inches.

Rays, P. 16, V. 9, D. 12, A. 7, C. 18.—*Le Sueur.*

HISTORY.—"This fish I discovered," says Le Sueur, "in the state of Vermont; I have not seen it in any other state." Not having met with this fish, I can only give Le Sueur's account of it.

GENUS LEUCISCUS.—*Klein.*

Generic Characters.—The dorsal and anal fins short and without strong rays at the commencement of either: no cirri.

This genus embraces those fishes which are generally known in New England, by the names of Dace, Club and Shiner.

THE COMMON DACE.
Leuciscus pulchellus.—STORER.
Storer's Report on Fishes of Massachusetts, p. 91.

DESCRIPTION.—Upper part of the head and tail blackish; back approaching to olive; sides lighter; belly white; cheeks, gill covers and lower fins more or less ruddy; scales striated, exhibiting a most beautiful play of green, blue, golden and silvery reflections. A dark colored membrane visible at the junction of the scales, giving the sides of the fish a reticulated appearance; 49 scales on the lateral line, which begins near the upper part of the gill-opening, bends rapidly downward through 9 scales, and then pursues a straight course to the tail. Head and operculum smooth, the latter with cupreous reflections. Scales rather large and much crowded above the pectoral fins. Eyes small, pupil black, surrounded by a golden line which fades into gray on the iris. Mouth large; lips, tongue and palate fleshy; jaws toothless; two patches of pectinated teeth in the throat, with four teeth in each. Ventral fins under the front of the dorsal; the anal fin twice its length from the caudal; the two first rays short and closely applied to the third in the dorsal and anal fin. Swimming bladder in two sacks connected by a tube. Length of the specimen before me 17 inches—from the snout to the posterior part of the operculum 3½—to the vent 9½. Total length 4½ times the greatest depth.

Rays, Br. 3, P. 16, V. 8, D. 10, A. 10, C. 19.

HISTORY.—This fish is quite common in lake Champlain and its tributaries. It is readily caught with the hook, and the flavor of its flesh is agreeable, but it is so soft and filled with small bones that it is not much valued as an article of food. The length of those usually taken varies from 5 to 12 inches, but they sometimes grow to the length of 20 inches.

THE SHINER.

Leuciscus crysoleucas.—MITCHELL.

Trans. Lit. and Phil. Soc. of N. Y., p. 459.
Fauna Boreali Amer. Fishes, page 122.
Storer's Report, Fishes of Mass., page 88.

DESCRIPTION.—Form ovate; body deep and thin, the depth contained 4 times in the total length. Color greenish above, lighter on the sides and yellowish white beneath; a very broad indistinct yellowish or cupreous stripe along the side to the middle of the tail. The fins of a dull yellow color, with the extremities of the dorsal, caudal and anal fins and the first ray of the pectoral more or less black; cheek and operculum with yellow and silvery reflections; scales rather large, radiated, crossed by concentric undulations, or striæ; the whole side exhibiting blue, green, cupreous, yellow and silvery reflections, according to the direction of the light. Eyes large; iris bright yellow. Head and gill covers smooth, mouth in front of the eyes, small, toothless, and directed upwards. The lateral line commences near the upper part of the gill opening, bends downwards and passes along nearly parallel to the curve of the abdomen, to the tail, being only one third as far from the belly as from the back at the ventral fin Swimming bladder in two sacks. Length of the pectoral fins to their height as 2 to 7; ventrals before the dorsal with slender bracts above their base; dorsal fin medial, its length being to the height of the anterior part as 1 to 2; the anal fin commences under the termination of the dorsal, its length being to the height of the anterior part as 7 to 6; tail large and forked. Length of the specimen before me 4.6 inches; depth 1.1.

Rays, Br. 3, P. 17, V. 8, D. 10, A. 15, C. 19.

HISTORY.—This fish is quite common, particularly in the small ponds and coves along the shore of lake Champlain, and about the mouths of our large streams, where it is found associated with perch, bull-pouts and mud fishes.

THE BROOK MINNOW.

Leuciscus atronasus.—MITCHELL.

Trans. Lit. and Phil. Soc. p. 460. Storer's Report on Fishes of Mass., p. 92.

DESCRIPTION.—Body rather thick and deep through the abdomen; head a little flattened above, and narrowed towards the snout. Color above brownish olive spotted with black; beneath white with cupreous and silvery reflections, and sometimes red; a dark band passes round the nose, crosses the eye, passes along the sides and through the middle of the tail, which is forked; above this band is usually a yellowish stripe; eyes middling size; iris bright yellow, where it is not darkened by the above mentioned dark band. The lateral line commences on the nape of the neck, passes obliquely downwards across the dark band on the side and along the lower margin of the band to the tail. Nostrils large, double and tubelar. Dorsal fin behind the ventrals and twice as high as it is long. Vent medial and under the posterior rays of the dorsal fin. Fins brownish yellow. Swimming bladder in two sacks connected by a tube. Length 2¼ inches; head a little more than one sixth of the total length.

Rays, Br. 3, P. 12, V. 7, D. 7, A. 7, C 19.

HISTORY.—This species is quite common in most of the streams in Vermont, and particularly so in those that fall directly into lake Champlain. It is an active, lively little fish, and on account of the stripes on its sides, the colors of which are changeable, according to the direction of the light falling upon them, it is one of our most beautiful fishes. When fully grown this fish is only from 2¼ to 3 inches long, and, though found in great numbers, its diminutive size renders it of no account as an article of food. It is chiefly sought to be used as bait for Pike and other large fishes.

The *Exoglosson nigrescens*, described by Rafinesque in the Journal of Academy Nat. Sci., Phil., I—422, which he says he found in lake Champlain, and several others of this family, which I know to exist in our waters, I have thought it best to omit, because I cannot speak of them with confidence without further examination.

GENUS HYDRARGYRA.—*Le Sueur.*

Generic Characters.—Ventral fins 6 rayed; teeth in the jaws and throat; those of the jaws conic and recurved; none in the palate; jaws protractile; lower jaw longer than the upper one; one dorsal fin, situated nearer the tail than the head, opposite to the anal fin; scales on the opercula and body; head flat, shielded above with large scales, the centre scale largest.

THE MUD FISH.
Hydrargyra fusca.

DESCRIPTION.—Color above dark olive, mottled with blackish; sides mottled or variegated with brown, green and golden, with faint indications of yellowish bars; belly dull brownish, bronzy yellow; fins dusky yellow; sides yellowish at the base of the tail, crossed by a vertical black bar, with a brownish, crescent-shaped line along the base of the caudal rays, making, with a vertical line, the form of the letter D. Form thick and plump; head slightly flattened above; upper jaw shorter than the lower, and broadly truncated; lower jaw curved upward and rounded; mouth slightly cleft; teeth in both jaws and front part of the vomer, small, crowded, and incurved; four patches of short, conical teeth in the throat. Eyes moderately large, pupil black, iris yellow, cornea very prominent and clear. Scales on the body, head, cheeks and operculum; those on the back part of the head largest. Tail fully rounded, a little shorter than the head, which is a little more than one-fifth the total length of the fish. Ventral fins small, medial, and slightly in advance of the beginning of the dorsal; anal fin under the posterior part of the dorsal and about as high as long; the dorsal nearly twice as long as high, and about its length from the caudal. The dorsal and anal have their first rays short and closely applied to the second ray; outer rays of the caudal also very short. Length of the longest of 12 specimens before me 4½ inches; greatest depth .8; thickness .5.

Rays, Br. 4, P. 15, V. 6, D. 14, A. 10, C. 16.

HISTORY.—These fishes exist in considerable numbers in the marshes and coves along the margin of lake Champlain, and of the rivers which fall into it. They are very tenacious of life, and live longer than most fishes without water. During droughts, as the waters subside and recede from the coves, they have the power, by a springing motion, of transporting themselves from one little puddle to another. They also have the power of partially burying themselves and living in the mud and among the moist grass-roots, after the other small fishes associated with them are all dead for the want of water. In these situations vast numbers of them are devoured by birds, muskrats, and foxes. In severe droughts, like that of 1841, the quantity of small fishes which die in consequence of the drying up of the coves, is exceedingly great. In one small cove, which I visited on the 24th of September, 1841, I found *Mud Fishes* and other small fishes dead in piles, in the low places which had become dry. One small portion of the cove, still covered with water and leaves to the depth of 4 or 5 inches, was literally filled with fishes struggling together for existence. This portion amounted to about one square rod, and in this space there could not have been much less than a barrel of fishes. They consisted of pickerel, yellow perch, shiners, bull pouts and mud fishes, but mostly of the two last. My feelings were really pained at the sight, and moved by compassion for the poor fishes, I heartily wished for rain, which, on the next day, came in abundance, to the joy, not only of the fishes and their sympathizers, but of the whole country.

II.—ESOCES, OR PIKE FAMILY.
GENUS ESOX.—*Linnæus.*

Generic Characters.—Snout elongated, broad, depressed, and obtuse; sides of the lower jaw with long acute teeth; intermaxillaries, palate, vomer and tongue studded with small teeth; a single dorsal fin, situated far back and over the anal fin.

THE COMMON PIKE.
Esox estor.—LE SUEUR.

Journal Acad. Nat. Sci., Phil., 1-419.
Esox lucius, Rich. Fauna Boreali, p 124.

DESCRIPTION.—Body thick, somewhat four-sided; back nearly straight from the head to the dorsal fin, and parallel to the abdomen. Color of the back blackish green; sides lighter, with violet and silvery reflections and several longitudinal rows of rounded and oblong yellowish spots; belly pearly white. Head one fourth the total length, flattened or concave on the upper part, and of a dark bottle green color; large pores on the head and lower jaw; upper jaw broad, flattened and thinned down to an edge at the extremity; lower jaw reflected and longer than the upper; tongue truncated at the extremity; teeth on the tongue, vomer, palatine bones and jaws, of different sizes, and either straight or hooking inwards; eyes lateral, close to the crown, and mid-way between the gill opening and end of the snout; pupil surrounded by a golden line and grayish iris. Scales small, often emarginate, and towards the back marked with bright lines in the form of the letter V. Lateral line nearly straight, nearer the back than belly, and formed by a deep notch in every 3d or 4th scale; usually several irregular rows of these notched scales on the sides resembling lateral lines. *Fins* all marked with brownish and yellow, and usually more or less ruddy except the dorsal; pectoral and ventral fins small; the posterior attachment of the ventrals medial; vent under the front part of the dorsal, and anal fin under the posterior part; tail forked. Preoperculum irregular, narrow in the middle; operculum quadrangular, scaly on the upper part; suboperculum narrow, and a little longer than the operculum; interoperculum small and mostly concealed. Length of the specimen before me 17 inches—to the pectorals 4, ventrals 8, anal 11½.

Rays, Br. 15, P. 13, V. 10, D. 18, A. 16, C. 19.

HISTORY.—This species is very common in lake Champlain and all its larger tributaries. It is generally known in Vermont by the name of *Pickerel.* About the north end of the lake and in Canada generally it is called the *Pike,* on account of its resemblance to the English Pike. Indeed the resemblance is so close that Dr. Richardson regards them as identical, and has described our Pike in his Fauna Boreali Americana under the name of the foreign species, *Esox lucius,* but they are generally regarded by naturalists as distinct species. This fish grows to a large size, frequently exceeding 30 inches in length, and weighing 10 or 12 pounds. It is very voracious, and devours great numbers of reptiles and small fishes. It is taken both with the hook and seine, and is considered a very good fish for the table. The fishermen say that there is another fish of this family in lake Champlain, which they call the *Maskalongè.* If so, it is probably the fish which Richardson *(Fauna Boreali, p. 127)* calls *E. estor,* Maskinonge. I lately received one which was sent me as a Maskalonge, but which proved to be only a plump specimen of the Common Pike.

THE PICKEREL.

Esox reticulatus.—LE SUEUR.
Journal Academy Nat. Sci., I—414.
Storer's Report, Fishes of Mass., p. 97.

DESCRIPTION.—Color variable from greenish brown to brilliant golden, but in all cases marked with irregularly distributed longitudinal lines; beneath white. Snout obtuse; gape of the mouth great; lower jaw longer than the upper; teeth in front of the lower jaw small, on the sides large and pointed. Eyes moderate in size, pupil black, iris yellow; nostril double; fins greenish; the pectoral and anal reddish after death; dorsal fin longer than the anal; pectorals commence on a line with the 16th branchial ray; vent large, 2 lines in front of the anal fin; from the dorsal fin to the commencement of the caudal 2 inches. Length of the specimen from which the above description was made 16 inches; head about one fourth the length of the body; width of the head in front of the eyes equal to half its length.

Rays, B. 17, D. 18, P. 13, V. 11, A. 17, C. 19.—*Storer.*

HISTORY.—This is the Common Pickerel on the east side of the Green Mountains in Vermont, as the preceding species is on the west side. It is found in Connecticut river and most of its larger tributaries, and it has multiplied exceedingly in several ponds to which it has been transported by the inhabitants in the neighborhood. This is the Common Pickerel of Massachusetts and the other New England states.

III.—SILURIDÆ OR CAT-FISH FAMILY.

GENUS PIMELODUS.—*Lacepede.*

Generic Characters.—Body covered with a naked skin; no lateral armature; jaws and often palatine bones furnished with teeth, but there is no exact row of teeth on the vomer parallel to that on the upper jaw. The form of the head varies exceedingly, as well as the number of cirri. Two dorsal fins, the second adipose.

THE BULL POUT.

Pimelodus vulgaris.

Silurus catus, Mitch. Trans. Lit. Phi. Society of New York, page 433.

DESCRIPTION.—Body without scales, covered with a mucous skin, tapering and cylindrical; head large, broad, depressed, color above dark, approaching to black; sides dark olive, or fuliginous, the color rubbing off or becoming lighter after being taken from the water; belly dirty white, often tinged with red; fins dark, often purplish; mouth broad; under jaw longest, and a broad band of small conical teeth in each; cirri 8, 4 in a row upon the under lip, the two outer ones nearly twice as large as the middle ones, one still larger at each angle of the mouth, and a small one at each nostril; the first dorsal ray and the first ray in each pectoral fin a strong spine, with the point free and sharp. A bony process projects backward over the base of the pectoral fin. Tail slightly rounded. Length of the specimen before me 12½ inches, width of the head 2.3, depth of the body 1.8, thickness 1.6.

Rays, B. 7, P. 1|7, V. 8, D. 1|6—0, A. 20, C. 17.

HISTORY.—This fish, which is quite plentiful in lake Champlain, is here generally known by the name of Bull Pout. Those taken from the lake are usually from 9 to 13 inches in length. For the table they require skinning like the Eel; but, though their flesh is tender and well flavored, there is so much waste in dressing, because of the great size of the head, that very little account is made of them as an article of food. This fish I suppose to be the species described by Dr. Mitchell under the name of *Silurus catus*, but whether it is the *Pimelodus catus* of Le Sueur, I have no means of judging, never having seen his description.

THE HORNED POUT.
Pimelodus nebulosus.—LE SUEUR?
Memoires du Mus. d'Hist Nat., V—149. Storer's Report, page 102.

DESCRIPTION.—Color dark olive, or fuliginous, darkest on the head and back, yellowish or cupreous on the sides, approaching to ruddy white on the belly; fins mostly ruddy at the base and brownish towards the extremity; head flattened above; upper jaw rather longest; both jaws furnished with numerous small conical teeth; 8 cirri about the head, 2 short ones at the nostrils, 4 longer ones on the chin, and 2 much longer, being 1.1 inch, extend backward from the angles of the mouth, and terminate in a fine filament. Spine of the 1st dorsal articulated, and free at the point; spines of the pectorals also free at the point, and strongly serrated interiorly; adipose fin over the posterior part of the anal. Tail nearly even. Length of the specimen before me 4½ inches, width of the head .8. Body much flattened vertically towards the tail.

Rays, B. 7, P. 1|7, V. 8, D. 1|5, A. 20, C. 17.

HISTORY.—This fish is common in Connecticut river, and in many of its larger tributaries. The specimen from which my description was drawn was taken in Connecticut river at Barnet It is there called the Pout, or Horned Pout. Having had an opportunity to compare only this one small specimen from Connecticut river with the Bull Pout found in lake Champlain, I am not prepared to say with confidence that they do not both belong to the same species; but as this specimen differs from the lake fish in having its body more flattened towards the tail, in having its upper jaw longest instead of shortest, in having the cirri at the angles of the mouth proportionally longer and the adipose fin more distant from the tail, I have introduced them as distinct species.

THE CAT FISH.
*Pimelodus * * * * *.*

DESCRIPTION.—Color dark smoky brown approaching to black above; cupreous or fuliginous on the sides; belly dull ruddy white; skin scaleless and smooth; fins dull smoky brown, more or less ruddy below. Head slopes gradually from the nape of the neck to the snout, which, as well as the head, is narrower and more pointed than the preceding species; the body also is more elongated; 8 cirri in the usual situations, all blackish excepting the two middle ones on the under lip which are flesh-colored, and not more than half as large as the two outer ones; those at the angle of the mouth very long, reaching beyond the pectorals half way to the ventral fins; those at the nostrils smallest. Mouth narrow, with the upper jaw overlapping the lower; teeth small, conical and numerous. Bony spine in the pectoral fin very strong, with about 20 sharp teeth on the posterior edge, and a strong bony process lying over the base of the fin; first dorsal mid-way between the pectorals and ventrals, twice as high as long, spine more slender than in the pectorals; height of the adipose fin 1 inch, situated over the posterior half of the anal, which is long and slightly rounded; tail rather deeply forked with spreading, pointed lobes; lateral line indistinct. Length of the specimen before me, which was caught in Winooski river,

18 inches; from the snout to the pectoral 2½; to the first dorsal 4½; width of the head 2.4, longest cirri 4.3.

Rays, B. 8, P. 1|7, V. 8, D. 1|6—0, A. 25, C. 18.

HISTORY.—When I prepared my list of fishes at the beginning of this chapter, I supposed our Cat Fish to be the *P. cœnosus* of Richardson. Upon re-examination, since that list was printed, I find our fish does not agree with his description, and I am now satisfied that it does not belong to that species. It is probably one of the eight species described by Le Sueur in the *Memoires du Museum d'Histoire Naturelle*, at Paris, but not having access to that work, I am unable to designate the species, or to say with certainty that it is embraced among those there described. This species is only occasionally taken in the vicinity of Burlington, but is regarded as very good fish for the table. In some parts of lake Champlain it is said to be quite plentiful.

IV.—SALMONIDÆ—SALMON FAMILY.

GENUS SALMO.

Generic Characters.—Head smooth; body covered with scales; two dorsal fins, the first supported by rays, the second fleshy, without rays: mouth large; sharp teeth on the jaws and tongue; branchial rays usually about ten; ventral fins opposite the centre of the first dorsal one.

THE SALMON.

Salmo salar.—LINNÆUS.

DESCRIPTION.—Color bluish silvery above, lighter on the sides and white beneath; black blotches upon the sides, much more numerous above the lateral line, for the most part surrounding the outline of the scales, leaving the color of the body unchanged; the spots upon the scaleless head are unbroken, and of a deeper color. Length of the head equal to one fifth the length of the fish; head sloping, darker colored above than the back of the specimen. Gill covers light silvery colored. Eyes small, pupil black, irides silvery; diameter of the eye equal to one fourth the distance between the eyes. Nostrils nearer the eyes than the extremity of the snout. Upper jaw longest, receiving into a notch at its middle the prominent tip of the lower jaw; both jaws, the palatine bones, vomer and tongue armed with sharp incurved teeth; lateral line nearly straight. The first dorsal fin commences on the anterior half of the body, height of its first rays equals its length; dark colored, with longitudinal rows of black blotches upon its base; length of the adipose fin equals one third its height: pectorals arise in front of the posterior angle of the gill covers; length equals one fourth their height; ventrals on a line opposite the middle of the dorsal, having on their sides a large axillary scale; anal fin white, higher than long; caudal dark brown, forked.

Rays, D. 12, P. 15, V. 9, A. 10, C. 19. —*Storer*.

HISTORY.—The Salmon, formerly very plentiful in nearly all the large streams in this state, is now so exceedingly rare a visitant that I have not been able to obtain a specimen taken in our waters, from which to make a description for this work. They have entirely ceased to ascend our rivers, and only straggling individuals are now met with in lake Champlain. I have heard of only one being taken here during the past summer, and that I did not see. The causes which have been principally operative in driving these fishes from our waters have already been mentioned. When the country was new, according to Dr. Williams, there was a regular and abundant migration of these fishes to and from our waters, in spring and autumn.* They came up Connecticut river about the 25th of April, and proceeded to the highest branches. Shortly after they appeared in lake Champlain and the large streams which fall into it. So strong is their instinct for migration, that, in ascending the streams, they forced their passage over cataracts of several feet in height, and in opposition to the most rapid currents. They were sometimes seen to make six or seven attempts before they succeeded in ascending the falls. When thus going up in the spring they were plump and fat, and of an excellent flavor; and from the beginning of May to the middle of June they were taken in great numbers. When they arrived in the upper parts of the streams they deposited their spawn. Towards the end of September they returned to the ocean, but so emaciated and lean as to be of little account as an article of food. In the spring, salmon were often taken weighing from 30 to 40 pounds.

THE NAMAYCUSH, OR LONGE.
Salmo namaycush.—PENNANT

DESCRIPTION.—Form resembling the

* History of Vermont, vol. 1, page 147.

Salmon; head flattened and slightly convex between the eyes; greatest depth contained about five times in the total length. Color dark bluish brown above approaching to black on the head; sides thickly spotted with roundish, yellowish gray spots on a dark brownish gray ground, the spots unequal, but usually about the size of a small pea; belly yellowish white; fins dark brown mottled with yellowish white; the pectorals, ventrals and anal slightly tinged with orange yellow. Lateral line plain, prominent and nearly straight. Scales small and thin, but much larger than on the Brook Trout. *Eyes* midway between the tip of the snout and the nape, and twice as near the former as to the hind edge of the gill cover, the measurement being made from the centre of the pupil; iris yellowish. *Nostrils* nearer the eye than the tip of the snout, double, orifices nearly equal, the anterior having a raised margin. *Jaws* equal, strong, and armed with incurved, sharp, conical teeth; similar teeth on the front part of the vomer, on the palate bones, and two rows on the tongue, with a deep groove between them. *Preoperculum* but little curved, and nearly vertical, suboperculum large and finely grooved. The *dorsal fin* medial, higher than long, and the ventral situated nearly under the middle of it; adipose fin club-shaped and nearly over the posterior ray of the anal; the anal higher than long, the anterior part being three times the height of the posterior; tail forked, with pointed lobes. Length of the specimen before me 23½ inches—to the posterior edge of the operculum 5¼—to the beginning of the dorsal 10½—to the vent 15—weight 4 pounds.

Rays, B. 12, P. 15, D. 11, V. 9, A. 11, C. 19.

HISTORY.—This species of Trout bears considerable resemblance to the *Salmo trutta*, or Salmon Trout, of Europe, and being mistaken for that fish by the first European settlers of this country, it has since usually borne the name of *Salmon Trout*. In the northern parts of this state and in the eastern townships in Canada, it is at present extensively known by the name of *Longe*. In Pennant's Arctic Zoology, and by the fur traders at the northwest, its more common appellation is *Namaycush*, or *Namaycush Salmon*. It is called by Dr. Mitchell the Great Lake Trout, and he describes it under the scientific name of *Salmo amethystus*.* This magnificent trout equals or surpasses the Common Salmon in size, and is found in most of the lakes and large ponds in the northern parts of North America. In the great lakes at the northwest it is often taken weighing from 30 to 60 pounds, and according to Dr. Mitchell, it has been taken at Michilimackinac of the enormous weight of 120 pounds. This fish was formerly common in lake Champlain and in several ponds in the western part of the state, but, like the Salmon, it is now rarely caught in those waters. It is, however, still found in considerable plenty in several ponds in the northern part of Vermont, particularly in Orleans county. Bell-water pond in Barton, and several ponds in Glover, Charleston, &c., are much celebrated on account of the fine Longe which they afford. These usually vary from half a pound to 10 pounds, but are often much larger. Individuals are said to have been taken recently in Glover weighing 25 pounds, and in Charleston exceeding 40 pounds.

This fish passes most of the time in the deepest parts of the lakes and ponds, but according to Dr. Richardson, resorts to the shallows to spawn in October. It is a very voracious fish, and is sometimes termed the tyrant of the lakes. It is taken with the hook and line, and is also speared by torch light. Its flesh is of a reddish yellow color, and is very much esteemed as an article of food. Roasting is said to be the best method of cooking it. "The Canadian voyageurs are fond of eating it raw, in a frozen state, after scorching it for a second or two over a quick fire, until the scales can be easily detached, but not continuing the application of heat long enough to thaw the interior."*

THE BROOK TROUT.
Salmo fontinalis.—MITCHELL.

DESCRIPTION.—Color above brown, with darker markings, fading into white or yellowish white on the belly; sides with numerous roundish yellow spots of unequal size, but usually about the bigness of a small pea; and also very small bright red spots commonly situated within the yellow ones. These red spots are extremely variable, being very few in some specimens and numerous in others. The caudal and first dorsal fin transversely banded or mottled with black. Head one seventh the total length, darker colored than the back. Eyes large, iris silvery. Teeth hook inward, on the jaws, tongue, palatine bones and vomer; those on the tongue largest. Jaws equal. Scales very

---

* Jour. of the Acad. Nat. Science, Philadelphia, Vol. 1, page 410.

* Richardson's Fauna Boreali Americana, vol. III, page 180.

minute. Lateral line straight. First dorsal fin on the anterior half of the body; adipose fin small, brownish yellow margined with black, and behind the anal; pectorals under the posterior part of the operculum; ventrals under the middle of the first dorsal; first ray of the anal, ventral and pectoral fins white; the second or third ray usually black, the rest of the fin reddish. Tail slightly forked.

Rays, Br. 11, P. 13, V. 8, A. 10, D. 10, C. 19.

HISTORY.—The Brook Trout is more generally diffused over the state than any other species of fish; there being scarcely a brook, or rill of clear water, descending from our hills and mountains in which it is not found. When the country was new they also abounded in the larger streams, where they often grew to the weight of two or three pounds. But they have been diminished by the causes already mentioned, and have been sought after with such eagerness as the most delicious article of food of the fish kind, that they are now seldom taken in our streams exceeding half a pound in weight, and much the greater number of them weigh less than a quarter of a pound. In many of the ponds they are still taken of of a larger size, but their flavor is thought to be less delicious than that of those taken in running water, especially in ponds with muddy bottoms. The rapidity with which this and other species of fishes multiply under favorable circumstances was exemplified in an astonishing manner at an early day, in Tinmouth, in this state. 'A stream which was about 20 feet wide, and which, like other streams, contained trout and suckers of the ordinary size and number, had a dam built across it for the purpose of supplying water for a saw mill. This dam formed a pond, which covered, by estimation, about 1000 acres, where the trees were thick and the soil had never been cultivated. In two or three years, the fish were multiplied in this pond to an incredible number. At the upper end, where the brook fell in, the fish were to be seen in the spring running over one another, so embarrassed by their own numbers as to be unable to escape from any attempt made to take them. They were taken by the hands at pleasure, and swine caught them without difficulty. With a small net the fishermen would take half a bushel at a draught, and repeat their labors with the same success. Carts were loaded with them in as short a time as people could gather them up when thrown upon the banks; and it was customary to sell them in the fishing season for a shilling a bushel. While they thus increased in numbers they also became more than double their former size. This great increase of fishes is supposed to have been occasioned by the increased means of subsistence, in consequence of carrying the water over a large tract of rich and uncultivated land.' *

The trout is usually taken with the hook, and the bait universally used is the red earth worm, every where known by the name of *Angle Worm*. Fishing for trout is a favorite and common amusement, and parties frequently go 15 or 20 miles for the sake of indulging in it.

GENUS OSMERUS.—*Artedi.*

Generic Characters.—Body elongated, covered with small scales; two dorsal fins, the first with rays, the second fleshy without rays; ventral fins under the front part of the first dorsal; teeth long on the jaws and tongue, two distinct rows on the palatine bones, but none on the vomer, except at the most anterior part; branchial rays eight.

THE SMELT.
Osmerus eperlanus.—ARTEDI.
Yarrell's British Fishes, II—75, fig.
Journal Acad. Nat. Sci., Phil., I—230.
Fauna Boreali Amer., Fishes, page 185.
Storer's Report, Mass. Fishes, page 108.

DESCRIPTION.—Semi-transparent, color silvery, greenish above and white beneath; top of the head and edges of the jaws blackish; under jaw longest, with a keel-shaped projection near its extremity; teeth on the tongue and palate, and two rows on each jaw, mostly large and hooking inwards; mouth large; nostrils very large and nearer to the snout than to the eye. Eye rather large, iris silvery; lateral line straight. Scales of moderate size, thin and transparent. Fins slender and transparent; the dorsal, caudal, and upper edges of the pectoral brownish; all the rest white and delicate; height of the first dorsal twice its length; ventrals under the first rays of the dorsal; tail forked, with spreading, pointed lobes. Length of the longest of two specimens before me 9 inches, greatest depth 1¼ inch.

Rays, B. 8, P. 11, V. 8, D. 11, A. 15, C. 17.

HISTORY.—The Smelt is one of those migratory species of fishes, which pass a part of the time in salt water and a part

---

* Williams' History of Vermont, vol. 1, p. 149.

in fresh. Though not a constant visitant in our waters, he occasionally makes his appearance, and is sometimes taken in lake Champlain in very considerable numbers. The form of this fish is long and slender, and its bright silvery hue renders it very beautiful. It is sometimes taken with the hook, but more commonly with the net, and is very highly esteemed as an article of food. In Massachusetts, according to Dr. Storer's Report, 750,000 dozen of these fishes are taken annually in Watertown alone, and sent to Boston market.

Genus Coregonus.

Generic Characters.—Head small; mouth small and edentate, or furnished with very small teeth; scales large; length of the first dorsal fin less than the height of its anterior portion, second dorsal adipose and without rays; branchial rays seven or eight.

WHITE FISH, OR LAKE SHAD.
Coregonus albus.—Le Sueur.

Journal Academy Nat. Sci., Phil., I.-332.
Fauna Boreali Amer., Fishes, page 195, fig.
Boston Journal Natural History, III—477, pl. 28.

Description.—Form ovate, slightly tapering towards the tail; body deep and thick; head pointed, and with the mouth, very small; teeth in the jaws few, and so minute as scarcely to be perceptible to the sight or touch in the recent specimen; color silvery, bluish gray on the back, lighter on the sides, and pearly white on the belly, with a delicate iridescent play of colors throughout. Scales large, thin, pearly and very deciduous, arranged in about 20 longitudinal rows, giving the fish a slightly striped appearance; lateral line very nearly straight; fins small, brownish, often tinged with red; the dorsal mid-way between the snout and the extremity of the tail; the posterior rays of the dorsal and anal fins much shorter than the anterior, giving those fins a triangular appearance; adipose fin rather large; caudal forked and spreading; a long, slender bract above and partly behind the ventral fins. Length of the specimen before me, which is considerably larger than the average size and very fat, 22 inches, depth 6, thickness $2\frac{1}{4}$, and weight $5\frac{1}{2}$ pounds.

Rays, Br. 8, P. 15, V. 11, D. 14,—0, A. 14, C. 19.

History.—This fish, though the same as the celebrated White Fish of the western and northwestern lakes, is generally known in Vermont by the name of *Lake Shad*. Its Indian name at the northwest is *Attihawmeg*. This fish is quite common in lake Champlain, and, in some years, is taken in the months of May and June in considerable quantities with the seine. It is also found in many of the small lakes, in Lower Canada, connected with the St. Lawrence on the south side, notwithstanding the assertion of Dr. Richardson * that it does not exist in the St. Lawrence below the falls of Niagara. This is universally considered a most excellent fish, and nearly all are disposed to acquiesce in the opinion of Charlevoix, that, "whether fresh, or salted, nothing of the fish kind can excel it;" but few, I think, will agree with the Baron La Hontau, who says that it should be eaten without any kind of seasoning, because "it has the singular property that all kinds of sauce spoil it." In warm weather this fish should be either cooked, or salted, soon after it is taken, as it quickly becomes soft and is spoiled. It is excelent either boiled or fried. The mode of boiling at the northwest, according to Dr. Richardson, is as follows: "After the fish is cleansed, and the scales scraped off, it is cut into several pieces, which are put into a thin copper kettle, with water enough to cover them, and placed over a slow fire. As soon as the water is on the point of boiling the kettle is taken off, shook by a semi-circular motion of the hand backwards and forwards, and replaced on the fire for a short time. If the shaking be not attended to exactly at the proper moment, or be unskilfully performed, the fish, coagulating too suddenly, becomes comparatively dry to the taste, and the soup is poor." The stomach of this fish is remarkably thick, and when cleansed and cooked is esteemed a great luxury. The White Fish is very thick and fleshy, and on account of the smallness of the head, fins and intestines, the waste in dressing is less than in any other fish. The greater part of those taken in lake Champlain are from 15 to 20 inches in length, and weigh from 1 to 3 pounds, though smaller ones are often taken, and occasionally larger ones, weighing from 3 to 6 pounds. They are usually sold fresh as taken from the water, and the price varies from 6 to 10 cents a pound. The White Fish seems to subsist principally upon small molluscous animals. I have sometimes found more

* Fauna Boreali Americana, vol. III, page 196.

than 100 univalve and bivalve shells in the stomach of a single fish.

V.-CLUPIDÆ or HERRING FAMILY.

GENUS ALOSA.—*Cuvier.*

Generic Characters.—Body compressed; scales large, thin, and deciduous; head compressed; teeth minute, or wanting; a single dorsal fin; abdominal line forming a sharp keel-like edge, which in some species is serrated; upper jaw with a deep notch in the centre; gill rays 8.

THE COMMON SHAD.
Alosa vulgaris.—CUV.

McMurtrie, Cuvier, ii, 235. Yarrell's British Fishes, ii, 136. Storer's Report, Fishes of Massachusetts, page 116.

DESCRIPTION.—Color of the top of the head and back bluish; upper portion of the sides, including the opercula, cupreous; beneath silvery; whole body covered with large, deciduous scales, with the exception of the head, which is naked; eyes large; pupils black; irides silvery; diameter of the eye equal to the distance between the eyes; nostril nearer the eye than the snout; upper jaw notched in the centre; its lateral edges slightly crenated; abdomen serrated; a black blotch at the posterior angle of the operculum; dorsal fin on the middle of the back, shuts into a groove; height equal to two-thirds its length; pectorals silvery; height to the length as 3 to 1; ventrals opposite the middle of the dorsal; anal received into a groove; caudal deeply forked. Length of the head to the whole length of the body as 1 to 6. Usual weight from 1 to 4 pounds.

Rays, D. 19, P. 16, V. 9, A. 20, C. 20. —*Storer.*

HISTORY.—This excellent and valuable fish, which is common both to Europe and America, was formerly taken in Connecticut river in large quantities, particularly in the neighborhood of Bellows Falls. It is still taken plentifully in Merrimack river, and in many other streams which flow into the Atlantic ocean from N. England. I cannot learn that it has ever been taken in lake Champlain, but on account of some resemblance in form and appearance between this species and the *Coregonus albus,* or White Fish, the name of Shad, or Lake Shad, is here very generally applied to the latter.

GENUS HIODON.—*Le Sueur.*

Generic Characters.—The form of a herring; abdomen trenchant, but not serrated; one dorsal fin opposite to the beginning of the anal; hooked teeth on the jaws, vomer and tongue; head small; eyes very large and situated near the end of the snout; branchial rays eight or nine.

THE WINTER SHAD.
Hiodon clodalus.—LE SUEUR.

Hiodon clodalus et H tergisus. Le Sueur, Jour. Ac. Nat. Sci. Phil. I—364, fig.

DESCRIPTION.—Body deep and thin: back elevated and nearly straight; belly trenchant; dorsal fin quadrangular; ventrals with large branching rays, and a long bract over their base; anal fin long, with the anterior portion large and pointed, and nearly straight, or rounded with a depression between it and the posterior portion. Color towards the back bluish, with metalic reflections, pearly and silvery below; head small, greenish brown above, with bronze reflections on the sides; dorsal and caudal fins brown, the others lighter. Eyes far forward, large, round; pupil black; iris with yellow and pearly reflections. Nostrils large, double, and very near the end of the snout; lateral line nearly straight, nearer the back than the belly; tail deeply forked; scales rather large, brilliant, about 60 on the lateral line. Mouth oblique; jaws even when shut, but on account of the obliquity of the gape the lower jaw appears longest when the mouth is open; numerous small conical teeth in both jaws, on the vomer, palatine bones, and tongue, the latter largest and hooking inward. Length 13½ inches; depth 3½; diam. of the eye .7.

Rays, B. 8, P. 12, V. 7, D. 11, A. 30, C. 18.

HISTORY.—Le Sueur's account of the genus Hiodon was published in 1818, in the Journal of the Academy of Natural Sciences. In this paper he describes what he considers two species, to which he gives the name of *H. tergisus* and *H. clodalus,* but at the same time intimates a possibility that they may both belong to the same species. The difference upon which he constituted the two species, was in the form of the anal fins, the *H. tergisus* having the anterior portion of that fin rounded, with a depression between that and the posterior portion, and *H. clodalus* with the anterior portion pointed, and the line to the posterior angle nearly straight. I have before me two specimens, which were caught at the same time. One is 13½ inches long, and has the pointed and straight anal fin of Le Sueur's *H. clodalus,* and the other, 13 in. long, has the rounded, notched anal fin of his *H. tergisus.* In other respects scarcely any difference can be discovered, and I have no doubt that they both belong to the same species. This fish is often called the White Fish by the fishermen. It is considered a very good fish for the table, but is not taken in lake Champlain very plentifully.

GENUS LEPISOSTEUS.—*Lacepede.*

Generic Characters.—Both jaws with rasp-like teeth, having a row of longer, pointed ones on the margin; branchæ united on the throat by a common membrane, which has three rays on each side; scales of a stony hardness; dorsal and anal fins opposite to each other, and far back.

THE COMMON BILL FISH.
Lepisosteus oxyurus.—RAFINESQUE.

Ichthy: Ohiensis, p 74. Kirtland's Report, p 196. Boston Jour. Natural History, IV—16.
Lepisosteus huronensis, Fauna Boreali Americana, p 237.

DESCRIPTION.—Body long, cylindrical; back slightly arched in a regular curve; head flattened above and on the sides, encased in a bony covering, having distinct striæ, grooves and sutures, with the jaws, which are thickly set with teeth of different sizes, lengthened out into a slender, flattened beak; upper jaw reaches beyond the lower, with nostrils near its extremity; tongue fleshy, bilobate; roe green; eyes just behind the angle of the mouth, and near the articulation of the lower jaw. Color above brownish leaden, sometimes with an umber hue, darkest on the head, yellowish pearly white below; sides spotted with blackish towards the tail; pectoral and ventral fins brownish; dorsal, caudal and anal yellow and ruddy, spotted with black; dorsal fin commences over the posterior part of the anal; the attachment of the caudal oblique, fin rounded, with the outer rays armed with sharp, spiny scales. Body covered with thick, strong, hard, bony scales, of rhomboidal form, and regularly arranged in oblique rows. Upon the lateral line, which is straight, but indistinct towards the tail, there are 60 scales. Length of the specimen before me 3 ft. 4 in.; upper jaw to the angle of the mouth 7 in.; from the angle to the orbit 1.2 in; from the point of the bill to the middle of the gill opening 12, or just one third of the total length, measured through the middle of the caudal fin; ventrals midway between the point of the bill and extremity of the tail. Weight 6 pounds.

Rays, P. 11, V. 6, D. 8, A. 9, C. 12.
HISTORY.—This singular fish was described by Samuel Champlain, as an inhabitant of the lake now bearing his name, more than 200 years ago. He called it *Chausarou*, which was probably the Indian name. The Indians assured him they were often seen eight or ten feet long, but the largest he saw was only five feet long, and about the thickness of a man's thigh. It is considered a very voracious fish, and when any of them are taken, or seen in the water, the fishermen calculate upon little success in taking other kinds. Charlevoix tells us that he preys not only upon other fishes, but upon birds also; and that he takes them by the following stratagem: Concealing himself among the reeds growing on the marshy borders of the lake, he thrusts his bill out of the water in an upright position. The bird, wanting rest, takes this for a broken limb, or dry reed, and perches upon it. The fish then opens his mouth and makes such a sudden spring that the bird seldom escapes him. Charlevoix also assures us that the Indians regarded the teeth of this fish as a sovereign remedy for the headache, and that pricking with it where the pain was sharpest took it away instantly. The scales with which this fish is covered are so thick and strong, as to form a coat of mail, which is not easily pierced with a spear. They are taken only occasionally in the seine at the present day, but are said to be sometimes seen in considerable numbers lying in the marshy coves. Its flesh is rank and tough, and is not used for food. The usual length of those now taken, is from two and a half to three feet, though they are often much longer. The specimen, from which the preceding figure and description were made, was taken at the mouth of Winooski river, May 11, 1841. One of the largest specimens which I have seen was taken at the same place, June 16, 1838, and is now in my possession. It is 46 inches long, and when caught weighed 9½ pounds. This species is found in the great western lakes, and in the Ohio river, where this and several other species are known by the name of *Gar Fishes.*

THE STRIPED BILL-FISH.
Lepisosteus lineatus.

DESCRIPTION.—Color above light olive, with a dark line along the middle of the back, and dark roundish spots on the upper mandible and towards the tail. A broad dark bluish brown stripe commences on the side of the bill, passes backward through the eye, across the cheek and operculum, and along the side and through the middle of the tail to its extremity; below this, commencing on the lower jaw a little forward of the angle of the mouth, is a bright yellowish white stripe, which touches upon the lower side of the eye, passes through the base of the pectoral fin and vanishes near the tail; still lower is a grayish brown stripe, with a lighter one along the middle of the belly to the vent; fins yellowish, spotted with brown; under mandible black; eye close to the angle of the mouth, and directly behind it; pupil black, surrounded by a bright golden line; iris brown where covered by the brown stripe, but lighter on the upper and lower margin. Bill flatter and broader, proportionally, than in the *L. oxyurus*; teeth sharp, and of different sizes, 4 rows above and 2 below; upper jaw considerably longest, terminated in a knob on which the nostrils are situated, and which is articulated over the tip of the lower jaw; all the fins proportionally much longer and more slender than in the *L. oxyurus*, the dorsal and anal reaching the base of the caudal. Lateral line straight, passing along near the upper edge of the dark lateral stripe, containing 62 scales. Scales rhomboidal, arranged in oblique rows. Pectoral fins situated under the membranous prolongation of the gill cover; ventrals nearly medial; height of the dorsal 1 in., length .4, commences over the posterior part of the anal, and extends half its length beyond it; height of the anal fin 1 inch, length .5; the attachment of the tail oblique; tail contained about 6 times in the total length; the head, including the bill, a little more than 3 times. Length of the specimen before me 10.3 inches; lower jaw 2, upper 2.2, from the snout to the eye 2.3, to the posterior part of the gill cover 3.2, to the ventral fins 5, to the commencement of the anal 7, of the dorsal 7.3; longest rays of the caudal 1.7.

Rays, P. 12, V. 6, D. 8, A. 9, C. 12.

HISTORY.—The only specimen which I have seen of this fish was the one from which the preceding description and figure were drawn. It was taken in Burlington during the drought in August, 1841, in a small cove, whose communication with the Winooski river had been cut off by the subsiding of the water. This fish may be the young of the preceding species, but finding so many points of difference, I have thought it best to introduce a separate description.

ORDER III.—*Malacopterygii—Subbrachiati.*

Fishes of this order have their gills pectinated, or comb-like, and the ventral fins very near the pectoral, either before, beneath, or a very little behind.

I.—GADIDÆ, OR COD-FISH FAMILY.

GENUS LOTA.—*Cuvier.*

Generic Characters.—Body elongated, one anal and two dorsal fins; the second dorsal and the anal fin long; cirri more or less numerous.

THE LING OR METHY.
Lota maculosa.—LE SUEUR.

Rich. Fauna Boreali, p. 248. Kirtland's Report, 196. Bost. Jour. Nat. Hist. 1V—24. *Gadus maculosus*, Le Su. Jour. Acad. Nat. Sci., Phil., I—83.

DESCRIPTION.—Body thick; back nearly straight from the snout to the tail; abdomen capacious, and often flabby when not distended with food or spawn; head broad and much depressed; upper jaw longest, with the upper lip extending considerably beyond the jaw; snout pointed; orbit elliptical; eyes rather small and nearly round, pupil bluish black, iris grayish golden. Above varied with brownish, olive and fuliginous, darkest on the head; sides obscurely spotted with whitish; belly yellowish, rusty-white, with ruddy tinges; lateral line commences above the gill opening and runs a straight course to the middle of the tail: nostrils double, the anterior lengthened into short cirri; the cirrus depending from the tip of the under lip reddish brown; all the fins brownish with their margins blackish; ventral fins before the pectoral, slender and pointed; pectorals broad and rounded; first dorsal short; second dorsal commences nearly over the vent, and extends to the base of the caudal; whole outline of the caudal rounded; anal fin commences about an inch behind the beginning of the second dorsal, and terminates a little anterior to the termination of the dorsal; teeth small and card-like on the jaws, palate and throat; tongue fleshy and smooth. Length of the largest of three specimens before me 23 inches, head, to the upper part of the gill opening, 4, first dorsal 1.5, second dorsal 9.5, anal 8.3, height of the dorsals and anal 1, of the jugular and pectorals 3, cirrus on the lip 1.3; orbit .4 by .5, distance between the orbit 1.2; vent 1 inch nearer the snout than to the extremity of the tail.

Rays, B. 7, V. 6, P. 20, D. 10—74, A. 68, C. 40.

HISTORY.—This fish, which is quite common in lake Champlain and its tributaries, I have referred to Le Sueur's species the *Gadus maculosus*, as agreeing more nearly with his description than with any other to which I have access. There are, however, several differences between them. In Le Sueur's species the jaws are said to be equal; in ours, the upper jaw is uniformly longest;—in his the lateral line is said to be in the middle of the body; in ours, anterior to the vent, it is much nearer the back than the belly. Our fish bears considerable resemblance to the *Lota brosmiana* described by Dr. Storer in the Boston Journal of Natural History, vol. IV, page 58. But it differs from his description and figure in having the upper jaw longest, in having the snout more pointed and less orbicular, &c. Judging from the descriptions without specimens for comparison, I should say that our fish differs as much from either of the species referred to, as they differ from each other, and that they either constitute three distinct species, or are all varieties of the same species.

The Ling is held in very low estimation as an article of food, the flesh being tough and the flavor unpleasant. This fish is one of the greatest gormandizers found in our waters. If he can procure food, he will not desist from eating so long as there is room for another particle in his capacious abdomen He is frequently taken with his abdomen so much distended with food as to give him the appearance of the globe or toad-fish. The smallest of the three before me, when my description was made, being 16 inches long, was so completely filled with the fishes swallowed, that their tails were plainly seen in its throat by looking into its mouth. On opening it, I found no less than 10 dace, *L. pulchellus*, all about the same size, and none of them less than 4 inches long. Seven of these were entire, and appeared as if just swallowed. Upon the others, the digestive process had commenced.

THE EEL-POUT.
Lota compressa.—LE SUEUR.

Jour. Acad. Nat. Sci., I—84. Storer's Report, 134.

DESCRIPTION.—Color of the back and sides yellowish brown, variegated with darker brown spots; gill cover and snout darkest; abdomen whitish. Body in front of the first dorsal cylindrical, beginning to be compressed at the sides, at the extremity of the pectorals, gradually becoming more so towards the tail, so that the caudal rays appear a membranous prolongation of the body; body covered with minute scales, looking like cup-shaped depressions; lateral line straight, conspicuous. Head much compressed; eyes circular; nostrils double; a minute cirrus rises from the back of each anterior nostril, and from the tip of the chin; upper jaw longest; jaws and palate armed with minute teeth. First dorsal lighter than the body, situated the length of the head back of head, short; second dorsal long, reaching to the tail; anal, the same length as the dorsal; caudal rounded; most of the fins margined with black. Length of the specimen 6 inches, head 1. Rays could not be counted on account of the fleshy texture of the fin-membrane. —*Storer*.

HISTORY.—This fish is found in Connecticut river and its tributaries. Not having obtained a specimen of it, I have copied Dr. Storer's description. It was first described by Le Sueur, from a specimen obtained at Northampton.

ORDER IV.—MALACOPTERYGII—APODES.

Fishes of this order have long bodies, a thick skin, and no ventral fins.

MURÆNIDÆ, OR EEL FAMILY.
GENUS MURÆNA.—*Linnæus*.

Generic Characters.—Body cylindrical, elongated, covered with a thick and smooth skin; the scales very small, lubricated with copious mucous secretion; mouth with a row of teeth in each jaw, and a few on the anterior part of the vomer; pectoral fins close to a small branchial aperture; no ventral fins; dorsal fin, anal fin and caudal fin united.

THE COMMON EEL.
Murena vulgaris.

Murena anguilla, Lin. et Pen. *Anguilla acutirostris*, Yarrell, Brit. Fishes, II—284. *A. vulgaris* Trans. Lit. and Phi. Soc. N. Y., 1—360.

DESCRIPTION.—Specimen 31 inches in length; from the tip of the snout to the base of the pectorals 3.6, to the vent 13.3, to the commencement of the anal 13.8; circumference just before the eyes 2.3, one and a half inch from the tip of the

THE BLACK EEL.

upper jaw 3.7, at the base of the pectorals 5, at the commencement of the dorsal 6.5, of the anal 5.7, distance between the eyes .6, height of the pectorals 1.4, base .6. Body cylindrical; color above dark olive brown, extending down low upon the sides; belly white, or yellowish white, sometimes with a ruddy tinge; lateral line irregular, indistinct, and above the middle of the body, before the vent, behind it, medial and straight to the middle of the tail; jaws narrow and rounded at the end; lower jaw longest, tipped with brown; lips fleshy; a broad band of small, short teeth in each jaw and upon the vomer; eye over the angle of the mouth, pupil black, iris golden; nostrils near the eyes; a short fleshy cirrus on each side of the snout; small mucous pores on various parts of the head; gap of the mouth small; gill opening small and under the anterior origin of the pectoral fin, which is pointed; dorsal, caudal and anal fin united. Pectoral rays 12. Vent 3 inches nearer the snout than to the extremity of the tail.

HISTORY.—This is the common Eel in Vermont, on the west side of the Green Mountains, and also in Canada, where it is taken in very large quantities. When skinned and skilfully cooked it is an agreeable and nourishing article of food, and is by many considered one of our best fishes; some, however, find it difficult to surmount the prejudice occasioned by its slender snake-like appearance. The ordinary weight of those taken in our streams is from 1 to 3 pounds. By comparing the above description with the two following, it will be seen that this Eel differs very materially from those found in other parts of New England, particularly in the relative position of the pectoral fins. By comparing our Eel with the description and figure of the Sharp-nosed Eel, *Anguilla acutirostris*, in Yarrell's British Fishes, vol. II, p. 284, I find the agreement in the position of the fins, &c., so perfect, that I have little doubt that they belong to the same species, and that the Common Eel of the St. Lawrence and its tributaries is identical with the Common Eel of Great Britain. Between our fish and Yarrell's figure there are some slight differences. In the figure the head is too broad, and the middle rays of the pectoral fins are too short. In our fish the middle rays are longest, making the fin appear pointed.

THE BLACK EEL.

Murœna bostoniensis.—LE SUEUR.

Journal Acad. Nat. Science, Phil., I--87. Storer's Report, page 157.

DESCRIPTION.—Specimen 23 inches in length: from the tip of the snout to the base of the pectorals 8 inches; circumference of the body back of the head, at the commencement of the pectorals, 3.4 inches; at the commencement of the dorsal fin 3.4; around the head 3.2, at the distance of 1.5 from the snout; in front of the eyes 1.7; from the tip of the lower jaw to the anal fin 10½ inches; width of the body over the pectorals 1.2, pupil black, iris golden; width between the eyes .4; lateral line indistinct. Color grayish brown above; yellowish white beneath, with a tinge of red about the tail.—*Storer.*

HISTORY.—The Common Eel, found in Connecticut river, and in the streams and ponds in this state on the east side of the Green Mountains, I suppose to belong to this species. Not having obtained specimens of this and the following species, I can only give Dr. Storer's description of them. In some of the ponds this Eel grows to a very large size. They are frequently taken at the outlet of Barnard pond weighing 8 or 10 pounds.

THE SILVER EEL.

Murœna argentea.—LE SUEUR.

DESCRIPTION.—Specimen 23 inches in length; from the tip of the snout to the base of the pectorals 7½ inches; circumference of the body back of the head at the commencement of the pectorals 3½, around the head 1½ inch from the snout 3, in front of the eyes 1.4, at the origin of the dorsal 3½; from the tip of the lower jaw to the anal fin 9½; width of the body over the pectorals .7; width between the eyes .3. Lateral line exceedingly distinct, appearing to divide equally the darker colored back from the beautiful lighter silvery abdomen For the extent of 6 inches in front of the anal orifice, a well marked line or furrow resembling in appearance the lateral line.—*Storer.*

HISTORY.—The fish known by the name of Silver Eel on the east side of the Green Mountains in this state, I suppose to belong to this species, but I have had no opportunity for deciding the point by the examination of specimens.

II. CARTILAGINOUS FISHES.

1. STURIONIDÆ, OR STURGEON FAMILY.

Fishes of this Family have free branchæ, wide gill openings, an operculum, but no rays in the gill membrane.

THE ROUND-NOSED STURGEON.

Genus Acipenser.—*Linnæus.*

Generic Characters.—Body elongated, which, with the head, is provided with rows of radiated bony prominences; snout pointed, conical; mouth placed on the under surface of the head, tubular, and without teeth.

ROUND-NOSED STURGEON.
Acipenser rubicundus.—Le Sueur.

Description.—General color bluish gray above, white with brushes of ruddy beneath; all the fins of a brownish hue, and slightly ruddy, with the outer margin whitish; form rounded, elongated and tapering regularly to the caudal; head rounded; snout short and rounded; upper part of the head with a bony covering; three rows of small and slightly developed bony tubercles without spines extending the whole length of the body, one on the back, and one on each side along the lateral line. Plates or tubercles on the lateral line 31 or 32; also a few plates between the dorsal and anal, and the caudal; but there are no ventral rows as there are in the *oxyrhynchus* and most other species. Eyes rather small, prominent, iris dark golden; nostrils double and large; four equal cirri suspended in a transverse line between the mouth and end of the snout, but nearest the latter, being 2 in. from the snout and 2½ from the mouth; cirri 2½ inches long, round, the size of a goose-quill at the base, and tapering to a point; color brownish white excepting their points, which are red; mouth under side of the head, tubular, ovate, 3 in. by 2 in., and capable of 2 inches protrusion. All the fins thick. The anal commences 4½ in. behind the vent, and a little behind the middle of the dorsal. Color of the intestines dark; stomach a thick sack resembling a fowl's gizzard. Length of the specimen before me 4 ft. 2 inches; weight 26½ lbs. Length of the head to the total length as 1 to 5; distance between the eyes 4 in., from the eyes to the end of the snout 4½; from the nose to the commencement of the dorsal 37 inches.

History.—This fish is quite common in lake Champlain, and grows to a very large size. It is frequently taken in the seine measuring more than 6 ft. in length, and weighing 100 pounds or more. Its flesh, though not generally very much esteemed, if properly cooked is very good eating. When eaten fresh it is usually cut into slices and fried in butter, with suitable seasoning; but whether eaten fresh or salted, the skin should always be taken off before it is cooked, as the oil contained in that imparts a disagreeable flavor. The Indian method of capturing the Sturgeon in lake Champlain, according to Charlevoix (Travels, Vol. I—119), was as follows: 'Two men placed themselves in the two ends of a canoe. The one behind steered and the other stood up holding a dart in one hand, to which one end of a long cord was fastened, and the other end fastened to the canoe. When he saw a Sturgeon within his reach, he threw his dart and endeavored to strike where there were no scales. If the fish was wounded he darted off, drawing the canoe pretty swiftly after him, but usually died after swimming about 150 paces, and was then drawn in by the cord.'

THE SHARP-NOSED STURGEON.
Acipenser oxyrhynchus.—Mitchell.

Description.—Body elongated, tapering; form pentagonal, with the angles covered with rough, radiated bony plates, each having a saddle-like base and a spur-like process arising from its centre and hooking backward, and usually terminating in a sharp point; the rest of the skin roughened by small scabrous patches of bony matter, resembling the spiculæ of minute crystals; head encased in a bony covering, and lengthened into an acute, conical snout; mouth on the under side of the head, ovate, toothless, and protractile; four cirri depending in a cross row between the mouth and the end of the snout, a little nearest the latter. The operculum is a single radiated bony plate; eyes rather small, the anterior part of the orbit just midway between the point of the snout and the posterior margin of the operculum; nostrils before the eyes, double, lower orifice much largest. Color grayish brown above, yellowish white beneath. *Bony plates* 12 between the encasement of the head and the dorsal fin, one of which rests upon the base of the dorsal, and is usually without a spine; between the dorsal and the caudal is usually one large plate and two or three smaller ones; lateral plates variable, but generally 28; ventral plates from 8 to 10; the spur-like processes longest and most pointed in the smaller specimens; usual length from 2 to 3 feet.

THE BLUE LAMPREY.

HISTORY.—This fish is occasionally taken in lake Champlain, and is here known by the name of *Rock Sturgeon*. It seldom exceeds 3 feet in length or 20 pounds in weight, but is much more generally and highly esteemed as an article of food than the preceding species, some even ranking it as one of our best fishes for the table. This, like the preceding, should be skinned before it is cooked, and for the same reasons.

II.—CYCLOSTOMIDÆ, OR LAMPREY FAMILY.

Fishes of this family have their jaws fixed in an immoveable ring. Their branchiæ are fixed with numerous openings.

GENUS PETROMYZON.—*Linnæus.*

Generic Characters.—Body eel-shaped; mouth circular, armed with tooth-like processes; lips forming a continuous circle around the mouth; seven openings on each side of the neck, leading to seven branchial cells; no pectoral or ventral fins; dorsal, anal and caudal fins formed by an extension of the skin on those parts.

THE BLUE LAMPREY.
Petromyzon nigricans.—LE SUEUR.
Trans. Am. Phil. Soc. N. S.1. 385. Storer's Rep.197.

DESCRIPTION.—Color above dark bluish gray, beneath and fins dingy white; several rows of blackish dots about the head and neck. Anterior third of the body cylindrical; the posterior two-thirds flattened laterally, and very much so toward the tail; head slightly flattened above and terminated in an oblique, oval or circular mouth, which is armed within with numerous yellowish, spinous teeth, projecting from widened bases, and surrounded by a fleshy lip which is margined with a row of fine papillæ; a small white spot on the top of the head between the eyes, in front of which is a spiracle. The first dorsal commences in the middle of the fish, the separation between the dorsals merely a notch; the length of the first dorsal contained 4½ times in the second. Length of the specimen before me 5 inches,—head, to the eye, 1 inch, to the vent 3½, width of the mouth .4.

HISTORY.—The fresh water *Lampreys*, or *Lamprey-Eels*, as they are more commonly called, resemble, in their habits, the Blood-Sucker much more than the ordinary fishes. They obtain their subsistence principally by attaching themselves by their mouths to the bodies of larger fishes, and drawing nourishment from them by suction; for this purpose their mouth and tongue are admirably adapted, the latter acting in the throat like the piston of a pump, while the circular lips of the former adhere closely to the side of the fish, and by these means the softer parts of the larger fish are drawn into the mouth and swallowed by the parasite. When a Lamprey once fastens himself, in this manner, upon a large fish, he adheres with such force as to baffle all the efforts of the fish to rid himself of his unwelcome incumbrance. Fishes are frequently taken in the seine with Lampreys still adhering to them, and others with deep depressed wounds upon their sides, affording indubitable proof of their having been attached. The fresh water Lampreys seldom exceed 6 or 8 inches in length, and no account is made of them as an article of food.

GENUS AMMOCŒTES.—*Dumer.*

Generic Characters.—Form of the body, the branchial apertures and fins, like those of the Lampreys; upper lip semi-circular, with a straight, transverse under lip; mouth without teeth, but furnished with numerous short membranous cirri.

THE MUD LAMPREY.
Ammocœtes concolor.—KIRTLAND.
Boston Journal Nat. History, vol. III. p. 473, pl. 28.

DESCRIPTION.—Form nearly cylindrical for two-thirds the length, then gradually flattened to the extremity of the tail, where it is quite thin; color yellowish brown above, gradually becoming lighter towards the belly, but without the dividing line between the lighter and darker parts, mentioned by Le Sueur in his description of the *A. bicolor*. Eyes so minute as hardly to be seen by the naked eye; nostrils on a light colored disk on the upper part of the head in front of the eyes; upper lip longer than the lower, in the form of a horse-shoe, protractile and capable of being closed so as to conceal the lower one; small papillæ on the inside of the lips and fringes within the mouth. The branchial openings, seven in number, commence below and a little back of the eye, and extend backward, passing obliquely downward, the apertures appearing like short oblique slits. Sides with an annular, or ribbed appearance. The fin, which is of a dull yellowish color, commences near the middle of the back, passes round the tail and terminates just behind the vent. About three fourths of an inch from the commencement is a considerable depression in the fin for more than half an inch, but it does not amount to a division. The fin rays are white, minute and forked. The longest of three specimens before me 5.3 inches; from the snout to the posterior branchial opening 1.1, to the vent 4.1. Rays too small to be counted.

History.—This fish agrees very well with Kirtland's description excepting the depression in the dorsal, and that the broadest part of the dorsal is some distance behind the vent. During the drought in September, 1841, I found large numbers of these fishes, which had buried themselves in the mud at the bottom of the small coves along the banks of Winooski river, from which the water had evaporated. This fish is known in many places by the name of Mud-Eel, or Blind-Eel.

CHAPTER VI.

INVERTEBRAL ANIMALS OF VERMONT.

Preliminary Observations.

Invertebral animals are such animals as are destitute of a spine or back bone, and are so exceedingly numerous that, with the exception of the molluscous animals, we shall not even attempt to give a catalogue of them. The animals of this great division are extremely various in their structure, habits, and dispositions. Some have their bodies protected by a shelly covering, while others have their bodies and limbs surrounded by crustaceous plates, while, again, others have no other covering than a soft and tender skin. A few only of them have red blood, and none of them possess all of the five senses. In many cases the sexes are united in the same individual, and in some cases the species is continued by a process somewhat resembling vegetation. They all afford eminent manifestations of the wisdom and skill of the Creator; and, though generally regarded as insignificant and contemptible, many of them contribute largely to the comfort and interest of man, while a still greater number are employed in annoying and injuring him.

Section I.—Mollusca.

Fresh-Water and Land Shells.

Prepared expressly for this work,
By Charles B. Adams, A. M.,
Professor of Natural History, Middlebury College.

FAMILY PERISTOMIANA.

Genus Paludina.

Generic Characters.—Shell conoid; whorls convex, modifying the aperture, which is ovate or nearly orbicular, with the margins united. Operculum thin, corneous, concentric. Animal with the head short; rostrum small and truncate; tentacles slender, with the eyes on an enlargement at their base; foot broad, thin.

Paludina decisa.—Say.

Description.—Shell ovate-conic, with revolving rows of bristly filaments when young, smooth when mature, green; apex truncate; whorls six, convex; suture deep; spire a little longer than the aperture, which is pyriform; umbilicus very small. Length 1.25 inch; breadth 0.75 inch; divergence of the spine 58°.

Remarks.—This species is very common in ponds and streams, and is found near the water's edge partly buried in mud or sand. Sometimes they are found crawling at the distance of a few feet from the water. They are viviparous, and produce their young in May. These, at birth, are furnished with a shell about an eighth of an inch in diameter, globular, and of a pale horn color, and are nearly transparent. In the progress of growth, the shell becomes proportionally more elongate, and the part which was formed previous to birth is invariably broken off. They are very rarely found heterostrophe. One such individual, of the size of a pea, was found in Otter Creek, in Middlebury.

Paludina integra.—SAY.

DESCRIPTION.—This species so much resembles the preceding, that a formal description is unnecessary. Its apex is not truncated, so that, with a greater divergence of the spire, it is, nevertheless, longer than that shell. It is also thicker, and the whorls are less convex. This shell is common in the western states, but it is extremely rare in Vermont, only three or four specimens having been obtained in lake Champlain. Length 1.3 inch; breath 0.75 inch; divergence of the spire, 63°.

Paludina porata.—SAY.

DESCRIPTION.—Shell conic, horn color; whorls four and a half, convex; suture rather deep; apex subacute, spire as long as the aperture, the labium of which is appressed to the penultimate whorl; umbilicus rather large. Length 0.27 in.; breadth 0.19 inch; divergence of the spire 72°.

REMARKS.—This species is found plentifully in streams and in lake Champlain. It is sometimes brownish or greenish.

Paludina lustrica.—SAY.

DESCRIPTION.—-Shell ovate-elongate, horn color; whorls four and a half, convex; suture rather deep; apex very obtuse; spire as long as the aperture, which is ovate-orbicular, with the labium not appressed to the penultimate whorl, and sometimes scarcely touching it; umbilicus small. Length 0.16 inch; breadth 0.11 inch; divergence of the spire 47°.

REMARKS.—This small species is common in lake Champlain. It differs from the preceding in the obtuseness of the apex, less divergence of the spire, and small umbilicus; also in the labium, which is quite distinct from the penultimate whorl, so that the shell much resembles a valvata.

GENUS VALVATA.

Generic Characters.—Shell discoid or conoid; whorls cylindrical; aperture orbicular, not modified by the penultimate whorl; margins continuous, distinct from the penultimate whorl. Operculum orbicular, concentric. Animal with the foot bilobed before; head proboscidiform; tentacles very long, slender, obtuse, cylindrical; eyes sessile behind the tentacles, with a branchial filament resembling a third tentacle.

Valvata tricarinata.—SAY.

DESCRIPTION.—Shell depressed, conic, thin, green, obsoletely striate; suture well impressed; whorls three or four, rendered subquadrangular by the revolving carinæ, of which two appear on the spire, and three on the last whorl; these are very much raised, rounded, equi-distant, the inferior bordering the umbilicus, which is broad and deep.— Length 0.13 inch; breadth 0.22 inch; divergence of the spire 90°, sometimes much greater.

REMARKS.—This shell, very curious on account of its carinæ, is common in lake Champlain, and in some of our streams. Varieties occur in which the middle carina is obsolete, or in which none are very distinct.* Other varieties have the spire less elevated, or even in the plane of the last whorl.

Valvata sincera.—SAY.

DESCRIPTION.—-Shell globose-discoid, obsoletely striate, brownish-green; whorls three and a half, accurately rounded, rapidly enlarging to the aperture; suture deeply impressed; spire but little elevated; apex obtuse; umbilicus deep, about two-thirds as wide as the last whorl; margin of the aperture touching the penultimate whorl. Length 0.1; breadth 0.2 inch; divergence of the spire about 135°

REMARKS.—This shell is much like the *var. simplex* of the preceding species. The umbilicus is usually a little larger, but the most striking characteristic is the rapid enlargement of the whorls, the last being more than three times the diameter of the penultimate. The divergence of the spire is never so small as in that species, but like that is sometimes much more than in the type of the species, even to 180°.

FAMILY MELANIANA.
GENUS MELANIA.

Generic Characters.—Shell turrited; aperture entire, ovate, effuse; columella thickened, arcuate. Operculum horny, subspiral. Animal oviparous; foot short; rostrum truncate; tentacles filiform, with the eyes outside, at or near their base.

Melania depygis.—SAY. Var.

DESCRIPTION.—-Shell elongate-conic, yellowish horn-color, with a broad rufous band on the whorls of the spire, with a second similar band on the lower third of the last whorl; upper whorls carinate on the lower side; whorls eight or nine; spire twice as long as the aperture. Length 0.53 inch; breadth 0.22 inch; divergence of the spire 33°.

---

* *Var. simplex.*--GOULD.

FRESH WATER AND LAND SHELLS.

REMARKS.—This species is interesting, as the only representative in New England of a family whose species are so numerous in the Southern and Western states. Here it is found only on our western border in lake Champlain, where but a few specimens have been obtained. It has some claims to be regarded as a new species, differing much in its proportions from the type of Say's species. But since specimens from Ohio vary much in their proportions, we have not been satisfied that it is a distinct species.

FAMILY LIMNÆANA.
GENUS LIMNÆA.

Generic Characters. Shell thin, oval or elongate; spire elevated, more or less acute; aperture longer than wide; margins not continuous; columella with a single oblique fold. No *operculum*. Animal hermaphrodite, spiral; head depressed; tentacles flattened, triangular, short, with the eyes at their base, on the inner front side; foot thin, oval, shorter than the shell.

Limnæa megasoma.—SAY.

DESCRIPTION.-Shell large, ovate, brown, with coarse incremental striæ; whorls five, convex; last whorl very large, inflated; * suture deep; spire two-thirds as long as the aperture, which is large. Length 2 inches; breadth 1.2 inch; divergence of the spire 58°.

REMARKS.—This large and noble species was originally discovered in the North West Territory, in latitude 48°. Subsequently it has been found only in Burlington. It is very rare in cabinets, but quite recently the author of this work discovered a large number in Burlington, at a low stage of the water.

Limnæa appressa.—SAY.

DESCRIPTION.—Shell large, thin, horn color, elongate; whorls seven; upper ones planulate, lower ones convex, last one much enlarged and obtusely shouldered above; suture not much impressed; spire long, slender; apex acute; aperture long-oval; margin thin and sharp; columellar fold strong. Length 1.75 inch; breadth 0.75 inch; divergence of the spire above 33°, below 40°.

REMARKS.-This species has been found for the most part with the preceding at Burlington. Its claims to be regarded as distinct from the *L. stagnalis*, of Europe, are very slight.

* Whorls inadvertently made to revolve the wrong way in our figure.

Limnæa gracilis.—JAY.

DESCRIPTION.—Shell very long and slender, pale horn color; whorls four and a half, very oblique, slightly and regularly convex; suture not much impressed; aperture more than half as long as the spire, long-oval; labium entirely separate from the penultimate whorl, moderately reflected, with a large rima behind it, as strong as the labrum. Length 1 inch; breadth 0.18 inch; divergence of spire 18°.

REMARKS.—This extremely rare species was discovered by Prof. Benedict, in Lake Champlain, at Crown Point. One or two specimens have been found on the Vermont side of the lake. The shell is remarkable for its length, which is nearly six times the breadth, although the whorls are very few. The development of the labium is also very remarkable. No other species can be compared with this.

Limnæa pallida.—ADAMS.

DESCRIPTION.—Shell moderately elongate, ovate-fusiform, very pale horn color, semi-transparent, not very thin, with fine irregular striæ of growth, whorls five and a half, moderately convex; suture well impressed; spire four-fifths as long as the aperture, acutely conic; apex sub-acute; body whorl not much enlarged, somewhat produced below; columellar fold moderate; umbilicus large. Length 0.48 inch; breadth 0.22 inch; divergence of the spire 45°.

REMARKS.—This species is rather common in lake Champlain, clinging to rocks and stones. It has not yet been found in any other region except in Andover, Ms. It is sometimes nearly white. It differs from *L. desidiosa* in having its columella much less tortuous, and its aperture less elongated below the fold.

Limnæa elodes.—SAY.

DESCRIPTION.—Shell brown horn-color; whorls seven, convex; suture well impressed; spire longer than the aperture, conic, sub-acute; last whorl somewhat ventricose; labium appressed closely to the penultimate whorl; columella prominent, with a very strong fold. Length 1.2 inch; breadth 0.55 inch; divergence of the spire 45°.

REMARKS —*Limnæa umbrosa*, SAY, is probably only a variety of this species, its principal difference consisting in the feebleness of its columellar fold, which is, in this species, of a variable character. This variety is much more abundant in Vermont than the type of *L. elodes*. This species differs from *L. desidiosa* chiefly in not having the columella produced in a straight line below the fold; from *L. pallida* in the less proportional size of the last whorl, and greater convexity of the whorls; from *L. palustris* of Europe chiefly in the greater convexity of the whorls and less acumination of the spire. By some it is regarded as a variety of the latter.

Limnæa desidiosa.—SAY.

DESCRIPTION.—Shell brown horn color, elongate-ovate; whorls nearly six, slightly convex; suture distinct; spire about as long as the aperture, which is lengthened below; columellar fold feeble; labium appressed; columella produced below the fold in a straight line. Length 0.55 inch; breadth 0.25 inch; divergence of the spire 45° to 55°.

REMARKS.—This species is very common, and is subject to great variation of form, frequently being elongated, and resembling *L. elodes.* Other individuals are short, as in Say's figure (Am. Conch.,) and the upper part of the last whorl is inflated and more or less shouldered, while the lower part is produced as is usual. This variety approaches *L. umbilicata* of Mass., which has the umbilicus larger, and the lower part of the last whorl abbreviated, inflated, and globular.

Limnæa caperata.—SAY.

DESCRIPTION.—Shell ovate, brown, with minute revolving raised lines, which are in some very distinct, and in others mostly obsolete; whorls nearly six, convex; suture distinct; spire about as long as the aperture, conic, acute; columella reddish, slightly folded, thickened, and reflected over an umbilicus. Length 0.45 inch; breadth 0.24 inch; divergence of the spire 57°.

REMARKS.—This species is well characterized by the revolving raised lines, which will generally be seen around the umbilical region, when obsolete elsewhere. The last whorl and the aperture are more regularly rounded than in the preceding species.

GENUS PHYSA.

Generic Characters.—Shell heterostrophe, shining, otherwise like Limnæa; operculum wanting; animal with long, slender tentacles; having the eyes at their base on the inner side.

Physa ancillaria.—SAY.

DESCRIPTION.—Shell ovate, yellowish brown, sometimes of a bay color; whorls four, flattened; suture not impressed; spire less than one-fifth of the length of the aperture; apex acute; last whorl very large; aperture acute and narrow above, wide below; outer lip often thickened within; columella produced in a right line below its fold. Length 0.65 inch; breadth 0.48 inch; divergence of the spire 110°.

REMARKS.—This species, seldom found plentifully, is not uncommon in lake Champlain. It is there found of a deep bay color.

Physa heterostropha.—SAY.

DESCRIPTION.—Shell ovate, brown; whorls five, slightly convex; suture slightly impressed; apex acute; aperture acute and somewhat narrowed above; columella produced in a right line; outer lip often thickened within. Length 0.75 inch; breadth 0.45 inch; divergence of the spire varying in different shells from 65° to 70°.

REMARKS.—This species is abundant in various parts of this state. Its young are not easily distinguished from those of the preceding species.

Physa gyrina.—SAY.

DESCRIPTION.—Shell long-ovate, yellowish brown; whorls five, slightly convex; suture moderately impressed; apex acute; aperture less acute above than the preceding species; columella a little curved below; outer lip often thickened within. Length 0.55 inch; breadth 0.75 inch; divergence of the spire 50°.

REMARK.—This species is very rare in this state.

Physa hypnorum.—DRAP.

DESCRIPTION.—Shell elongate, yellowish brown; whorls six, moderately convex; suture well impressed; apex acute; spire nearly as long as the aperture, which is regularly narrowed to the tip; columella oblique, in its lower part turned backwards and upwards; outer lip not thickened within. Length 0.58 inch; breadth 0.25 inch; divergence of the spire 45°.

REMARKS.—This species, described by Say as *P. elongata,* does not differ from the European shell, whose name we have prefixed to it. It is found in swamps and in small sluggish streams.

The above four species of Physa differ chiefly in the ratio of the spire to the aperture, and in the divergence of the former, which depends on the ratio of the length and breadth so far as it is uniform in different parts of the spire. The gradation in these characters is parallel, as may be seen by a comparison of their measurements.

GENUS PLANORBIS.

Generic Characters.—Shell with the revo-

FRESH WATER AND LAND SHELLS.

lutions of the spire in a plane, and subsequently visible on both sides; aperture lunated by the intrusion of the penult whorl; operculum none; animal long, rolled up like the shell; head saddle-shaped; tentacles long, contractile, with the eyes at their inner base.

Planorbis lentus, *P. corpulentus*, and *P. trivolvis*, of SAY, are undoubtedly varieties of one species, to all of which the following description will apply.

DESCRIPTION.—Shell brown, sometimes greenish, coarsely striate across the whorls, of which there are four and a half; inner whorls sharply carinate on the left side; suture very deep, except between the inner whorls of the left side, where it is not depressed below the carina; inclination of the shell to the left from a perpendicular 15° to 20°; aperture extending beyond the plane of the left side, sometimes beyond that of the right side, narrowing from the right to the left, with about three quarters of the height of the penult whorl moderately intruding. Greatest breadth 1.1 inch; least breadth 0.36 inch; height of aperture 0.58 inch.

REMARKS.—Sometimes the carination of the left side extends through all the whorls. The extension of the aperture on the right side is of a very variable character, especially at different ages, and in some localities the growth is very exuberant. A remarkable example of the latter case occurred in Otter Creek, just below the falls in Middlebury, where great numbers of large and beautiful specimens were obtained in the spring of 1839, although they have since entirely disappeared.

Planorbis campanulatus.—SAY.

DESCRIPTION.-Shell brownish or greenish yellow, finely striate; whorls four and a half, narrow, sub-carinate on the left side; inner whorls on this side scarcely depressed below its plane, exhibiting the apex distinctly; cavity of the right side very profound; inclination from a perpendicular to the left about 20°; aperture abruptly campanulate, oblique, including the lower two-thirds of the height of the penult whorl. Greatest breadth 0.59 inch; least breadth 0.45 inch; height 0.27 inch.

REMARKS.-This species resembles some small varieties of the preceding; but is distinguished by the abruptly campanulate aperture, and the narrowness of the outer whorl, which in this species is scarcely wider than the penult whorl, while in that species, owing to the rapid enlargement of the whorls from the centre, the last greatly exceeds all the others.

Planorbis bicarinatus.—SAY.

DESCRIPTION.—Shell brown, or greenish horn color; irregularly striate across, with very slight revolving striæ; whorls three, carinate on both sides, but more acutely on the left side; suture generally coincident with the carinæ except in the last semi-volution on the right side; concavities of both sides equally deep, that of the right wider; inclination to the left about 20°; aperture large, angulated by the left carina, embracing four-fifths of the length of the penult whorl. Greatest breadth 0.62 inch; least breadth 0.44 in.; height of aperture 0.31 inch.

REMARKS.—This species inhabits both quiet and running waters in ponds and streams of every size. It is very common.

Planorbis armigerus.—SAY.

DESCRIPTION.—Shell brownish horn color, feebly striate, shining; whorls four, subcarinate on the left side; right side slightly concave, left side deeply umbilicated; suture distinct and well impressed on both sides; inclination to the left about 40°; aperture nearly orbicular, slightly intruded upon by one-fourth of the height of the penult whorl, very far within armed with six teeth, of which two are on the inner side, one on the middle, elevated, lamellar, oblique, tortuous, large, the other just below it very small, nearly conical; four on the outer side, of which the two left are large, elevated, lamellar, oblique, converging outwardly, the two on the right small, subconic, but little elevated. Greatest breadth 0.34 inch; least breadth 0.29 inch; height of aperture 0.13 inch.

REMARKS.—This species is remarkable and singular in the genus for its teeth, which have been elevated by Haldeman to a generic character. It is common among dead leaves in still water. In swamps which are dried in the summer, it then takes refuge in the moist earth and leaves.

Planorbis exacutus.—SAY.

DESCRIPTION-.Shell extremely thin and fragile, brown, sometimes encrusted with a blackish substance, meniscoid; whorls four, carinate on the left side; inner whorl on the right side slightly depressed; left side deeply umbilicated; last whorl much broader than all the others, convexly compressed on both sides to an extremely acute, medial carina; inclination to the left about 60°; aperture large, cordiform. Greatest breadth 0.24 inch; least breadth 0.19 inch; height 0.055 inch.

REMARKS.—This species is more compressed than any other native Planorbis, the breadth being usually almost four times the height; the regular double convex form is also remarkable; also its tenuity, a full grown specimen weighing only .05 of a grain.

Planorbis parvus.—SAY.

DESCRIPTION.—Shell brownish horn color, feebly striate,shining; whorls three and a half or four, moderately increasing; both sides concave, but the left more than the right; last whorl subcarinate in the middle; inclination to the left about 40°; aperture subelliptical, slightly modified by the intrusion of two thirds of the height of the penult whorl; greatest breadth 0.25 inch; least breadth 0.2 inch; height 0.07 inch.

REMARKS.—This species is found plentifully in a great variety of stations.

Planorbis deflectus.—SAY.

DESCRIPTION.—Shell horn color; finely striate; whorls four; last whorl well rounded, indistinctly carinate below; right side convex, flattened at the apex; left side deeply concave; suture deep; inclination to the left about 45°; aperture round-ovate; greatest breadth 0.17 inch; least breadth 0.13 inch; height 0.06 inch.

REMARKS.—The shell above described is *P. elevatus*, ADAMS, which is probably the young of Say's species. It is very nearly allied to the preceding, but differs in the elevation of the spire on the right side, and deeper concavity of the left, and in the absence of a medial carina; the last whorl is also often abruptly deflected downwards.

Planorbis hirsutus.—GOULD.

DESCRIPTION.—Shell horn-color, striate; epidermis green, with raised revolving hirsute lines; whorls three and a half, last one strongly carinate in mature shells, less so in the young, and in the former often abruptly deflected downwards near its termination; right side with a small narrow concavity; left side sometimes generally concave, sometimes like the right; inclination to the left about 40° to 50°, increasing with age; aperture nearly orbicular, scarcely modified by the intrusion of the penult whorl. Greatest breadth 0.31 inch; least breadth 0.25 inch; height 0.1 inch.

REMARKS.—The mature shell resembles *P. deflectus*, but is distinguished by the medial carina of the outer whorl. It very nearly resembles *P. albus* of Europe, and probably is not specifically distinct.

FAMILY COLIMACEA.
GENUS SUCCINEA.

Generic Characters.—Shell ovate or ovate-conic, umber-colored; aperture large, longer than wide; outer lip sharp, never reflected; columella not folded, thin; operculum wanting; animal with four tentacles, with the eyes at their summit as in Helix.

Succinea obliqua.—SAY.

DESCRIPTION.—Shell ovate, striate; whorls three, oblique; spire half as long as the aperture; last whorl very large and convex; aperture ovate, nearly as broad above as below, somewhat oblique.—Length 0.97 inch; breadth 0.55 inch; divergence 70°.

REMARKS.—In the New England states this shell is generally of a deep umber color, but in Ohio it is pale. It is found in moist grounds, under stones and wood. The animal is beautifully mottled with dark purple on a cream-colored ground. It goes into winter-quarters in October, forming a thin transparent epiphragm. The shell which we have described may be *S. campestris*, SAY, or more probably the latter is only a variety of *S. obliqua*.

Succinea ovalis.—SAY.

DESCRIPTION.—Shell ovate, somewhat conic, striate; whorls three; spire less than one-third as long as the aperture, small, conic; last very large, elongate, patulous; aperture very large, exhibiting much of the interior of the spire, ovate. Length 0.61 inch; breadth 0.3 inch; divergence 64°.

REMARKS.—This species is common about the margins of water. It is extremely fragile.

Succinea avara.—SAY.

DESCRIPTION.—Shell small, ovate, conic, striate; whorls three, very convex, with the suture very deeply impressed; spire conic, five-sevenths as long as the aperture, which is not large, ovate.—Length 0.3 inch; breadth 0.17 inch; divergence 67°.

REMARKS.—The shell which Say describes under the name of *S. vermeta* is probably the adult of this species. The aperture is proportionally larger in the young, as is also true of *S. obliqua*. When young a viscid substance attaches dirt to the shell, which becomes clean when mature.

GENUS BULIMUS.

Generic Characters.—Shell ovate, or oblong-ovate, with the last whorl larger than the penult ; aperture longer than wide ; with the margins not continuous ; columella smooth, sometimes truncate. No operculum. Animal of the form of the shell, with four tentacles, of which the larger are oculiferous. The number of species in this genus, including the sub-genus Achatina, exceeds two hundred. But not more than six or eight are known in the United States, and only one in New England.

Bulimus lubricus.—Drap.

DESCRIPTION.—Shell oblong ovate, brown, shining ; whorls six, moderately convex ; suture well impressed ; spire twice as long as the aperture, which is ovate ; labrum a little thickened within, making a little more than a right angle with the columella, which is truncate. Length 0.26 inch ; breadth 0.1 inch ; divergence 45° in the upper part of the spire, below it is much less.

REMARKS.—This species, being common over a large part of Europe, is supposed by some to have been introduced thence into this country. It is remarkable, on this supposition, that it should have spread as far as the lake of the Woods and lake Winnipeg. As the divergence below the middle is very slight, the shell, when half grown, is nearly as wide as when mature.

GENUS PUPA.

Generic Characters.—Shell cylindrical ; apex obtuse ; aperture parallel to the axis of the shell, rounded below, more or less biangular above ; margins reflected, separated by a lamina appressed on the columella. No operculum. Animal with the form of the shell ; with four tentacles, of which the larger two are oculiferous at their summit, and the others are very minute.

Although a large portion of the exotic species exceed a half inch and many an inch in length, the native species are all minute, and some of them are the least of all our shells.

Pupa milium.—GOULD.

DESCRIPTION.—Shell ovate, brown, shining, with slight incremental striæ not discernible without a microscope ; whorls five, convex ; suture well impressed ; apex very obtuse ; aperture horizontally truncate above by the penult whorl, indented on the outer lip, with six teeth, of which one is at the indenture of the labrum, two very small teeth are in the lower part of the aperture, on the left side is a larger tooth double at its base, and at right angles to this are two on the horizontal margin ; the umbilicus is large. Length 0,06 inch ; breadth 0,03 inch.

REMARKS.—This species, the least of all which have been described in this country, was originally discovered in Middlebury. Its weight is 0.005 of a grain. It lives under moist decaying leaves, and at the foot of limestone ledges. None but a naturalist would find it.

Pupa ovata.—SAY.

DESCRIPTION.—Shell brown, ovate, tapering above the penultimate whorl ; whorls five, convex, with a distinct suture ; aperture small, ovate, with an indenture on the right side ; with six primary teeth, of which two are on the transverse lip, viz. a large one on the middle, and a small one to its right ; two are on the left and two on the right side ; sometimes a very small tooth is found on the left part of the transverse lip. Length, 0.08 inch ; breadth 0.05 inch.

REMARKS.—In color this species resembles *P. milium*, but is easily distinguished by its size and proportions, and the arrangement of the teeth. *P. modesta*, Say, for which this species has sometimes been mistaken, is described as having only four teeth.

Pupa badia.—ADAMS.

DESCRIPTION.—Shell reddish brown, cylindrical, very obtusely tapering in the two upper whorls ; whorls seven, moderately convex, with a well impressed suture ; aperture orbicular, less than one third of the length of the shell, with the margin slightly reflected, and the submargin contracted, with a single rather small tooth on the penultimate whorl ; umbilicus moderate. Length 0.14 ; breadth 0.07 inch.

REMARKS.—This rare species was discovered by Prof. Benedict at Crown Point, where, only, it has yet been found. Its aperture is wider, and umbilicus less than in *P. marginata*, DRAP. of Europe, but it may be only a variety. It is easily distinguished by its mahogany color.

Pupa armifera.—SAY.

DESCRIPTION.—Shell oblong ovate, of a dingy white, striate ; whorls seven, a little convex, with a moderately impressed suture ; apex very obtuse ; aperture subovate, with six teeth, of which the larger on the transverse lip is obliquely elongated, and nearly meets the labrum above ; one is on the left side, and four are below and on the right side ; of the latter, the first and fourth are the least, and are sometimes wanting. Length 0.17 inch ; breadth 0.09 inch.

REMARKS.—This is the largest species of Pupa found in the United States, and by its color is distinguished from all which approximate to it in size. It occurs plentifully at Crown Point under stones in very dry situations. A few dead specimens have been found in Bridport, on the margin of lake Champlain, which may have been drifted from the opposite side.

Pupa albilabris.—WARD. Inedit.

DESCRIPTION.—Shell brown, finely striate, long-ovate, tapering above the penult whorl; whorls six, convex, with a well impressed suture; aperture a little less than half as long as the spire, without teeth, with a reflected, white, thick, flattened margin; umbilicus moderate.—Length 0.18 inch; breadth 0.07 inch.

REMARKS.—This species is well known as Say's *cyclostoma marginata*. As the latter specific name is preoccupied in the genus Pupa, to which it belongs, it has received the name under which we have described it. A very few specimens only have been found alive at Crown Point, and one dead on the Vermont shore of the lake.

Pupa contracta.—SAY.

DESCRIPTION.—Shell white, ovate, tapering above the body whorl; whorls five, convex, with a well impressed suture; aperture sub-triangular, with the transverse lamina raised, and forming with the labrum a continuous lip, much contracted in the throat, with three teeth, one on the transverse lip, large, prominent, and sinuous, another on the right side, where the throat is most contracted, and the third is merely a convexity caused by the fold of a large umbilicus. Length 0.1 inch; breadth 0.06 inch.

REMARKS.—This species is easily recognized by its elevated transverse lip. It is found under wood or stones in moist pastures.

Pupa Tappaniana.—WARD. Inedit.

DESCRIPTION.—Shell very small, pale horn color, translucent, tapering above the penultimate whorl; whorls a little more than five, convex, with a well impressed suture; aperture sub-orbicular, (the penult whorl cutting off about one-third of the circle,) about one-third of the length of the shell; margin sharp, with a narrow contraction in the sub-margin, beneath which is a thickening within, on which are the labial teeth; teeth eight, five primary and three secondary; of the former the largest is on the penultimate whorl, the next largest on the left side of the aperture; at the base, beginning at the left hand, is a primary, then a secondary, a primary, a secondary, a primary, and another secondary, extending nearly to the upper extremity of the right margin: the last three primaries are not constant in size; umbilicus open. Length 0,08 inch; breadth 0,05 inch.

REMARKS.—This species is easily distinguished from the preceding by its teeth.

Pupa exigua.—SAY.

DESCRIPTION.—Shell white, shining, elongate, tapering above the penultimate whorl; whorls six, convex, with a well impressed suture; aperture ovate, with the upper lip oblique, margin reflected and thickened, teeth two, of which the larger is on the oblique lip, and the other, which is small, is on the left side; umbilicus distinct. Length 0,08 inch; breadth 0,03 inch.

REMARKS.—This shell is easily distinguished by its neat, shining appearance, and graceful form. It is more common than any other species of this genus in Vermont, and is found under stones and logs in moist places.

GENUS HELIX.

Generic Characters.—Shell orbicular or globose, usually convex or conoid above, but sometimes flattened; aperture wider than long, semi-elliptic or lunate, contiguous to the axis of the shell, with the outline interrupted by the intrusion of the penult whorl. No operculum. The animal, commonly called a *snail*, has four tentacles, of which the posterior pair are larger and oculiferous.

Helix albolabris.—SAY.

DESCRIPTION.—Shell globose-conic, with a light brown, sometimes reddish epidermis, with five parallel oblique incremental striæ, and very minute revolving lines; whorls five and a half, convex, with a well impressed suture; aperture contracted by the labrum, which is white, flat, broadly reflected, and extends beneath to the centre of the shell, covering the umbilicus, which is open only in the young. Greatest breadth 1.35 inch; least breadth 1 inch; height 0.8 inch; divergence of the spire 135°.

REMARKS.—This species is found very commonly in most parts of Vermont. On the islands called the Four Brothers,

FRESH WATER AND LAND SHELLS.

in lake Champlain, it is abundant, in company with *Succinea obliqua*. The reddish variety is rare. The size of mature specimens is sometimes less than an inch in their greatest diameter. During the day, except in damp weather, they are confined to their retreats under logs and stones. Their eggs are white, nearly globular, and about 0.2 inch in diameter. The young shell does not receive the reflected lip until of its full size.

Helix thyroidus.—SAY.

DESCRIPTION.–Shell globose-conic, with a light brown, sometimes reddish epidermis, with five parallel oblique incremental striæ ; whorls five, convex, with a well impressed suture ; aperture contracted by the labrum, which is widely reflected, flat, white, next the aperture, yellowish externally ; inner margin with an oblique tooth ; umbilicus partly covered by the reflected labrum, exhibiting only one volution. Greatest breadth 0.95 inch ; least breadth 0.7 inch ; height 0.47 inch ; divergence 140°.

REMARKS.—This species is extremely rare in Vermont, but is more common in the western states. It might, at first, be confounded with the preceding, but is distinguished by the tooth on the inner margin of the aperture, the partially open umbilicus, and the yellow color of the outside of the labrum.

Helix dentifera.—BINNEY.

DESCRIPTION.—Shell depressed, with a yellowish horn-colored epidermis, with fine parallel oblique incremental striæ ; whorls five, with the suture distinct but not deep ; aperture contracted by the lip, which is white, and broadly reflected ; inner lip with a large tooth, long and parallel with the lower margin ; umbilicus none. Greatest breadth 0.9 inch ; least breadth 0.6 inch ; height 0.44 inch ; divergence 135°.

REMARKS.—This very rare species has been found only by Dr. Binney on the east side of the Green Mountains.

Helix palliata.—SAY.

DESCRIPTION.—Shell depressed, with a dark reddish brown epidermis, which is thickly covered, when in a perfect state of preservation, with acute hair-like projections ; with numerous fine oblique incremental striæ; whorls five, flattened, with a distinct suture ; aperture much contracted and made three-lobed by the teeth ; labrum white and broadly reflected ; teeth three, of which one is long and curved, nearly covering the pillar lip ; two are on the inner margin of the labrum ; one above is acute and prominent, and the other below is long and lamellar ; the labrum is continued over the umbilical region in a white callus. Greatest breadth 0.9 inch ; least breadth 0.6 inch ; height 0.48 inch ; divergence about 160°.

REMARKS.—This species, which is not rare in the western states, is seldom found in Vermont. It is easily distinguished from *H. tridentata* by the want of an umbilicus.

Helix monodon.—RACKETT.

DESCRIPTION.-Shell globose-conic, with a brown hirsute epidermis, with minute incremental striæ ; whorls six, with a distinct suture ; aperture contracted by a deep groove behind the tip, which is white, reflected, flattened, covering more or less of the umbilicus, which is deep but not wide ; inner lip with a compressed elongated tooth, parallel with the lower part of the margin. Greatest breadth 0.45 inch ; least breadth 0.42 inch ; height 0.26 inch ; divergence 135°.

REMARKS.—In this description we have included *H. fraterna*, SAY, a variety in which the umbilicus is entirely covered by the labrum. As this is a variable character, and the other characters present no distinction, we cannot separate them. Rackett's name has the priority both of Say's description of the variety and of Ferussac's use of the same name for another species. This is common on hill sides in rather dry places. Specimens vary in respect of size and the elevation of the spire.

Helix concava.—SAY.

DESCRIPTION.—Shell depressed, a little convex above, with fine oblique incremental striæ ; epidermis pale greenish horn color ; whorls five, flattened above, elegantly rounded below, the outer one dilating towards the aperture, with a well impressed suture ; labrum partially reflected below, simple above ; inner lip with a thin callus, which connects the extremes of the labrum ; umbilicus wide and deep, exhibiting all the volutions. Greatest breadth 0.75 inch ; least breadth 0.6 inch ; height 0.33 inch ; divergence about 155°.

REMARKS.—This species is rare in Vermont, but more common in the western states. West of the Rocky Mountains it is of a much greater size, exceeding an inch in diameter.

Helix pulchella.—MULL.

DESCRIPTION.—Shell much depressed, pale horn color, nearly transparent, finely striate, with a colorless epidermis ; whorls three and a half, convex, with a deep suture, the last one much larger than the preceding; aperture nearly orbicular, dilated; labrum much thickened, white, reflected, scarcely interrupted by the intrusion of the penultimate whorl; umbilicus large. Greatest breadth 0.095 inch; least breadth 0.078 inch; height 0.05 inch; divergence 160°.

REMARKS.—This species is remarkable for its wide geographical distribution. It is common in Great Britain and a large part of Europe, and in this country is found as far south as South Carolina, as far west as Council Bluffs, and as far east as Maine. It is very abundant in some parts of Vermont. It is the *H. minuta* of Say.

Helix Sayii.—BINNEY.

DESCRIPTION.—Shell depressed globose, with numerous fine oblique incremental striæ; epidermis very light brown, shining; whorls five and a half, convex, with a well impressed suture; labrum white, narrow, reflected, with a small rounded tooth on the inner edge below; inner lip with a small oblique tooth on the middle; umbilicus not very wide but deep and exhibiting all the volutions. Greatest breadth 1 inch; least breadth 0.8 inch; height 0.55 inch; divergence 135°.

REMARKS.—This species was originally described by Say with the name of *H. diodonta*, but as this name had been preoccupied, Dr. Binney proposed that of *H. Sayii*. The species is rare in Vermont. It is easily recognized by its narrow lip and two small teeth, of which, however, the one on the inner margin is sometimes wanting.

Helix tridentata.—SAY.

DESCRIPTION.—Shell depressed, a little convex above, with crowded oblique incremental striæ; epidermis brown; whorls five, a little flattened above, with a distinct suture; aperture three-lobed, contracted by a groove behind the labrum, which is white, reflected, flattened, furnished with two acute prominent teeth; inner lip with a prominent, oblique and slightly curved tooth; umbilicus rather wide, deep.

REMARKS.—This species is widely distributed, having been found in Florida, and in the western states. In the former region it is very small, in the latter very large. In Vermont it is of an intermediate size.

Helix labyrinthica.—SAY.

DESCRIPTION.—Shell small, elevated conic above, flattened below, with very coarse, regular, oblique incremental striæ, so crowded that the intervening spaces are rounded ribs, which are obsolete beneath; epidermis brown, sometimes inclining to horn color; whorls six, convex, with a well impressed suture; labrum thickened, reflected, and usually reddish brown; inner margin with two compressed, perpendicular, parallel teeth, which are prolonged into the throat of the aperture, resembling the track of a rail road; but the lower tooth is smaller, and sometimes obsolete; umbilicus narrow and not deep. Greatest breadth 0.1 inch; least breadth 0.08 inch; height 0.08 inch; divergence 135° in the upper third, half as much below.

REMARKS.—This beautiful little shell is at once distinguished by its peculiar teeth. The aperture is sometimes of an elegant red color. It is found under leaves in the forests, and at the foot of limestone ledges. It occurs as far west as Council Bluffs.

Helix indentata.—SAY.

DESCRIPTION.—Shell much depressed, convex above, shining, of a pale horn color, nearly transparent, with distant, nearly equi-distant impressed transverse lines, of which there are 25 to 30; there is often an impressed line parallel with and immediately below the suture; whorls four and a half, slightly convex, with a distinct impressed suture, and rapidly enlarging; aperture large; labrum sharp, terminating beneath at the centre of the shell, where is a deep indentation rather than umbilicus. Greatest breadth 0.18 inch; least breadth 0.15 inch; height 0.08 inch; divergence 160°

REMARKS.—This species resembles *H. arborea*, SAY, but is distinguished by its distant impressed lines, by the enlargement of the last whorl, and the want of an umbilicus. It is rare.

Helix arborea.—SAY.

DESCRIPTION.—Shell somewhat depressed, convex above, shining, of a pale horn color or brown, nearly transparent, with very fine crowded incremental striæ; whorls nearly five, convex, with a well impressed suture; aperture a little modified by the intrusion of the penult whorl; labrum sharp; umbilicus deep, about three fourths as wide as the last whorl. Greatest breadth 0.3 inch; least breadth 0.26 inch; height 0.15 inch; divergence 135°.

REMARKS.—This very common species is found both in a dry and in a wet station. In the former, the shell and the animal are of a pale horn color, and smaller. In the latter the shell is brown, and the animal nearly black. The dimensions above given are of a large specimen of the latter variety. The species is very

FRESH WATER AND LAND SHELLS.

widely distributed through the United States and Missouri Territory.

Helix electrina.—GOULD.

DESCRIPTION.—Shell much depressed, convex above, shining, of a pale horn color, sometimes yellowish or brownish, nearly transparent, with numerous very fine inequidistant impressed lines or striæ of growth; whorls three and a half, slightly convex, with a well impressed suture, and an impressed line immediately below the suture, and parallel with it; the last whorl rapidly enlarging; aperture large, slightly modified by the intrusion of the penult whorl; labrum sharp; umbilicus narrow and deep. Greatest breadth 0.2 inch; least breadth 0.16 inch; height 0.1 inch; divergence 165°.

REMARKS.—This species much resembles *H. indentata* above, but has the striæ much more numerous, and usually one whorl less; beneath the resemblance to *H. arborea* is equally striking, but the umbilicus is not so wide. Without examination of both sides, it is very liable to be confounded with one or the other of the above species. It has been found in Missouri, Ohio, Massachusetts, New York and Vermont.

Helix inornata.—SAY.

DESCRIPTION.—Shell much depressed, convex above, shining, with very fine oblique incremental striæ; epidermis brown horn color; whorls five, slightly convex, with a distinct but not deep suture; the last whorl much larger than the preceding; aperture very wide, much modified by the intrusion of the penultimate whorl, with an opaque white deposit within, which is a little distant from the sharp labrum; the latter extends nearly to the centre of the shell, projecting into the small umbilicus. Greatest breadth 0.55 inch; least breadth 0.47 inch; height 0.27 inch; divergence 165°.

REMARKS.—A single specimen only of this species has been found in Vermont, in Middlebury. It closely resembles *H. cellaria*, Mull.

Helix fuliginosa.—GRIFFITH.

DESCRIPTION.-Shell globose-conic, with very minute irregular oblique striæ of growth; epidermis dark smoky brown; whorls four and a half, convex, with a well impressed suture; the last whorl much larger than the preceding; aperture nearly orbicular, not much modified by the intrusion of the body whorl, with a very thin deposit on the inside; umbilicus deep, moderately wide. Greatest breadth 0.85 inch; least breadth 0.8 inch; height 0.5 inch; divergence 135°.

REMARKS.—This species is not common. It resembles the preceding, but differs in size, color, form of the aperture, and greater width of the umbilicus. It is the *H. lucubrata* of Say, a name perhaps entitled to preference, since that of Griffith, although previously in use in cabinets, was not published until after Say's name had appeared in print.

Helix multidentata.—BINNEY.

DESCRIPTION.—Shell much depressed, conoid above, shining, reddish brown, translucent, with very fine, somewhat regular impressed lines or striæ of growth; whorls seven, narrow, convex, often with a very small impressed line revolving just above the suture, which is deep; the whorls increasing but slightly in diameter; aperture narrow, very much modified by the intrusion of the penult whorl; labrum sharp; teeth in rows, far within the aperture, on its outer and lower half; the rows are curved, with the convexity towards the aperture, and contain from 4 to 6 closely approximate teeth, appearing through the shell, under a magnifier, like glass beads; the number of rows varies from two to four, of which one only is visible from the aperture; the umbilicus is very narrow and deep. Greatest breadth 0.12 inch; least breadth 0.11 inch; height 0.06 inch; divergence 150°.

REMARKS.—This elegant little species was discovered by Dr. Binney in Strafford, and has since been found in Middlebury, also in New York, at Malone. It has so little resemblance to any other species, that comparison is unnecessary.

Helix minuscula.—BINNEY.

DESCRIPTION.—Shell depressed, whitish horn color, with microscopic incremental striæ; whorls more than four, very convex, with a deep and very conspicuous suture; last whorl not much larger than the preceding; aperture nearly circular, not much modified by the intrusion of the penult whorl; labrum sharp; umbilicus very large. Greatest breadth 0.08 inch; least breadth 0.07 inch; height 0.03 inch; divergence about 150°.

REMARKS.—This little species has been found in Ohio and in this state. In size and color it is like *H. pulchella*, but in the other characters is at once distinguished.

Helix lineata.—SAY.

DESCRIPTION.—Shell very much depressed and discoid, with parallel equidistant raised revolving lines; epidermis green; whorls four and a half, very convex, narrow, with a deep suture, last whorl very little enlarged; aperture lunate, very much modified by the intrusion

FRESH WATER AND LAND SHELLS.

of the penult whorl; labrum sharp; umbilicus concave, very broad and deep, exhibiting very distinctly all the volutions to the apex; far within the aperture may often be seen a pair of conical teeth on the inner side of the outer whorl, one on the middle, the other below; sometimes one is obsolete; often a second and sometimes a third pair may be seen through the sides of the shell much farther within. Greatest breadth 0.14 inch; least breadth 0.13 inch; height 0.06 inch; divergence never less than 160°, usually 170°.

REMARKS.—Above, this shell resembles H. *multidentata*, in the depression of the spire and narrowness of the whorls, but in the other characters is very different. No other native species has such revolving minute carinæ. It has been found in the northern and middle states.

Helix striatella.—ANTH.

DESCRIPTION.—Shell depressed-convex, with very much crowded deep incremental striæ; epidermis reddish or yellowish brown; whorls four, convex, with a well impressed suture, moderately increasing in diameter; aperture nearly circular, slightly modified by the intrusion of the penult whorl; labrum sharp; umbilicus not so wide as the last whorl, deep, distinctly exhibiting the volutions to the apex. Greatest breadth 0.25 inch; least breadth 0.22 inch; height 0.12 inch; divergence 140° to 150°.

REMARKS.—This species is quite common in Vermont. It resembles H. *perspectiva*, SAY, a species, which has not been found in the New England states. The latter has one or two more whorls, the umbilicus much wider, and the striæ much coarser It is also a larger shell. This species does not appear to differ from the European shell, H. *ruderata*, STUDER. Comparing specimens from Stiria with those of Vermont, we are unable to detect any difference. But as some naturalists are not convinced of their identity, we have retained the name of the American author, although the European name has the priority of many years.

Helix alternata.—SAY.

DESCRIPTION.—Shell depressed-convex, with acute, raised, equi-distant obliquely curved striæ, which render the shell scabrous; epidermis horn color, variegated with rufous spots and bars obliquely arranged; whorls six, convex, with a well impressed suture; aperture very oblique, nearly circular, brilliant, sometimes pearly within; labrum sharp; umbilicus broad and deep, exhibiting all the volutions; beneath, the colored bars are more regular, and converge into the umbilicus: they are interrupted by a colorless zone a little below the middle of the last whorls. Greatest breadth 1 inch; least breadth 0.87 inch; height 0.59 inch; divergence 125° to 135°.

REMARKS.—This species has been found throughout most of the territory of the United States. It is very common in this state, living under stones and logs on hill-sides in rather moist but not wet places. When young, its outline is carinated. It resembles the H. *radiata*, of Europe, but cannot be mistaken for any other American species.

Helix chersina.—SAY.

DESCRIPTION.—Shell elevated and conic above, convex and shining beneath, striæ of growth excessively minute; epidermis brownish amber-colored; whorls six, very convex, with a deep suture, not increasing much, so that the last is but little larger than the penultimate whorl; aperture very wide, reaching to the axis beneath, much modified by the intrusion of the penultimate whorl; labrum sharp; umbilical region indented. Greatest breadth 0.115 inch; least breadth 0.105 inch; height 0.09 inch; divergence 90°

REMARKS.—This and H. *labyrinthica* are distinguished from other native species of Helix by the elevation of the spire, and are very distinct from each other in most characters other than size and form. The species is not very rare in this state, and having been found in Georgia and the North West Territory, is, no doubt, widely dispersed. From its minute size it is liable to escape detection.

FAMILY LIMACIANA.
GENUS VITRINA.

Generic Characters.—Shell with a depressed, convex, obtuse spire, with but few whorls, of which the last is extremely large; the aperture is very large, wider than long, interrupted by the penult whorl; umbilicus wanting. The shell is extremely thin and transparent, and is capable of containing only a part of the animal. No operculum. The animal is much too large to enter the shell, resembling a Helix. It is long, mostly straight, with the posterior part distinct, spiral, protected by the shell; with four tentacles, of which the anterior pair is very short.

Vitrina pellucida.—DRAP.

DESCRIPTION.—Shell globose-discoid, shining, with the incremental striæ ex-

FRESH WATER AND LAND SHELLS.

cessively minute, transparent, and nearly colorless; whorls two and a half, scarcely convex, with the suture but little impressed, sometimes with a slightly impressed line revolving near the suture; aperture elliptic, not much modified by the intrusion of the penultimate whorl; labrum thin and sharp; inner lip slightly reflected. Greatest breadth 0.24 inch; least breadth 0.18 inch; height 0.12 inch; divergence about 160°.

REMARKS.—This species, well known over a large part of Europe, was observed first on this continent by Mr. Say, who remarks that it "was first found near Coldwater Lake, in lat. 48¾ N., under stones, fallen timber, &c. It afterwards occurred, in similar situations, until we approached Lake Superior, when it was no more seen." This side of Lake Superior it has been found only at Rogers' rock, near the N. E. extremity of Lake George, within the space of a square rod. As it occurred so near to Vermont, and will very probably be found within its limits, we have included it among our species. It does not appear to differ from the European shell, except in the want of a greenish tinge.

Genus LIMAX.

Generic Characters.— Animal without a shell, oblong, convex above, furnished with a leathery shield over the anterior dorsal region; beneath with a flattened longitudinal foot; with four tentacles, of which the posterior pair are larger and oculiferous; with the branchial cavity beneath the shield, opening on the right side.

The species of this and of kindred genera are commonly *slugs*, or *snails*, from their resemblance to the inhabitants of snail shells. In turning over stones and logs or boards, they are often seen.

Limax campestris.—BINNEY.

DESCRIPTION.—"Color usually of various shades of amber, without spots or markings, sometimes blackish; head and tentacles smoky. Body cylindrical, elongated, terminating in a very short carina at its posterior extremity, mantle oval, fleshy, but little prominent, with five concentric lines; back covered with prominent, elongated tubercles and furrows; foot narrow, whitish; respiratory foramen on the posterior dextral margin of the mantle; body covered with a thin watery mucus. Length about one inch."

REMARKS.—This species is smaller than *L. agrestis*, LINN. "The tuberosities of the surface are more prominent in proportion to their size, are not flattened or plate like, and are not separated by darker colored anastomosing lines, the intervening lines being of the same color as the general surface." It is found under wood and stones in various situations.

Genus TEBENNOPHORUS.—*Binney*

Generic Characters.—"Mantle covering the whole superior surface of the body; pulmonary cavity anterior, orifice on the right side towards the head; orifice of the rectum contiguous to and a little above and in advance of the pulmonary orifice; organs of generation united, orifice behind and below the superior tentacle of the right side; without testaceous rudiment, terminal mucous pore, or locomotive band of the foot."

Tebennophorus Caroliniensis.—BOSC.

DESCRIPTION.—Body whitish, with brownish or blackish spots arranged in three ill defined, longitudinal, anastomosing bands, with small spots between; inferior margin cream colored; foot whitish; superior tentacles knobbed at the extremity, with the eyes on the upper part of the knob; "cuticle covered with irregular, vermiform glands, anastomosing with each other, and having a general tendency to a longitudinal direction, with shallow furrows between, lubricated with a watery mucus." Length, when fully extended, upwards of three inches.

REMARKS.—This species inhabits forests, in damp, shaded places, about decaying wood. In the cabinet of Middlebury college are two specimens, which were taken from the nest of the brown hawk, (*Falco fuscus*, GM.)

Genus PHILOMYCUS.—*Rafinesque.*

Generic Characters.—Animal resembling the preceding, but entirely destitute of a mantle.

Philomycus dorsalis.—BINNEY.

DESCRIPTION.—"Color of upper surface ashy, with a shade of blue, an uninterrupted black line extending down the centre of the back; superior tentacles black, about one eighth of the length of the body; lower tentacles blackish, very short; body cylindrical and narrow, terminating posteriorly in an acute point; base of foot white, very narrow, its separation from the body not well defined; upper surface covered with elongated and slightly prominent glandular projections, the furrows between indistinct; respiratory orifice very minute, situated on the right side, about one eighth of an inch behind the insertion of the superior tentacle." Length nearly an inch.

REMARKS.—This species is found in the forests, in the soil about decaying wood. It is probably not very common.

FAMILY CALYPTRACIANA.
Genus Ancylus.

Generic Characters.—Shell thin, oblong-elliptic, obliquely conic; apex acute, curved backwards; aperture elliptic; margins sharp. Animal covered, not concealed, by the shell, with two compressed tentacles and the eyes on the inner part of the base; foot elliptic, not so wide as the body.

Ancylus parallelus.—Haldeman.

Description.—Shell nearly transparent, oblong-ovate; epidermis thin, horn color; sides straight, slightly divergent forwards; apex subacute, moderately elevated, with two fifths of the length of the shell behind, leaning to the right. Length 0.25 inch, width 0.15 inch, height 0.08 inch.

Remarks.—This species is found in streams and ponds in many parts of the New England states. It was supposed to be Say's *A. rivularis*, not on account of any resemblance between the two shells, but from the meagerness of the description. From some remarks of this learned naturalist, comparing *A. rivularis* with *A. tardus*, it seems probable that the former is not an elongate species.

Ancylus tardus.—Say.

Description.—Shell nearly transparent, elliptical; epidermis thin, horn color; sides somewhat curved; apex subacute, elevated, a little behind the middle, leaning backwards but scarcely to the right. Length 0.25 inch, width 0.16 inch, height 0.13 inch.

Remarks.—This is at once distinguished from the preceding by its proportions. *A. rivularis* differs in having the apex more on one side, and one end distinctly wider than the other.

FAMILY NAIADES.
Genus Anodonta.

Generic Characters.—Shell equivalve, inequilateral, transverse; hinge toothless; the two muscular impressions remote; ligament long. The shell is usually very thin. Animal with the lobes of the mantle entirely separate.

Anodonta Benedictensis.—Lea.

Description.—Shell ovate-trapezoidal, thin; epidermis coarsely striate, yellowish or greenish brown, usually with two or three dark green rays posteriorly, in old shells of a very dark color, obscuring the rays; beaks rather small, wrinkled, approximate; discs moderately inflated; anterior side two thirds to one half as long as the posterior; hinge margin straight; anterior and posterior margins straight and divergent above, below abruptly rounded into the basal margin, which is moderately curved throughout, except in old shells, in which it is straight or even incurved in the middle. Dimensions of two specimens: No. 1, length 4.5 inches, height 2.75 inches, width 1.7 inch; No. 2, length 3.87 inches, height 2.5 inches, width 1.5 inch.

Remarks.—It will be seen in the above measurements, that the proportionate length is subject to considerable variation, which affects only the posterior side, and in part is a sexual distinction. This species is abundant in lake Champlain, but is not found elsewhere. It is much larger than any other Anodonta in this state.

Anodonta marginata.—Say.

Description.—Shell ovate, widest below the beaks, thin; epidermis yellowish and greenish brown, with very irregular striæ of growth; beaks rather prominent, with numerous small wrinkles; discs moderately inflated, flattened; anterior side about two fifths as long as the posterior; hinge margin curved; posterior margin slightly curved in a descent of one third of the length of the shell, then rapidly rounding into the basal margin, which is nearly straight at and behind the middle; anterior margin regularly rounded. interior bluish. Length 3.8 inches, height 1.6 inch, width 1.15 inch.

Remarks.—This species may be most easily distinguished from the *A. undulata* by the greater size and very minute wrinkles of the beaks, and the flattening of the umbo. It has been found in Otter Creek at Wallingford. If it be not the *A. marginata* of Say, that species cannot now be recognized. It has been found more abundantly in Massachusetts by Dr. Gould, on whose authority I have given it this name.

Anodonta fluviatilis —Dillwyn.

Description.—Shell oblong-ovate, widest behind the beaks, thin; epidermis smooth, yellowish, and brownish green, olivaceous posteriorly and above, where are a few obscure dark rays; beaks quite small, with numerous small wrinkles; discs moderately inflated, convex; anterior side between a third and a fourth as long as the posterior; hinge margin straight, rising into a wing posteriorly; posterior margin very obliquely descending to a truncate extremity; inferior margin nearly straight; anterior margin regularly rounded; interior surface bluish, iridescent. Length 2.4 inches, height 1.25 inch, width 0.9 inch.

REMARKS.—A few small specimens of this species have been found in Middlebury. In Massachusetts and further south it attains a much greater size. It is very similar to the preceding, but is distinguished by its wing, small beaks, and convex disc. It more nearly resembles *A. cygnea* of Europe.

Anodonta undulata.—SAY.

DESCRIPTION.—Shell oblong ovate, widest behind the beaks, not thin, with coarse and fine striæ of growth; epidermis yellowish, brownish, or blackish green, with numerous irregular dark green rays, which are obscured when the general color is dark; beaks quite prominent, much undulated; discs moderately inflated, convex; anterior side usually less, sometimes more than one third as long as the posterior; hinge margin nearly straight; posterior margin descending in a curve through a third of the length of the shell, then abruptly rounded into the inferior, which is slightly curved or straight; anterior margin regularly rounded; interior bluish, but often covered with a light salmon colored nacre, with a dark blue or brown margin; hinge with obsolete teeth. Dimensions of two specimens: No. 1, length 2.75 inches, height 1.4 inch, width 0.85 inch. No. 2, length 2.65 inches, height 1.45 inch, width 1.1 inch.

REMARKS.—This species is found in small streams and in lake Champlain.— When the epidermis is of a light color and the rays conspicuous, it is a very beautiful shell. More frequently it is dark, and the appearance unattractive. It is intermediate between this genus and the next.

GENUS ALASMODONTA.

Generic Characters.—Shell as in Anodonta, but furnished with a stout, striated, and simple or divided cardinal tooth in each valve: also the shell is usually thicker. Animal as in Anodonta.

Alasmodonta arcuata.—BARNES.

DESCRIPTION.—Shell very long ovate, arcuate; epidermis black, or brownish black, with very distinct striæ of growth, very much developed at the margin; beaks small, depressed, much eroded; discs moderately inflated, flattened; anterior sides more than one-fourth as long as the posterior; hinge margin regularly curved into the posterior, which descends at first very obliquely, and is then irregularly rounded into the basal margin; this is incurved, and the anterior is regularly rounded; interior with a brilliant, thick nacre, iridescent posteriorly. Length 4.9 inches; height 2.2 inches; width 1.35 inch.

REMARKS.—This species has been found at Burlington. It has been considered identical with *Unio margaritiferus* of Europe, but that shell is shorter, and has the beaks more central and elevated. It yet more nearly resembles the *Unio sinuatus* of Europe, which is higher and has the beaks more central. Perhaps it may not be distinct from the latter. The young have the basal margin straight. It is found throughout New England.

Alasmodonta rugosa.—BARNES.

DESCRIPTION.—Shell ovate; epidermis with irregular incremental striæ, which are mostly fine, greenish brown; beaks small, not prominent, undulate; discs flattened, with two ridges extending posteriorly in slightly curved lines, between and above which the surface is crowded with numerous crowded wrinkles, which, for the most part, run posteriorly and upwards; anterior side much depressed, about one-third as long as the posterior; hinge margin arcuate behind the teeth, otherwise nearly straight, ascending posteriorly; posterior margin descending in a straight line to the upper umbonial angle; extremity truncate between the umbonial angles; inferior margin nearly straight; anterior margin regularly rounded; inner surface often with a light salmon-colored deposit. Length 4.1 inches; height 2.3 inches; width 1.25 inch.

REMARKS.—This species is common in the western states, where it attains a greater size. Lake Champlain and the streams west of the Green Mountains appear to be the most eastern limit of its habitation.

Alasmodonta undulata.—SAY.

DESCRIPTION.—Shell ovate, epidermis smooth, blackish or greenish brown, with obscure darker rays; beaks large and prominent, with large and deep undulations; discs much inflated and convex, with a ridge more or less obtuse extending posteriorly; anterior side small, one-sixth to one-third as long as the posterior; hinge margin sinuous or simply curved;

pòsterior margin descending obliquely in a straight or slightly curved line, rounded below; inferior margin slightly curved; anterior margin regularly rounded; inner surface bluish, sometimes with a light salmon-colored nacre anteriorly or throughout. Dimensions of two specimens: No. 1, length 2.2 inches; height 1.4 inch; width 1.08 inch. No. 2, length 2.06 inches; height 1.2 inch; width 0.9 inch.

REMARKS.—This species is rather common in the northern middle states.— When young the epidermis is of a lighter color, the rays are more conspicuous, and the shell is shining and beautiful.

GENUS UNIO.

Generic Characters.—Shell as in Alasmodonta, but is also furnished with very long lamellar lateral posterior teeth, usually one on the right valve entering between two on the left. Very rarely the right valve has one entering between two on the left. The cardinal teeth are often double, sometimes triple. Animal as in Anodonta.

Unio alatus.—SAY.

DESCRIPTION.—Shell ovate-triangular, moderately thick; epidermis olive, or brownish green, with numerous fine and some coarse striæ of growth; beaks small, not prominent, in the young shell exhibiting small wrinkles; discs moderately inflated posteriorly, compressed anteriorly, with one or two small posterior angles above; anterior side small, one-fourth to one-fifth as long as the posterior; hinge margin straight, very much elevated behind into a triangular connate wing, the posterior margin of which is incurved; the remainder of the posterior and the anterior margins are regularly rounded; inferior margin nearly straight; inner surface usually purplish red, rarely very pale red, sometimes of a rich reddish salmon color; cardinal teeth rather small. Dimensions of two specimens: No. 1, length 5.3 inches; height 3.85 inches; width 1.75 inch. No. 2, length 5.9 inches; height 3.85; width 2.2 inches.

REMARKS.—No. 2 is a very old shell. In such the wing is nearly obsolete, and consequently the form is more ovate. This species is very abundant in Lake Champlain, east of which it has never been found. In the western states it is common.

Unio gracilis.—BARNES.

DESCRIPTION.—Shell ovate-triangular, rather thin; epidermis straw-color, coarsely striate near the margins, otherwise smooth and shining; beaks small, not prominent, smooth; discs considerably inflated, convex, with two or three slight ridges proceeding posteriorly above; anterior side small, compressed, about one-third as long as the posterior; hinge margin nearly straight, much elevated posteriorly into a triangular connate wing, of which the posterior margin is incurved; other margins regularly rounded, the basal moderately; inner surface iridescent, bluish, pink above; cardinal teeth very small. Length 5 inches; height 3.5 inches; width 1.6 inch.

REMARKS.—This species has the form and size of the preceding, but is easily distinguished by the color of the epidermis, of the nacre, greater inflation, and thinness. It is common in lake Champlain, and, like *U. alatus*, is not found any farther to the eastward, but is common through the western states.

Unio compressus.—LEA.

DESCRIPTION.—Shell oblong-ovate, not thick; epidermis grass-green, or olivaceous, with numerous irregular yellowish rays, with distinct striæ; beaks small, pointed, much wrinkled; discs moderately inflated posteriorly, scarcely convex; anterior side three-sevenths to three-eighths as long as the posterior; hinge margin straight, rising posteriorly into a slightly elevated wing, which is often more or less connate; posterior margin descending obliquely in a straight line to a somewhat rounded truncate extremity; inferior margin somewhat rounded; anterior margin regularly rounded; interior bluish, sometimes tinged with pale brownish yellow; cardinal teeth much compressed, on the left valve deeply and broadly bifid, or even trifid; of the lamellar teeth of the left valve one is very small. Length 2.85 inches; height 1.6 inch; width 0.8 inch.

REMARKS.—This species also is found in the western states, and has its eastern limit in the streams west of the Green Mountains. It is much larger in the west.

Var. plebeius.—ADAMS. Epidermis olivaceous, rays obscure; wing scarcely elevated; lamellar teeth very small, with the three divisions of the left cardinal very remote. Length 4.3 inches; height 2.3 inches; width 1.25 inch. This variety is found in a small brook in Middlebury.

FRESH WATER AND LAND SHELLS.

Unio complanatus.—LEA.

DESCRIPTION.—Shell oblong, rather thick; epidermis blackish or greenish brown, sometimes yellowish, with numerous irregular green rays; striæ of growth rather coarse; beaks rather prominent, small; discs compressed, sometimes considerably inflated, but always flattened; anterior side from one-fifth to one-third as long as the posterior; hinge margin nearly straight; posterior margin a little curved, oblique; inferior margin straight, sometimes a little incurved or excurved; anterior margin well rounded; nacre purplish red, pink, sometimes light salmon color, rarely white; lamellar teeth nearly straight; cardinal teeth double. Dimensions of three specimens: No. 1, length 3.9 inches; height 2 inches; width 1.4 inch. No. 2, length 3 inches; height 1.53 inch; width 0.8 inch. No. 3, length 3.05 inches; height 1.53 inch; width 1.36 inch.

REMARKS.—This species is subject to great variations of form, of which the most remarkable in this state is that of a gibbous variety in lake Champlain. No. 3 is an example; No. 2 exhibiting on the contrary a very compressed form. This species is the most common of the Naiades in this, as in the other New England states. Immense numbers cover the shores of lake Champlain.

Unio siliquoideus.—BARNES.

DESCRIPTION.—Shell ovate, not very thick; epidermis yellowish or somewhat greenish brown, with numerous irregular green rays, shining; striæ of growth usually rather fine; beaks small, rather prominent, wrinkled; discs convex, tumid; anterior side a little more or less than one-third as long as the posterior; inferior margin sometimes curved, sometimes straight; other margins rounded; nacre clear white, sometimes light salmon color; cardinal teeth equally bifid in the left valve, unequally in the other; lateral teeth a little curved, not long. Dimensions of three specimens: No. 1, length 2.7 inches; height 1.9 inch; width 1.3 inch. No. 2, length 2.43 inches; height 1.3 inch; width 0.85 inch. No. 3, length 3.05 inches; height 1.65 inch; width 1.4 inch.

REMARKS.—This species, although always ovate, varies much in the ratios of the three dimensions. To illustrate this, the above measurements are taken from examples of the greatest extremes; No. 1, of height; No. 2, of length; and No. 3, of width. The largest individuals are about 4 inches in length. According to Mr. Lea this species is *U. luteolus*, LAMARCK, and the latter name has the right of priority; but according to others, Lamarck's species above quoted is *U. cariosus*, SAY. We therefore, provisionally, give the preference to the name affixed by Mr. Barnes.

Unio ventricosus.—BARNES.

DESCRIPTION.—Shell short, ovate, not very thick; epidermis usually pale yellowish brown, with green rays, of very unequal width, sometimes numerous, often obsolete, except on the corselet; smooth and shining; beaks large and prominent, wrinkled; umbones very tumid, with a more or less distinct angle extending to the bottom of the posterior margin; discs convex; anterior side about half as long as the posterior; hinge margin sinuous; posterior extremity irregularly rounded, in the females high and truncate, in the males somewhat tapering and produced; inferior margin more or less rounded; anterior extremity depressed, well rounded; nacre white; cardinal teeth not large, deeply bifid; lamellar short, distant from the beaks. Dimensions of three specimens: No. 1, length 5.5 inches; height 3.3 inches; width 2.3 inches. No. 2, length 3.35 inches; height 2.35 inches; width 1.77 inch. No. 3, length 3.8 inches; height 2.3 inch.; width 1.83 inch.

REMARKS.—The variations of form are for the most part those of sex, as exhibited in the above measurements. Nos. 1 and 3 are males, No. 1 being unusually large. No. 2 is a female. This species is not rare in lake Champlain, which is its most eastern limit. It is common in the western states.

Unio rectus.—LAMARCK.

DESCRIPTION.—Shell very long ovate, thick; epidermis olivaceous above or throughout, usually yellowish brown below, but nearly covered with dark, broad, more or less confluent, green rays; beaks rather prominent, smooth; discs moderately inflated, scarcely convex; anterior side about one third as long as the posterior; hinge margin slightly curved; posterior extremity sub-rostrate; inferior somewhat curved, straight, or in females incurved; anterior margin rounded; nacre white, pink above; cardinal teeth pink, double, both divisions stout on the left valve, also the inner one on the right. Length 5.75 inches; height 2.3 inches; width 1.55 inch.

REMARKS.—This species is common in the western states, and has its most eastern limit in lake Champlain, where it is rare. The females are much higher in the posterior half, in consequence of a development of the inferior margin.

FAMILY CONCHACEA.
Genus Cyclas.

Generic Characters.—Shell small, thin, globose-elliptic, hinge with two minute cardinal teeth in each or in one valve, which are sometimes obsolete, with compressed lateral teeth on each side. Animal with the mantle posteriorly prolonged into two siphons, which have no retractor muscle; foot very thin and long.

Cyclas elegans.—ADAMS.

DESCRIPTION.—Shell subglobular, rhombic-orbicular, equilateral, finely and elegantly striated; epidermis rather light olive green, with two straw-colored concentric zones, of which the exterior is marginal; beaks not prominent, slightly undulate; umbones very thin; within bluish; lateral teeth large and strong; cardinal teeth rudimentary. Length 0.43 inch; height 0.36 inch; width 0.26 inch.

REMARKS.—This species was discovered in Weybridge, in a swamp, near the site of an old Indian encampment. It has also been found at Burlington. It is remarkable for its shining and elegantly striated surface, and for its inflation, which continues far over the disc, and terminates abruptly near the margin. *C. rhomboida*, SAY, resembles it, but has coarse striæ, no yellow zones, and the discs are less inflated. This is a rare species, and the most beautiful of the genus in our knowledge.

Cyclas similis.—SAY.

DESCRIPTION.—Shell subelliptic, nearly equilateral; epidermis dark brown or yellowish and greenish brown; striæ of growth coarse, deep; umbones not much inflated, broad; disc rather tumid; anterior and posterior margins subrectilineal and divergent; inferior and superior margins rounded; within bluish; cardinal teeth small; lateral teeth compressed, strong. Length 0.68 inch, height 0.5 inch, width 0.4 inch.

REMARKS.—The form of the young differs much from that of the adult. It is rectangular, longer than high, and much compressed. This species differs from the preceding in the coarseness of the striæ; the discs near the margin are less tumid, and the form is much less quadrilateral, and the young, although quadrilateral, are longer and much more compressed. Sometimes there are in this species also yellow zones.

Cyclas rhomboida.—SAY.

DESCRIPTION.—Shell rhombic, nearly equilateral, very coarsely striate; epidermis yellowish horn color; beaks not prominent, nor undulate; umbones prominent; discs moderately tumid; anterior and posterior margins nearly straight, divergent; superior and inferior margins moderately curved; within white; cardinal teeth rudimentary, lateral teeth strong. Length 0.46 inch, height 0.38 inch, width 0.27 inch.

REMARKS.—This species is very nearly allied to the preceding, but the difference is constant. That shell is longer, and the umbones less elevated. The young of this species, although rectangular, are more tumid, which is the cause of the difference in the umbones of mature shells. This species is very plentiful in lake Champlain, and is the only one which occurs in the open waters of the lake in its southern part.

Cyclas partumeia.—SAY.

DESCRIPTION.—Shell ovate-globose, higher behind, nearly equilateral, very thin, translucent, rather finely striate; epidermis shining, straw color, or bluish horn color; beaks not prominent; umbones moderately tumid; discs much inflated and quite regularly convex; posterior and hinge margins nearly straight; other margins much rounded; cardinal teeth small; lateral teeth much developed, compressed. Length 0.3 inch, height 0.25 inch, width 0.17 inch.

REMARKS.—This species inhabits stagnant water, and even swamps which are dried during the autumn. The young are less tumid, very regularly elliptical, and of a light honey yellow. In Massachusetts this species attains a greater size. It resembles *C. cornea* of Europe, which, however, is wider, has the umbones more prominent, and both sides of equal height. *C. similis* is longer, much larger, and more coarsely striate.

Cyclas calyculata.—DRAP.

DESCRIPTION.—Shell rhombic orbicular, higher behind, nearly equilateral, extremely thin and fragile, translucent, with very fine striæ; epidermis shining, bluish horn color, or lemon yellow; beaks swollen, and very prominent, resembling knobs; umbones moderately tumid; discs with a small degree of convexity; posterior and hinge margins nearly straight, making an obtuse angle; anterior and inferior margins rounded; anterior much shorter than the posterior margin; cardinal teeth extremely minute; lateral teeth small, compressed; inner surface colored like the exterior. Length 0.35 inch, height 0.29 inch, width 0.17 inch.

REMARKS.—This species has been found in a swamp in Middlebury, and in Putt's swamp, on the west side of lake Champlain. It has also been found in Maine. The very young are tumid and elliptic, and of a lemon yellow. Some were found in an embryo state in the early part of July. Its dimensions are, length 0.07 inch, height 0.055 inch, width 0.04 inch. The shell of the parent did not exceed 0.002 inch in thickness. The species is easily distinguished by the prominence of the beaks. There seems to be no ground for separating our shell from the European species, whose name we have prefixed.

Cyclas minor.—MIGHELS AND ADAMS.

DESCRIPTION.—Shell ovate, tumid, inequilateral, oblique, very finely striate; epidermis straw color, shining; beaks prominent, two fifths of the difference from one extremity to the other; umbones and discs tumid; posterior and hinge margins slightly rounded; the other margins much rounded; both cardinal and lateral teeth well developed. Length 0.18 inch, height 0.15 inch, width 0.11 inch.

REMARKS.--This species inhabits swamps and is the least of all the native species of this genus. It differs from *C. dubia*, SAY, in having the beaks less removed from the centre, and the posterior and dorsal margins more rounded.

APPENDIX.

Limnæa expansa.—HALDEMAN.

This species is said by the describer to have been found in Vermont, on the authority of Dr. Gould, who received it from a third person as a Vermont shell.

Auricula bidentata.—SAY.

This species, referred by its describer to the genus Melampus, was given to Dr. Gould by some one who professed to have found it in Vermont. As this species has not otherwise been found out of the reach of salt water, we cannot, without better authority, regard it as a native of this state.

Amnicola.

Dr. Gould and Mr. Haldeman have proposed a sub-genus of *Paludina* under this name. It includes of the shells of this state, *Paludina porata* and *P. lustrica*.

Amnicola pallida.—HALD.

On the cover of No. 4 of the Monog. Limniad. Mr. H. has described with this name one of the species just named, but the description is not sufficiently exact to determine to which of them it must be referred. That the shell in question is one of them is inferred from the fact that Mr. H. received them from the writer of this article.

SECTION II.—INVERTEBRATA.

Annulata, Crustacea, Arachnides, and Insects.

The above are four of the classes into which Cuvier's third great division of the animal kingdom is subdivided. The animals belonging to the first 3 classes, which are found in Vermont, are of very little importance, and only a few of them are generally known. We shall pass over them all with only a few remarks.

Annulata.

These are small, insignificant animals, with elongated bodies, consisting of segments, and having red blood. Some of them are protected by a shelly tube, which they never leave during life, and breathe by means of branchiæ at one extremity of the body. These constitute the order Tubicola. Others have their organs and branchiæ disposed longitudinally along the body. These last belong to the order dorsibranchiata. Our brooks and ponds furnish several animals belonging to the above orders, but they have not been properly examined. The third order of Annelides are denominated Abranchiatæ, on account of their having no apparent external organs of respiration. The horse leech, *Hirudo sanguisuga* L., which is so common in marshes and muddy places in this state, belongs to this order. It grows to a much larger size than the medicinal leech, *H. medicinalis* L., and is sometimes used for the same purposes; but its teeth are more blunt, and the wound produced by them is said in some cases to be dangerous. A specimen before me, which was taken in Burlington, is a very dark olive green above, and the same color, but a little lighter beneath, with a few small spots of black. When not in motion he lies in an oval form, and is about 3 inches long, and 1¼ inch wide, but when moving he stretches himself to the length of 6 or 7 inches. The animal is furnished with a flattened disc at each extremity, fitted for adhering to bodies by what is called suction, and its locomotion is performed by reaching forward its anterior extremity, fixing the disc, and then bringing forward the posterior, which is fixed in like manner, and the anterior again thrust forward. In this manner it ascends the side of a perpendicular pane of glass without difficulty, but when at rest it usually adheres by the whole under surface.

The little animal commonly called the *Hair Snake* also belongs to this order, and to the genus Gordius. These are very common in the still waters and mud in all parts of the state. They are usually about the size of a large horsehair, and are from one to 6 or 8 inches in length. In color they vary from pure white to nearly black, and hence we probably have several species. The vulgar notion that they originate from hairs which fall from horses and cattle, and become animated in the water, would seem to be too absurd to need contradiction; and yet, absurd as it is, people are to be found who believe it.

Another, and, indeed, the most common animal belonging to this class in Vermont, is the earth worm, *Lumbricus terrestris*, L., called here the *Angle worm*, on account of the great use made of it for bait in fishing. Its body is cylindrical, of a reddish color, and grows to the length of 5 or 6 inches, with the size of a common goose quill. It is destitute of teeth, eyes, and limbs. It traverses the ground in all directions, and seems to subsist chiefly upon the rich soil, which it swallows. It comes to the surface of the ground during the night, and in wet weather, but descends during the day and in dry weather, so as to be in contact with the moist earth.

Crustacea.

This class embraces the crabs, lobsters, and the like. They usually have a crustaceous covering, which is more or less hard, with articulated limbs, and distinct organs of circulation. They breathe by means of branchiæ, which vary much in form and situation, being in some cases on the abdomen, and in others on the bottom of the feet. The animals of this class are very numerous, but they are confined principally to the ocean, and to tropical climates. The following is the only one found in Vermont, which we shall describe.

THE FRESH WATER LOBSTER,
Astacus Bartonii. Bosc.

DESCRIPTION.—General color greenish brown or dark olive; legs 10, the three anterior ones on each side each terminated by two claws forming a kind of forceps; anterior forceps large, strong, toothed, orange colored at the point and edges and besprinkled with spots formed by indentations. Tail terminated by 5 fan-like plates, forward of which, upon the under side, are two rows, with three in each, of small fringed fins, and still further forward are 4 bony limbs which fold inward towards the abdomen; horns, or feelers, 6, two of which are 3 inches long, the others much shorter. Limbs edged with sparse, downy hairs; body and limbs covered with shell, with numerous articulations. Length of the specimen before me 4½ inches.

This singular little animal is so exact a miniature of the large salt water Lobster that some have supposed it to be the young of that species, or rather a dwarfed variety of it. But it is evidently a distinct species, and though it lives and continues to grow for many years, it very seldom exceeds 4 or 5 inches in length. It is very common in many of the small streams in the western parts of the state. It is sometimes eaten, and by some is esteemed a luxury. It is often called the Craw Fish.

Arachnides.

The principal animals in Vermont which belong to this class are the Spiders, of which we have, probably, about 100 species. The Spiders belong to the genus *Aranea* of Linneus. And though usually called insects, they differ very materially from the proper insects in their form and habits, and constitute a very interesting family, but we are neither prepared nor have we room to go into particulars respecting them. Their classification is based to a considerable extent upon the arrangement of their eyes, which are usually eight in number.

Insects.

Insects constitute the most numerous division of the animal kingdom. European naturalists have computed that there are on an average 6 insects to one plant. This computation is probably too high for our country, but, estimating only two thirds of that number to a plant, as we have about 1000 plants, it will give us 4000 species of insects. The number of known species of New England insects is now about 3000, of which the greater part are found in Vermont. How many remain to be examined and described is, of course, unknown, but the number is, doubtless, very considerable. The word *Insect* comes from the Latin word *Insecta*, and is applied to these small animals on

TRANSFORMATION OF INSECTS.

account of their appearing to be intersected, or divided into sections. Most insects are subject to several changes of form and habit called *metamorphoses*, and in this consists their most remarkable peculiarity. Their existence is made up of four principal stages, viz: the egg, the larva, the chrysalis, and the perfect animal. Directed by instinct, the parent insect is sure to deposit its eggs in the place most favorable for the support of the young, which are in due time to be hatched from them. From these the larvæ are at length produced in the form of maggots, worms, or caterpillars. In this state, which is entirely dissimilar to the parent in form and mode of life, they feed voraciously and grow rapidly, often attaining a weight and bulk much greater than that of the perfect insect. At length they cease to feed, become stationary and encased in a shelly covering, which is often surrounded by a cocoon formed of silky fibres. This is what is called the chrysalis or *pupa*. After remaining for a while in this condition, the shell is burst and thrown off, and the insect emerges in its perfect state, usually provided with wings and often exhibiting the most brilliant and beautiful colors. In this state only is it capable of propagating its species. But it, in general, continues in this state only a short period, just long enough to lay its eggs and die. Most insects feed much more sparingly in their perfect than in their larva state, and some do not feed at all in their perfect state.

This Butterfly measured 1.7 inch in length, and the spread of its wings was just 6 inches. The color of the body belts on the abdomen and portions of the wings was a dark brick-red. General color of the wings different shades of brown beautifully variegated with white, blue, and violet. A roundish black spot, containing a lunated light blue spot near the extremity of each outer wing,&c. This individual was a female, and in the course of the seven days which it lived it laid about 200 eggs.

On the 17th of August, 1840, a caterpillar was picked up in the door-yard, of which the above is a figure. It was 3.5 inches long and 0.75 inch in diameter. Its color was light pea-green. Upon its body were six rows of spines, two on each side, which were blue and pointed, and two on the back, the four anterior ones terminated by balls of the size of small pin-heads, which were red, and covered with small black thorns; all the rest yellow with black points. Being placed under a glass vessel, it immediately commenced spinning, and, before the next day, had completely enveloped itself in a cocoon, precisely similar to the one above described. This remained in a chamber during the winter, and in the spring of 1841, we had from it another butterfly, answering exactly to that figured above.

These details are introduced merely to illustrate the metamorphosis which insects generally experience, and to show the manner in which many of them are preserved through the winter. Others, however, pass the winter in the larva state, in the ground, and still more are preserved in the egg, while some live through the winter in their perfect state.

While much pains have been taken,

The Cocoon, of which the above is a figure, was found on a pine plain in Burlington, upon a small bush, as above represented, in March, 1840. The Cocoon was composed of strong brown silk, and measured 3.5 inches in length and 1.5 in thickness. After being kept about three weeks, or till the 20th of April, in a warm room, a large butterfly, of which the following is a figure, came out of it, by making an opening in the upper end.

and legislative enactments have been resorted to for the destruction of the larger kinds of noxious animals, insects have for the most part been regarded as too insignificant to deserve notice, while the damage sustained on account of the ravages of insects is probably three times as great, on an average, as that produced by all the vertebral animals together. We have been paying liberal bounties for the destruction of catamounts, wolves, bears, and foxes, while the wheat fly, from which we were sustaining far greater damage than from all those larger animals, has hardly received any attention. We have even paid a bounty for the destruction of crows, while in consequence of that destruction our fields were suffering from the ravages of grubs, which the crows are designed to check. Crows may do some mischief in the spring by pulling up corn, but it is believed to be more than counterbalanced by the good which they do, principally by the destruction of vermin. We are of opinion that all birds, without a single exception, are to be regarded as friends to the farmer and gardener, kindly provided by Providence to prevent the undue multiplication of noxious insects, and we cannot too severely reprobate the barbarous practice in which boys are permitted to indulge, of shooting birds for amusement. It is a practice which should be discountenanced by every friend of his country—by every friend of humanity.

Some insects are most injurious in their perfect state. Of these are the various kinds of bugs, which feed upon vines, &c. But far the greater part do most mischief while in the larva state. Of these are the various kinds of caterpillars, which are the larvæ of butterflies and moths,—the weevil, which is the larva of the wheat fly,—the maggots which cause the fruit to fall off prematurely, and which are the larvæ of curculio and other insects,—the borers, which are the larvæ of beetles, bugs, &c.

The Borer, which at present appears to be doing most injury in this state, is the larva of the *Clitus pictus*, which feeds upon the Locust tree, *Robinea pseudoacacia*. It commenced its ravages in the southern part of the state, about ten or twelve years ago. It made its appearance at Middlebury, where it destroyed nearly all the locust trees, about 1835. A year or two after this it had proceeded northwardly as far as Vergennes, and in 1840 it had reached Burlington, but did little injury that year. About the first of June, 1841, its operations began to show themselves, and were continued till the beginning of August, in which time many of the fine locust trees in this town were entirely spoiled, and others more or less injured. During the month of August they were in the chrysalis state, and consequently inactive. About the first of September they emerged from that state, and during the first half of that month the perfect insects were seen in large numbers, often paired, depositing their eggs upon the locust trees in the crevices of the bark, which were in due time hatched. The same operations have been repeated during the past summer, and now (Sept. 6, 1842,) the insects are busily engaged in depositing their eggs for a new generation. The following is a figure of this insect:

Clitus pictus.

The color of this insect is black, with the wing cases crossed by 5 or 6 irregular bright yellow bars, and there are about the same number of yellow bars upon the abdomen. The color of the legs is reddish umber. Length of the female .8 inch;—the male smaller. The color of the larva, or Borer, is yellowish white.

The Cucumber-Bug, *Galeruca vittata*, is one of our most troublesome insects in gardens. It usually makes its appearance upon cucumber, squash and melon vines early in June, or about the time the leaves begin to expand. Various means have been resorted to for the purpose of preventing its depredations, but from two years' experience we are inclined to believe that sprinkling the plants occasionally with ground plaster of Paris, is the most simple and effectual remedy.

The Cock-chafer, or May Beetle, *Melolontha quercina*, is often plentiful, and does considerable mischief by the destruction of the first leaves and blossoms upon our fruit trees. During the day they lie concealed, but come forth from their retreats and commit their depredations in the evening. The larva of this beetle is the large white grub, which is so often seen in rich grounds and in turfs. This insect continues four years in the larva, or grub form, and often does extensive damage by eating the roots of grass, corn and other vegetables. At the end of the fourth year it descends deep into the earth, constructs its cocoon from which the beetle is hatched in its perfect form the following spring. This is the large beetle which so often enters houses in the evening, attracted by the light within.

Although a large proportion of insects are more or less injurious, there are also others from which man derives very considerable benefit. Among the most valuable of these in this state, may be reckoned the Honey Bee and the Silk Worm, which furnish us with most exquisite articles of food and clothing. But of the great majority of insects scarcely any thing is known either of good or evil.

CHAPTER VII.

BOTANY OF VERMONT.

SECTION I.

Catalogue of Vermont Plants.

By WM. OAKES, of Ipswich, Massachusetts.

Preliminary Observations.

THE State of Vermont, in the richness and beauty of its vegetation, is scarcely equalled by any of the New England States. It owes this, no doubt, to the fertility of its soil, the moisture of its climate, and its situation on the ridges and western borders of the mountains. Its ranges of mountains, stretching the whole length of the State from north to south, intercept and often exhaust the summer clouds and rains, which generally come from the west, so that the destructive droughts, which are so often felt in New Hampshire and the other New England States, are almost unknown in Vermont. The State excels in the number and variety of its Forest Trees, possessing, with the exception of eight, all the known species of New England. The following is the list of

THE NATIVE FOREST TREES OF VERMONT.

Lime Tree, or Bass Wood. *Tilia Americana.*
Wild Black Cherry. *Cerasus serotina.*
Sugar Maple. *Acer saccharinum.*
White Maple. *Acer dasycarpum.*
Red Maple. *Acer rubrum.*
White Ash. *Fraxinus acuminata.*
Red Ash. *Fraxinus pubescens.*
Black Ash. *Fraxinus sambucifolia.*
Sassafras. *Laurus Sassafras.*
Tupelo, or Sour Gum. *Nyssa multiflora.*
Red Mulberry. *Morus rubra.*
Hornbeam. *Carpinus Americana.*
Iron Wood. *Ostrya Virginica.*
White Beech. *Fagus sylvestris.*
Red Beech. *Fagus ferruginea.*
Chestnut. *Castanea vesca, var. Americana.*
White Oak. *Quercus alba.*
Swamp White Oak. *Quercus bicolor.*
Overcup White Oak. *Quercus macrocarpa.*
Black Oak. *Quercus tinctoria.*
Red Oak. *Quercus rubra.*
Rock Chestnut Oak. *Quercus montana.*
Scarlet Oak. *Quercus coccinea.*
Large White Birch. *Betula papyracea.*
Small White Birch. *Betula populifolia.*
Black Birch. *Betula lenta.*
Yellow Birch. *Betula excelsa.*
Balsam Poplar. *Populus balsamifera.*
Heart-leaved Balsam Poplar. *Populus candicans.*
Cotton Poplar. *Populus Canadensis.*
Vermont Poplar. *Populus monilifera.*
Large Aspen. *Populus grandidentata.*
American Aspen. *Populus tremuloides.*
Button Wood. *Platanus occidentalis.*
Common Elm. *Ulmus Americana.*
Slippery Elm. *Ulmus fulva.*
Northern Cork Elm. *Ulmus racemosa.*
Hoop Ash, or Hackberry. *Celtis occidentalis.*
Butternut, or Oilnut. *Juglans cinerea.*
Shellbark Hickory. *Carya squamosa.*
Pignut Hickory. *Carya porcina.*
Bitter Pignut Hickory. *Carya amara.*
White Pine. *Pinus Strobus.*
Red Pine, or Norway Pine. *Pinus resinosa.*
Pitch Pine. *Pinus rigida.*
Double Spruce. *Pinus nigra.*
Single Spruce. *Pinus alba.*
Balsam Fir. *Pinus balsamea.*
Hemlock Spruce. *Pinus Canadensis.*
American Larch, or Hackmatack. *Pinus pendula.*
Arbor Vitæ, or "White Cedar." *Thuja occidentalis.*
Red Cedar. *Juniperus Virginiana*

52 species.

SMALL TREES.   VERMONT PLANTS RARE IN OTHER STATES.

Besides the above, there are several trees of small size.
Striped Maple. *Acer Pennsylvanicum.*
Mountain Maple. *Acer montanum.*
Choke Cherry. *Prunus Virginiana.*
June Berry. *Amelanchier Canadensis.*
Mountain Ash. *Sorbus Americana.*
Wild Yellow Plum, or "Canada Plum." *Prunus Americana.*

And also many large shrubs, which sometimes become small trees.

The Stag's Horn Sumac. *Rhus tuphina.*
The Poison Sumac, or Dogwood. *Rhus venenata.*
The Hawthorns. *Crataegus coccinea, &c.*
The Witch Hazel. *Hamamelis Virginiana.*
The High Laurel. *Kalmia latifolia.*
Several species of Willow and Alder.
Several species of *Cornus, Viburnum, &c.*

The Forest Trees of New England not found in Vermont are,

The Tulip Tree. *Liriodendron Tulipifera.*
Sweet Gum. *Liquidambar Styraciflua.*
Black Walnut. *Juglans nigra.*
White Hickory. *Carya alba.*
White Cedar of Middle States. *Cupressus thyoides.*
Chestnut Oak. *Quercus Castanea.*
Post Oak. *Quercus obtusiloba.*
Cotton Tree. *Populus heterophylla.*

There are three species found in Vermont, and not elsewhere in N. England.
The Overcup White Oak. *Quercus macrocarpa.*
The Northern Cork Elm. *Ulmus racemosa.*
The Heart-leaved Balsam Poplar. *Populus candicans.*

The *Overcup White Oak* belongs to the states of the West, and has not been found even in New York. It was found in 1829, by Dr. Robbins, in many towns on the western border of the state from St. Albans to Bennington. It is distinguished by the great size of the acorn, and the fringed border of the cup.

The *Northern Cork Bark Elm* was first found in the state of New York, and was described by Mr. Thomas, in Silliman's Journal, in the same year (1829) that it was found by Dr. Robbins in Bennington and Pownal. It is easily distinguished from the other New England species by the broad plates of cork on its branches.

Three fine species of *Poplar*, the two *Balsam Poplars*, and the magnificent *Vermont Poplar*, Populus monilifera, are scarcely found unless cultivated, in any other of the New England states. Neither of these three Poplars, nor the *Cotton Poplar*, have been found native in New York by the Botanists of that State, according to the late Report and Catalogue of Dr. Torrey. (According to the younger Michaux, the *Cotton Poplar* is found native in the west of New York.)

The *Vermont Poplar*, and the *Heart-leaved Balsam Poplar*, which Dr. Robbins found wild in many parts of Vermont, were not seen native in North America by either the elder or younger Michaux, and do not appear to have been previously seen in a wild state by any Botanist in the United States.

List of VERMONT PLANTS not found in any other New England state.

Anemone Pennsylvanica,
" Hudsoniana,
Corydalis aurea,
Nasturtium natans,
Sisymbrium teres,
Draba arabisans,
Sinapis arvensis. Introduced
Cerastium nutans,
Flœrkea proserpinacoides,
Ceanothus ovalis,
Lathyrus ochroleucus,
Phaca Robbinsii,
Zizia integerrima,
Symphoricarpus racemosus,
Viburnum pubescens,
Valeriana sylvatica,
Aster ptarmicoides,
Solidago humilis,
Pterospora andromedea,
Justicia Americana,
Shepherdia Canadensis,
Euphorbia platyphylla,
Quercus macrocarpa,
Populus candicans,
" monilifera,
Ulmus racemosa,
Listera convallarioides,
Calypso bulbosa,
Trillium grandiflorum,
Zannichellia palustris,
Carex eburnea,
Equisetum variegatum,
Aspidium aculeatum,
Pteris gracilis.

Besides the species in the above list, many of which are among the rarest and most interesting plants of the U. S. there, are a great number of species common in the west of Vermont, and of Massachusetts and Connecticut, which are entirely unknown in the eastern parts of New England. Among these we may mention the *Ginseng*, the *Golden Corydalis*, the curious and beautiful species of *Dielytra*, and the *Spring Beauty*, Claytonia Caroli-

Of the four beautiful species of *Lady,s* most delicate and brilliant blossoms. ground in the woods with its cheerful and niana, which in early spring spangles the

NUMBER OF PLANTS.

Slipper, only two, *Cypripedium acaule* and *arietinum*, are found in the eastern part of New England.

Four species of *Trillium* are also found in Vermont, of which one, the magnificent *Great flowered Trillium*, is found nowhere else in New England. In the eastern part of Massachusetts, no species is found except *Trillium cernuum*.

Vermont is peculiarly rich in Orchideæ. The rare and beautiful *Calypso* has been found no where else in the United States, and *Listera convallarioides* in no other New England state. All the species of New England are found in Vermont, except two, *Tipularia discolor* and *Orchis rotundifolia*.

Of the beautiful order of Ferns, Vermont contains two species not found elsewhere in New England, *Pteris gracilis* and *Aspidium aculeatum*, and several fine species which are wanting or rare in the east of New England, are common in Vermont. It has all the species of New England except *Lygodium palmatum* and *Woodwardia onocleoides*.

On the other hand Vermont is wanting in a great number of plants common in the south and east of New England. Of course it is destitute of all the species peculiar to the sea shore, and of all the numerous and beautiful "Weeds" of the Sea. The elegant *Tulip Tree*, common in the southwest of New England, the splendid *Rosebay*, and the fragrant *Magnolia*, are not found in Vermont. In the whole there are more than 500 New England species which it does not possess, of which we will only mention *Berberis vulgaris, Silene Pennsylvanica, Tephrosia Virginiana, Rhexia Virginica, Liatris scariosa, Clethra alnifolia, Euchroma coccinea, Anagallis arvensis, Hypoxis erecta, Aletris farinosa, Lilium superbum, Poa Eragrostis,* and *Baptisia tinctoria*.

The number of known phænogamous plants of New England, with the addition of the Ferns, is nearly or quite 1500, excluding a great number of nominal species generally admitted. The number of plants of Vermont of the same Orders, in the present catalogue, is 929. The whole number of species of the same orders existing within the limits of the state, is doubtless as many as 1100 or 1200, so that there is still a very ample field for the discovery of additional species. Many species, indeed, exist on the very borders of Vermont, in New Hampshire and Massachusetts, which we have no authority for inserting as natives of the state, and have not admitted into the catalogue, although we have no doubt that they are also Vermont plants.

WESTERN PART OF VERMONT.

We must not forget to mention that the vegetation of the eastern part of Vermont is greatly inferior in beauty and variety to that of the western border. The pines and firs prevail more at the east, and the species of forest trees are not so numerous. While the west has nearly every plant of the east, the east is destitute of a vast number of those of the west. Among the species of Vermont plants wanting at the east, we may mention the *Vermont Poplar*, both the *Balsam Poplars*, the *Cotton Poplar*, the *Northern Cork Elm*, the *Overcup White Oak, Viola Canadensis* and *rostrata, Dielytra Canadensis, Uvularia grandiflora, Asplenium angustifolium, rhizophyllum,* and *Ruta muraria,* &c., besides others to be immediately noticed.

The western ridge of the Alleghany mountains, which at the head of lake Champlain ceases to exist, is broken and interrupted in the state of New York opposite the southwestern border of Vermont, and thus an indirect and difficult entrance is opened to some of the plants of the west and northwest. The western border of Vermont thus appears to become the eastern limit of a considerable number of plants, of which the following is a pretty complete list.

Anemone Pennsylvanica,
Corydalis aurea,
Symphoricarpus racemosus,
Justicia Americana,
Flœrkea proserpinacoides,
Ceanothus ovalis,
Nasturtium natans,
Viburnum pubescens,
Zannichellia palustris,
Carex eburnea,
Lathyrus ochroleucus,
Ulmus racemosa,
Quercus macrocarpa,
Aster ptarmicoides,
Pterospora andromedea,
Pteris gracilis,
Zizia integerrima,
Lonicera hirsuta,
Polanisia graveolens,
Trillium grandiflorum,*

Many of the above species, though not found more eastwardly in the United States, may possibly extend farther to the east along the banks of the St. Lawrence.

The summits of Mansfield and Camel's Hump Mountains, the highest mountains in the state, have been pretty thoroughly examined by Dr. Robbins, Mr. Tuckerman, and Mr. Macrae. These mountains, though destitute of trees at their very summits, from the violence of the winds

* Found in New Brunswick, according to Hooker.

which sweep over them, do not probably quite reach the true limits of trees, and possess only a few of the alpine plants of the White Mountains, which are about 80 miles distant to the eastward.* The only truly alpine species found on these mountains are, perhaps, *Juncus trifidus*, and *Hierochloa alpina*. Other species, almost alpine, are *Poa alpina, Empetrum nigrum, Salix Uva-ursi, Bartsia pallida, Lycopodium Selago*,&c.

The materials upon which the present Catalogue is founded, are the following.

The Catalogue of the plants of Middlebury, published in 1821 in Professor Hall's "Statistical Account of the town of Middlebury," and which was subsequently republished in the first edition of the present work, with the addition of the common cultivated plants, and about 30 indigenous and naturalized species, some of which were probably collected in other parts of the state, making in the whole 569 indigenous and naturalized species. The author of this Catalogue was Dr. EDWIN JAMES, the well known botanist in Long's Expedition to the Rocky Mountains. It was probably made almost entirely from his own collections, and though literally a mere list of names, it bears the marks every where of the great accuracy and research of its author, then a young botanist. It is still the only authority for several rare species.

The collections made by JAMES W. ROBBINS, M.D., of Uxbridge, Mass., who in the year 1829 examined with the greatest care and success the whole western border of Vermont, from Massachusetts to Canada. Dr. Robbins entered the state at Pownal, on the 20th of May, and passing slowly along the western border to the Canada line, examined the large islands of lake Champlain, and afterwards visited Camel's Hump Mountain, leaving the state at Windsor on the 10th of June. On the 20th of July he again entered the state at Guildhall, and after examining the southern border of lake Memphremagog, and the towns in that vicinity, he visited Mansfield Mountain. From thence he proceeded to Burlington and Colchester, where he first discovered the remarkable botanical region at High Bridge and Winooski falls, so rich in rare and interesting plants, and after examining the shores of the lake and the islands of South and North Hero, he visited the mouth of Otter Creek, and, proceeding along the western range of towns from Shoreham to Pownal, left the state at Brattleboro' on the 23d of August. Dr. Robbins found and collected a vast number of rare and interesting species, a large part of which were additions to the Flora of New England, and many of them were also new to the United States.

The collections of JOHN CAREY, Esq., of the city of New York, well known to Botanists by his contributions to the Flora of Torrey and Gray, who resided at Bellows Falls during the five years preceding 1836, and who also made frequent visits to the northeastern counties of the state. Though Mr. Carey's examinations were principally confined to the eastern part of the state, which is very inferior as a botanizing region to the western border, yet he collected very many rare and interesting plants, among which we may mention *Calypso bulbosa*, *Listera convallarioides*, and *Equisetum variegatum*. Mr. Carey has also added to the catalogue a large number of common species, especially Grasses and Cyperaceæ.

The collections of W. F. MACRAE, Esq. of Montreal, Canada, who, while resident at Burlington a few years ago, as a student in the University of Vermont, examined with great zeal the Botany of that vicinity, and besides the more common plants of that region, collected many rare and interesting species, among which were *Pteris gracilis*, and *Draba arabisans*, the first new to New England, the last collected there only by Michaux. Mr. Macrae also, in 1839, in company with EDWARD TUCKERMAN, JR., Esq., the author of several valuable papers on the Lichens of New England, visited Camel's Hump and Mansfield mountains, where, besides other rare species, they collected, on the sides of Mansfield, *Aspidium aculeatum*, found in the United States only by Pursh, and by him in the same region. Mr. Tuckerman has also communicated other species collected by him in various parts of Vermont.

Several very interesting species were added to the Flora of Vermont by the late J. CHANDLER, M. D., of Bennington, Vt., who also accompanied Dr. Robbins during a part of his first tour, and several are given on the authority of ISAAC BRANCH, M. D., of Abbeville District, S. C., JEREMIAH BURGE, M. D., of Drewsville, N. H., M. M. REED, M.D. of Jacksonville, Ill., and P. T. WASHBURN, Esq. of Ludlow, Vt.

All the rarer species collected by Dr. Robbins, and many of the common ones, are ascertained from specimens received from him—the remainder rest on the authority of his journals in my possession, which were made daily during his tour. From his thorough acquaintance with the

---

* Height of Mansfield mountain 4,279 feet, and of Camel's Hump 4,183 feet, above tide water.

CATALOGUE OF PLANTS.

plants of New England, and our mutual knowledge of each other's species, derived from long intercourse and interchange of specimens, I believe that very few if any mistakes have occurred as to the species received from him.

I have received specimens from Dr. Chandler of all the plants given on his authority, and Dr. Robbins saw and examined the species derived from Drs. Branch, Burge, and Reed, in the herbaria of those gentlemen.

I have also seen specimens from Mr. Macrae, of nearly all the species given on his authority.

I have seen only a few specimens from Mr. Carey, but have not hesitated to depend on his known accuracy, and intimate intercourse with Drs. Torrey and Gray.

In preparing the Catalogue, I have generally followed, especially as to the nomenclature of the species, the truly excellent North American Flora of Torrey and Gray, now published as far as Vol. 2, No. 2, which corresponds with the first part of the Catalogue as far as the genus *Bidens*, inclusive. As to the remaining part, I have preferred such names and synonyms as are most certain and familiar to American Botanists, not always following my own opinions, as such a catalogue affords no room for their explanation and support. Owing to the excellent materials at my disposal, the Catalogue is doubtless as complete as that of any state of the Union yet published, and I hope that it will be found useful and acceptable to Botanists.

CATALOGUE OF PLANTS.

[The sign § is prefixed to such species as have been introduced and naturalized.]

CLASS I. EXOGENS, OR MONOCOTYLEDONOUS PLANTS.

ORDER RANUNCULACEÆ. *The Crowfoot Tribe.*

Clematis, *Linn.* Virgin's Bower.
   Virginiana, L. Borders of thickets &c., in moist soil. Aug.
   verticillaris, DC. Shady ledges. Rather rare. May, June.
Anemone, *Haller.* Wind Flower.
   nemorosa, L. Woods, &c. May.
   Virginiana, L. On dry rocky hills, &c. June, July.
      var. *alba*. Castleton, *Branch, Robbins.* Colchester, Burlington, &c. *Robbins.* By an accidental transposition, placed under *A. cylindrica*, in Hovey's Mag. Vol. 7, p. 18.
   cylindrica, Gray. Dry hills, &c. Bellows Falls, *Carey* Burlington, *Macrae.* July.
   Hudsoniana, Richardson. Torrey & Gray, Vol. Suppl. p. 658. A. *multifida.*
      var. *Hudsoniana*, DC. T. & G. I. p. 13. On the limestone ledges of the Winooski river, at Winooski falls, Colchester, and below High Bridge, Burlington, *Robbins.* May, June.
   Pennsylvanica, L. In stony places occasionally overflowed, on the banks of lake Champlain. Westhaven, South Hero, &c., *Robbins.* At Mallet's Bay, Sharpshin Point, and Winooski falls, Burlington, *Macrae.* June, July.
Hepatica, *Dillen.* Noble Liverwort.
   triloba, Chaix. *Anemone Hepatica*, L. Woods. April.
Ranunculus, L. Crowfoot.
   aquatilis, L. var. *capillaceus*, DC. Small streams. June—Sept.
   reptans, L. var. *filiformis*, DC. Overflowed borders of rivers and lakes. July, Aug.
   abortivus, L. Shady banks, &c. May, June.
   sceleratus, L. Ditches, &c. July, Aug.
   acris, L. Buttercups. Meadows, &c. June—Aug.
   bulbosus, L. Buttercups. Pastures on hills, &c. May, June.
   repens, L. Low moist grounds. June—Aug.
   Pennsylvanicus, L. Low moist grounds. July, Aug.
   recurvatus, Poir. Shady moist banks. June.
   Purshii, Richardson. R. *multifidus*, Pursh. Ponds and lakes. Castleton, *Chandler.* South Hero, Alburgh, Colchester, &c., *Robbins.* Middlebury, *Burge.* May, June.

CATALOGUE OF PLANTS.

Caltha, *L.* *Meadow Cowslip. Marsh Marigold.*
 palustris, L. Wet meadows and swamps. May, June.
Coptis, *Salisbury. Gold Thread.*
 trifolia, Salisb. Woods, in boggy soil. May.
Aquilegia, *Tourn. Columbine.*
 Canadensis, L. Rocky places. May, June.
Actaea, *L.*
 alba, Bigelow. *White Cohosh.* Rocky woods. May.
 rubra, Bigelow. *Red Cohosh.* Rocky woods. May.
Cimicifuga, *L.*
 racemosa, Elliott. *Actæa racemosa,* L. *Black Snakeroot.* Woods. Middlebury, *James.* Mansfield mountain, Shelburne and Sharpshin Points near Burlington—rare.—*Macrae.*
Thalictrum, *Tourn. Meadow Rue.*
 dioicum, L. Shady rocky banks. May.
 Cornuti, L. Moist grounds. July.

Order MENISPERMACEÆ. *The Moonseed Tribe.*

Menispermum, *Tourn. Moonseed.*
 Canadense, L. Woods, &c. Middlebury, *James.* St. Albans and South Hero, *Robbins.* Burlington, *Carey.* Vergennes, *Macrae.* June, July.

Order BERBERIDACEÆ. *The Barberry Tribe.*

Leontice, *L.*
 thalictroides, L. *Blue Cohosh.* Woods. May.
Podophyllum, *L. May Apple.*
 peltatum, L. Woods in rich soil. Castleton, *Branch.* May.

Order CABOMBACEÆ.

Brasenia, *Schreber.*
 purpurea. Hydropeltis purpurea, Michx. *Brasenia peltata,* Pursh. In water. In Minaud's pond, Rockingham, *Carey.* In Colchester pond, *Macrae.* July.

Order CERATOPHYLLACEÆ.

Ceratophyllum, *L. Hornwort.*
 echinatum? Gray. In ponds and rivers. Near the mouth of Winooski river, and in lake Memphremagog, *Robbins.*

Order NYMPHÆACEÆ. *The Water-Lily Tribe.*

Nymphæa, *Tournefort.*
 odorata, Aiton. *White Water-Lily.* Ponds and rivers. July, Aug.
Nuphar, *Smith.*
 advena, Aiton. *Yellow Water-Lily.* Ponds and rivers. June, July.
 lutea. var. *Kalmiana,* Torr. & Gr. *N. Kalmiana,* Pursh. Ponds and rivers. July.

Order SARRACENIACEÆ.

Sarracenia, *Tourn.*
 purpurea, L. *Side-saddle Flower. Forefather's Cup.* Sphagnous bogs. June.

Order PAPAVERACEÆ. *The Poppy Tribe.*

Sanguinaria, *Dillenius. Blood-root.*
 Canadensis, L. Woods, &c. May.
Chelidonium, *Tourn.*
 § majus, L. Road-sides, and about houses. June—Sept.

Order FUMARIACEÆ. *The Fumitory Tribe.*

Dielytra, *Borckh.*
 cucullaria, DC. Woods, &c. May.
 Canadensis, DC. *Squirrel Corn.* Woods. St. Albans, *Robbins.* In the southwest of Vermont, *Oakes.* May.
Adlumia, *Raf.*
 fungosa. Corydalis fungosa, Ventenat. *Adlumia cirrhosa,* Raf. Rocky woods. Middlebury, *James, Burge.* Castleton, Burlington, and Westhaven, *Robbins.* Ludlow, *Washburn.* July—Sept.

CATALOGUE OF PLANTS.

Corydalis, *DC.*
  aurea, Willd. Rocky woods. Castleton, *Chandler.* Burlington, *Macrae.* May, June.
  glauca, Pursh. Rocks and ledges. May, June.

Order CRUCIFERÆ. *The Cruciferous Tribe.*

Nasturtium, *R. Br.*
  palustre, DC. Wet places. July, Aug.
  natans, DC. *var. Americanum,* Gray, T. & G. I. p. 75. In shallow water on the borders of Otter Creek below Vergennes, abundant for several miles, *Robbins.* July, Aug.
Barbarea, *R. Br.*
  vulgaris, R. Br. *Winter Cress.* Road-sides, &c., generally in moist soil. June.
Arabis, *L. Wall Cress.*
  hirsuta, Scop. *A. sagittata,* DC. *Turritis hirsuta,* L. Rocks. June.
  laevigata, DC. *Turritis lævigata,* Muhl. Rocks. June.
Cardamine, *L.*
  rhomboidea, DC. *C. rotundifolia var.,* Tor. & Gray. Wet meadows. Castleton, *Robbins.* May, June.
  hirsuta, L. *C. Pennsylvanica,* Muhl. Brooks &c. June, July.
  pratensis, L. *Lady's Smock. Cuckoo Flower.* Wet meadows. Whiting and Alburgh, *Chandler.* St. Albans, *Robbins.* May, June.
Dentaria, *L. Toothwort.*
  diphylla, Michx. *Pepper Root.* Woods. May.
  laciniata, Muhl. Woods. Castleton, *Robbins.* May.
Sisymbrium, *Allioni.*
  § *officinale,* Scop. *Hedge Mustard.* Road-sides and about houses. June—Aug.
  teres, Torr. & Gray, I. p. 93. *Cardamine teres,* Michx. Vermont, on Lake Champlain, *Michaux.* No botanist except Michaux has ever collected this species.
Sinapis, *L. Mustard.*
  § *nigra,* L. *Black Mustard.* Old fields, &c. June—Aug.
  § *arvensis,* L. Road sides, old fields,&c., called "*Charlock,*" which it resembles. Charlotte and Alburgh, *Robbins.* About Burlington, *Macrae.* May, June.
Draba, *L.*
  arabizans, Michx. On rocks. On Lake Champlain, *Michaux.* At Sharpshin Point, Burlington, and on the north side of Juniper Island, *Macrae.* May.
Cochlearia, *L.*
  § *Armoracia,* L. *Horse-radish.* Banks of rivers, and about houses, in moist soil. June. This well known species is also thoroughly naturalized in Massachusetts, often in places distant from habitations.
Camelina, *Crantz.*
  § *sativa,* Crantz. Old fields, flax fields, &c. Ferrisburgh, *Robbins.* Bellows Falls, *Carey.*
Lepidium, *L. Pepperwort, or "Pepper Grass."*
  Virginicum, L. Sandy fields and roadsides. June, July.
Capsella, *Vent. Shepherd's Purse,*
  § *Bursa-pastoris,* Mœnch. Gardens and fields. April—Sept.
Raphanus, *L.*
  § *Raphanistrum,* L. *Charlock. Wild Radish.* Cultivated grounds. South Hero, *Robbins.* June, Sept.

Order CAPPARIDACEÆ. *The Caper Tribe.*

Polanisia, *Raf.*
  graveolens, Raf. On the gravelly banks of Lake Champlain, above high water. July, Aug.

Order POLYGALACEÆ. *The Milkwort Tribe.*

Polygala, *L. Milkwort.*
  verticillata, L. Dry Soils. At Bellows Falls, *Tuckerman, Carey.* July—Sept.

Senega, L. Seneca Snake-root. Dry rocky woods and banks. June.
polygama, Wait. *P. rubella*, Willd. Dry fields and borders of woods. July, Aug.
paucifolia, Willd. Pine woods and sphagnous swamps. May, June.
ambigua, Nuttall. Dry fields, &c. Pownal, *Robbins*. July, Aug.

Order VIOLACEÆ.   *The Violet Tribe.*

Viola, *L.* *Violet.*
palmata, L. Woods and shady banks. Pownal, *Robbins*. May.
cuculiata, Ait. Wet meadows and woods. May.
sagittata, Ait. var. *ovata*, T. &. G. I. p. 138. *V. ovata*, Nutt. Dry hills, &c. May.
rotundifolia, Michx. Woods. May.
blanda, Willd. Wet meadows and woods. May.
Muhlenbergii, Torrey. Moist woods. May, June.
rostrata, Pursh. Woods. May, June.
pubescens, Ait. Woods. May, June.
Canadensis, L. Woods. May, June.

Order DROSERACEÆ.   *The Sundew Tribe.*

Drosera, *L.* *Sundew.*
rotundifolia, L. Sphagnous bogs. June—Aug.
longifolia, L. Sphagnous bogs. June—Aug.
Parnassia, *Tourn.* Grass of Parnassus.
Caroliniana, Michx. Wet meadows, &c. Aug., Sept.

Order CISTACEÆ.   *The Rock-rose Tribe.*

Helianthemum, *Tourn.*
Canadense, Michx. Dry sandy pastures, &c. Pownal, *Robbins*. Bellows Falls, *Carey*. Burlington, *Macrae*. June.
Lechea, *L.* *Pin Weed.*
major, Mich. Dry pastures, &c. Middlebury, *James*. July, Aug.
minor, Lam. Dry hills, &c. Middlebury, *James*. Burlington, *Macrae*. Bellows Falls, *Carey*. July, Aug.

Order HYPERICACEÆ.   *The St. John's Wort Tribe.*

Hypericum, *L.* *St. John's Wort.*
pyramidatum, Ait. *H. ascyroides*, Willd. Banks of rivers. Burlington, *Bigelow*. Near Rutland, *Robbins*. On Black river, Springfield, *Carey*. On White river, between Royalton and Hartford, *Oakes*. July, Aug.
§ *perforatum*, L. Common St. John's Wort. Grass fields, pastures, &c. July, August.
corymbosum, Muhl. Shady banks, &c. July, Aug.
ellipticum, Hooker. Moist meadows, &c. Middlebury, *Burge*. Westford and Ferrisburgh, *Robbins*. Burlington, *Tuckerman*. Bellows Falls, &c., *Carey*. July, Aug.
mutilum, L. *H. parviflorum*, Willd. Wet soils. July, Aug.
Canadense, L. Wet soils. July, Aug.
Elodea, *Adans.*
Virginica, Nutt. Swamps, &c. Middlebury, *James*. Burlington, *Macrae*. July Aug.

Order ILLECEBRACEÆ.   *The Knot-grass Tribe.*

Spergula, *Bartl.*
§ *arvensis*, L. Old fields, &c. June, Oct.
Anychia, *Michx.*
dichotoma, Michx. Dry hills, &c. Pownal, *Robbins*. July, Aug.

Order CARYOPHYLLACEÆ.   *The Pink Tribe.*

Mollugo, *L.*
verticillata, L. Sandy soils. Bellows Falls, *Carey*. July—Sept.
Arenaria, *L.* *Sandwort.*
stricta, Michx. Rocks. June.

CATALOGUE OF PLANTS.

Grœnlandica, Spring. *A. glabra*, Bigel. non Michx. On the summits of Mansfield mountain and Camel's Hump, *Robbins, Tuckerman,* and *Macrae.* July, Aug. (Identical with *A. glabra* of Michaux, *Macrae.*)
§ *Serpyllifolia*, L. Sandy fields. Burlington, *Tuckerman.* May—July.
lateriflora, L. Moist woods. Middlebury, *Burge.* Fairhaven, *Robbins.* June.

Stellaria, *L.*
§ *media*, Smith. *Chickweed.* Gardens, &c. April—Nov.
longifolia, Muhl. Bellows Falls, *Carey.* June.
borealis, Bigel. Swamps, and on mountains. June, July.

Cerastium, *L. Mouse-ear Chickweed.*
§ *vulgatum*, L. Roadsides, &c. June.
nutans, Raf. Moist shady places. Middlebury, *Burge.* Danby and Rutland, *Robbins.* May.

Silene, *L. Catchfly.*
antirrhina, L. Dry fields, &c. On the rocks about Winooski falls, Colchester. *Robbins.* Bellows Falls, *Carey.* June.
§ *noctiflora*, L. Old fields, &c. Bellows Falls, *Carey.* Burlington, *Macrae.* July.

Agrostemma, *L.*
Githago, L. *Corn Cockle.* Cultivated fields, &c. June.

Order PORTULACEÆ. *The Purslane Tribe.*

Portulaca, *L.*
oleracea, L. *Purslane.* Gardens, &c. July, Aug.

Claytonia, *L.*
Caroliniana, Michx. *Spring Beauty.* Woods. April, May.

Order LINACEÆ. *The Flax Tribe.*

Linum, *L. Flax.*
§ *usitatissimum*, L. *Common Flax.* Old fields, &c. July.
Virginianum, L. Dry woods, &c. Pownal, *Robbins.* June—Aug.

Order GERANIACEÆ. *The Geranium Tribe.*

Geranium, *L.*
maculatum, L. Woods. June.
Carolinianum, L. Dry soils. Bellows Falls, *Carey.* Burlington, *Oakes.* June.
Robertianum, L. Shady ledges, &c. June—Sept.
§ *dissectum*, L. Hills. Castleton, *Robbins.* June, July. Exactly the European plant, and found also by Dr. Robbins at Augusta, Me., and Uxbridge, Mass.

Order BALSAMINACEÆ. *The Balsam Tribe.*

Impatiens, *L. Balsam.*
pallida, Nutt. Moist shady grounds. Pownal, *Oakes.* At the base of Mansfield mountain, Westhaven, Jericho, &c., *Robbins.* Guildhall, *Carey.*
fulva, Nutt. Moist grounds. Aug. Sept.

Order LIMNANTHACEÆ.

Flœrkea, *Willd.*
proserpinacoides, Willd. Wet banks, and margins of streams, &c. Castleton, *Robbins.* May.

Order OXALIDACEÆ. *The Wood-sorrel Tribe.*

Oxalis, *L. Wood-sorrel.*
acetosella, L. Mountain woods. June, July.
stricta, L. Cultivated grounds. June—Sept.

Order XANTHOXYLACEÆ.

Xanthoxylum, *L.*
Americanum, Miller. *X. traxineum*, Willd. *Prickly Ash.* On rocky hills and banks. Middlebury, *James.* Ferrisburgh, Shoreham, Grand Isle, Shelburne, St. Albans, and Arlington, *Robbins.* April, May.

CATALOGUE OF PLANTS.

ORDER ANACARDIACEÆ. *The Cashew Tribe.*

Rhus, *L. Sumac.*
  typhina, L. Stag's horn Sumac. Hills. June.
  glabra, L. Smooth Sumac. Hills, &c. July.
  copallina, L. Mountain Sumac. Hills and pastures. July.
  venenata, DC. R. *vernix*, L. in part. Poison Sumac. Poison Dogwood.— Swamps. June.
  Toxicodendron, L. Poison Ivy. Woods and along fences. June.
  aromatica, Ait. Dry hills and banks. Shoreham, *Dr. Hill.* Westhaven and Pownal, *Robbins.* May.

ORDER MALVACEÆ. *The Mallow Tribe.*

Malva, *L. Mallows.*
  § *rotundifolia*, L. Road-sides and about houses. June—Sept.
Sida, *L.*
  Abutilon, L. Waste places, cultivated grounds, &c. Pownal, *Robbins.* Aug., Sept.

ORDER TILIACEÆ. *The Linden Tribe.*

Tilia, *L. Linden, or Lime Tree.*
  Americana, L. Bass Wood. Woods. July.

ORDER VITACEÆ. *The Vine Tribe.*

Vitis, *L. Vine.*
  Labrusca, L. Fox Grape. Woods and thickets. June.
  æstivalis? Michx. Summer Grape. Banks of rivers, &c. On the alluvial banks of the Winooski, near High Bridge, Colchester, *Robbins.* Rocks at Sharpshin Point, Burlington, *Macrae.* Bellows Falls, *Carey.* June.
  riparia, Michx. Thickets on the banks of rivers. Bellows Falls, *Carey.* June.
  cordifolia, Michx. Frost Grape. Winter Grape. Borders of thickets, &c. June.
Ampelopsis, *Michx.*
  quinquefolia, Michx. Common Creeper. Woods, &c. July.

ORDER ACERACEÆ. *The Maple Tribe.*

Acer, *L. Maple.*
  Pennsylvanicum, L. A. *striatum*, Michx. Striped Maple. Woods. May, June.
  spicatum, Lam. A. *montanum*, Ait. Mountain Maple. Woods. June.
  saccharinum, L. Sugar Maple. Woods. May. var. *nigrum.* A. *nigrum*, Michx. Black Sugar Maple. Woods. May.
  dasycarpum, Ehrh. White Maple, Soft Maple. Banks of rivers. April.
  rubrum, L. Red Maple. Swamps, &c. April, May.

ORDER CELASTRACEÆ.

Staphylea, *L. Bladder-nut.*
  trifolia, L. Rocky banks, &c. Middlebury, *James.* Pownal, *Robbins.* May.
Celastrus, *L.*
  scandens, L. Wax-work. False Bitter-Sweet. Borders of woods, fences, &c. June.

ORDER RHAMNACEÆ. *The Buck-Thorn Tribe.*

Rhamnus, *L. Buck-thorn.*
  alnifolius, L'Her. Sphagnous swamps. Castleton, Whiting, Craftsbury, &c., *Robbins.* Hubbardton, *Chandler.* Danville, *Carey.* May, June.
Ceanothus, *L.*
  Americanus, L. New Jersey Tea. Dry woods, pastures, &c. July
  ovalis, Bigel. Dry open sandy woods, &c. Burlington, June.

ORDER LEGUMINOSÆ. *The Pea and Bean Tribe.*

Vicia, *L. Vetch.*
  sativa, L. Common Vetch. Tare. Old fields, &c. July.
  Cracca, L. Old fields, &c. Middlebury, *Burge.* June, July.

CATALOGUE OF PLANTS.

Lathyrus, *L.*
  maritimus, Bigel. *Pisum maritimum*, L. *Shore Pea.* On the sandy shore of lake Champlain, Burlington, *Macrae.* June, July.
  palustris, L. Wet meadows, &c. June.
    var. *myrtifolius*, L. *myrtifolius*, Muhl. "In Vermont. *Torrey & Gray.*"
  ochroleucus, Hooker. *L glaucifolius*, Beck. On the banks of lake Champlain, in dry soil, in North and South Hero, *Robbins.* June, July.
Apios, Boerhaave.
  tuberosa, Mœnch. *Glycine Apios*, L. *Ground Nut.* Moist shady places. Aug.
Amphicarpæa, *Elliott.*
  monoica, Elliott. *Glycine monoica*, L. Woods. July.
Trifolium, *L. Clover. Trefoil.*
  § *arvense*, L. Dry sandy soil. July, Aug.
  § *pratense*, L. *Red Clover.* Meadows, fields, &c. June—Sept.
  repens, L. *White Clover.* Meadows, fields, woods, &c. May—Oct.
Melilotus, *Tourn. Melilot.*
  officinalis, Willd. *Yellow Melilot.* Middlebury, *James.* June—Aug.
Medicago, *L.*
  § *lupulina*, L. *Nonesuch.* Fields, &c. South Hero. *Robbins.* June, Aug.
Phaca, *L.*
  Robinsii, Oakes, in Hovey's Mag., May, 1841. On a limestone ledge in Burlington, on the banks of Winooski river, a quarter of a mile below High Bridge, *Robbins.* May, June.
Desmodium, *DC.* Hedysarum, *L.*
  nudiflorum, DC. Dry woods. Aug.
  acuminatum, DC. Dry woods. Aug.
  Canadense, DC. Woods and by fences. July, Aug.
  canescens, DC. Dry soil. Pownal, *Robbins.* Aug.
  paniculatum, DC. Dry woods. Ferrisburgh, *Robbins.* Aug.
  Dillenii, Darlington. Dry woods. Bellows Falls, *Carcy.* Aug.
Lespedeza, *Michx.* Hedysarum, *L.*
  violacea, Pers. Dry woods. Rockingham, *Carey.* Aug.
  hirta, Ell. Dry fields, banks, &c. Colchester, *Robbins.* Aug.
  capitata, Michx. Dry pastures, &c. Bellows Falls, *Carey.* August.
Lupinus, *L. Lupine.*
  perennis, L. *Wild Lupine.* Sandy woods and fields. June.
Cassia, *L.*
  Marilandica L. *Wild Senna.* Orwell, *Dr. Hill.* Bellows Falls, *Carey.* Aug.

ORDER ROSACEÆ. *The Rose Tribe.*

Prunus, *Tourn. Plum.*
  Americana, Marshall. *P. nigra*, Ait. *Canada Plum. Wild Yellow Plum.*— Woods. May.
Cerasus, *Juss.* Prunus, *L Cherry.*
  pumila, Michx. *Sand Cherry.* Rocky or sandy shores. May.
  Pennsylvanica, Loisel. *C. borealis*, Michx. *Wild Red Cherry.* Woods. May.
  serotina, DC. *C. Virginiana*, Michx. *Wild Black Cherry.* Fields, woods, &c. June.
  Virginiana, DC. *P. obovata*, Bigel. *Choke Cherry.* Fields, woods, &c. June.
Spiræa, *L.*
  salicifolia, L. *Meadow Sweet.* Low grounds. July, Aug.
  tomentosa, L. *Hardhack.* Low grounds. July, Aug.
Geum, *L. Avens.*
  strictum, Ait. Low grounds. July.
  Virginianum, L. Fields, &c. June, July.
  rivale, L. *Water Avens.* Bogs. June.
Waldstenia, *Willd.*
  fragarioides, Tratt. *Dalibarda fragarioides*, Michx. Woods. June.
Agrimonia, *Tourn. Agrimony.*
  Eupatoria, L. Woods and pastures. July.
Potentilla, *L. Cinquefoil.*
  fruticosa, L. Bogs. July—Sept.
  Canadensis, L. *P. simplex*, Michx. *Five Finger.* Woods. May, June.
    var. *pumila.* *P. Pumila*, Poir. Pastures, &c. May—Aug.

CATALOGUE OF PLANTS.

Norvegica, L. Old fields, &c. June—Aug.
tridentata, Ait. On the Alpine summits of Mansfield mountain and Camel's Hump, *Robbins*. July, Aug.
arguta, Pursh. *P. confertiflora*, Torrey. Rocky Hills. Pownal, Castleton, *Robbins* Bellows Falls, *Carey*. May, June.
anserina, L. Overflowed places. June, July.
argentea, L. Dry hills, &c. Bellows Falls, *Carey*. Burlington, *Macrae*. Pownal, *Robbins*. June.

Comarum, *L.*
palustre, L. Bogs. Burlington, *Robbins*. Charleston, *Carey*. July.

Fragaria, *Tourn.* *Strawberry.*
Virginiana, Ehrh. *Wild Strawberry.* Woods and meadows. May.
vesca, L. Common "*English*" *Wood Strawberry.* Woods, especially on mountains. May.

Dalibarda, *L.*
repens, L. Woods, especially on mountains. June—Aug.

Rubus, *L.* *Bramble.*
odoratus, L. *Flowering Raspberry.* Shady rocky banks. June—Aug.
strigosus, Michx. *Red Raspberry.* About woods. May, June.
occidentalis, L. *Thimble-berry. Black Raspberry.* By fences, &c. May, June.
villosus, Ait. *High Blackberry.* Borders of woods and fields. June.
Canadensis, L. *R. trivialis*, Pursh. *Low Blackberry.* Fields, &c. June.
hispidus, L. *R. sempervirens* and *setosus*, Bigelow. Woods. June.
triflorus, Richardson. *R. saxatilis*, Michx. Swamps and woods. June.

Rosa, *Tourn.* *Rose.*
Carolina, L. Borders of swamps, &c. July.
lucida, Ehrh. Pastures, &c. June.
blanda, Ait. On rocks. Bellows Falls, *Carey*. Burlington, *Macrae*. On the ledge near High Bridge, Burlington, with *Phaca Robbinsii*, *Oakes*. June.
§ *rubiginosa*, L. *Sweet Briar.* Thickets, pastures, &c. June, July.

Cratægus, *L.* *Hawthorn.*
coccinea, L. Borders of thickets, &c. May, June.
tomentosa, L. var. B., Torrey & Gray, I—466. Thickets, &c. Bellows Falls, *Carey*. May, June.
punctata, Jacq. Borders of woods, &c. Ferrisburgh, Charlotte, Colchester, &c, *Robbins*, May, June.

Pyrus, *L.*
arbutifolia, L. f. var. *erythrocarpa*. Dry woods. June.
var. *melanocarpa*. Chokeberry. Swamps. June.
Americana, DC. *Sorbus Americana*, Willd. *Mountain Ash.* Woods, especially on mountains. June.

Amelanchier, *Medic.* DC. Mespilus, *L.* Aronia, *Pers.* *Juneberry.*
Canadensis, T. & G. 1—473. *Mespilus Can.* L. *Pyrus Botryapium*, L. fil.
var. *Botryapium*, T. & G. Woods, &c. May, June
var. *oblongifolia*, T. & G. Woods, &c. May, June.
var. *rotundifolia*, T. & G. Rocky banks of rivers, &c. May, June.
var. *oligocarpa*, T. & G. Near the summits of Camel's Hump and Mansfield mountain, *Robbins*, *Tuckerman*, and *Macrae*. In a swamp at Guildhall, *Carey*. June.

ORDER LYTHRACEÆ. *The Loosestrife Tribe.*

Decodon, *Gmelin.*
verticillatum, Elliott. *Lythrum vert.*, L. Borders of ponds, &c. Colchester, *Robbins*.

ORDER ONAGRACEÆ. *The Evening Primrose Tribe.*

Epilobium, *L.* *Willow Herb.*
angustifolium, L. *E. spicatum*, Lam. Burnt woods, &c. July, Aug.
coloratum, Muhl. Wet places. July, Aug.
palustre, L. var. *albiflorum*, Lehm. *E. lineare*, Muhl. *E. squamatum*, Nuttall. Swamps. Aug.

Œnothera, *L.* *Evening Primrose.*
biennis, L. Old fields, &c. July, Aug.

CATALOGUE OF PLANTS.

pumila, L. Old fields, &c. June—Sept.
Circaea, *Tourn. Enchanter's Nightshade.*
 Lutetiana, L. Woods, &c. July.
 alpina, L. Old woods, on fallen mossy trunks, &c. July, Aug.

Sub-Order HALORAGEÆ.

Proserpinaca, *L.*
 palustris, L. Ditches, borders of ponds, &c. July, Aug.
Myriophyllum, *Vaill. Water Milfoil.*
 spicatum, L. In ponds, &c. July.

Order CUCURBITACEÆ. *The Gourd Tribe.*

Sicyos, *L. Single-seeded Cucumber.*
 angulatus, L. Cultivated grounds and river banks. Aug.
Echinocystis, *Torrey & Gray*, 1, 542.
 lobata, T. & G. *Momordica echinata*, Willd. *Hexameria echinata*, T. & G. in New York State Cat. p 137. Alluvial banks of rivers. On the Hoosic, Pownal, Vt., *Oakes.* On the Winooski, below High Bridge, Colchester, *Robbins.* Aug.

Order GROSSULARIACEÆ. *The Currant and Gooseberry Tribe.*

Ribes, *L. Currant and Gooseberry.*
 Cynosbati, L. Rocky woods, &c. May.
 lacustre, Poiret. Rocky mountain woods. May, June.
 prostratum, L'Herit. *R. rigens* and *trifidum*, Michx. Mountain woods. May.
 floridum, L'Herit. *Wild Black Currant.* Woods. Bridgewater, *Thompson.* May.
 rubrum, L. *Red garden Currant.* Swamps. St. Johnsbury, *Carey.* Also on the rocky banks of the Winooski, *Oakes.* May, June.

Order CRASSULACEÆ. *The House-leek Tribe.*

Penthorum, *L.*
 sedoides, L. Low moist places. July, Aug.

Order SAXIFRAGACEÆ. *The Saxifrage Tribe.*

Saxifraga, *L. Saxifrage.*
 Virginiensis, Michx. Rocks. May.
 Pennsylvanica, L. Wet meadows and swamps. May, June.
Mitella, *L.*
 diphylla, L. *False sanicle.* Woods. May.
 nuda, L. *M. cordifolia*, Lam. *M. prostrata*, Michx. Shady bogs. May, June.
Tiarella, *L. Mitre Wort.*
 cordifolia, L. Woods. May, June.
Chrysosplenium, *Tourn. Golden Saxifrage.*
 Americanum, Schweinitz. *C. oppositifolium*, Michx. &c. not L. Wet boggy soil. May, June.

Order HAMAMELACEÆ. *The Witch Hazel Tribe.*

Hamamelis, *L. Witch Hazel.*
 Virginiana, L. Woods, &c. Oct., Nov.

Order UMBELLIFERÆ. *The Umbelliferous Tribe.*

Hydrocotyle, *Tourn. Marsh Penny Wort.*
 Americana, L. Swamps, &c. July, Aug.
Sanicula, *Tourn. Sanicle.*
 Marilandica, L. Woods. June.
Cicuta, *L.*
 maculata, L. *Water Hemlock.* Moist meadows, &c. July, Aug.
 bulbifera, L. Borders of swamps, &c. August.
Sium, *L. Water Parsnip.*
 latifolium, L. Wet places. July, Aug.
Cryptotænia, *DC.*
 Canadensis, DC. *Sison Canadense*, L. Shady banks, &c. July.
Zizia, *Koch.*
 aurea, Koch. *Smyrnium aureum*, L. Meadows, &c. July.
 integerrima, DC. *Smyrnium integerrimum*, L. Shady banks, &c. June.
Thaspium, *Nutt.*
 cordatum, Torrey & Gray, 1, 615. Middlebury, *James.* June.

CATALOGUE OF PLANTS.

Conioselinum, *Fisch.*
   Canadense, T. & G. 1, 619. *Selinum Can.*, Michx. *Cnidium Can.*, Spreng. Cedar swamps and wet woods. Fairhaven, and at the base of Mansfield mountain, *Robbins.* Burlington, *Macrae.* July.

Archangelica, *Hoffm.*
   atropurpurea, Hoffm. *Angelica triquinata*, Michx. *Angelica.* Low grounds. July.

Pastinaca, *Tourn.*
   § sativa, L. *Common Parsnep.* By fences, &c. June, July.

Heracleum, L. *Cow Parsnep.*
   lanatum, Michx. By fences, &c. June, July.

Osmorhiza, *Raf.*
   longistylis, DC. *Sweet Cicely.* Woods. May, June.
   brevistylis, DC. Woods. May, June.

Conium, L. *Hemlock.*
   § maculatum, L. *Poison Hemlock.* Road sides, &c. July, Aug.

ORDER ARALIACEÆ. *The Aralia Tribe.*

Aralia, L.
   nudicaulis, L. *Wild Sarsaparilla.* Woods. May, June.
   racemosa, L. *Spikenard.* Woods and shady banks. July.
   hispida, L. Burnt woods, &c. July.

Panax, L. *Ginseng.*
   quinquefolium, L. *Common Ginseng.* Woods. July.
   trifolium, L. *Dwarf Ginseng.* Moist woods. May.

ORDER CORNACEÆ. *The Dogwood Tribe.*

Cornus, L. *Dogwood.*
   alternifolia, L. Woods. June.
   circinata, L. Woods, &c. Middlebury, *James.* Castleton, Colchester, and Burlington, *Robbins.* Bellows Falls, *Carey.* June.
   stolonifera, Michx. *Cornus alba*, Wang. Banks of streams, &c. May, June.
   paniculata, L'Her. Borders of woods, &c. June.
   sericea, L. Low grounds, &c. July.
   florida, L. *Common Dogwood.* Woods. Castleton, *Robbins.* May, June.
   Canadensis, L. Woods. May.

ORDER CAPRIFOLIACEÆ. *The Honeysuckle Tribe.*

Linnæa, *Gronov.*
   borealis, Gronov. *Linnæa.* Old woods. June, July.

Symphoricarpus, *Dillenius.*
   racemosus, Michx. *Snowberry.* Rocky banks. On Grand Isle and South Hero, at the "Point of Rocks" in Shoreham, and at Fort Cassin, *Robbins.* On the extremity of Sharpshin Point, Burlington, *Macrae.* July, Aug.

Lonicera, L. *Honeysuckle.*
   hirsuta, Eaton. Rocky woods. Middlebury, *James.* Castleton, *Branch.* Pownal, *Robbins.* June.
   parviflora, Lam. Rocky banks, &c. June.
   ciliata, Muhl. Shady ledges, &c. May, June.
   cœrulea, L. *Xylosteum villosum*, Michx. Bogs, &c. May, June.

Diervilla, *Tourn.*
   trifida, Mœnch. *D. Canadensis*, Willd. Rocky woods. July.

Triosteum, L.
   perfoliatum, L. *Feverwort.* Rocky woods, &c. Bennington, *Robbins.* May, June.

Sambucus, *Tourn.* *Elder.*
   Canadensis, L. *Common Elder.* Along fences, &c. July.
   pubens, Michx. *Red-berried Elder.* Woods and mountains. May.

Viburnum, L.
   nudum, L. *V. pyrifolium*, Pursh. *V. cassinoides.* L. Moist woods, &c. June.
   Lentago, L. Moist thickets. June.
   dentatum, L. *Arrow wood.* Moist thickets. June.
   pubescens, Pursh. Dry rocky banks. Middlebury, *James.* Shoreham, Castleton, and Westhaven, *Robbins.* Sharpshin Point and a high rock behind it, Burlington, *Macrae.* June.

CATALOGUE OF PLANTS

acerifolium, L. Rocky woods. June.
Opulus, L. *var. Americanum*, Ait. *V. Oxycoccus*, Pursh. *Cranberry Bush.*
  Woods, &c. May, June.
  var. eradiatum, Oakes in Hovey's Mag., May, 1841. *V. pauciflorum*, La
    Pylaie. T. &. G. 2, 17. Near the summit of the
    Mansfield mountain, *Tuckerman* and *Macrae*. July.
Lantanoides, Michx. *Hobble Bush.* Old woods. May, June.

Order RUBIACEÆ. *The Madder Tribe.*

Houstonia, *L.*
  cœrulea, L. Wet pastures, &c. May, June.
  longifolia, Michx. Dry woods. July.
Galium, *L.* *Bedstraw.*
  Aparine, L. *Goose grass. Cleavers.* Shady banks. June.
  trifidum, L. *G. tinctorium*, L. *G. obtusum*, Bigel. Low grounds. June, July.
  asprellum, Michx. Moist thickets. July.
  triflorum, Michx. Woods. June, July.
  pilosum, Ait. Dry pastures, &c. Pownal. *Robbins.* June.
  circaezans, Michx. Woods. June, July.
    var. lanceolatum, Torr. & Gray. *G. lanc.* Torrey. Woods. Castleton,
      Branch. Middlebury, *Burge.* Essex, *Robbins.* Bellows
      Falls, *Carey.*
    var. montanum, T. & G. 2, 24. *G. Littellii*, Oakes in Hovey's Magazine,
      May, 1841. On the sides of Camel's Hump mountain,
      Robbins. Notch of Mansfield mountain, *Tuckerman* and
      Macrae. July, August. A pubescent var. grows on
      Sharpshin Point, Burlington, *Macrae.*
Cephalanthus, *L.* *Button Bush.*
  occidentalis, L. Small ponds and wet places. July, August.
Mitchella, *L.* *Checker-Berry.*
  repens, L. Woods. June, July.

Order VALERIANACEÆ. *The Valerian Tribe.*

Valeriana, *Tourn.* *Valerian.*
  sylvatica, Herb. Banks. Cedar and other swamps. Fairhaven and Craftsbury.
    Robbins. May—July.

Order DIPSACEÆ. *The Teasel Tribe.*

Dipsacus, *L.* *Teasel.*
  § *sylvestris*, L. *Wild Teasel.* Waste grounds. Castleton, *Reed.*

Order COMPOSITÆ.

Vernonia, *Schreber.*
  Noveboracensis, Willd. *Iron-weed.* Low grounds. Middlebury, *James.* Aug.
Eupatorium, *L.*
  perfoliatum, L. *Thorough wort.* Bogs and wet grounds. Aug.
  ageratoides, L. f. Shady banks, &c. August, Sept.
  purpureum, L. *E. verticill.* and *maculatum*, L. Moist grounds. Aug., Sept.
Nardosmia, *Cass.* L.
  palmata, Hook. *Tussilago palmata*, Ait. Swamps. Fairhaven, *Robbins.*
    April, May.
Tussilago, *Tourn.*
  ? § *Farfara*, L. *Colts-foot.* Banks of streams, and moist banks. Pownal, *Oakes.*
    Danby, Castleton, Grand Isle, Arlington, &c., *Robbins.*
    Burlington, *Tuckerman.* Rockingham, *Carey.* April, May.
Aster, *L.* *Starwort.*
  conyzoides, Willd. Dry open woods, &c. Pownal and Arlington, *Robbins.*
    July, August.
  lævis, L. Borders of woods, &c. Bellows Falls, &c., *Carey.* Aug., Sept.
  undulatus, L. Dry woods, &c. Burlington, *Macrae.* Bellows Falls, *Carey.*
    August, Sept.
  corymbosus, Ait. Woods and shady banks, Aug., Sept.
  cordifolius, L. Woods, &c. Sept.
  multiflorus, Ait. Dry hills, pastures, &c. Pownal, *Robbins.* Aug., Sept.

CATALOGUE OF PLANTS.

 dumosus, L. var. *strictior*, T. & G., 2, 128. Borders of woods, &c.
 Tradescanti, L. var. *fragilis*, T. & G., 2, 129. Rocky banks of the Winooski, Colchester and Burlington, *Robbins*. Aug., Sept.
 miser, L. var. *hirsuticaulis*. T. & G., 2, 131. Borders of thickets, &c. Bellows Falls, *Carey*. Burlington, *Macrae*. Aug., Sept.
 simplex, Willd. Wet grounds. Bellows Falls, *Carey*. August, Sept.
 præaltus, Poir. Moist woods, &c. Bellows Falls, *Carey*. August, Sept.
 puniceus, L. Low moist grounds. August, Sept.
 Novæ-Angliæ, L. Moist grounds, &c. Middlebury, *James*. Sept.
 acuminatus, Michx. Woods. August, Sept.
 ptarmicoides, T. & G., 2, 160. *Chrysopsis alba*, Nutt. *Heleastrum album*, DC. Rocky hills, Pownal, *Robbins*. August.
 linariifolius, L. Dry sandy pastures, &c. August, Sept.
 umbellatus, Miller. Moist thickets. August, Sept.
 macrophyllus, L. Dry woods. Sept.

Erigeron, L. *Flea-bane.*

 Canadense, L. Old fields, &c. July—Oct.
 bellidifolium, Muhl. Poor Robert's Plantain. Borders of woods, &c. May, June.
 Philadelphicum, L. *E. purpureum*, Ait. Banks of rivers. Putney, *Reed*. Burlington, *Robbins*. Bellows Falls, *Carey*. June.
 strigosum, Muhl. *E. Philadelphicum*, and *E. integrifolium*, Bigel. Fields, &c. June—Aug.
 annuum, Pers. *E. heterophyllum*, Muhl. *E. strigosum*, Bigel. Old fields, &c. July, August.

Solidago, L. *Golden Rod.*

 Canadensis, L. About fences and woods. August, Sept.
 gigantea, Ait. Borders of woods, &c. Bellows Falls, *Carey*. August, Sept.
 juncea, Ait. *S. arguta*, Torr. and Gray. Borders of woods, &c. Burlington, *Carey*.
 neglecta, Torrey & Gray. Moist woods, &c. Fairhaven, *Robbins*. Aug., Sept.
 altissima, L. Low grounds, &c. August, Sept.
 nemoralis, Ait. Dry fields and hills. August, Sept.
 odora, Ait. Woods. August, Sept.
 bicolor, L. Dry woods. August, Sept.
 cæsia, L. Woods. Bellows Falls, *Carey*. Sept.
 flexicaulis, L. *S. latifolia*, L. Shady banks and woods. Sept.
 virgaurea, L. Bigel. *S. thyrsoidea*, E. Meyer. T. & G., 2, 207. Woods on the sides of Killington Peak and of Mansfield Mountain. *Robbins*. August.
 squarrosa, Muhl. Dry banks and woods. Castleton, Essex and Colchester, *Robbins*. August, Sept.
 lanceolata, L. Low grounds, &c. August, Sept.
 humilis, Pursh, 2, 543. On limestone rocks at Winooski falls, Colchester, and also on the ledge with *Phaca Robbinsii*, Burlington, *Robbins*. August.

Inula, *L.*

 Helenium, L. *Elecampane.* Road sides. August.

Xanthium, L. *Cocklebur.*

 Strumarium, L. var. *Canadense*, Torrey and Gray. Road sides, &c. Middlebury, *James*. South Hero, *Robbins*. Burlington, *Carey*. August.

Ambrosia, *L.*

 Artemisiæfolia, L. *A. elatior*, L. *Bitter Weed.* Old fields, &c. Aug., Sept.
 trifida, L. Low grounds. Pownal, *Robbins*. August, Sept.

Rudbeckia, L.

 laciniata, L. Low grounds, &c. August, Sept.

Helianthus, L. *Sun flower.*

 divaricatus, L. Sandy woods, &c. August, Sept.
 decapetalus, L. Moist places and woods about Burlington and Colchester, *Macrae*. August, Sept.

Bidens, L. *Bur Marigold.*

 frondosa, L. Moist fields, &c. August, Sept.

CATALOGUE OF PLANTS.

 chrysanthemoides, Michx. Wet grounds. Bellows Falls, *Carey.* Aug., Sept.
 cernua, L. Wet grounds. August, Sept.
 Beckii, Torrey. Lakes, ponds, &c. In Lake Champlain, near Benson, *Chandler.* August, Sept.
 connata, Muhl. Moist grounds. Middlebury, *James.* August, Sept.
Anthemis, *L.*
 cotula, L. *May weed.* Road sides, &c. July—Sept.
Achillea, *L.*
 § *Millefolium*, L. *Yarrow, Milfoil.* Pastures, &c. July, August.
Chrysanthemum, *L.*
 leucanthemum, L. *Whiteweed.* Pastures and grass fields. June—Aug.
Artemisia, *L. Wormwood.*
 § *Absinthium*, L. *Common Wormwood.* Road sides, &c. Naturalized abundantly in Danby, Barre, Williamstown, Mount Tabor, Dorset, Pownal, &c., *Robbins.* Aug.
 § *vulgaris*, L. *Mugwort.* Road sides, &c. In Castleton, *Branch.* Middlebury, *Burge.* In North Hero, St. Albans, Georgia, Danby, &c. *Robbins.* Hubbardton, *Chandler.* Swanton, *Carey.* Colchester, *Oakes.* July, August.
Tanacetum, *L. Tansy.*
 § *vulgare*, L. *Common Tansy.* Road sides, &c. August.
Gnaphalium, *L. Cudweed.*
 decurrens, Ives. Fields and pastures. Near Mansfield Mountain, *Robbins.* Highgate, *Tuckerman.* Bellows Falls, *Carey.* Burlington and Colchester, *Oakes.* August, Sept.
 polycephalum, Michx. *Life everlasting.* Fields and pastures. August, Sept.
 uliginosum, L. Low grounds. August, Sept.
Antennaria, *R. Br.*
 margaritacea, R. Br. *Gnaphalium marg.* L. Pastures, &c. August, Sept.
 plantaginea, R. Br. *Gnaph. plant.* L. Pastures, &c. April, May.
Senecio, *L. Groundsel.*
 Balsamitæ, Muhl. Rocky banks. June.
 obovatus, Muhl. Dry rocky banks, &c. Bennington and Pownal, *Robbins.* May, June.
 aureus, L. Bogs, &c. June.
  var. lanceolatus, Oakes, in Hovey's Mag. May, 1841. In a cedar swamp at Brownington, *Robbins.* July.
 hieracifolius, L. *Fireweed.* Low grounds, &c. Aug.
Cirsium, *Tourn. Thistle.*
 § *lanceolatum*, Scop. *Cardus lanceolatus*, L. Road-sides, &c. July—Sept.
 discolor, Spreng. *Cnicus discolor*, Muhl. Fields and woods. Aug.
 pumilum, Spreng. *Cnicus odoratus*, Muhl. *Carduus pumilus*, Nutt. Pastures. Essex, *Robbins.* Bellows Falls, *Carey.* Sept., Oct.
 muticum, Michx. *Cnicus glutinosus*, Big. Moist woods. August, Sept.
 § *arvense*, Scop. *Cnicus arvensis*, Hoff. *Canada Thistle.* Fields, meadows, roadsides, &c. July, Sept.
Onopordon, *L. Cotton Thistle.*
 § *Acanthium*, L. Dry pastures, &c. Williston and Grand Isle, *Robbins.*
Arctium, *L.*
 Lappa, L. *Burdock.* Waste places. July—Sept.
Lactuca, *Tourn. Lettuce.*
 elongata, Muhl. Along fences, &c. July.
  var. sanguinea. L. sanguinea, Big. Dry pine woods. July, Aug.
Leontodon, *L.*
 Taraxacum, L. *Dandelion.* Fields, gardens, &c.
Sonchus, *L. Sow thistle.*
 oleraceus, L. *Common Sow thistle.* Gardens, &c. August, Sept.
  var. spinulosus,. S. spinulosus, Bigel. *S. oleraceus E.* Smith E. H., 3, 344. Pluk. t. 61, f. 5. Waste grounds, &c. Bellows Falls, *Carey.* Common in the east of Massachusetts, and apparently a starved variety of *S. oleraceus*, though the ochenia are also smoother than in the common variety.
 floridanus? L. *S. acuminatus*, Bigelow. Moist woods. August, Sept.
Hieracium, *L. Hawk-weed.*

CATALOGUE OF PLANTS.

   venosum, L. Dry open woods, &c. June.
   Marianum, Willd. Dry woods, &c. Aug.
   Canadensis, Michx. *H. Kalmii*, Bigelow, &c. Borders of woods. Aug.
   paniculatum, L. Dry woods. Aug.
Krigia, *Schreber*.
   Virginica, Willd. Dry sandy pastures, &c. Middlebury, *James*. May—July.
Prenanthes, *Vaill*.
   altissima, L. Shady banks, &c. August, Sept.
   alba, L. Woods, &c. August, Sept.

ORDER LOBELIACEÆ. *The Lobelia Tribe.*

Lobelia, *L.*
   Kalmii, L. Moist rocks and bogs. Brownington and Colchester, *Robbins*. Burlington, *Carey, Macrae, Oakes*. July, Aug.
   Claytoniana, Michx. *L. pallida*, Muhl. Moist meadows. June.
   Cardinalis, L. *Cardinal Flower.* Wet places. August, Sept.
   inflata, L. *Indian Tobacco.* Fields, road-sides, &c. Aug.

ORDER CAMPANULACEÆ. *The Bell Flower Tribe.*

Campanula, *L. Bell Flower.*
   rotundifolia, L. *Hare-bell.* Rocky banks, &c. June, July.
   amplexicaulis, Michx. *C. perfoliata*, L. Dry ledges, &c. Middlebury, *James.* Fairhaven, *Chandler.* June, July.
   aparinoides, Pursh. Wet meadows, &c. June, July.

ORDER ERICACEÆ. *The Heath Tribe.*

Andromeda, *L.*
   polifolia, L. Sphagnous bogs, especially on the edges of ponds. May, June.
   paniculata, L. *Pepper bush.* Swamps, &c. Pownal, *Robbins.* Bellows Falls, *Carey.* Ludlow, *Washburn.* June, July.
   calyculata, L. Bogs, &c. May.
Arbutus, *L.*
   Uva-ursi, L. *Bear berry.* Rocky hills, &c. April, May.
Gaultheria, *L.*
   procumbens, L. *Partridge Berry.* Dry woods. June, July.
Rhododendron, *L. Rosebay.*
   nudiflorum, Torr. *Azalea nudiflora*, L. *Wild Honeysuckle.* Swamps and moist woods. Middlebury, *James.* Pownal, *Oakes.* Fairhaven and Georgia, *Robbins.* Bellows Falls, *Carey.* Ludlow, *Washburn.* June.
   viscosum, Torrey. *Azalea viscosa*, L. Swamps. Middlebury, *James.* July.
   Canadense, Torrey. *Rhodora Can.*, L. Bogs, &c. Brattleboro', *Robbins.* Guildhall, *Carey.* May, June.
Kalmia, *L.*
   latifolia, L. *Calico bush. High Laurel.* Rocky hills, &c. Rockingham, *Carey.* June, July.
   angustifolia, L. *Sheep Laurel. Low Laurel.* Moist places. June, July.
   glauca, Ait. Sphagnous bogs. May, June.
Epigæa, *L.*
   repens, L. *Ground Laurel.* Sandy woods and on mountains. April, May.
Ledum, *L.*
   latifolium, L. *Labrador Tea.* Bogs. On the summits of Camel's Hump and Mansfield mountains, *Robbins* and *Tuckerman.* May, June.
Vaccinium, *L.*
   frondosum, L. *Dangleberry.* Woods. Middlebury, *James.* June.
   resinosum, Ait. "*Huckleberry,*" or *Black Whortleberry.* Dry woods, &c. May, June.
   corymbosum, L. *High Blueberry.* Swamps, &c. May, June.
   Pennsylvanicum, Lam. *V. virgatum*, Ait. Big. *Low Blueberry.* Dry woods. Essex, *Robbins.* May, June.
   tenellum, Ait. Big. *Low Blueberry.* Dry woods, pastures, &c. On the summits of Camels Hump and Mansfield mountains, *Robbins, Macrae,* and *Tuckerman.* May, June.

CATALOGUE OF PLANTS.

Canadense, Richardson. *Low Blueberry.* Pastures, swamps, &c. Bellows Falls, *Carey.* Fairhaven, *Oakes.* May, June.
uliginosum, L. On the summits of Mansfield and Camel's Hump mountains. *Robbins, Tuckerman,* and *Macrae.* June.
Vitis Idæa, L. *Cowberry.* With the preceding, *R., T.* and *M.* June, July.
macrocarpon, Ait. *Common Cranberry.* Bogs, &c. June.
oxycoccus, L. *Small Cranberry.* Bogs. On the summit of Mansfield mountain, *Robbins.* June, July.

Lasierpa, *Torr.*
hispidula, Torr. *Gaultheria serpyllifolia,* Pursh. Old pine woods and swamps. May, June.

Pyrola, *L. Winter green.*
rotundifolia, L. Woods. July.
chlorantha, Swartz. *P. asarifolia,* Torrey. Not of Michx. Old pine woods, &c. June, July.
elliptica, Nutt. Dry woods. July.
secunda, L. Old Pine woods, &c. June, July.
uniflora, L. Rare. In a cedar swamp, Brownington, *Robbins.* In Pine woods, Burlington and High Bridge, *Macrae.* In Charleston, with calypso borealis, *Carey.* July.
umbellata, L. *Pipsissewa.* Dry woods. July.
maculata, L. Dry woods. Middlebury, *James.* July.

Monotropa, *L.*
uniflora, L. *Indian Pipe.* Woods. July.

Hypopithys, *Dillen. Pine sap.*
lanuginosa, Nutt. *Monotropa lanuginosa,* Michx. Woods. July, Aug.

Pterospora, *Nutt.*
andromedea, Nutt. Dry rocky pine woods, near High Bridge, Colchester, *Robbins,* and Burlington, *Oakes.* Shady rich soil on the rocks of Sharpshin Point, Burlington, *Macrae.* July.

Order AQUIFOLIACEÆ. *The Holly Tribe.*

Nemopanthes, *Raf.*
Canadensis, Raf. *Ilex Canadensis,* Michx. Swamps, &c. May.

Prinos, *L.*
verticillatus, L. *Black Alder. Winter Berry.* Swamps. Middlebury, *James.* July.

Order OLEACEÆ. *The Olive Tribe.*

Fraxinus, *L. Ash.*
sambucifolia, Lam. *Black Ash.* Moist woods, Middlebury, *James.* Lyndon, *Carey.* In Vermont, *Tuckerman.* May.
acuminata, Lam. *F. Americana,* Michx. f. *White Ash.* Woods. May.
pubescens, Walter. *F. tomentosa,* Michx. f. *Red Ash.* Woods, &c. In Castleton, *Chandler.* In Burlington, and in Grand Isle, *Robbins.* May.

Order APOCYNACEÆ. *The Dog's-bane Tribe.*

Apocynum, *L. Dog's-bane.*
androsaemifolium, L. Borders of woods, by fences, &c. June, July.
hypericifolium, Ait? Pursh. Gravelly banks of ponds and rivers. June, July.

Order ASCLEPIADACEÆ. *The Milkweed Tribe.*

Asclepias, *L. Milkweed.*
Syriaca, L. *Common Milkweed.* Along fences, &c. July.
phytolaccoides, Pursh. Woods, &c. July.
incarnata, L. Low grounds. July, August.
obtusifolia, Michx. Dry sandy soil. July.
quadrifolia, Jacq. Rocky woods. June.
tuberosa, L. *Pleurisy-Root.* Sandy fields, &c. Pownal, *Robbins.* Bellows Falls, *Carey.* July, August.
debilis, Michx. Shady dell near Burlington, *Macrae.* July.

CATALOGUE OF PLANTS.

Order GENTIANACEÆ. *The Gentian Tribe.*

Gentiana, *L. Gentian.*
   saponaria, L. *Soap-wort Gentian.* Moist thickets, &c. August, Sept.
   quinqueflora, L. Woods. Castleton, *Reed.* Pownal, *Robbins.* Rockingham, *Carey.* August.
   crinita, Frœl. Wet meadows. Sept., Oct.
Centaurella, *Michx.*
   Virginica. Sagina Virginica, L. *Centaurella paniculata*, Michx. *C. autumnalis*, Pursh. Swamps, &c. Rockingham, *Carey.* August, Sept.
Menyanthes, *L.*
   trifoliata, L. *Buckbean.* Bogs, &c. Burlington and Georgia, *Robbins.* Derby, *Carey.* Colchester, *Macrae.* May, June.

Order CONVOLVULACEÆ. *The Bindweed Tribe.*

Convolvulus, *L. Bind weed.*
   sepium, L. Moist borders of thickets, &c. July.
   spithameus, L. Dry sandy plains. July.
Cuscuta, *L. Dodder.*
   Americana, L. Low grounds. August.

Order BORAGINACEÆ.

Lithospermum, *L. Gromwell.*
   § *officinale*, L. Dry pastures, &c. Sudbury and Benson, *Chandler.* Middlebury, St. Albans, and South Hero, *Robbins.* Burlington, *Macrae, Oakes.* June, July.
   § *arvense*, L. *Corn Gromwell.* Old wheat fields, &c. May.
Lycopsis, *L.*
   § *arvensis*, L. Road sides, &c., in dry soil. Pownal, *Reed.*
Echinospermum, *Lehm.*
   § *Lappula*, Lehm. *Myosotis Lappula*, L. Road sides, &c. July, Aug.
   Virginianum, Lehm. Borders of thickets, road sides, &c. Bellows Falls, *Carey.* July.
Cynoglossum, *L. Hound's Tongue.*
   § *officinale*, L. Road sides, &c. May, June.
   Virginianum, L. Woods. Rare. June.

Order HYDROPHYLLACEÆ.

Hydrophyllum, *L.*
   Virginianum, L. Woods. June.
   Canadense, L. Woods. At the base of Mansfield mountain, and frequent in the south west of Vermont, *Robbins.* June.

Order LABIATÆ. *The Mint Tribe.*

Lycopus, *L. Water Horehound.*
   sinuatus, Ell. *L. Europœus*, Pursh., not of Linn. Low grounds. Aug.
   Virginicus, L. Low grounds. Aug.
Mentha, *L. Mint.*
   § *Piperito*, L. *Peppermint.* Ludlow, *Washburn.*
   borealis, Michx. ? Tor. Manual, Bigel. Wet grounds. Aug.
   Canadensis, L. ? Torrey, Manual. Banks of rivers, &c. On the Hoosic, at Pownal, *Oakes.*
   § *viridis*, L. *Spearmint.* Moist meadows, about springs, &c. July, Aug.
Monarda, *L. Horsemint.*
   fistulosa, L. *M. allophylla*, Michx. *M. oblongata*, Ait. Dry rocky woods. At Middlebury, *James.* July, Aug.
Blephilia, *Raf.*
   hirsuta, Raf. *Monarda hirsuta*, Pursh. In Castleton, *Branch.* In a wet meadow, Craftsbury, *Robbins.* In moist woods, Chester, *Oakes.* July, August.
Pycnanthemum, *Michx.*
   incanum, Michx. *Mountain Mint.* Rocky woods. Cavendish, *Macrae.* Aug.
   lanceolatum, Pursh. Borders of thickets, &c. Pownal, *Robbins.* Bellows Falls, *Carey.* July, Aug.

CATALOGUE OF PLANTS.

muticum, Pursh. Pastures, &c. Pownal, *Robbins.* July, Aug.
Collinsonia, *L.*
 Canadensis, L. *Horse-weed.* Shady banks, &c. Middlebury, *James.* Arlington, *Robbins.* July, Aug.
Hedeoma, *Pers.*
 pulegioides, Pers. *Penny-royal.* Pastures, &c. Middlebury, *James.* Bellows Falls, *Carey.*
Melissa, *L. Balm.*
 § *Clinopodium*, Benth. *Clinopodium vulgare*, L. Rocky banks. July.
Prunella, *L.*
 vulgaris, L. *Self Heal.* Pastures, &c. June—Sept.
Scutellaria, *L. Scullcap.*
 lateriflora, L. Common Scullcap. Low grounds. Aug.
 galericulata, L. Moist places. Aug.
 parvula, Michx. *S. ambigua*, Nutt. Sharpshin Point, Burlington, *Macrae.* July.
Lophanthus, *Benth.*
 nepetoides, Benth. *Hyssopus nepetoides*, L. Thickets and along fences. Middlebury, *James.* Rutland, *Branch.* Pownal, Bennington, and Arlington, *Robbins.* July, Aug.
Nepeta, *L.*
 § *Cataria*, L. *Catnep.* Roadsides, &c. July, Aug.
 § *Glechoma*, Benth. *Glechoma hederacea*, L. *Ground Ivy. Gill.* On cultivated grounds, &c. May, June.
Leonurus, *L.*
 § *Cardiaca*, L. *Motherwort.* Roadsides, &c. July, Aug.
Stachys, *L. Hedge Nettle.*
 aspera, Michx. Old fields, &c. Grand Isle and South Hero, *Robbins.* Burlington, *Macrae* and *Tuckerman.* July Aug.
Galeopsis, *L. Hemp Nettle.*
 § *Tetrahit*, L. Roadsides, &c. July, Aug.
 § *Ladanum*, L. Waste places, &c. Bellows Falls, *Carey.* July.
Teucrium, *L. Germander.*
 Canadense, L. Low grounds. South Hero, *Robbins.* Bellows Falls, *Carey.* Red Rocks, Burlington, *Macrae.* July, Aug.

ORDER SOLANACEÆ. *The Night Shade Tribe.*

Solanum, *L. Night Shade.*
 § *Dulcamara*, L. *Bitter-sweet.* Roadsides, &c. July, Aug.
 § *nigrum*, L. Cultivated grounds. July, Aug.
Physalis, *L. Ground Cherry.*
 viscosa, L. Dry fields, &c. Pownal, *Robbins.* June, July.
Datura, *L.*
 § *Stramonium*, L. *Thorn Apple.* Waste grounds. July—Sept.
Hyoscyamus, *L. Henbane.*
 § *niger*, L. Roadsides, &c. Panton, *Burge.* Mount Independence, *Dr. Hill.* June, July.

ORDER SCROPHULARIACEÆ· *The Figwort Tribe.*

Verbascum, *L. Mullein.*
 § *Thapsus*, L. Common *Mullein.* Old fields, &c. July, Aug.
Veronica, *L. Speedwell.*
 § *serpyllifolia*, L. Meadows and Pastures. May, June.
 scutellata, L. Ditches, &c. June.
 Beccabunga, L. *Brooklime.* In grounds wet by springs, &c. June.
 Anagallis, L. *Water Speedwell.* Ditches, &c. Middlebury, *Burge.* June, July.
 peregrina, L. Cultivated grounds. Middlebury, *James.* May, June.
 § *arvensis*, L. Old fields, &c. May, June.
 Virginica, L. Moist bank on Mr. U. H. Penniman's grounds, with *Trillium grandiflorum*, Colchester, *Oakes.* Aug.
Linaria, *Tourn. Toad Flax. Snap Dragon.*
 § *vulgaris*, Mœnch. *Antirrhinum Linaria*, L. Roadsides, &c. Manchester, *Robbins.* July—Sept.

Canadensis, Spreng. Moist bare soils. Bellows Falls, *Carey.* July, Aug.
Scrophularia, *L. Figwort.*
   Marilandica, L. Along fences, &c. Middlebury, *James.* Colchester, *Robbins.* July, Aug.
Mimulus, *L. Monkey Flower.*
   ringens, L. Wet grounds. Aug.
Gratiola, *L. Hedge Hyssop.*
   aurea, Muhl. Borders of Ponds, &c. Middlebury, *James.* August, Sept.
Lindernia, *L.*
   Pyxidaria, L.
     var. *dilatata*. *L. dilatata*, Muhl. Moist open grounds. Middlebury, *James.* Brattleboro' and West Haven, *Robbins.*
     var. *attenuata*. *L. attenuata*, Muhl. Craftsbury and Cambridge, *Robbins.* July, Aug.
Chelone, *L. Snake-head.*
   glabra, L. Borders of swamps, &c. August, Sept.
Pentstemon, *L'Her.*
   pubescens, Ait. Rocky hills, &c. Middlebury, *James.* Castleton, *Chandler.* Benson, *Prof. Woodward.* Pownal, *Robbins.*
Gerardia, *L.*
   tenuifolia, Vahl. Dry soil. Pownal and Brattleboro', *Robbins.* Bellows Falls, *Carey.* Aug.
   flava, L. Dry woods. Near Bellows Falls, *Carey.* Aug.
   pedicularia, L. Dry woods, &c. Pownal, *Robbins.* Bellows Falls, *Carey.* August.
   quercifolia, Pursh. Woods. Castleton and Pownal, *Robbins.*
Pedicularis, *L. Lousewort.*
   Canadensis, L. Borders of woods, &c. May, June.
Castilleja, *Mutis.* Bartsia, *L.*
   pallida, Kunth. *Bartsia pallida*, L. On the north side of Mansfield mountain, near the summit, *Tuckerman* and *Macrae.* July.
Melampyrum, *L. Cow Wheat.*
   Americanum, Michx. Woods. June—Aug.

Order OROBANCHACEÆ. *The Broom-Rape Tribe.*

Orobanche, *L. Broom-rape.*
   Americana, L. Woods. On White Creek, *Chandler.* Sharpshin Point, Burlington, *Macrae.*
   uniflora, L. Woods. June.
Epiphegus, *Nutt. Beech Drops.*
   Virginiana. Orobanche Virginiana, L. *Epiphegus Americanus*, Nutt. Woods, under beech trees. Sept.

Order VERBENACEÆ. *The Vervain Tribe.*

Verbena, *L. Vervain.*
   hastata, L. Low grounds, roadsides, &c. July, Aug.
   urticifolia, L. Roadsides, &c. July.
Phryma, *L.*
   leptostachya, L. Woods and shady banks. Middlebury, *James·* South Hero and Arlington, *Robbins.* Bellows Falls, *Carey.* Burlington, *Oakes.* July.

Order ACANTHACEÆ

Justicia, *L.*
   Americana, Vahl. *J. pedunculosa*, Michx. In water. "At Ferrisburgh." *Dr. Paddock's* herbarium in the Museum of the University at Burlington, the specimen thus ticketed, seen by Dr. Robbins.

Order LENTIBULACEÆ.

Utricularia, *L. Bladder-wort.*
   vulgaris, L. In ditches, ponds, &c. Aug.
   cornuta, Michx. Bogs, &c. Vermont, *Carey.* July, Aug.

CATALOGUE OF PLANTS.

Order PRIMULACEÆ. *The Primrose Tribe.*

Trientalis, *L.*
  Americana, Pursh. Wet woods and swamps. May, June.
Lysimachia, *L.* *Loose-strife.*
  thyrsiflora, L. Swamps. Castleton, *Chandler.* Burlington, *Macrae.*
  stricta, Ait. Low grounds, &c. July.
  quadrifolia, L. Woods. June, July.
  ciliata, L. Borders of woods, &c. July.
  hybrida, Michx. Wet grounds. Ferrisburgh and South Hero, *Robbins.* July.
Samolus, *L.* *Water Pimpernel.*
  Valerandi, L. Borders of rivers. Middlebury, *James.* July—Sept.

Order PLANTAGINEÆ. *The Plantain Tribe.*

Plantago, *L.* *Plantain.*
  § *major*, L. *Common Plantain.* About houses, fields, &c. June—Sept.

Order AMARANTHACEÆ. *The Amaranth Tribe.*

Amaranthus, *L.*
  § *hybridus*, L. Gardens, &c. Aug.
  Blitum ? L. Cultivated and waste grounds. Pownal, *Robbins.* Aug.

Order CHENOPODIACEÆ. *The Goosefoot Tribe.*

Chenopodium, *L.* *Goosefoot.*
  § *album*, L. Gardens, fields, &c. July, August.
  § *Botrys*, L. *Jerusalem Oak.* Sandy banks of Lake Champlain, &c. Alburgh, *Robbins.* Middlebury, *James.* Burlington, *Oakes.* Bellows Falls, *Carey.* July, August.
  § *hybridum*, L. Waste grounds. August.
  § *rubrum*, L. Cultivated grounds. Bennington, *Robbins.* August.
Blitum, *L.*
  § *capitatum*, L. *Strawberry Blite.* Road sides, &c. Hubbardton, *Branch.* Newport, *Robbins.* North Troy, *Carey.* June.

Order PHYTOLACEÆ.

Phytolacca, *L.*
  decandra, L. *Poke.* Waste places, &c. July—Oct.

Order POLYGONACEÆ. *The Buckwheat Tribe.*

Polygonum, *L.* *Knotweed.*
  § *aviculare*, L. *Knot-grass.* About houses, &c. June—Oct.
  Virginianum, L. Rocky woods. Arlington and Castleton, *Robbins.* Waterbury, *Macrae.* July, August.
  Hydropiper, L. *Water Pepper.* Low grounds, ditches, &c. August.
  mite, Pers. Wet places. West Haven, *Robbins.* Castleton, *Chandler.* July, August.
  § *Persicaria*, L. Gardens, &c. July—Sept.
  amphibium, L. var. *natans*, Michx. Floating in water.
    var. *emersum*, Michx. Margin of ponds, &c. Aug., Sept.
  Pennsylvanicum, L. Low grounds, &c. July, August.
  sagittatum, L. *Scratch-grass.* Low grounds. August, Sept.
  arifolium, L. Swamps, &c. August, Sept.
  scandens, L. Fields, &c. July, August.
  cilinode, Michx. Woods, &c. July, August.
  § *convolvulus*, L. Road sides. July, August.
  § *Fagopyrum*, L. *Buckwheat.* Old fields, &c. July, August.
Rumex, *L.* *Dock.*
  § *crispus*, L. *Curled Dock.* Cultivated grounds. July, August.
  § *obtusifolius*, L. Cultivated grounds. June, July.
  verticillatus, L. In water. July.
  § *Acetosella*, L. *Sheep Sorrel.* Pastures and cultivated grounds. May—July.

CATALOGUE OF PLANTS.

Order LAURACEÆ. *The Cinnamon Tribe.*

Laurus, *L.*
   sassafras, L. *Common Sassafras.* Woods, &c. Pownal, *Robbins.* May.
   Benzoin, L. *Fever Bush.* Swamps, &c. Bellows Falls, *Carey.* May.

Order ELEAGNACEÆ. *The Oleaster Tribe.*

Shepherdia, *Nutt.*
   Canadensis, Nutt. Rocky banks of Lake Champlain, &c. May.

Order THYMELACEÆ. *The Mezereon Tribe.*

Dirca, *L. Leather-wood.*
   palustris, L. Moist woods. April, May.

Order SANTALACEÆ. *The Sanders-wood Tribe.*

Nyssa, *L.*
   multiflora, Walt. *N. sylvatica*, Michx. f. *N. villosa*, Willd. *Tupelo, or Sour Gum.* Woods and swamps. Craftsbury, *Robbins.* June.
Comandra, *Nutt.*
   umbellata, Nutt. *Thesium umb.*, L. Borders of woods, &c. June.

Order ARISTOLOCHIACEÆ. *The Birthwort Tribe.*

Asarum, *Tourn.*
   Canadense, L. *Wild Ginger.* Rocky woods. May.

Order EMPETRACEÆ. *The Crowberry Tribe.*

Empetrum, *L.*
   nigrum, L. *Crowberry.* Summit of the Mansfield and Camel's Hump Mountains, *Robbins*, *Tuckerman* and *Macrae*. June, July.

Order EUPHORBIACEÆ. *The Spurge Tribe.*

Acalypha, *L. Three-seeded Mercury.*
   Virginica, L. Fields and road sides. Middlebury, *James.*
Euphorbia, *L. Spurge.*
   § *Helioscopia*, L. Waste ground, &c. In Addison county, *Burge.* July, Aug.
   § *platyphylla*, L. *E. obtusata?* Pursh. Road sides, &c. Benson, *Chandler.* Vergennes, South Hero, and Grand Isle, *Robbins*, Aug.
   maculata, L. Sandy fields, &c. July—Sept.
   hypericifolia, L. Dry sandy fields, &c. Burlington, *Tuckerman.* Aug., Sept.

Order URTICACEÆ. *The Nettle Tribe.*

Urtica, *Tourn. Nettle.*
   pumila, L. Shady places. July, August.
   § *dioica*, L. Road sides, &c. July.
   Canadensis, L. Shady, moist woods, &c. July, August.
Parietaria, *Tourn. Pellitory.*
   Pennsylvanica, Muhl. Shady rocks. Fair Haven, *Robbins.* Extremity of Sharpshin Point, Burlington, *Macrae.* July.
Bœhmeria, *Willd.*
   cylindrica, Willd. Swamps, &c. Bellows Falls, *Carey.* July, Aug.
Cannabis, *Tourn. Hemp.*
   § *sativa*, L. Waste places. June, July.
Humulus, *L. Hop.*
   § *Lupulus*, L. Borders of thickets, &c. Middlebury, *Burge.* Castleton, *Robbins.* August.
Morus, *Tourn. Mulberry.*
   rubra, L. *Red Mulberry.* Banks of rivers, woods, &c. Pownal, *Oakes.* May.

Order AMENTACEÆ.

Sub-Order Cupuliferæ.

Carpinus, *L. Hornbeam.*
   Americana, Michx. Woods. May.
Ostrya, *Scop. Hop Hornbeam.*
   Virginica, Willd. *Carpinus ostrya*, Michx. f. t. *Iron-wood.* Woods. May.

BOTANY OF VERMONT.

CATALOGUE OF PLANTS.

Corylus, *Tourn.* *Hazel Nut.*
   Americana, Walt. *American Hazel Nut.* Thickets, &c. April.
   rostrata, Ait. *Beaked Hazel Nut.* Shady banks, &c. April.
Fagus, *Tourn.* *Beech.*
   sylvestris, Michx. and Michx. f. t. *White Beech.* Woods. May.
   ferruginea, Ait? Michx. f. t. *Red Beech.* Woods. May.
Castanea, *Gært.* *Chestnut.*
   vesca, Gært. var. *Americana,* Michx. *Chestnut.* Woods. July.
Quercus, *L.* *Oak.*
   tinctoria, Bartram. *Black oak.* Woods. May.
   rubra, L. *Red oak.* Woods. May.
   ilicifolia, Wang. *Q. Banisteri,* Michx. *Scrub Oak.* Barren plains, &c. Bellows Falls, *Carey.* May.
   macrocarpa, Michx. *Over-cup White Oak.* Woods, &c. Burlington, Colchester, St. Albans, Grand Isle, South Hero, Shoreham, West Haven and Bennington, *Robbins,* This is perhaps *Q. olivæformis* of Dr. James' catalogue.
   alba, L. *White Oak.* Woods. May, June.
   bicolor, Willd. *Q. Prinos discolor,* Michx. fil. *Swamp White Oak.* Wet woods. May.
   montana, Willd. *Q. Prinos monticula,* Michx. f. t. *Rock Chestnut Oak.* Rocky woods. Bennington, *Robbins.* May.
   chinquapin, Pursh. *Dwarf Chestnut Oak.* Dry hills, &c. Pownal, *Robbins.* May.
   coccinea, Wangenheim. *Scarlet Oak.* Woods. May.

Sub-Order Betuleæ. *The Birch Tribe.*

Betula, *Tourn.* *Birch.*
   populifolia, Ait. *Small White Birch.* About barren fields, woods, &c. May.
   papyracea, Ait. *Large White Birch. Canoe Birch.* Woods. May.
   lenta, L. *Black Birch. Sweet Birch. Cherry Birch.* Woods. May.
   excelsa, Ait. *B. lutea,* Michx. f. *Yellow Birch.* Woods. May.
Alnus, *Willd.* *Alder.*
   serrulata, Willd. *Common Alder.* Swamps, &c. April.
   glauca, Michx. f. sylv. t. Swamps, &c. April.
   crispa, Hook. *Betula Alnus crispa,* Ait. Near the summit of Camel's Hump and Mansfield mountains, *Robbins.* June.

Sub-Order Salicineæ. *The Willow Tribe.*

Salix, *Tourn.* *Willow.*
   candida, Willd. Pursh. In a sphagnous swamp on the borders of Lake Bombazin, Hubbardton, *Robbins.* April.
   Muhlenbergiana, Willd. Dry woods, &c. Bellows Falls, *Carey.* April, May.
   pedicellaris, Pursh. Bogs and swamps. Burlington, *Robbins. Macrae.* May.
   conifera, Wang. Wet thickets, &c. April.
   rostrata, Richardson. Borders of thickets, &c. Bellows Falls, *Carey.* April, May.
   nigra, Marshall. Banks of streams, &c. May.
   lucida, Muhl. Borders of swamps, &c. May.
   cordata, Muhl. Low wet grounds. April, May.
   rigida, Muhl. Low wet grounds, &c. Bellows Falls, *Carey.* April, May.
   grisea, Willd. Borders of swamps, &c. April, May.
   § *vitellina,* L. Road sides, &c. May.
   Uva-ursi, Pursh. On the summit of Mansfield Mountain, *Robbins.* June.
Populus, *Tourn.* *Poplar.* (According to Michaux's Sylva.)
   balsamifera, Michx. Michx. f. Sylv. t. *Balsam Poplar.* Woods and banks of rivers, &c. Pownal, *Oakes.* Westhaven, *Robbins.* April.
   candicans, Ait. Michx. f. Sylv. t. *Heart-leaved Balsam Poplar.* South Hero, Grand Isle, Cambridge, Jericho, &c., *Robbins.* Burlington, *Macrae, Oakes.* April.
   Canadensis, Michx. f. Sylv. t. *Cotton Wood. Cotton Poplar.* Banks of rivers, &c. On the Hoosic, Pownal, *Oakes.*
   monilifera, Ait. Michx. f. Sylv. t. *Vermont Poplar.* Banks of rivers, lakes, &c. In Orwell, *Branch, Chandler.* In Pownal, Brattleboro',

CATALOGUE OF PLANTS.

North Hero, South Hero, Alburgh, Johnson, and Hydepark, *Robbins*. Burlington, *Oakes*. April.
tremuloides, Mich. Michx. f. Sylv. t. *American Aspen*. Woods. April.
grandidentata, Michx. Michx. f. Sylv. t. *Large Aspen*. Woods. April, May.

SUB-ORDER MYRICEÆ. *The Gale Tribe.*

Comptonia, *Banks.*
  asplenifolia, Ait. *Sweet Fern.* Dry hills and plains. April, May.

SUB-ORDER PLATANEÆ. *The Plane Tribe.*

Platanus, *L. Plane Tree.*
  occidentalis, L. *Button Wood. Sycamore.* Banks of rivers, &c. May.

ORDER ULMACEÆ. *The Elm Tribe.*

Ulmus, *L. Elm.*
  Americana, L. *Common Elm.* Woods, banks of rivers, &c. April.
  fulva, Michx. *Slippery Elm.* Woods, banks of rivers, &c. April.
  racemosa, Thomas in Sill. Journal, 1829. *Northern Cork Elm.* Moist woods, &c. Bennington and Pownal, *Robbins.*
Celtis, *L. Hackberry.*
  occidentalis, L. *Hoop Ash.* Woods, &c. Burlington, *Robbins.* May.

ORDER JUGLANDACEÆ. *The Walnut Tribe.*

Juglans, *L. Walnut.*
  cinerea, L. *Butter Nut. Oil Nut.* Woods, &c. May, June.
Carya, *Nuttall. Hickory.* Juglans, *L.*
  alba, Nutt. Juglans alba, L. *J. squamosa*, Michx. f. not *J. alba*, Willd, Bigel. *Shell-bark or Shag-bark Hickory.* Woods. May, June.
  porcina, Nutt. *J. porcina*, Michx. f. Sylv. t. *J. glabra*, Muhl., Bigelow. *Pig Nut.* Woods. Middlebury, *James.* May, June.
  amara, Nutt. *J. amara*, Michx. f. Sylv. t. *Bitter Pig Nut.* Woods. Colchester, *Robbins.* Burlington, *Carey, Macrae.* May, June.

CLASS II. GYMNOSPERMS.

ORDER CONIFERÆ. *The Fir Tribe.*

Pinus, *L. Pine.*
  resinosa, Ait. *P. rubra*, Michx. f. Sylv. t. *Red Pine.* "Norway Pine," a bad name, as it is not found in Norway. Dry barren woods. June.
  rigida. Pitch Pine. Woods, in poor soil. June.
  Strobus, L. *White Pine.* Woods and swamps. June.
  nigra, Ait. *Black or Double Spruce.* Woods and swamps. May, June.
  alba, Ait. *White or Single Spruce.* Woods and swamps. May, June.
  balsamea, L. *Balsam Fir. Silver Fir.* Mountain woods, &c. June.
  var. Fraseri. P. Fraseri, Pursh. Near the summits of Mansfield and Camel's Hump Mountains, *Robbins, Tuckerman,* and *Macrae.* Essex, *Macrae.*
  Canadensis, L. *Hemlock Spruce.* Rocky woods, &c May, June.
  pendula, Ait. Larix Americana, Michx. *American Larch. Hackmatack. Tamarack.* Woods and swamps. May, June.
Thuya, *Tourn. Arbor Vitae.*
  occidentalis, L. *American Arbor Vitae.* "*White Cedar.*" In swamps and rocky woods. May.
Juniperus, *L. Juniper.*
  Virginiana, L. *Red Cedar. J. prostrata*, James? Dry rocky woods, &c. May.
  communis, L. *Common Juniper.* Dry rocky pastures, &c. May.
Taxus, *Tourn. Yew.*
  Canadensis, Willd. *American Yew. Ground Hemlock.* Swamps, &c. May.

ORDER CALLITRICHACEÆ.

Callitriche, *L.*
  vernalis, L. *C. autumnalis*, L. *C. terrestris*, Raf. In water, and on moist soil on the margins of ponds, &c. May—Sept.

CLASS III. ENDOGENS OR MONOCOTYLEDONS.

Order IRIDACEÆ. *The Iris Taibe.*

Sisyrinchium, *L.* *Blue-eyed Grass.*
   anceps, Cavan. Meadows. Burlington, *Macrae.*
   var. mucronatum. Dry soil. Burlington, *Macrae.* Bellows Falls, *Carey.* June.
Iris, *L.*
   versicolor, L. Blue Flag. Wet meadows, &c. July.

Order HYDROCHARACEÆ. *The Frog-bit Tribe.*

Udora, *Nutt.*
   Canadensis, Nutt. *Elodea Canadensis*, Michaux. *Serpicula occidentalis*, Pursh. In water. Middlebury, *James.* At the mouths of Winooski river and Otter Creek, and in lake Memphremagog, *Robbins.* August.
Valisneria, *Micheli.*
   spiralis, L. *V. Americana*, Michx. In lakes and slow flowing water. Middlebury, *James.* At the mouth of Winooski river, in Castleton river, in lake Champlain near the mouth of the Lamoille, in Shoreham, and in the Connecticut at Brattleboro', *Robbins*, August, Sept.

Order ORCHIDACEÆ. *The Orchis Tribe.*

Orchis, *L.*
   Sect. 1. Orchis.
   spectabilis, L. Woods. May, June.
   Sect. 2. Habenaria, *Willd.*
   orbiculata, Pursh. Woods. Leaves flat on the ground. June, July.
   Hookeriana, *Habenaria Hookeriana*, Torrey. Woods. June.
   blephariglottis, Willd. "Sphagnous margin of a closely shaded pond in North Troy, *Carey.* Aug.
   hyperborea, L. *H. Huronensis*, Spreng. Swamps, &c. Base of Mansfield mountain, and Burlington, *Macrae.* July.
   var. dilatata. *O. dilatata*, Pursh. Swamps, &c. July.
   psycodes, L. not of Bigelow, &c. *O. fimbriata*, Ait. Wet meadows, &c. July, August.
   grandiflora, Bigelow Wet meadows, &c. July.
   lacera, Michx. *O. psycodes*, Willd, Big. &c., not of L. Bogs, &c. Middlebury, *James.* July.
   ciliaris, L. Swamps, &c. Middlebury, *James.* Aug.
   obtusata, Pursh. High mountains and sphagnous swamps at the North. In Charleston, with the Calypso, *Carey.* Brownington, *Robbins.* July.
   viridis, Swartz. *O. bracteata*, Muhl. Woods. May, June.
   tridentata, Muhl. On the east side of Mansfield mountain, *Macrae.* July.
   flava, L. *Habenaria herbiola*, Brown. Burlington, *Macrae.*
Microstylis, *Nutt.* Malaxis, *Swartz.*
   ophioglossoides, Nutt. Woods. July, August.
   monophyllos, Lindl. *M. brachypoda*, Gray. *Ophrys monophyllos*, L. In Vermont, probably near Castleton, *Chandler.* July.
Liparis, *Richard.* Malaxis, *Swartz.*
   liliifolia, Richard. Hills near Bellows Falls, *Carey.* June, July.
   Læselii, Richard. *Malaxis lorreana*, Barton. Boggy soil, &c. July.
Aplectrum, *Nuttall.*
   hyemale, Nutt. *Cymbidium hyemale*, Muhl. Woods. Middlebury, *James.* Near Castleton, *Chandler.*
Corallorhiza, *Haller.*
   innata, R. Brown. *C. verna*, Nutt. Sphagnous swamps. May, June.
   multiflora, Nutt. Pine woods, &c. August, Sept.
   odontorhiza, Nutt. Woods. Bellows Falls, *Carey.* Sept.
Arethusa, *L.*
   bulbosa, L. Bogs. Hubbardton, *Robbins.* Near Burlington, *Macrae.*

CATALOGUE OF PLANTS

Pogonia, *Juss.*
   ophioglossoides, R. Brown. Bogs. Near Burlington, *Robbins, Macrae.* July.
   verticillata, Nutt. Woods. Near High Bridge, Colchester, *Robbins, Oakes.* May, June.

Triphora, *Nuttall.*
   pendula, Nutt. In a dry wood of beech, birch, &c., on a hill south of Fair Haven village, *Chandler.* August.

Calopogon, *R. Brown.*
   pulchellus, R. Brown. Bogs. July.

Spiranthes, *Richard.* Neottia, *Swartz.*
   cernua, Richard. Moist grounds, &c. August, Sept.
   gracilis, Hook. *N. gracilis*, Big. Dry woods. Colchester, *Robbins.* Burlington, *Macrae.* July.
   æstivalis, Rich. *Neottia æstivalis*, Lam. *N. cernua, var. latifolia*, Torrey. Moist woods, banks of rivers, &c. Burlington, *Macrae.* Bellows Falls, *Carey.* June.

Goodyera, *R. Brown.*
   pubescens, R. Brown. Woods. July, August.
   repens, R. Brown. Old woods. July.

Listera, *R. Brown.*
   cordata, R. Brown. On high mountains and in sphagnous swamps. Fairhaven, *Chandler.* Near the summit of Mansfield Mountain and Camel's Hump, *Robbins, Tuckerman* and *Macrae.* North Troy, *Carey.* June, July.
   convallarioides, Nutt. In Charleston, with Calypso borealis, *Carey.*

Calypso, *Salisbury.*
   bulbosa. Cypripedium bulbosum, L. *Calypso borealis*, Salisbury. In a dark sphagnous wood or swamp on the line between Charleston and Morgan, the entrance to which is opposite the house of Mr. Charles Cummings. *Carey.*

Cypripedium, *L. Lady's Slipper.*
   pubescens, Willd. *C. parviflorum*, Ait. *Yellow Lady's-Slipper.* Dry woods and in swamps. May, June.
   acaule, Ait. *Red Lady's-Slipper.* Dry woods, and also in swamps. May, June.
   spectabile, Swartz. *White Lady's-Slipper.* Swamps. June, July.
   arietinum, Ait. Dry woods and sphagnous swamps. In the cedar swamp at Fair Haven, *Chandler, Robbins.* In Grand Isle, and in dry woods near High Bridge, Colchester, *Robbins.* Burlington, *Carey, Macrae*, and *Oakes.*

ORDER PONTEDERIACEÆ.

Pontederia, *L.*
   cordata, L. *Pickerel-weed.* In water. July, August.

Schollera, *Schreber.*
   graminifolia, Muhl. Middlebury, *James.* In Otter Creek near its mouth, *Robbins.* In Castleton River, *Chandler.* July, August.

ORDER MELANTHACEÆ. *The Colchicum Tribe.*

Veratrum, *Tourn. White Hellebore.*
   viride, Ait. Swamps, &c. June.

ORDER TRILLIACEÆ.

Trillium, *L.*
   erythrocarpum, Michx. *T. pictum*, Pursh. Woods and swamps. May.
   erectum, L. Woods. May.
   grandiflorum, Salis. Woods, shady banks and swamps in the west of Vermont, from Pownal to Alburgh, *Robbins.* May, June.
   cernuum, L. Woods. Castleton, *Branch, Robbins.* May.

Medeola, *L. Indian Cucumber.*
   Virginica, L. Woods, &c. June, July.

ORDER LILIACEÆ. *The Lily Tribe.*

Lilium, *L. Lily.*
   Philadelphicum, L. *Wild Red Lily.* Pastures, &c. July.
   Canadense, L. *Wild Yellow Lily.* Moist meadows. July.

BOTANY OF VERMONT.

CATALOGUE OF PLANTS.

Erythronium, *L. Dog's-tooth Violet.*
 Americanum, Smith. Moist grounds, &c. May, June.
Allium, *L. Onion and Garlic.*
 tricoccum, Ait. *Wild Onion or Leek.* Woods. July.
Convallaria, *L. Lily of the Valley. Solomon's Seal.*
 pubescens, Willd. Woods. May, June.
 bifolia, L. Woods. May.
 stellata, L. Moist meadows and banks. May, June.
 trifolia, L. Sphagnous swamps and bogs. May, June.
 racemosa, L. Rocky woods, &c. June.
 borealis, Torr. *Dracæna borealis*, Ait., not *C. umbellulata*, Michx. Woods. June.
Streptopus, *Michx.*
 roseus, Michx. Woods, especially on mountains. May, June.
 amplexifolius, var. Americanus, Gray. *Uvularia amplexifolia*, L. *S. distortus*, Michx. Mountain woods. On the sides of Mansfield and Camel's Hump, *Robbins*, *Macrae*, and *Tuckerman*. Newport and Danville, *Carey*. June, July.
Uvularia, *L. Bellwort.*
 grandiflora, Smith. Woods. May.
 sessilifolia, L. Woods. May.

Order ALISMACEÆ. *The Water Plantain Tribe.*

Sagittaria, *L. Arrow-head.*
 sagittifolia, L. Ditches, ponds, &c. July, August.
Alisma, *L.*
Plantago, *L. Water Plantain.* In water. July, August.

Order JUNCEÆ. *The Rush Tribe.*

Luzula, *DC.* Juncus, *L. Wood Rush.*
 campestris, DC. Woods, pastures, &c. May.
 pilosa, Willd. Woods and swamps. May.
 parviflora. L. melanocarpa, Desv. *Juncus. parviflora*, Retz. At the base of Mansfield Mountain, *Robbins*. On the Chin of Mansfield and on Camel's Hump, *Macrae* and *Tuckerman*. June, July.
Juncus, *L. Rush.*
 effusus, L. Wet meadows, &c. June, July.
 filiformis, L. On the summits of Camel's Hump and Mansfield Mountains, and on the shore of Lake Champlain at Ferrisburgh, *Robbins*. June, July.
 nodosus, L. Wet meadows, &c. June.
 tenuis. Willd. Low grounds, &c. June, July.
 acuminatus, Michx. Margins of ponds, &c. Burlington, *Macrae*, *Oakes*.
 bufonius, L. Low grounds, &c. July.
 trifidus, L. On the summit of Mansfield Mountain, *Robbins*, and of Camel's Hump, *Tuckerman* and *Macrae*. June.

Order RESTIACEÆ.

Eriocaulon, *L. Pipewort.*
 septangulare, With. *E. pellucidum*, Michx. Borders of ponds, generally in the water. Seymour's pond, Morgan, and Minaud's pond, Rockingham, *Carey*. August, Sept.

Order SMILACEÆ. *The Smilax Tribe.*

Smilax, *L.*
 rotundifolia, L. *Green Briar.* Woods and thickets. June.
 herbacea, L. *S. peduncularis*, Muhl. Borders of woods, &c. June.

Order ARACEÆ. *The Arum Tribe.*

Arum, *L.*
 Dracontium, L. *Dragon-root.* Moist grounds. Shoreham, *Robbins*. May, June.
 triphyllum, L. *Indian Turnip.* Shady banks and swamps. May, June.
Peltandra, *Rafinesque.*
 Virginica, Raf. *Calla Virginica*, Michx. In water on the borders of ponds and rivers. Colchester pond, *Robbins*. June, July.
Calla, *L.*
 palustris, L. Swamps. Middlebury, *James*. Fair Haven and Whiting, *Robbins*. Bellows Falls, and Guildhall, *Carey*. July.

CATALOGUE OF PLANTS.

Symplocarpus, *Salisbury.* *Skunk Cabbage.*
   fœtidus, Nutt. *Pothos fœtida*, L. Wet meadows and swamps. April.
Acorus, *L.*
   Calamus, L. *Sweet Flag.* Wet meadows, &c. June.

ORDER TYPHACEÆ. *The Cat's-tail Tribe.*

Typha, *Tourn.* *Cat's tail.* *Reed Mace*,
   latifolia, L. Ditches, pools, &c. July.
Sparganium, *Tourn.* *Burr Reed.*
   ramosum, L. In ditches, &c. June, July.
   simplex, Hudson. *S. Americanum*, Nutt. Borders of streams, &c. July.

ORDER FLUVIALES.

Najas, *L.*
   Canadensis, Michx.
      var. *fragilis.* *Caulinia fragilis*, Willd. Middlebury, *James.*
      var. *flexilis.* *Caulinia flexilis*, Willd. In water three feet deep at the mouth of Otter Creek. Ferrisburgh, *Robbins.* July, Aug.
Zannichellia, *Micheli.*
   palustris, L. In shallow water, in Lake Champlain, at South Hero. *Robbins.*
Potamogeton, *L. Pondweed.*
   natans, L. Ponds, and slow flowing waters. July, August.
   heterophyllum, Schreber. Ponds, and slow streams. August.
   diversifolium, Barton. Ponds, &c. In Lake Champlain at South Hero, *Robbins.* July.
   perfoliatum, L. Ponds, &c. August.
   lucens, L. Ponds, &c. August.
   compressum, L. Rivers, ponds, &c. July, August.
   pauciflorum, Pursh. *P. gramineum*, Michx. Ponds, &c. July, August.
   pectinatum, L. Ponds, &c. July.

The species of Potamogeton as above are all according to Torrey's Flora of the Northern States, vol. I, p. 196.

ORDER JUNCAGINACEÆ. *The Arrow-Grass Tribe.*

Scheuchzeria, *L.*
   palustris, L. Sphagnous swamps and bogs. In Georgia, *Chandler.* At the southern end of Colchester Pond, *Robbins.* In North Troy, with Orchis blephariglottis, *Carey.* June.

ORDER PISTIACEÆ. *The Duckweed Tribe.*

Lemna, *L. Duckweed.*
   polyrhiza, L. Ditches, &c.
   minor, L. Ditches, &c. At North Hero, *Robbins.*
   trisulca, L. Ditches, ponds, &c. At North Hero, *Robbins.*

ORDER CYPERACEÆ. *The Sedge Tribe.*

Dulichium, *Richard.*
   spathaceum, Rich. Borders of ponds, &c. July, August.
Cyperus, *L.*
   diandrus, Torr. var. *castaneus*, Torr. Margins of ponds, &c. August.
   strigosus, L. Low moist grounds. August.
   repens, Elliot. *C. phymatodes*, Muhl. Wet sandy soil. In South Hero, West Haven, and on the banks of Otter Creek, Ferrisburgh, *Robbins.* August.
   filiculmis, Vahl. *C. mariscoides*, Ell. Dry sands August.
   inflexus, Muhl. *C. uncinatus*, Pursh. Sandy shores of rivers and lakes. Aug.
Eleocharis, *R. Brown.* Scirpus, *L.*
   palustris, R. Brown. Wet places, ditches, &c. May, June.
   obtusa, Schultes. *Scirpus capitatus* of American authors, not of Linnæus. Ditches and margins of ponds. June, July.
   acicularis, R. Brown. Margins of ponds, &c. June.
   tenuis, Schultes. Margins of ponds, &c. June.
Scirpus, *L. Club Rush.*
   lacustris, L. *S. acutus*, Muhl. *Bulrush.* In water on the borders of lakes, ponds, &c. July.

CATALOGUE OF PLANTS.

triqueter, L. Wet places, borders of rivers, &c. July.
atrovirens, Muhl. Moist meadows, &c. July.
brunneus, Muhl. Swampy grounds. Pownal, *Robbins*. August.
Eriophorum, Michx. Wet meadows, ditches, &c. August.
Eriophorum, *L. Cotton Grass.*
   alpinum, L. Bogs. Brownington, *Robbins*. Danville, *Carey*. May, June.
   vaginatum, L. Bogs. June, July.
   Virginicum, L. Bogs. July, August.
   polystachyon, L. Bogs. May, June.
   angustifolium, Reichard. *E. gracile*, Roth. Bogs. May, June.
Isolepis, *R. Brown.* Scirpus, *L.*
   capillaris, Rœm. and Sch. Dry sands. Bellows Falls, *Carey*. August.
Rhyncospora, *Vahl.*
   glomerata. Moist pastures, &c. Bellows Falls, *Carey*. July, August.
Carex, *Micheli. Sedge.*
   disperma. Dewey. Sphagnous swamps.
   rosea, Schk. Woods and shady banks.
   cephalophora, Muhl. Woods, &c.
   sparganioides. Mahl. Moist shady banks, &c.
   stipata, Muhl. Wet meadows.
   bromoides, Schk. Moist woods, &c.
   vulpinoidea, Michx. *C. multiflora*, Muhl. Moist pastures, &c
   paniculata, var. teretiuscula, Wahl. Bogs.
   siccata, Dewey. Moist banks, &c. Burlington, *Macrae*.
   trisperma, Dewey. Bogs and swamps.
   Deweyana, Schw. Woods, &c.
   tenuiflora, Wahl. Cedar and other swamps. In Salem, in a shady swamp near a small pond at the head of Lake Memphremagog, also in Burlington, *Robbins*. On the western side of the great cedar swamp at Fair Haven, *Oakes*.
   stellulata, Good. *C. scirpoides*, Schk. *C. sterilis*, Willd. Wet meadows and swamps.
   curta, Good. Swamps.
   scoparia, Schk. Wet meadows.
     var. *lagopodioides*. *C. lagopodioides*, Willd. Wet meadows.
   festucacea, Schk. Moist woods and meadows.
   aurea, Nutt. Moist rocky ledges, &c. Pownal, Burlington and Colchester, *Robbins*. Bellows Falls, *Carey*.
   saxatilis, L. Summits of Mansfield and Camel's Hump mountains, *Robbins*, *Tuckerman* and *Macrae*.
   cespitosa, L. Wet meadows, &c.
   acuta, L. Wet meadows.
   crinita, Lam. Wet shady banks, &c.
   leucoglochin, Ehr. *C. pauciflora*, Willd. Bogs, especially at the north, and on mountains. At Colchester pond, *Robbins*. At North Troy, with Orchis blephariglottis, *Carey*.
   polytrichoides, Muhl. Swamps, &c.
   pedunculata, Muhl. Woods, &c.
   squarrosa, L. In a low wet wood on the margin of Otter Creek, Ferrisburgh, *Robbins*.
   gracillima, Schw. Wet meadows and woods. Burlington, *Carey*. Colchester, *Macrae*.
   vestita, Willd. Borders of woods, &c. Middlebury, *James*.
   Pennsylvanica, Lam. *C. varia* and *marginata*, Muhl. Woods.
   Emmonsii, Dewey. *C. alpestris*, Torr. and Schw. *C. Davisii*, Dewey. Bellows Falls, *Carey*.
   oligocarpa, Schk. Woods.
   laxiflora, Lam. Woods, &c. Castleton, *Robbins*.
   granularis, Muhl. Moist shady rocks. Burlington, *Oakes*.
   eburnea, Boott. *C. alba*, Dewey. Limestone rocks. On the rocks at High Bridge, Colchester, and at Grand Isle, South Hero, West Haven and Pownal, *Robbins*.
   anceps, Muhl. Woods and shady banks.
   plantaginea, Lam. Woods.
   sylvatica, Huds. Woods, especially on mountains.

CATALOGUE OF PLANTS.

flava, L. Wet meadows. Sutton, *Carey.*
intumescens, Rudge. *C. folliculata* of Schk., not of Linn. Wet woods.
lupulina, Muhl. Wet meadows and woods.
tentaculata, Muhl. Wet meadows.
retrorsa, Schw. Swamps, &c.
bullata, Schk. Wet meadows, &c. South Hero, *Robbins.*
vesicaria, L. *C. ampullacea*, Dewey. *C. utriculata*, Boott. Wet meadows, &c.
lacustris, Willd. Borders of ponds, &c.
scabrata, Schw. Swamps, &c.
hystericina, Muhl. Wet meadows.
Pseudo-cyperus, L. Ditches and margins of ponds.
longirostris, Torrey. Shady ledges, &c. On the sides of Camel's Hump, and at Castleton, *Robbins.* Rocky banks of Saxton's river, near Bellows Falls, *Carey.*
limosa, L. Bogs, especially at the the north.
miliacea, Muhl. Moist banks, &c.
pallescens, L. Wet meadows, &c.
umbellata, Schk. Rocky hills, &c. Summit of Mansfield mountain, *Robbins*

ORDER GRAMINEÆ. *The Grass Tribe.*

(*Mostly according to Torrey's Flora of the Northern States, Vol. I.*)

Agrostis, *L. Bent Grass.*
 § *vulgaris*, Smith. *Red-top.* Meadows, pastures, &c. June—Aug.
 § *alba*, L. Meadows, pastures, &c. June—Aug.
lateriflora, Michx. Moist meadows, sides of hills, &c. August, Sept.
sobolifera, Muhl. Rocky shady hills, &c. August, Sept.
tenuiflora, Willd. Rocky shady hills, &c. July, Aug.
sylvatica, Torrey. Dry rocky hills, &c. August.
canina, L.
 var. *alpina*, Oakes. *Agrostis rupestris*, Gray in Sill. Jour., vol. 42. On the summit of Camel's Hump mountain, *Robbins, Tuckerman* and *Macrae.* July. This variety is common on the White Mountains, and is connected with the common variety, which is abundant in Essex county, Massachusetts, by several intermediate forms, found at the base and on the sides of the White Mountains.

Cinna, *L.*
arundinacea, Willd. Wet woods, &c. August, Sept.
Polypogon, *Desfontaines.*
racemosus, Nutt. *P. glomeratus*, Willd. Wet meadows, &c. Aug., Sept.
Brachyelytrum, *P. de Beauv.*
aristatum, P. de B. *Muhlenbergia erecta*, Roth. Woods, &c. June, July.
Alopecurus, *L. Fox-tail Grass.*
 § *pratensis*, L. Moist meadows, &c. Bellows Falls, *Carey.* May, June.
geniculatus, L. Wet meadows, &c. June.
Phleum, *L. Cat's-tail Grass.*
 § *pratense*, L. *Herd's Grass*, *Timothy.* Fields, &c. July, August.
Phalaris, *L. Canary Grass.*
 § *Canariensis*, L. Pastures, &c. Cavendish, *Macrae.* July.
Milium, *L. Millet Grass.*
effusum, L. Woods, &c. Banks of Saxton's river, Bellows Falls, *Carey.* July.
pungens, Torr. Dry rocky woods, &c. May.
Piptatherum, *P. de Beauv.*
nigrum, Torr. Shady ledges, &c. August.
Oryzopsis, *Michx.*
asperifolia, Michx. Woods, especially on mountains. May, June.
Panicum, *L. Panic Grass.*
 § *Crus-Galli*, L. Cultivated grounds, &c. July—Sept.
clandestinum, L. *P. pedunculatum*, Torrey. Woods. July.
latifolium, L. Sandy woods, &c. July.
dichotomum, L. *P. nitidum*, Lam. Low grounds. July.
depauperatum, Muhl. *P. rectum*, Roemer and Shultes Sandy soils. Bellows Falls, *Carey.* Burlington, *Macrae*, July.
xanthophysum, Gray. Sandy woods, &c. Burlington, *Carey.* June, July.
capillare, L. Sandy fields and cultivated grounds. August, Sept.

CATALOGUE OF PLANTS.

Setaria, *P. de Beauvois.*
   § *viridis*, P. de B. *Panicum viride*, L. Cultivated grounds, &c July, August.
   § *glauca*, P de B. *Panicum glaucum*, L. Cultivated grounds, &c July, Aug.
Digitaria, *Haller.*
   § *sanguinalis*, Scop. Cultivated grounds, &c. August, Sept.
   glabra. Sandy fields, &c. Castleton, Colchester, West Haven, and Ferrisburg, *Robbins.* August, Sept.
Paspalum, *L.*
   ciliatifolium, Michx. Dry fields, &c. Bellows Falls, *Carey.* Aug.
Aristida, *L.*
   dichotoma, Michx. Barren fields, &c. Pownal, *Robbins.* Aug.
Calamagrostis, *Roth.* Arundo, *L.*
   Canadensis, P. de Beauv. *Arundo Canadensis*, Michx. *Calamagrostis Mexicana*, Nutt. Wet meadows, &c. July.
Anthoxanthum, *L.* *Sweet-scented Vernal Grass.*
   § *odoratum*, L. Meadows and pastures. Middlebury, *James.* May, June.
Aira, *L.* *Hair Grass.*
   flexuosa, L. Dry rocky woods. June.
   cespitosa, L. *Aira aristulata*, Torrey. On the moist rocky banks of rivers. On the Connecticut, at Guildhall, *Robbins.* July.
Trisetum, *Pers.*
   striatum, Michx. *T. purpurascens*, Torrey. *Avena striata*, Michx. Rocky woods. Castleton, Georgia, and Woodstock, *Robbins.* May, June.
   molle, Trinius. *Avena mollis*, Michx. On dry limestone rocks, at High Bridge and Winooski falls, Colchester, *Robbins.* June.
Hierochloa, *Gmelin.*
   alpina, Roem. and Sch. On the summit of Mansfield mountain, *Tuckerman* and *Macrae.* July.
Arundo, *L.* *Reed Grass.*
   Phragmites, L. In water on the borders of ponds, &c. In lake Memphremagog, *Robbins.* Aug.
Danthonia, *DC.*
   spicata, P. de B. Dry barren woods, pastures, &c. June, July.
Festuca, *L.* *Fescue Grass.*
   § *duriuscula*, L. Dry pastures, &c. June.
   tenella, Willd. Dry sandy fields, &c. Bellows Falls, *Carey.* June, July.
   § *elatior*, L. Grass fields, &c. Middlebury, *James.* June.
   § *pratensis*, Huds. Grass fields, &c. Bellows Falls, *Carey.* June, July.
Glyceria, *R. Brown.*
   fluitans, R. Br. Stagnant water, Burlington, *Carey.* June.
Poa, *L.* *Meadow Grass.*
   § *annua*, L. Cultivated grounds, &c. May—Aug.
   dentata, Torrey. Ditches and wet places. July, Aug.
   aquatica, L. Wet meadows, &c. July, Aug.
   § *pratensis*, L. Grass fields, roadsides, &c. June, July.
   compressa, L. Sandy fields, and in woods, &c. June.
   serotina, Ehrh. Wet meadows. July.
   nemoralis, L. Woods. May, June.
   nervata, Willd Wet meadows. June, July.
   obtusa, Muhl. Wet meadows, &c. Bellows Falls, *Carey.* Aug.
   Torreyana, Sprengel. *P. elongata*, Torr. not of Willd. Woods. At the base of Mansfield mountain, *Robbins.* Morgan, near the line of Charleston, *Carey.* Burlington, *Macrae.* July.
   Canadensis, Torr. *Briza Can.*, Michx. Wet meadows and swamps. July.
   hirsuta, Michx. Sandy and gravelly beach of Connecticut river, at Bellows Falls, *Carey.*
   alpina, L. Summit of Mansfield mountain, *Robbins.* July.
   reptans, Michx. Wet sandy shores of rivers and lakes. On the banks of the Otter Creek, Ferrisburgh, and of the Winooski, Colchester, *Robbins.* July, Aug.
Tricuspis, *P. de Beauv.*
   seslerioides, Torr. *Poa quinquefida*, Pursh. Sandy soil Middlebury, *James.* Aug.
Dactylis, *L.* *Orchard Grass.*
   § *glomerata*, L. Grass fields, &c. Bellows Falls, *Carey.* June.

Bromus, *L. Brome Grass.*
§ *secalinus*, L. *Chess or Cheat.* Cultivated grounds. July.
ciliatus, L. Woods, &c. July.
purgans, L. Woods, shady banks, &c. Castleton and Brattleboro', *Robbins.*
Secale, *L. Rye.*
§ *cereale*, L. Old fields and on rocks, &c. June.
Elymus, *L. Lyme Grass.*
Canadensis, L. and *var. glaucifolius.* Rocky river banks, &c. July, Aug.
striatus, Willd. E. *villosus*, Torrey, Flora. Dry rocky banks, &c. Middlebury, *James.* July, Aug.
Hystrix, L. Rocky woods. Middlebury, *James.* West Haven, *Robbins.*
Triticum, *L. Wheat.*
§ *repens*, L. Couch Grass. "*Witch Grass.*" Cultivated grounds,&c. June, July.
Spartina, *L. Cord Grass.*
cynosuroides, Willd. Banks of rivers, &c. Bellows Falls, *Carey.* Aug.
Andropogon, *L. Beard Grass.*
scoparius, Michx. Dry fields, &c. Pownal, *Robbins.* Bellows Falls, *Carey.* Burlington, *Macrae.* August, Sept.
furcatus, Muhl. Dry rocks and fields. Colchester, *Robbins.* Bellows Falls, *Carey.* Aug. Sept.
nutans, L. Dry fields, &c. Pownal and Brattleboro', *Robbins.* Bellows Falls, *Carey.* Burlington, *Macrae.* Aug. Sept.
Leersia, *Solander.*
Virginica, Willd. Wet woods, &c. Aug.
oryzoides, Swartz. Ditches, &c. Aug. Sept.
Zizania, *L. Wild rice.*
aquatica, Lambert. In shallow water in rivers and lakes. Burlington and S. Hero, *Robbins.* Aug.

CLASS IV. ACROGENS.

Order EQUISETACEÆ. *The Horsetail Tribe.*

Equisetum, *Tourn. Horsetail.*
limosum, L. Bogs, borders of ponds, &c. June.
sylvaticum, L. Moist woods and shady banks. May.
hyemale, L. Wet woods and banks. June.
variegatum, Schleich. Interstices of rocks on the shores of the Connecticut river, near low water mark, Bellows Falls, *Carey.*
scirpoides, Michx. Moist woods and banks. June.

Order FILICES. *The Fern Tribe.*

Polypodium, *L.*
vulgare, L. Shady rocks, &c.
Dryopteris, L. Woods and swamps.
Phegopteris, L. Woods and shady banks.
var. *connectile.* P. *connectile*, Michx.
var. *hexagonopterum.* P. *hexagonopterum*, Michx.
Aspidium, *Swartz. Shield Fern.*
acrostichoides, Swartz. Rocky woods, &c.
Goldianum, Hooker. Woods. In Orleans county, *Carey.*
Thelypteris, Swartz. A. *noveboracense*, Willd. Moist woods.
cristatum, Sw. A. *Lancastriense*, Sw. Moist woods near Burlington, *Macrae.*
marginale, Sw. Rocky woods.
dilatatum, Sw. Woods.
aculeatum, Sw. Woods about the "Notch" at the north base of Mansfield mountain. *Macrae* and *Tuckerman.*
Cistopteris, *Bernhardi.*
fragilis, Bernh. *Aspidium tenue*, Sw. Moist rocks, &c.
bulbifera, Bernh. *Aspid. bulb.* Willd. Shady rocks, generally on limestone.
Dicksonia, *L'Heritier.*
pilosiuscula, Willd. Moist pastures, shady woods, &c.
Woodsia, *R. Brown.*
Ilvensis, R. Br. On rocks. Fairhaven, &c., *Robbins.* On the summit of Mansfield mountain, *Tuckerman* and *Macrae.*

CATALOGUE OF PLANTS.

obtusa, Torr. *Aspid. obtusum*, Swartz. W. *Perriniana*, Hooker and Greville. Rocks. Bellows Falls, *Carey*.
Asplenium, *L. Spleenwort.*
 rhizophyllum, L. Shady limestone rocks.
 angustifolium, Michx. Woods. Middlebury, *James.*
 ebeneum, Ait. Rocky ledges.
 Trichomanes, L. Steep rocky ledges.
 thelypteroides, Michx. Woods and shady banks. Bellows Falls, *Carey*. In Colchester, on the eastern side of High Bridge, *Oakes*. Ludlow, *Washburn.*
Ruta muraria, L. *Wall rue Spleenwort.* In the crevices of limestone rocks, facing the woollen factory at Winooski falls, near Burlington, *Robbins* and *Macrae*. At the place of the former bridge, near High Bridge, Colchester, also at Pownal and West Haven, *Robbins.*
Filix-fæmina, Bernh. *Aspidium Filix-fæmina*, Sw. *Aspidium asplenioides*, Sw. *A. angustum*, Willd. Woods.
Woodwardia, *Smith.*
 Virginica, Sw. Bogs. At Colchester pond, *Robbins.*
Pteris, *L. Brake.*
 aquilina, L. *Common Brake.* Dry woods, &c.
 atropurpurea, L. Crevices of limestone rocks. Near High Bridge and at Winooski falls, and at Pownal and West Haven, *Robbins.*
 gracilis, Michx. On rocks overhanging the "Devil's Den," Burlington, *Macrae.*
Adiantum, *Tourn. Maidenhair.*
 pedatum, L. Woods.
Struthiopteris, *Willd.*
 Germanica, Willd. *S. Pennsylvanica*, Willd. Woods, and low grounds.
Onoclea, *L.*
 sensibilis, L. Moist woods and banks.
Ophioglossum, *L. Adders' Tongue.*
 vulgatum, L. Bellows Falls, *Carey*. Burlington, *Macrae.*
Osmunda, *L. Flowering Fern.*
 Claytoniana, L. *O. interrupta*, Michx. Moist grounds, &c.
 cinnamomea, L. Moist grounds, &c.
 regalis, L. *O. spectabilis*, Willd. Moist grounds, &c.
Botrychium, *Swartz. Moonwort.*
 fumarioides, Willd. Pastures, &c.
 var. *dissectum*, Oakes. *B. dissectum*, Muhl. Rockingham, *Carey.*
 Virginianum, Sw. *B. gracile*, Michx. Woods.
 simplex, Hitchcock. At Sutton, near the village, on the road leading to Burke, *Carey.*

Order LYCOPODIACEÆ. *The Club-Moss Tribe.*

Lycopodium, *L. Club-Moss. Winter-green.*
 clavatum, L. Dry woods.
 complanatum, L. Woods.
 obscurum, L. *L. dendroideum*, Michx. *Ground Pine.* Woods.
 annotinum, L. Woods, especially near the mountains.
 rupestre, L. On dry rocks. Georgia, *Robbins*. Fair Haven, *Chandler.*
 Selago, L. Summits of Mansfield and Camel's Hump mountains, *Robbins*, *Tuckerman* and *Macrae.*
 lucidulum, Michx. Woods.

INDEX TO THE GENERA IN THE PRECEDING CATALOGUE.

Acalypha,	196	Agrostemma,	181	Amaranthus,	195	Anthemis,	189	Aralia,	186
Acer,	182	Agrostis,	204	Ambrosia,	188	Anthoxanthum,	205	Arbutus,	190
Achillea,	189	Aira,	205	Ampelopsis,	182	Anychia,	180	Archangelica,	186
Acorus,	202	Alisma,	201	Amphicarpæa,	183	Apios,	183	Arctium,	189
Actæa,	178	Allium,	201	Andromeda,	190	Aplectrum,	199	Arenaria,	180
Adiantum,	207	Alnus,	197	Andropogon,	206	Apocynum,	191	Arethusa,	199
Adlumia,	178	Alopecurus,	204	Anemone,	177	Aquilegia,	178	Aristida,	205
Agrimonia,	183	Amelanchier,	194	Antennaria,	189	Aralis,	179	Artemisia,	189

INDEX TO GENERA.—Continued.

Arum,	201	Cuscuta,	192	Kalmia,	190	Osmunda,	207	Secale, 206
Arundo,	205	Cynoglossum,	192	Krigia,	190	Ostrya,	196	Senecio, 189
Asarum,	196	Cyperus,	202	Juglans,	198	Oxalis,	181	Setaria, 205
Asclepias,	191	Cypripedium,	200	Juncus,	201	Panax,	186	Shepherdia, 196
Aspidium,	206	Dactylis,	205	Juniperus,	198	Panicum,	204	Sicyos, 185
Asplenium,	207	Dalibarda,	184	Justicia,	194	Parietaria,	196	Sida, 182
Aster,	187	Danthonia,	205	Lactuca,	189	Parnassia,	180	Silene, 181
Barbarea,	179	Datura,	193	Lasierpa,	191	Paspalum,	205	Sinapis, 179
Betula,	197	Decodon,	184	Lathyrus,	183	Pastinaca,	186	Sisymbrium, 179
Bidens,	188	Dentaria,	179	Laurus,	196	Pedicularis,	194	Sisyrinchium, 199
Blephilia,	192	Desmodium,	183	Lechea,	180	Peltandra,	201	Sium, 185
Blitum,	195	Dicksonia,	206	Ledum,	190	Penthorum,	185	Smilax, 201
Bœhmeria,	196	Dielytra,	178	Leersia,	206	Pentstemon,	194	Solanum, 193
Botrychium,	207	Diervilla,	186	Lemna,	202	Phaca,	183	Solidago, 188
Brach'elytrum	204	Digitaria,	205	Leontice,	178	Phalaris,	204	Sonchus, 189
Brasenia,	178	Dipsacus,	187	Leontodon,	189	Phleum,	204	Sparganium, 202
Bromus,	206	Dirca,	196	Leonurus,	193	Phryma,	194	Spartina, 206
Calamagrostis,	205	Draba,	179	Lepidium,	179	Physalis,	193	Spergula, 180
Calla,	201	Drosera,	180	Lespedeza,	183	Phytolacca,	195	Spiræa, 183
Callitriche,	198	Dulichium,	202	Lilium,	200	Pinus,	198	Spiranthes, 200
Calopogon,	200	Echinocystis,	185	Linaria,	193	Piptatherum,	204	Stachys, 193
Caltha,	178	Echinosper'um	192	Lindernia,	194	Plantago,	195	Staphylea, 182
Calypso,	200	Eleocharis,	202	Linnæa,	186	Platanus,	198	Stellaria, 181
Camelina,	179	Elodea,	180	Linum,	181	Poa,	205	Streptopus, 201
Campanula,	190	Elymus,	206	Liparis,	199	Podophyllum,	178	Struthiopteris, 207
Cannabis,	196	Empetrum,	196	Listera,	200	Pogonia,	200	Symphoricarp. 186
Capsella,	179	Epiphegus,	194	Lithospermum	192	Polanisia,	179	Symplocarpus, 202
Cardamine,	179	Epigæa,	190	Lobelia,	190	Polygala,	179	Tanacetum, 189
Carex,	203	Epilobium,	184	Lonicera,	186	Polygonum,	195	Taxus, 198
Carpinus,	196	Equisetum,	206	Lophanthus,	193	Polypodium,	206	Teucrium, 193
Carya,	198	Erigeron,	188	Lupinus,	183	Polypogon,	204	Thalictrum, 178
Cassia,	183	Eriocaulon,	201	Luzula,	201	Pontederia,	200	Thaspium, 185
Castanea,	197	Eriophorum,	203	Lycopodium,	207	Populus,	197	Tiarella, 185
Castilleja,	194	Erythronium,	201	Lycopsis,	192	Portulaca,	181	Tilia, 182
Ceanothus,	182	Eupatorium,	187	Lycopus,	192	Potamogeton,	202	Tricuspis, 205
Celastrus,	182	Euphorbia,	196	Lysimachia,	195	Potentilla,	183	Trientalis, 195
Celtis,	198	Fagus,	197	Malva,	182	Prenanthes,	190	Trifolium, 183
Centaurella,	192	Festuca,	205	Medeola,	200	Prunella,	193	Trillium, 200
Cephalanthus,	187	Flœrkea,	181	Medicago,	183	Prinos,	191	Triosteum; 186
Cerastium,	181	Fragaria,	184	Melampyrum,	194	Proserpinaca,	185	Triphora, 200
Cerasus,	183	Fraxinus,	191	Melilotus,	183	Prunus,	183	Trisetum, 205
Ceratophyllum	178	Galeopsis,	193	Melissa,	193	Pteris,	207	Triticum, 206
Chelidonium,	178	Galium,	187	Menispermum	178	Pterospora,	191	Tussilago, 187
Chelone,	194	Gaultheria,	190	Mentha,	192	Pycnanthe'um	192	Typha, 202
Chenopodium,	195	Gentiana,	192	Menyanthes,	192	Pyrola,	191	Udora, 199
Chrysosplen.	185	Geranium,	181	Microstylis,	199	Pyrus,	184	Ulmus, 198
Chrysanthem.	189	Gerardia,	194	Milium,	204	Quercus,	197	Urtica, 196
Cicuta,	185	Geum,	183	Mimulus,	194	Ranunculus,	177	Utricularia, 194
Cimicifuga,	178	Glyceria,	205	Mitchella,	187	Raphanus,	179	Uvularia, 201
Cinna,	204	Gnaphalium,	189	Mitella,	185	Rhamnus,	182	Vaccinium, 190
Circæa,	185	Goodyera,	200	Mollugo,	180	Rhododendron	190	Valeriana, 187
Cirsium,	189	Gratiola,	194	Monarda,	192	Rhus,	182	Valisneria, 199
Cistopteris,	206	Hamamelis,	185	Monotropa,	191	Rhynchospora	203	Veratrum, 200
Claytonia,	181	Hedeoma,	193	Morus,	196	Ribes,	185	Verbascum, 193
Clematis,	177	Helianthemum	180	Myriophyllum,	185	Rosa,	184	Verbena, 194
Cochlearia,	179	Helianthus,	188	Najas,	202	Rubus,	184	Vernonia, 187
Collinsonia,	193	Hepatica,	177	Nardosmia,	187	Rudbeckia,	188	Veronica, 193
Comandra,	196	Heracleum,	186	Nasturtium,	179	Rumex,	195	Viburnum, 186
Comarum,	184	Hieracium,	189	Nemopanthes,	191	Sagittaria,	201	Vicia, 182
Comptonia,	198	Hierochloa,	205	Nepeta,	193	Salix,	197	Viola, 180
Conioselinum,	186	Houstonia,	187	Nuphar,	178	Sambucus,	186	Vitis, 182
Conium,	186	Humulus,	196	Nymphæa,	178	Samolus,	195	Waldsteinia, 183
Convallaria,	201	Hydrocotyle,	185	Nyssa,	196	Sanguinaria,	178	Woodsia, 206
Convolvulus,	192	Hydrophyllum	192	Œnothera,	184	Sanicula,	185	Woodwardia, 207
Coptis,	178	Hyoscyamus	193	Onopordon,	189	Sarracenia,	178	Xanthium, 188
Corallorhiza,	199	Hypericum,	180	Onoclea,	207	Saxifraga,	185	Xanthoxylum, 181
Cornus,	186	Hypopithys,	191	Ophioglossum,	207	Scheuchzeria,	202	Zannichellia, 202
Corydalis,	179	Impatiens,	181	Orchis,	199	Schollera,	200	Zizania, 206
Corylus,	197	Inula,	188	Orobanche,	194	Scirpus,	202	Zizia, 185
Cratægus,	184	Iris,	199	Oryzopsis,	204	Scrophularia,	194	
Cryptotænia,	185	Isolepis,	203	Osmorhiza,	186	Scutellaria,	193	Total, 393

*** Having been obliged, contrary to expectation, to work off the preceding Catalogue without awaiting the return of the *proof sheets* from the author, some typographical errors, &c., have occurred, for the correction of which see the Errata at the end of the volume.

Section II.
Trees and Fruits.

To the preceding full, and very perfect catalogue of Vermont Plants, kindly furnished for this work by Wm. Oakes, Esq., of Ipswich, Massachusetts, we here subjoin a brief account of our most important Forest Trees, a list of which has already been given on page 173, and also a few words respecting our Shade Trees, Fruits, &c., which is all our limits will admit.

BASSWOOD, OR LIME TREE.
Tilia Americana.

This tree is found in all parts of the state, and under favorable circumstances grows to the height of 70 or 80 feet with a proportional diameter. In newly cleared lands the stumps and large roots of the basswood are apt to send forth shoots which grow with great rapidity. To prevent the growth of these the bark is sometimes stripped from the stumps, or they are seared by building a fire around them. The inner bark of this tree is sometimes macerated in water and formed into ropes. The wood is white and tender, but is valuable for very many purposes. It is sawed into planks and boards, and is used for chair seats, trunks, and in the manufacture of a variety of other articles.

BLACK CHERRY.
Cerasus serotina.

This is our largest species of cherry tree, and sometimes, though rarely, exceeds 50 feet in height and 15 inches in diameter. It is scattered, but very sparingly, over the greater part of the state. It is sometimes called *Wild Cherry;* and also *Cabinet Cherry*, from the use made of it by cabinet makers. But it is more generally called *Black Cherry*, and this name may be derived either from the color of the bark or the ripe fruit. The perfect wood is of a dull light red color, which deepens with age. It is compact, fine grained, brilliant, and not liable to warp when perfectly seasoned. It is extensively used for almost all species of furniture, and sometimes rivals mahogany in beauty, but it has been sought for with so much eagerness, that there is very little now remaining in our forests large enough to be sawn into boards. The bark of this tree is aromatic, has an agreeable bitter taste, and is often used as a tonic.

THE SUGAR MAPLE.
Acer saccharinum.

The Sugar Maple is one of our most common and valuable forest trees. It grows to a larger size than any other species of maple, and its wood, when seasoned, is much heavier and harder. Hence it is often called *Rock Maple* or *Hard Maple*. Its ordinary height is about 60 feet, with a diameter of from 2 to 3 feet. The wood, when first cut, is white, but by exposure assumes a rosy tinge. Its grain is fine and close, and when polished has a silky lustre. It is strong and heavy, but when exposed to moisture soon decays, on which account it is little used either in civil or naval architecture. When thoroughly seasoned it is used by wheelwrights for axletrees and by sleigh makers for the runners of common sleds. It is also used by chair makers and cabinet makers in many kinds of their work. The wood of this tree exhibits two accidental forms of arrangement of the fibre, of which cabinet makers take advantage for manufacturing beautiful articles of furniture. The first consists of undulations, forming what is called *Curled Maple*. The second, which occurs only in old trees, appears to arise from an inflection of the fibre from the circumference towards the centre, producing spots, which are sometimes contiguous, and at others a little distance apart. This is what is called *Bird's-Eye Maple*, and the more numerous the spots, the more beautiful and more esteemed is the wood. Like the curled and striped maple, it is used for inlaying mahogany. It is also made into bedsteads, portable writing desks, and a variety of other articles, for which purposes it is highly valued. The sugar maple is the most valuable wood for fuel found in the state. Its ashes are very abundant and rich in alkali. Its charcoal is of the most valuable kind. Its wood may easily be distinguished from the other kinds of maple by its weight and hardness. Valuable as this tree is on account of its wood, and for being one of our most beautiful and flourishing ornamental shade trees, its value is greatly increased on account of the sugar extracted from it. When the country was new, nearly all the sweetening consumed in the state was obtained from the sugar maple, and although the proportional quantity has been diminished by the destructton of the maple forests, our people have become so sensible of its value, both for fuel and for its sugar, that they are taking much pains to preserve groves of the second growth. It is a tree which grows rapidly, and considerable quantities of sugar are now made from trees which sprung from the seed since the settlement of the state was commenced. The quantity of sugar manufactured in the state in 1840, was 4,647,934 lbs.

The quality of the sugar made in the state is very unequal. While some of it is black, dirty and disagreeable, there is much made which is no wise inferior in color or flavor to the very best West India sugar; and this depends entirely upon the manner and care with which it is manufactured. The dark color, the clamminess and disagreeable taste of much of our maple sugar, are owing chiefly to three causes. 1. The neglect to scald the buckets, &c., used for catching the sap, and to keep the sap clear from all impurities. 2. Allowing the kettles to become so much heated at the top as to cause the syrrup to burn upon them, and afterwards to be dissolved and mingled with the syrrup. 3. Allowing syrup to remain too long in iron kettles. It should in no case be allowed to stand in the kettle over night. If these causes be guarded against and the syrrup be well settled, well cleansed, and done down without being burnt, there can be no failure of having good sugar. To make white coarse grainod sugar, it should be done so that only about three fourths of it will grain. It should then be poured into a tub, and remain unstirred till the graining has ceased. The molasses should then be drained or poured off, and the sugar will be found to be very beautiful. It may be still further whitened by spreading upon the sugar a clean white cloth, and covering it for a few days with moist dough, made of Indian meal. The sugar made from this tree, in addition to its excellent qualities, has two important recommendation. It is the production of our own state, and it is never tinctured with the sweat, and the groans, and the tears, and the blood of the poor slave.

THE WHITE MAPLE.
Acer dasycarpum.

This tree so nearly resembles the Red Maple, that it is very generally confounded with it in Vermont, both being called Soft Maple. The name of White Maple may be derived either from the leaf or from the wood. The color of the under side of the leaf is a beautiful silvery white, and the wood is also very white, and of a fine texture; but it is softer and lighter than either of the other species of maple. It is sometimes used in the manufacture of furniture, for inlaying mahogony, cherry and walnut, but it is liable to change its color. Wooden bowls are sometimes made of it, but ash and poplar are preferable when they can be had. Sugar is sometimes made from the sap of this and the Red Maple, but the same quantity of sap does not yield more than half as much sugar as that of the sugar maple. Like the Red Maple, the extract from the inner bark of this tree produces a black preceptale with copperas, and is sometimes used for coloring.

THE RED MAPLE.
Acer rubrum.

This tree is found in most parts of the state, but in no part is so plentiful as the Sugar Maple. Its flowers appear in April, long before the leaves, and are the first indications which the forests exhibit of the returning spring. They are small, of a deep red color, and hence the name, *Red Flowering Maple.* This tree is most common in low moist lands, and on the banks of streams and ponds, but is sometimes met with at considerable elevations on our hills and mountains. Its usual height is about 50 feet, with a diameter from 20 to 30 inches. The wood is lighter and more porous than that of the sugar maple, but when seasoned under shelter it makes excellent fuel, and is valuable for various other purposes. It is easily wrought in the lathe, and is much used for yokes, the handles of agricultural implements, wooden dishes and other domestic wares. In old trees, the grain is sometimes undulated constituting as in preceding species, what is called *Curled Maple*. This is wrought into various articles of furniture, which for richness and lustre, often equals the finest mahogany. It is also used for the stocks of guns. From the inner bark of this tree an extract of a purple color is obtained, which is darkened by the addition of a little copperas or alum and sometimes used for writing ink, and also for dying black.

WHITE ASH.
Fraxinus acuminata.

This tree is thinly scattered over nearly the whole state, and seems to delight in cool situations. It is most frequently met with near the banks of streams, and on the acclivities surrounding ponds and swamps. In these situations it frequently attains the height of 70 or 80 feet, with a diameter of from two to three feet. It is universally known by the name of *White Ash*, and this name may be derived either from the color of the bark, the sapwood, or the under surface of the leaves, all of which are white. By the light color of the bark it is readily distinguished from the other species. The wood of this tree is highly esteemed for its strength, suppleness, and elasticity, and is applied with advantage to a great variety of uses. It is always selected by carriage makers for fills or shafts, the felloes of wheels, and the frames of carriage bodies. It is also used for chairs, scythe snaths and rake handles; for hoops, sieves, boxes, wooden bowls, and a variety of other domestic wares; also for the staves of casks, blocks for pullies, and on account of its strength and elasticity, it is considered superior to any other kind of wood for oars.

RED ASH.
Fraxinus pubescens.

The Red Ash is a handsome tree which grows to the height of about 60 feet. The bark on the trunk is of a deep brown color, and the wood differs from that of the White Ash in being redder, but it possesses most of the other properties of the White Ash, and is, in general, applied to the same purposes.

BLACK ASH.
Fraxinus sambucifolia.

The Black Ash requires a moister soil than the White Ash, and is commonly found growing on low lands, and in and about swamps; and hence it is sometimes called *Swamp Ash*. The perfect wood is of a brownish complexion, and by malling may be separated into thin narrow strips, which are employed for bottoming chairs, making baskets, riddles, &c. The saplings of this tree are much used for hoop-poles.

SASSAFRAS.
Laurus sassafras.

This interesting and valuable tree is found, but sparingly, in the southwestern parts of the state, and this seems to be its most northern limit. On account of its small size and scarcity, little account is made of the wood, but it is highly valued for its medicinal properties. For more than 200 years it has maintained its reputation as an excellent sudorific, and it is employed to advantage in cutaneous affections and chronic rheumatism. The bark of the roots contains the greatest quantity of the peculiar extract of this tree. The dried leaves and young branches contain a large amount of mucilage.

THE TUPELO, OR SOUR GUM.
Nyssa multiflora.

This tree, which is here usually called *Pepperidge*, is found sparsely scattered through the southern and western parts of the state, but no where in large quantities. It grows to the height of near 50 feet, with a diameter of 15 or 20 inches. The limbs usually descend low upon the trunk, which continues of nearly uniform size for some distance. The wood of this tree holds a middle place between the hard and soft wood trees. The most remarkable peculiarity of this tree consists in the arrangement of its woody fibres, which are so united into bundles and twisted and braided together, that it is nearly impossible to split it. Hence it is often employed for the naves of wheels, and other articles, which are liable to split when made of common materials.

THE RED MULBERRY.
Morus rubra.

Vermont is near the northern limit of the growth of this tree, and here it grows very sparingly. At the south it is said to attain the height of 60 or 70 feet, and the wood is employed for many useful purposes, but here neither its size nor its numbers render it of much consequence.

HORNBEAM, OR BLUE BEECH.
Carpinus Americana.

This tree is not common excepting in the western part of the state, where it is generally known by the name of *Blue Beech*. It seldom exceeds twenty feet in height or 4 or 5 inches in diameter. The bark is smooth and undivided, and sets very close to the wood, the surface of which is usually irregularly furrowed. The wood is white, compact and fine grained, but the tree is so small and rare that little account is made of it.

IRON WOOD.
Ostrya Virginica.

The body of this tree, while small, is much used for levers in rolling logs, and hence it is frequently called *Lever Wood*. It is also called *Hop-Hornbeam*, from the resemblance of the fruit to that of the hop. The growth of this tree is very slow, as may be seen by the great number of concentric annual layers contained in a tree of only a few inches in diameter. It never constitutes the principal part of the forest, but is thinly scattered among the other trees in almost all parts of the state. It seldom exceeds 40 feet in height or 10 inches in diameter. The wood is white, compact, fine grained, and very heavy. It is used for making the cogs of wheels, for mallets, and for various other purposes. When seasoned it makes the very best of fuel, but its slow growth and limited quantity prevents its being an object of much regard.

RED BEECH.
Fagus ferruginea.

The Red Beech is found in all parts of the state, and in some places is so much multiplied as to form almost entire forests of considerable extent. Its usual height when full grown is from 60 to 70 feet, with a diameter of 2 or 2½ feet. The wood is valuable for fuel and in the arts. That of the second growth in open lands is strong, compact, fine grained and heavy. As it is not liable to warp when well seasoned, it is very suitable for the backs of cards, and is generally chosen for that purpose. It is also used for shoe lasts, for the wood of joiners' planes and other tools, and for the handles of various kinds of instruments. For fuel it is but little inferior to the sugar maple, if it be seasoned and kept under shelter from rains and moisture, but if exposed it is soon injured, and the sap wood soon rots. The fruit of this tree is usually abundant, and as swine eat it with avidity the early settlers of the state relied much upon beechnuts for fattening their hogs. As beechnuts are injured by the fall rains, those which are designed for preservation should be gathered as soon as ripe, and should be thinly spread in a dry place till they are thoroughly seasoned. They are often eaten, but are not very highly esteemed. A rich oil may be extracted from the nut.

THE WHITE BEECH.
Fagus sylvestris.

The two kinds of Beech are distinguished chiefly by their wood and durability. In the White Beech the greater part of the tree is sap-wood and very perishable, while in the Red Beech the sap wood is thin, and the heart, or perfect wood exceedingly compact and durable. The White Beech also grows to a greater height, and its trunk is freer from limbs than that of the Red Beech.

CHESTNUT.
Castanea vesca, var. Americana.

The Chestnut in Vermont is confined mostly to the south western parts, and to the towns lying along the bank of Connecticut river in the counties of Windham and Windsor. The basis of the soil in which it there flourishes is an argillaceous slate. According to the journal of Samuel Champlain, he found this tree growing on the shore of the lake which bears his name, in 1609. The wood is durable, and where it exists in sufficient quantities, it is used for posts and rails for fences, for shingles, and for staves of dry casks. For posts, trees from six to ten inches in diameter are employed, and they are generally charred on their surface before they are set in the ground. Chestnut rails are said to last more than 50 years. The wood being filled with air snaps as it burns, and on that account is not much esteemed for fuel; but coal made of it is excellent.

THE WHITE OAK.
Quercus alba.

The growth of the White Oak is confined principally to the southern and western parts of the state, and even there was never very much multiplied. The original growth sometimes attained the height of 70 feet, with a diameter of three or four feet, but the old trees have been nearly all cut down, and only a second growth, which has sprung up since the country was settled, now remains. The wood of this tree is more valuable than that of any other of the American oaks. It is of a reddish white color, and is very strong and durable. When perfectly seasoned it is much used by carriage makers, and is preferred to any other wood for the frames of coaches, waggons, and sleighs, and also for the felloes, spokes and naves of wheels. The wood of the stocks of young trees is very tough and elastic, and is susceptible of minute division; and hence it is much used for baskets, the hoops of sieves, and for whip, pail and axe handles. It also makes the best of staves for casks, and is the most valuable wood for ship-building. The bark of the White Oak is much used in medicine on account of its astringent properties. It is taken internally in the form of a decoction, or powder, for intermittent fevers, and is applied externally to wounds and ulcers as a styptic and antiseptic. Inhaled in the form of an impalpable powder, it is said to cure the phthisic in its advanced stages. For medical purposes the inner bark on small branches is to be chosen.

RED OAK.
Quercus rubra.

This oak, though not very abundant in Vermont, is more plentiful and widely diffused in the state than the preceding species. The wood is reddish and very coarse grained, and is of little value compared with that of the White Oak. It is used principally for staves and heads of casks. The bark is used in tanning leather.

The other species of oak, mentioned on pages 173 and 174, are found in Vermont only in small quantities.

LARGE WHITE BIRCH.
Betula papyracea.

This tree is quite common, and often attains the height of 60 or 70 feet. It is often called *Canoe Birch*, from the circumstance of its bark often being employed by the Indians in the construction of canoes. They also manufacture the bark into baskets and boxes. Divided into thin sheets it has been used as a substitute for paper. In new settlements large plates of the bark of this tree were sometimes used for covering the roofs of houses. The wood of this tree is lighter, when seasoned, and less valuable than that of the Yellow Birch and Black Birch.

THE BLACK, OR CHERRY BIRCH.
Betula lenta.

This tree is called Cherry Birch, from its resemblance to the wild cherry. It is also sometimes called *Sweet Birch*, or *Spice Birch*, on account of its agreeable aromatic smell and taste. It grows best in a deep loose soil, and sometimes reaches the height of 80 feet, with a diameter, at the bottom, of more than three feet. It is not so abundant as the following species, but the wood is more highly valued by the cabinet makers, being finer grained and susceptible of a higher polish. When freshly cut the wood has a light rosy hue, which deepens by exposure to the light. It is much used in the manufacture of bedsteads, tables, sofas, armed chairs, and a variety of other articles, and with age assumes very much the appearance of mahogany.

THE YELLOW BIRCH.
Betula excelsa.

The Yellow Birch is common in all parts of the state, generally preferring a rich moist soil. It ranks as one of our largest trees, often attaining the height of 70 or 80 feet, with a diameter of three or four feet. It is remarkable for the color and arrangement of its epidermis or outer bark, which is of a golden yellow color, and which frequently divides itself into narrow strips, rolled backwards at the ends and attached in the middle, giving to the tree a ragged appearance. The bark and young shoots have an agreeable aromatic odor and spicy taste. The wood of this tree is very valuable. It ranks next to the sugar maple in excellence as an article of fuel, and is used for various other purposes. It is sawed into joists, planks and boards, and is used by the cabinet maker for bedsteads, tables, and numerous other articles of household furniture. It is also made into yokes for oxen, and ox-sleds. The saplings are used for hoop-poles, and from these most of the brooms were made which were used by the early settlers. The bark is used in tanning leather. Russia leather is said to owe its peculiar odor, and its power of resisting moisture and the attack of worms and insects, to an oil used in currying, which is extracted from the paper-like bark of the birch. Hence its value for book-binding. The oil is obtained by heating the bark in closed earthern or iron vessels.

BUTTONWOOD, OR SYCAMORE.
Platanus occidentalis.

The Buttonwood is usually found growing along the banks of streams and margins of lakes and ponds, and, although, in Vermont, it does not, in magnitude, exceed some other trees, it is said in some parts of our country to grow to a greater size than any other tree in the United States. We have accounts of button wood trees in the western part of the state of New York and on the Ohio river, measuring more than 40 feet in circumferance at the height of five feet from the ground. This tree, though generally known by the name of Buttonwood in New England, is called in other places by various other names. In Virginia it is sometimes called *Water Beech*. At the west it is frequently called *Sycamore*, or *Plane Tree*, and in Louisiana and Canada it bears the name of *Cotton Tree*. The wood of this tree in seasoning, becomes of a dull red color, and is susceptible of a bright polish. It is but little used by cabinet makers, in the form of boards, on account of its liability to warp, but it answers well for bedsteads, and requires only to be polished and varnished, without paint, to make a very neat article.

WHITE ELM.
Ulmus Americana.

With the exception of the white pine, we have no tree which grows to a greater size, or which appears more graceful and majestic than the White Elm. This tree is found, though not very plentifully, in all parts of the state, and is sometimes seen towering to the height of 100 feet, with a diameter at the base of more than 4 feet. The wood is of a dark brown color, and is wrought for several valuable purposes. It is often sawed into planks, and has been considerably used for the naves of wheels. During a part of the year the bark of this tree is very easily detached, and this, after being soaked in water and rendered supple by pounding, was formerly much used for bottoming common chairs. For fuel, the elm is inferior to several other kinds of wood, but its ashes are strongly impregnated with alkali, and no wood yields a greater quantity. The young of the elm is much admired, and much employed as a shade tree around our yards and dwellings, and seems to be preferable to the locust, inasmuch as it thrives in all parts of the state, and is not, like the locust, liable to be destroyed by the Borer.

RED, OR SLIPPERY ELM.
Ulmus fulva.

This tree, though found in most parts of the state, is less abundant than the preceding species, and of less magnitude, seldom exceeding 60 feet in height, with a diameter of 2 feet. The wood is of a reddish color, and is less compact than that of the white elm. It makes excellent and durable rails, into which it is easily split, but this last property renders it unsuitable for the naves of wheels. It is, however, said to answer a good purpose for the blocks of pullies. The inner bark of this tree is an important article of *materia medica*. Macerated in water it yields a thick and abundant mucilage, which makes a refreshing drink much used in colds, coughs and fevers. The bark, when dried and reduced to flour, is said to make excellent puddings.

BUTTERNUT.
Juglans cinerea.

The Butternut is common in most parts of the state, and is known in some places by the name of Oil-nut, which it derives from the oily nature of its fruit. It thrives best on a dark cold soil, and often measures three or four feet in diameter, although it seldom exceeds 60 feet in height. The roots of the Butternut usually extend horizontally, with little variation in size, and but a few inches below the surface of the ground, often to the distance of 30 feet or more, which makes it a troublesome tree, when growing upon or adjacent to lands designed for tillage. The wood of this tree is light, and of a reddish color, and, though it has little strength, it possesses, in a good degree, the property of durability The timber is little used for frames of buildings, but is sometimes sawed into boards and clapboards. It is also used for posts in fences, for corn shovels, wooden dishes, troughs for catching the sap of the sugar maple, and for panels for coaches and chaises. For all these purposes it answers well, as it is not liable to split, and receives paint in a superior manner. The extract of the bark of this tree is used for a cathartic. Its operation is said to be sure, and unattended, in the most delicate constitutions, with pain or irritation.

SHELLBARK HICKORY.
Carya squamosa.

This tree, though no where greatly multiplied, is by no means uncommon, particularly in the neighborhood of lake Champlain. It is usually found on moist lands, and often about swamps and in places which are liable to be inundated in high water. The wood possesses the characteristic properties of the hickories generally, being very elastic and tenacious. It also possesses their common defect of soon decaying and being very liable to be eaten by worms. The wood is straight grained and easily split, and, being also easily wrought when green, is made into ax handles and whip handles, which are much esteemed on account of their smoothness, suppleness and strength.

THE NORWAY PINE.
Pinus resinosa.

The Norway Pine, though originally plentiful in some places in Vermont, was never so abundant as the following species, and, though a large and lofty tree, does not equal the white pine in size and height, seldom exceeding 3 feet in diameter or 80 feet in height. This tree is often called *Red Pine* and sometimes *Yellow Pine* from the color of its bark. The wood is fine-grained, compact, and on account of the resin it contains much heavier than that of the white pine, and for many purposes is more valuable. It is employed in architecture in various ways and is much esteemed for floors in dwelling houses. It is becoming scarce. Leaves in twos.

THE WHITE PINE.
Pinus strobus.

The white pine is much the most lofty tree which grows in our forests and the most valuable for its timber. Dr. Williams states the height of this tree to be 247 feet,[*] but it is probable that a very few only have obtained that height in Vermont. The tallest trees which have fallen under our own observation have not exceeded 170 feet. While the pine forests were standing, trees measuring from 140 to 180 feet were not uncommon, and they have often measured more than 6 feet in diameter at the base.

This species of pine was originally very abundant in all the western parts of the state, particularly in the neighborhood of Lake Champlain, and was found in considerable quantities along the bank of the Connecticut and most of our smaller rivers. But in consequence of the indiscriminate havoc of our forest trees by the early settlers, and of the common use of this tree for timber, boards and shingles for buildings and other domestic uses, together with the great demand for it, for exportation, our forests of white pine have mostly disappeared, and boards and shingles of good quality are becoming scarce and difficult to be obtained. The leaves are in fives.

THE PITCH PINE.
Pinus rigida.

This pine is always found upon light sandy lands and seldom exceeds 50 or 60 feet in height. It is remarkable for the great number of its limbs, which usually occupy two thirds of the trunk and render the wood extremely knotty. A large proportion of the trunk consists of sap wood, and for architectural purposes it is much less valuable than either of the preceding species. When sufficiently free from knots it makes firm and durable floors, and for fuel it is much esteemed by bakers and by glass and brick-makers. From the knots and resinous stocks of this tree lamp black is manufactured. The leaves are in threes.

DOUBLE SPRUCE.
Pinus nigra.

This tree is found in all parts of Vermont, and is so greatly multiplied on many of our hills and mountains, as to constitute almost entire forests of considerable extent. The usual height of this tree is from 60 to 80 feet, with a diameter of from 1½ to 2 feet. It seems to prefer a cool gravelly or sandy soil, and is most common upon northern or northwestern declivities. It is found, though of diminutive size, on the very summits of our mountains, and to this tree, more than any other, are we indebted for the name of our state, *Verd-Mont*, it being the most plentiful evergreen upon our mountains. The wood of the Double Spruce is distinguished for strength, lightness and elasticity, and is extensively used for frames of houses and other buildings. It is also sawed into boards and clapboards, which, though harder to plane and more liable to split in nailing, are, for many purposes, little inferior to pine, and for some purposes are preferred. It likewise makes good shingles. In the interior parts of the state houses, barns and other buildings are very often made entirely of spruce. The young branches of this tree, boiled in water, and the decoction sweetened with molasses or maple sugar, makes what is called *spruce beer*, which is said, in long sea voyages, to be a sure preventive of the scurvy. The wood is not of much value for fuel. It contains little resin, except what exudes and forms concretions in the seams of the bark, and is called spruce gum.

SINGLE SPRUCE.
Pinus alba.

This Spruce is much less plentiful in Vermont than the preceding species, to which, in most respects, it bears a strong resemblance, and is applied to the same uses.

---

[*] Hist. Vt. Vol. 1. p. 87. The author of Memoirs of Dr. Wheelock, late president of Dartmouth College, states that he measured a white pine which grew on the plain where that College now stands, and found it 270 feet from the butt to the top. Memoirs p. 56.

THE SILVER, OR BALSAM FIR.
Pinus balsamea.

The fir tree flourishes best in a cold, moist, sandy loam, and hence it is most commonly found growing on the north side of our mountains and about the margin of cold springy swamps. It sometimes, though rarely, reaches 50 feet in height, and its diameter seldom exceeds 12 or 15 inches. Where this tree stands alone, and developes itself naturally, its branches, which are numerous and thickly garnished with leaves, diminish in length in proportion to their height, and thus form a round pyramid or cone of remarkable regularity and beauty. The wood is very white, but its texture is coarse and open. It is sometimes used for staves in making casks,, and answers well for dry casks, but is not so good for holding liquids. It is also sawn into boards for making boxes, and is used for rafters, joists, &c., in frames. The balsam, for which this tree is somewhat celebrated, is obtained from the blisters or tumors on the bark. It may be collected with considerable facility with a teaspoon. For this purpose an incision is made in the lower part of the blister with the point of the spoon, and the pressure required in the operation causes the balsam to flow into the spoon, from which it is transferred to phials. The balsam is nearly colorless, has the consistency of honey, and is of an acrid penetrating taste. It is commonly known in this state by the name of *fir balsam*, but is said to be sold in many places abroad under the improper name of *Balm of Gilead.* It is of some celebrity as a medicine, particularly in pulmonary complaints and sprains of the chest and stomach, for which it is taken, a few drops at a time, internally. It is also in repute for its healing properties when applied to external wounds and sores.

THE HEMLOCK.
Pinus Canadensis.

The Hemlock is found in all parts of the state, and in most parts in abundance. It flourishes best in a sandy loam at the foot of hills and on lands slightly inclining. In such situations the trees are often from three to four feet in diameter. The size of the body of this tree is nearly uniform for about two thirds of its length. In very old trees the large limbs are often broken off four or five feet from the trunk by the weight of the snows lodged upon them, giving to the trees a decrepid and unsightly aspect. The wood of this tree, though abundant, is unfortunately coarse-grained, and inferior to most of the other evergreens for architectural purposes. It is, however, extensively used for frames and joists of buildings, for the timbers and planks of bridges, for the floors of barns, for lining boards, lath boards, &c. The logs are used for building dams, wharves and breakwaters, and they are bored and much used for aqueducts. The bark of the hemlock is extensively used in Vermont in tanning leather.

AMERICAN LARCH.
Pinus pendula.

This tree is generally known in Vermont by the name of *Tamarack*, but is sometimes called *Larch*, and sometimes *Hachmatack*. It seems to delight in a cold wet soil, and in this state it is most commonly met with in cold swamps. In the southern and eastern part of the state this tree is extremely rare, but in the western and northern parts it is much more common, and in some swamps is found in considerable quantities. A short distance further north, in Canada, it becomes still more abundant. With us this tree seldom exceeds 80 or 100 feet in height, with a diameter of about 2 feet; but to the north it attains a greater magnitude, and in the neighborhood of Hudson's Bay it is said to emulate our white pine, rising to the height of nearly 200 feet. This tree sheds its leaves in autumn, though its appearance in summer might lead one to suppose it to be an evergreen. The wood is strong and durable, and makes valuable timber for frames of buildings. It is also used for posts in fences, and for staves of dry casks. Although it snaps considerably, it is much superior to the evergreens for fuel.

MOUNTAIN ASH, OR MOOSEMISSA.
Sorbus Americana.

This beautiful little tree is very common upon our hills and mountains, and by transplanting is found to thrive well in all parts of the state. It seldom exceeds 25 feet in height, or 4 or 5 inches in diameter. It is generally known by the name of *Mountain Ash*, but is not unfrequently called *Moosemissa*. No use is made of the wood, but the bark affords an agreeable bitter, and is considerably used as a tonic. But this tree is chiefly valued as an ornamental shade tree, and its beautiful white blossoms, its pinnated globrous leaves, and bunches of red berries, which remain upon the tree during the winter, make it much admired for that purpose.

WHITE CEDAR, or ARBOR VITAE.
Thuya occidentalis.

This tree is found growing only in swamps, and along the rocky banks of streams and ponds, and is universally known in Vermont by the name of *White Cedar*. It was originally very abundant in the northern and western parts of the state, and is still found in many places in considerable quantities. The wood of this tree is nearly white, with a slight tinge of red. It is very light, soft, fine-grained and somewhat odorous. For durability it ranks next, among our forest trees, to the red cedar, and is extensively used for posts and rails for fences.

RED CEDAR.*
Juniperus Virginiana.

Red Cedar formerly existed in some quantities along the banks and islands of lake Champlain, but on account of the eagerness with which it has been sought for posts and other purposes, it has now become exceedingly scarce. Trees were formerly found 30 or 40 feet in height and 10 or 12 inches in diameter, but few now remain which are more than 10 or 12 feet high, and their growth is so very slow that there seems to be little prospect of a supply by reproduction.

The perfect wood of this tree is of a bright reddish tint and hence it is called *Red cedar*. The wood is compact, fine grained and very light, though heavier and stronger than the White cedar. It contains an essential oil, which exhales considerable odor, and which serves as a protection both against insects and moisture. The recent chips and splinters of this wood are often placed in drawers with woollen cloths and beneath carpets, and they are found to be a very sure protection against moths. The wood is also much used in making black lead pencils. But the quality which renders the Red cedar most valuable is its durability; and for this it excels every other wood found in the state. There are red cedar posts which have been standing in the common fences in Burlington and other places for 50 years, and which are now, excepting the mere surface, as sound as when set. These are eagerly sought out and preferred to new posts of any other kind, for fences, where great durability is desired.

HOOP ASH, OR HACKBERRY.
Celtis occidentalis.

This tree is found very sparingly in Vermont. In favorable situations, at the south and southwest, it grows to the height of 70 or 80 feet, and with the disproportional diameter of not more than 18 or 20 inches. The wood is neither strong nor durable, but where plentiful, as it is easily split, it is much employed for the rails of rural fences.

For some notice of the Northern Cork Elm, *Ulmus racemosa*, and the Poplars, see page 174.

Shade Trees. There are few if any of the forest trees which we have described, which are not more or less employed for shade, or ornament, about our yards and dwellings; but there are some which seem to be much more suitable than others for this purpose. Among these are the sugar maple, the elm, and the moosemissa, or mountain ash. To the growth of these, the soil and climate of most parts of the state are well suited, and they are all transplanted without difficulty. The larch too makes a beautiful shade tree, and so do several of our evergreens; but their transplanting is attended with much more difficulty. The best time for transplanting trees generally is believed to be

---

* Our cut was made from a young villous branch, which differs materially from that of the old tree.

in the spring, just before the appearance of the leaves.

Besides the native forest trees which have been used for shade and ornament, several exotics have been introduced for the same purposes. A little more than 30 years ago the Lombardy poplar, *populus dilatata*, was brought into the state, and was, for a time, extensively propagated, and much admired. Its growth was extremely rapid, and the appearance of the young tree was very pretty, but it was soon found that these were its only recommendations, which were more than counterbalanced by several positive objections. The wood was found to be soft and brittle, and nearly useless for fuel or any other purposes. As the barren and fertile flowers of this poplar grow on separate trees, and as none but trees bearing barren flowers have been introduced into this country, no seed is brought to perfection, and being propagated wholly by shoots, its growth, though rapid, was soon found to be feeble and sickly. Before the trees attained any considerable magnitude, the top branches would begin to die and fall off, rendering them unsightly, and giving them, while young, the appearance of decrepitude and decay, and littering the grounds and walks with limbs and rubbish. These circumstances, and the disgusting worms bred among their foliage, gradually lessened them in the public estimation, and for many years past no pains have been taken to propagate them. Many of the old trees have been cut down, and those which remain are generally in a decaying, dilapidated condition, and the prospect now is that they will, in a few years, become extinct.

The locust tree, *Robinia pseudo-acacia*, is one of our most beautiful and agreeable shade trees, and is very much prized, particularly in the western part of the state. It thrives best on the light, warm soil, which was originally covered with forests of white pine, but either the soil or the climate of our mountain towns is unfavorable to its growth; and hence it is not often met with in the central parts of the state. Fears are now entertained that all our locust trees will be destroyed by the Borer.

Fruit Trees. For many years after the settlement of this state was commenced, very little attention was paid to the cultivation of fruit trees. Apple orchards, it is true, were early planted in many places, and in some cases a few plumbs, cherries and perhaps pears, but they were generally suffered to produce their natural fruit, and very little effort was made to improve it by pruning and cultivation. But for a few years past much more attention has been given to this subject, and many choice varieties of these fruits have been introduced and extensively propagated by grafting and budding.

APPLE. *Pyrus malus.*—This is our most important and abundant fruit, and is found to flourish in all parts of the state. In the older parts the orchards became very extensive, the trees large, and immense quantities of apples were produced. These were mostly manufactured into cider, in consequence of which much more cider was made than could well be consumed, in its crude state, even when it was customary for all to drink it as freely, or more so, than water, and the price abroad did not warrant the expense of transportation. Distillation was therefore resorted to, and large quantities of cider brandy were manufactured. The farmers generally having large orchards could each make, without inconvenience, from half a barrel to two or three barrels of this liquor, and when they had it in their houses, as it did not seem to have cost them much, they felt themselves at liberty to use it very freely; and to this single circumstance may be traced the temporal and perhaps everlasting ruin of many of our previously thrifty farmers. This cause of ruin and misery was in the full tide of operation when the first general movement was made in New England on the subject of temperance.

But after the spell was broken, which had so long bound down all our people to the use, or acquiescence in the use, of distilled spirits, and it was perceived that these liquors were not only unnecessary, but hurtful as a common drink, our farmers began to perceive that those large portions of their lands which were covered with apple orchards were not only yielding them no profit, but that which, under their present management, was doing them a real injury. From this time many endeavored to turn their apples to a better account, by feeding them to their cattle, and hogs, and horses, and for these purposes they were found to be valuable, but caution was necessary, that they should not be fed in too large quantities at a time, especially when the fruit was hard and sour. Many, whose orchards were extensive, cut down large portions of them, that the lands might be more profitably employed in the production of something else. At present our people appear more anxious to improve their fruit by grafting or inoculating choice varieties upon the trees they already have, than to enlarge their orchards; and their are few countries which are capable of producing a greater variety of fine apples than Vermont.

The *Pear Tree* does not grow so well in the northern and central mountainous parts of the state, but it flourishes in the southeastern and western parts, where many choice varieties are cultivated and bear well. A few *Quinces* and *Peaches* are raised, but very little attention has been paid to their cultivation. That as good peaches may be raised in Vermont as in any other place, we think will hardly be disputed by any who ate of those which grew in our own garden in Burlington during the past and present year. Our remarks respecting the pear tree will apply also to the *Plum*. In the northern parts of the state, the native, or Canada Plum is much cultivated. It bears plentifully, and the fruit is tolerable. Our plum trees generally are very uncertain bearers. After bearing profusely one year they often pass several years without producing any fruit. *Cherries* flourish well, and several varieties are cultivated.

The *Siberian Crab Apple* is cultivated in the northern parts of the state, where it flourishes well, and bears abundantly. With sugar this fruit makes an excellent marmalade.

Nuts. These are the walnut, chestnut, butternut, beech-nut, oak-nut or acorn, and hazle-nut. Of walnuts we have three kinds, but the pignuts are much the most common. The shell bark hickory is found in some parts, but is not very abundant. The chestnut thrives only in the southern part of the state. Butternuts are common in most parts, and some years they are produced in very great abundance. It is esteemed a luxury by many, and in plentiful years large quantities are gathered and dried. See page 215. The beechnut is the most plentiful nut found in the state, and it abounds in all parts. When the country was new the early settlers depended principally upon this nut for fattening their hogs. But it was in many places as necessary that they should be attended by a guard to protect them against the original proprietors, the bears, as it was that the first settlers should be guarded against the attacks of the Indians. See page 212. The hazlenut grows on a shrub four or five feet high; and, though quite common, but little account is made of it. The above are all indigenous, and grow in a wild state without cultivation. Acorns too were formerly plentiful in many parts of the state, and these, like the beechnut, were for swine and bears a favorite article of food.

Berries. Vermont produces a very considerable variety of berries, both wild and cultivated, and many of them are highly serviceable, not only for desserts, but as articles of food. One of the most important of these is the *currant*, of which we have four species. Of these the red, white, and black currant are largely cultivated in gardens, but the two former are most esteemed, and are much eaten, stewed or made into pies when green; and when ripe they are eaten raw, or in pies, or are preserved in sugar, and their juice mixed with clean sugar at the rate of one pound of the latter to a pint of the former, and boiled from 15 to 20 minutes in a tin or brass kettle, makes an exquisite jell, which may be kept in glass vessels for years without difficulty. The black currant has a peculiar musky taste and odor, and, though liked by some, is not so generally esteemed. Black currants are found in a wild state in our forests, and red currants are also found growing wild upon our mountains, the taste of which is much less agreeable than that cultivated in gardens.

Whortleberries. of the various kinds, are produced in great plenty in different parts of the state, particularly on the pine plains in the neighborhood of lake Champlain. In plentiful years, the quantities of these berries offered for sale in our villages along the western part of the state are very considerable. In 1841, which was remarkably productive in these berries, the quantity brought into the village of Burlington between the 25th of June and the 1st of September, could not have fallen much, if any, short of 200 bushels.

We have three kinds of *raspberries*, the red, black and white, all of which grow wild. The two latter are much improved by cultivation, and are considerably cultivated in gardens. The red raspberry is very abundant on most of our hills and mountains. *Gooseberries* are found growing wild in all parts, but the fruit is generally small. Several choice foreign varieties have been introduced into our gardens, where they are easily cultivated and brought to a high degree of perfection. They are a luxury, which, with very little trouble, every family might enjoy.

Blackberries, of two or three kinds, are common, and they are universally regarded as the most wholesome and delicious wild berry found in the state. A variety of this berry is occasionally found the color of which is a delicate yellowish white. It is sometimes cultivated in gardens, and, contradictory as the terms may seem, several have been able to assert, without contradiction, that they could en-

BERRIES.—MEDICINAL PLANTS.

tertain their visitants with a dessert of white black-berries.

The *barberry* bush grows well in most parts of the state, but so little use is made of the berry that no effort is made to multiply it. Two kinds of *cranberries*, the high and the low, are common in many of the swamps, and preserved in sugar they make an agreeable and wholesome sauce. Of *strawberries* there are several kinds. The wild, or woods strawberry, though a pleasant fruit, is not found in sufficient quantities to be an object of much regard. The common field strawberry is diffused over the whole state, and in its season affords considerable quantities of delicious fruit, though it seldom grows to a large size. Several varieties of foreign strawberries are cultivated in gardens. Some of these grow to a great size, and with proper attention a small plot of ground may be made to yield a very large quantity of choice fruit. The fox and frost *grapes* grow wild in most parts of the state, and several exotic grapes are successfully cultivated in gardens, and bear well. The large purple grape endures our climate and ripens its fruit without protection, and this is undoubtedly the most profitable for general cultivation. The more choice varieties must either be housed or buried to preserve them through the winter, and many of them require protection and artificial heat, in order to bring their fruit to maturity. In addition to the above, we have the hobbleberry, the mulberry, the checkerberry, the partridge berry, and some others which are eaten, and several kinds, as the sumac, elder, juniper, &c., which are used in medicine or the arts.

Medicinal Plants.—The native vegetables of Vermont already contribute somewhat to the Materia Medica of the country, and when the medicinal properties of our plants become better known, it is probable that the list of those which deserve to be employed in the healing art will be greatly increased. We are of the number of those who look with much more confidence to the vegetable than to the mineral kingdom, for antidotes to the various diseases and ills which flesh is heir to. Not that we would go to the lengths of some of our name, and banish all mineral substances from our pharmacopœia, but, being fully persuaded that for removing a great majority of diseases, the remedies derived from the vegetable kingdom are not only more effectual, but far more safe than those derived from the mineral kingdom, we would gladly see the medicinal properties of our plants more thoroughly investigated, their reputed virtues canvassed, and their proper places assigned them among the articles of our *materia medica*.

In the preceding account of our forest trees, we have briefly mentioned the medicinal purposes to which the parts of several of them are applied. We had intended in this place to notice a few of the many herbs and roots which are, or have been, of repute for their medicinal virtues, but we have not room. We would, however, remark that the Ginseng, *Panax quinquefolia*, was the first medicinal root which attracted much attention in this state, and is the only one which has been to any considerable extent an article of exportation. This root had long been regarded in China as a *panacea*, and was supposed to be indigenous only in that country and Tartary, till 1720, when it was discovered by the Jesuit *Lafitan*, in the forests of Canada. Such was the demand for the root in China, at that period, that it soon became a considerable article of commerce. Upon the settlement of this state the ginseng was found to grow here in great plenty and perfection, and it soon began to be sought with eagerness for exportation. For many years it was purchased at nearly all the retail stores in the state, and was sent to the seaports to be shipped to China. Those who dug the root sold it in its crude state for about 2 shillings or 34 cents per lb., and it was so plentiful in some places that digging it was a profitable business. The root is a mild, pleasant, and wholsome bitter, but it has never ranked very high as a medicine in this country, and its exportation and the clearing of the country has rendered it scarce.

Flowering Plants. This state is particularly rich, considering its northern situation and mountainous surface, in beautiful flowering plants. Several of these have already been noticed in the observations preliminary to the preceding catalogue. Among our most singular flowering shrubs may be mentioned the Witch Hazel, *Homomelis Virginica*. This shrub puts forth its modest yellow blossoms usually in October, after the leaves have been killed by the frost, but the seed is not matured till the following year.

Poisonous Plants, which are natives of Vermont, are not numerous. Enough, however, exist to render caution necessary in gathering herbs, either for food or medicine. A few poisonous plants have also been introduced, and to some extent naturalized. Of these may be mentioned the poison hemlock, which may be seen growing in many places by the roadsides.

CHAPTER VIII.

GEOLOGY AND MINERALOGY OF VERMONT.

When we commenced our undertaking four years ago, we had little doubt that there would be a Geological Survey of the state, under the patronage of the government, in season to enable us to embrace the results of it in the present work. In consequence of this expectation, we have devoted less attention to the geology of the state than to the other departments of our natural history; and, a survey not having been undertaken, as we anticipated, we must content ourselves for the present with only a few general remarks on these interesting subjects. Enough is already known to make it certain that our state ranks among the first in the Union in mineral resources, and by private and individual enterprise something has already been done towards turning these resources to account, as may be seen by reference to our account of Strafford, Bennington, Plymouth, and some other towns in part third. The few remarks which we shall offer will be presented under the heads of Rocks, Metals, and Minerals.

Rocks.

The ranges of rocks in this state, for the most part, extend through the state in lines parallel to the principal range of the Green Mountains. The greater part of the rocks are of primitive formation. The ranges, commencing on the west side of the state, according to Prof. Eaton, are nearly in the following order:—1. Old Red Sandstone in an interrupted range;—2. Graywacke;—3. Transition, or Metaliferous Limestone, alternating with Transition Argillite;—4. Transition, or Calciferous Sandstone;—5. Transition Argillite;—6. Primitive Argillite;—7. Sparry Limestone;—8. Granular Limestone;—9. Granular Quartz, containing hematitic iron ore and manganese, and lying at the foot of the Green Mountains on the west side;—10. Hornblende Rock;—11. Gneiss, with alternating layers of Granite;—12. Mica Slate, constituting the middle ridge of the Green Mountains, and extending, in many places, a considerable distance down the eastern side. Most of these ranges of rocks extend through the whole length of the state; On the east side of the Mountains the geological features are not so well defined, nor so well known. Although there are here indications of ranges nearly parallel with those on the west side, they are frequently interrupted and jumbled together; the different rocks often being arranged in alternating layers. The principal ranges of rocks in the central part of the state are nearly as exhibited in the following diagram of a vertical section passing from east to west, through Camel's Hump:

References.

1. Lake Champlain.
2. Camel's Hump.
3. Montpelier.
4. Connecticut River.

a. Sandstone.
b. Argillaceous slate.
c. Graywacke Sandstone.
d. Limestone.
e. Slates, Graywacke, Argillaceous, &c.
f. Mica Slate.
g. Quartz, Talcose Slate and Chlorite.
i. Argillaceous Slate.
k. Granite.
l. Lime.
m. Argillaceous Slate.

Granite. This rock shows itself very sparingly in the Green Mountain range, and on the west side of the mountains hardly exists at all, except in small rolled masses. On the east side of the mountains it occurs in many places in Windham and Windsor counties. In the northern part of Orange county, the southeastern part of Washington and southwestern part of Caledonia county, it constitutes the principal rock *in situ*. From this great granite region was obtained the material for building the State House. (*See Part III, p.* 9.). Orleans county abounds in huge granite boulders, which make excellent building stone.

Gneiss. This occurs in many places along the summits of the Green Mountain range and in the counties of Windham and Windsor, where it serves a good purpose for walls, under-pinnings, &c.

Mica Slate. This constitutes almost the entire middle range of the Green Mountains from Massachusetts to Cana-

ROCKS, METALS AND MINERALS.

da, and is met with more or less abundantly in all the counties on the east side of the mountains. It is of little value as a building stone, excepting for wall fences, but is found in many places suitable for covering stone bridges, for flagging stone, &c. In Halifax and some other places it is found of a quality suitable for common grave stones.

Argillaceous Slate. Several considerable ranges of this slate are found in Vermont extending from south to north. It is abundant along Connecticut river, and in Windham county it is extensively quarried at Dummerston and other places for roof and writing slate. A range of this slate extends north from White river through Montpelier, which, at Berlin and some other places, affords slate of a very good quality. A dark colored glazed variety of this slate extends along the eastern margin of lake Champlain, the seams of which are filled with calcareous spar.

Lime. The range of granular limestone, which enters the state at Pownal, and extends almost directly north to Canada, is the most important in the state. This range affords excellent marble, which is extensively wrought in many towns in the counties of Bennington, Rutland and Addison. Very beautiful marble is also found at Swanton. Throughout all the western parts of the state limestone, for the manufacture of lime of the best quality, is abundant. On the east side of the mountains, the best for the manufacture of lime is probably at Plymouth, near the head of Black river. (*See Part III, p.* 140.) Some of this limestone is found to receive a very good polish as it has been wrought to some extent for marble. The other most important localities are at Whitingham and in the southeastern part of Caledonia county. The lime on the east side of the mountains is not only more limited in quantity, but is darker colored, and otherwise inferior to that on the west.

Talcose Slate. This rock forms an interrupted range from Whitingham, on Massachusetts line, to Troy on Canada line. In this range are extensive beds of excellent steatite, or soap stone, which is, in many places, wrought into fire places, stoves, aqueducts, &c. The most important localities are at Grafton, Plymouth, Bridgewater, Bethel, Moretown and Troy. Talcose slate also abounds on the west side of the mountains in the county of Lamoille, and the eastern part of Franklin county.

Serpentine. Nearly in connection with the Talcose range, on the east side of the mountains, this rock shows itself in many places;—most extensively at Cavendish near Black river, and at Lowell near the source of Missisco river. At the former place, its connection with the limestone and steatite forms that most beautiful variety of marble called *Verd Antique.* (*See Part III, p.* 48.) At the latter place is found beautiful precious serpentine, and several varieties of amianthus and asbestos.

Metals.

Iron ore, in the form of oxydes, is found in greater or less quantities in almost all parts of the state. The most important beds of this ore which have been opened on the west side of the mountains are at Bennington, Tinmouth, Pittsford, Chittenden, Brandon, Monkton and Highgate, and on the east side of the mountains at Troy and Plymouth, for an account of which, see part third, under the respective names, particularly the latter. Sulphuret of iron is also abundant in many places. *See Strafford, in part third.*

Manganese is abundant in connection with the iron ore at Plymouth, Bennington, Chittenden, &c., and has already become a considerable article of exportation.

Lead ore has been found in small quantities at Thetford, Sunderland, Morristown, and some few other places. There is some prospect that the vein at Morristown may prove valuable. It is situated upon the top of a large hill, in the seam in talcose slate, the strata of which are nearly vertical, and extend from north to south. The seam at the surface of the rock, which is bare for some distance, is perhaps 18 inches wide, and can be traced north and south several rods. This seam is filled with a substance which seems to be mostly quartz, in which the sulphuret of lead, or galena, is scattered, being in masses from the size of a pin-head to that of a man's fist. The seam, which has been opened to the depth of several feet, is found to increase in width downward, and to become richer in ore, but whether it will repay the expense of working is at present problematical.

Copper ore is found sparingly at several places. At Strafford, where it has been found most plentifully, it has been smelted for the copper. (*See Part III, p.* 166.)

Silver is said to exist in a small proportion in the lead ore, but has been found here in no other connection.

Gold has been found in the lower part of Windham county, but in no other part of the state. In 1826 a lump of native gold was found in Newfane weighing 8 ounces, and in Somerset it has been found in small particles in connection with talcose slate.

LOCATION OF MINERALS.

Minerals.

We shall close this short chapter by indicating some of the principal localities of interesting minerals, many of which will be still further noticed in part third, under the names of the towns in which they are situated.

Actynolite.—Windham, Grafton, Newfane, Brattleboro', Norwich—the latter very beautiful

Agaric Mineral.-Lyndon, Groton, Manchester.

Aluminous Slate.-Pownal, Rockingham.

Amethyst.—Westminster, Ludlow.

Amianthus.—Weybridge, Mount Holly, Lowell, Barton.

Argillaceous Slate.—Common.

Asbestos.—Mount Holly, Lowell, Troy.

Augite.—Charlotte, Chester.

Bitter Spar.——Grafton, Bridgewater, Lowell.

Blende.—Orwell.

Calcareous Spar.—Vergennes, Shoreham, &c.

Calcareous Tufa.—Clarendon, Middlebury, Hubbardton, Manchester, Orwell.

Carbonate of Lime.—Common.

Chalcedony—Newfane.

Chlorite.--Grafton, Windham, Bethel, &c.

Chrysophrase.—Newfane.

Copper, (Carbonate Green)—Bellows Falls, *(Sulphuret,)* Strafford, Waterbury.

Copperas.—Strafford, Shrewsbury.

Cyanite.—Grafton, Bellows Falls, Norwich.

Diallage.—New Haven.

Dolomite.—Jamaica.

Epidote.—Middlebury, Chester, Berkshire, &c.

Feldspar.-Townshend, Thetford, Monkton, &c.

Fetid Limestone.—Shoreham, Bridport, &c.

Flint.—Orwell.

Fluate of Lime.—Putney, Rockingham.

Garnet——Bethel, Bridgewater, Norwich, &c.

Graphite, Plumbago, or Black Lead.—Hancock, Charlotte.

Hornblende.—Jericho, Ludlow, &c.

Hornstone.-Middlebury, Shoreham, Salisbury, Bennington, Orwell.

Jasper.—Middlebury, in rolled masses.

Kaolin.—Monkton, Brookline.

Lead, (Sulphuret) or Galena.—Sunderland, Thetford, Danby, Morristown.

Lime, Fluate.—Putney, Rockingham, Thetford.

Lime, Fetid Carbonate.—Bennington.

Lithomarge.—Bennington.

Macle.—Near Bellows Falls.

Manganese, Oxyde.--Bennington, Brandon, Monkton, Pittsford, Chittenden, Plymouth.

Marble.—Shaftsbury, Manchester, Dorset, Rutland, Middlebury, Swanton, Plymouth.

Marl.—Peacham, Barnard, Benson, Alburgh.

Mica.--—Chester, Craftsbury, Orange, Grafton, &c.

Novaculite, or Oil Stone.——Thetford, Memphremagog Lake.

Potstone.—Grafton, Newfane.

Potter's Clay.—Middlebury.

Prehnite.—Bellows Falls.

Quartz.— Common. *Fetid Q.*, Shrewsbury. *Greasy Q.*, Grafton, Hancock, New Haven, &c. *Quartz Chrystal,* Castleton, Vergennes, Waitsfield, St. Johnsbury, &c. *Milky Q.*, Stockbridge, Grafton, Middlebury. *Radiated Q.*, Hartford. *Smoky Q.*, Shrewsbury, Wardsborough. *Tabular Q.*, Windham.

Rubollite.—Bellows Falls.

Scapolite.—Brattleborough.

Schorl.—Grafton, Bridgewater, Brattleborough, Newfane, Dummerston, &c.

Serpentine, Precious.—Lowell, Ludlow, Troy, Cavendish, Windham.

Staurotide.—Rockingham, Vernon.

Steatite.—Grafton, Bethel, Moretown, Bridgewater, Troy, &c.

Stelactite.——Bennington, Dorset, Plymouth, Montpelier.

Sulphur.—Wilmington, Bridgewater.

Talc.—Grafton, Windham, Newfane, Ludlow, Bridgewater, Hancock, Montpelier, Fletcher, &c.

Titanium.—Whitingham.

Tourmaline.—Peacham.

Tremolite.—Bellows Falls, Wardsboro'.

Tufa Calcareous.—Orwell, Clarendon, Middlebury, &c.

Zinc.—Orwell.

Zoisite.—Rockingham, Wardsborough.

APPENDIX

TO

THOMPSON'S VERMONT.

NATURAL HISTORY

Topography.

When the History of Vermont, to which this is an Appendix, was published, in 1842, the boundary line between the United States and the British provinces was unsettled, and in dispute between the two governments; but in the latter part of the summer of that year, the matter was amicably arranged by a treaty, formed by Mr. Webster and Lord Ashburton, and ratified by the two governments. The northern boundary of the state was intended to be along the 45th parallel of latitude, and was supposed to be on that parallel till the survey of 1818 proved the 45th parallel to be some distance to the southward of what had been previously regarded as the northern boundary of the state, cutting off a strip through the whole width, varying from one-fourth of a mile to a few rods. By the treaty, the northern boundary of the state was established upon the old well known line, without reference to the 45th parallel. This line was marked in 1845, by cutting away the timber, where it passed through forests, and by putting up cast iron posts at short distances through its whole length.

The geological explorations and the railroad surveys, which have been made during the last ten years, have added much to our knowledge of the general topography of the state, and many objects of interest and value have been brought to light. Remeasurements have been made of several of our principal mountain summits, and their altitudes ascertained with greater accuracy, probably, than before, and a number of important peaks have, within that period, been measured for the first time.

In addition to these measurements of isolated mountain summits, there have been reconnoisances and surveys made, in almost every direction, through the state, for the location of the various railroads which have been built, or are now building. The profiles of these roads, together with the profiles of the canal routes, which had been surveyed previously, have furnished the means for giving a very tolerable exhibition of the elevation above the sea, of the principal places and most interesting objects in the state.

In the following list of altitudes, those of mountain summits are all derived from Barometrical measurements. The others are in part Barometrical; but they are derived principally from the various surveys for canals and railroads. Minute accuracy in these altitudes above the sea, cannot be expected. They are, however, believed to be a near approximation to the truth, and to show with sufficient exactness the relative elevation of the different places and objects.

APPENDIX TO THOMPSON'S VERMONT.

MOUNTAIN SUMMITS.

ALTITUDES ABOVE THE OCEAN.
Mountain Summits.

			Feet.
Chin,	Mansfield,	Thompson,	4348
Nose,	"	"	4044
South Peak,	"	"	3882
Camel's Hump,	Duxbury,	"	4083
Jay Peak,		Adams,	4018
Shrewsbury Peak,			4086
Killington	" Sherburne,	Partridge,	3924
Equinox,	Manchester,	"	3706
Ascutney,	Windsor,	"	3320
Snake Mt.,	Bridport,	Adams,	1310
Buck Mt.,	Waltham,	"	1035
Sugar Loaf,	Charlotte,	Thompson,	1003
Snake Hill,	Milton,	"	912
Cobble,	"	"	827

Passes over the Green Mountains.

Lincoln,		Adams,	2415
Granville,		"	2340
Peru,		"	2115
Sherburne,		Partridge,	1882
Walden,		De Witt Clinton,	1615
Mt. Holly,	(R. Road)	Gilbert,	1415
Roxbury,	(R. Road)		997
Williamstown,		Johnson,	908

Villages.

Burlington Town House,		Benedict,	202
" University,		"	367
Milton Falls,		Thompson,	298
Essex,		"	452
Jericho Corners,		"	604
Underhill Flat,		"	665
Williston,		"	402
Franklin,		"	430
St. Albans,		"	370
Highgate Springs,			160
Swanton,			160
E. Berkshire,			460
Winooski Falls,			203
Sheldon,			375
Richmond,			332
Waterbury,			425
Middlesex,			520
Montpelier, (Capitol,)			540
Northfield, (Depot,)			724
Braintree,			732
West Randolph,			678
Bethel,			556
Royalton,			476
White River Junction,			335
Windsor,			288
Bellows Falls,			225
Woodstock,			400
Brattleborough,			160
Bennington,			432
Manchester,			650
Rutland,			500

Villages (continued).

Castleton,	475
Ludlow,	985
Proctorsville,	895
Chester,	670
Brandon,	460
Middlebury,	390
Vergennes,	225
Norwich,	400
Newbury,	420
Barnet,	460
St. Johnsbury,	585
Lyndon,	735
Barton,	953
Derby Centre,	975
" Line,	1050
Craftsbury Common,	1158
Troy, south,	740
Irasburgh,	875
Hardwick Hollow,	720
Hydepark,	560
Cambridge,	410
Johnson,	460

Lakes and Ponds.

Champlain, Lake,*		90
Memphremagog "		695
Joe's Pond,	Cabot,	1544
Lyford's Pond,	Walden,	1692
Molley's Pond,	Cabot,	1626
Winooski Pond,	Peacham,	1410
Wells River Pond,	Groton,	1000
Crystal Pond,	Barton,	933
Mud Pond,	Sutton,	1183
Savanna Pond,	"	1210
Willoughby Lake,	Westmore,	1161
Elligo Pond,	Craftsbury,	893
Salem Pond,	Salem,	967
Pensioners Pond,	Charleston,	1140
Island Pond,	Brighton,	1182
Lake Connecticut, head of Con. River, in N. H.,		1589

Falls.

Great Falls, Marshfield, (head,)	1074
" " (foot,)	871
Nat. Bridge Falls, Waterbury, (foot,)	345
McIndoe's Falls, Barnet, (head,)	449
" " " (foot,)	436
20 Miles Rapids, Lunenburg, (head,)	822
" " Barnet, (foot,)	486
Guildhall Fall, Guildhall, (head,)	835

---

*The level of Lake Champlain is taken for a basis in many of the surveys, which have been made, for canals and railroads, and their profiles indicate the height of places above the lake. In estimating, from these, the heights above the ocean, for the accompanying tables, 90 feet are added. The mean height of the lake above the ocean is frequently stated at 94 or 95 feet, but from the data, to which I have had access, I am disposed to think that 90 feet is nearly the true height. The change of level of the lake, that is, the difference between the extreme high water and the extreme low water marks, amounts to eight feet.

CLIMATE AND METEOROLOGY.

A general account of the Climate and Meteorology of Vermont is given in Part I. page 9 to 23, to which the following tables and observations are now added:

Monthly and Annual Mean Temperatures at Burlington,—continued from the table on page 9, Part I.

MONTHS.	1842.	1843.	1844.	1845.	1846.	1847.	1848.	1849.	1850.	1851.	1852.	Mean, 11 Yrs.
	°	°	°	°	°	°	°	°	°	°	°	°
January,	22.30	28.02	9.91	21.36	19.77	20.97	24.17	15.06	23.74	19.60	14.36	19.93
February,	26.60	12.95	20.33	22.63	15.05	18.59	21.09	14.34	24.32	26.02	23.19	20.46
March,	35.80	25.66	31.00	34.09	33.89	25.73	29.03	31.66	30.47	33.35	28.50	30.83
April,	44.60	43.85	49.50	43.82	47.73	37.48	42.89	39.90	41.35	43.31	39.86	42.12
May,	53.50	53.92	58.50	53.81	57.60	56.40	58.86	51.59	51.64	54.13	56.16	55.10
June,	63.80	61 86	66.50	65.21	64.97	64.15	65.02	66.76	67.12	62.97	64.34	64.80
July,	70.00	64.16	67.10	68.40	69.51	71.03	68.39	72.74	70.03	67.40	71.08	69.08
August,	70.10	67.78	65.60	69.43	70.45	67.62	66.82	69.14	66.03	65.68	66.44	67.73
September,	57.30	59.59	59.90	58.12	64.75	58.80	56.41	58.02	59.69	60.58	59.42	59.32
October,	47.50	42.84	47.00	51.15	45.37	45.89	47.24	47.10	48.25	51.09	47.95	47.40
November,	34.40	31.56	34.10	39.26	41.26	39.84	34.81	43.29	40.38	31.74	35.58	36.39
December,	21.30	26.87	23.40	17.81	23.43	27.00	30.01	23.17	18.65	18.58	30.32	23.69
Annual Temp.	45.60	43.25	44.40	45.42	46.15	43.88	45.39	44.40	45.14	44.54	44.77	44.74

The above results were deduced from three daily observations, made at sunrise, 1 P. M. and 9 in the evening, by the Author. The location is in latitude 44° 29′ N. and longitude 73° 11′ W., and is one mile eastward from the shore of Lake Champlain, and elevated 256 feet above the lake, or 346 above the ocean.

EXTREMES OF TEMPERATURE.

Greatest and Least Heat in the Shade, and the Hottest and Coldest Day in each year since 1837—15 years.

Year.	Greatest Heat.	°	Greatest Cold.	°	Hottest Day.	Mean. °	Cold. D.	mean. °
1838	June 10 and July 20,	93	January 31.	–13	July 29,	83	Jan. 30,	–6
1839	July 20,	91	January 24,	–19½	August 20,	78	Jan. 23,	–4⅓
1840	July 16,	94	January 18,	–16	July 16,	81⅔	Jan. 16,	–7
1841	August 18,	96	January 4,	–10	August 18,	82	Jan. 4,	⅔
1842	July 19, Aug. 26,	93	January 13,	–11	July 19, Aug. 26,	79⅓	Jan. 13,	–4⅔
1843	June 22,	90	Feb 'ry 17,	–17	June 22,	76	Feb. 17,	0
1844	June 19,	88	January 28,	–24	June 19,	74⅔	Jan. 29,	13⅔
1845	July 12,	96	Decemb. 11,	–18	July 21,	80	Dec. 11,	–11⅓
1846	August 5,	96	Feb. 10 & 19,	–10	August 5,	81⅔	Jan. 18,	–5
1847	July 19,	98	Feb 'ry 16,	–14	July 19,	83	Jan. 31,	⅓
1848	July 12,21,22: Au. 12,	92	January 11,	–25	June 18,	80	Jan. 10,	–10⅔
1849	July 12, 13,	100	Feb 'ry 17,	–17	July 13,	87	Feb. 18,	–7
1850	June 19,	93	February 6,	–18	June 19,	81¾	Feb. 5,	–4¾
1851	September 10,	92	Feb 8, Dec. 26	–17	September 10,	79¾	Feb. 8,	–11½
1852	June 16,	97	January 15,	–17	June 15,	81½	Jan. 20,	–5½

By the above statement it will be seen, that, during the last fifteen years, the range of the Thermometer has been from 100° above to 25° below zero, equal to 125°; and that the warmest day was the 13th of July, 1849, and the coldest day, the 29th of January, 1844, and that the difference between the mean temperature of those two days was 100⅔°.

TEMPERATURE AT NEWBURY.   FALL OF WATER AT BURLINGTON.

ANNUAL MEAN TEMPERATURE AND WEATHER AT NEWBURY.

Year	THERMOMETER.				WEATHER.				
	Mean heat.	greatest.	least.	Range.	Fair.	Cloudy.	Rain.	Snow.	Sn. & Rain.
	°	°	°	°	°	°	°	°	°
1840	44.28	90	−33	123	169	196	96	32	6
1841	44.36	86	−22	108	146	219	78	46	4
1842	43.61	86	−19	108	163	202	101	49	7
1843	43.70	90	−25	115	157	208	77	57	5
1844	43.83	86	−26	112	160	206	106	31	2
1845	43.44	88	−28	116	142	223	100	47	5
1846	45.44	90	−20	110	149	216	87	42	4
1847	44.44	90	−22	112	144	221	106	42	6
1848	44.83	87	−32	119	165	201	107	56	6
1849	44.13	94	−21	115	161	204	99	61	2
10 yr.	44.20	94	−33	127	156	209	96	51	5

The materials for the above table are derived from Meteorological observations made at Newbury, by Mr. Johnson, of that place, and published in the Annual Report of the Regents of the University of New York for 1850. These observations were continued through a period of twenty-seven years, but the earlier observations were made without a thermometer, and embraced only the clearness of the sky, the rains and snows, the course of the winds, the progress of vegetation, aurora borealis, and other rare phenomena. The mean temperature in the above table, is derived from three daily observations, made at 6 A. M., noon, and 6 P. M. This mean is probably a little higher than it would have been if the observations had been made at sunrise, 1 P. M. and 9 P. M., as in the preceding table. By a comparison of the eight years, from 1842 to 1850, which are embraced in both tables, the mean annual temperature of Burlington appears to be about two-thirds of a degree warmer than Newbury, while the latitude of the place of observation in Burlington is 23′ greater, and its altitude above the ocean about 75 feet less than the place of observation at Newbury.

MONTHLY AND ANNUAL FALL OF WATER AT BURLINGTON.

Continued from page 12, Part I.

MONTHS.	1842.	1843.	1844.	1845.	1846.	1847.	1848.	1849.	1850.	1851.	1852.	Mean, 11 Yrs.
	Inches	Inches	Inches	Inches	Inches	Inches	Inches	Inches	Inches	Inches	Inches	Inches
January,	1.04	0.71	2.29	2.38	1.72	2.80	1.84	0.79	1.57	1.20	1.03	1.58
February,	3.75	1.43	0.73	2.52	1.47	1.85	0.90	0.41	1.79	1.90	1.69	1.52
March,	1.97	2.12	2.35	2.48	2.20	2.10	2.44	2.14	1.11	0.67	1.92	1.96
April,	2.52	0.82	1.43	2.22	0.91	3.15	1.09	0.47	2.41	1.67	1.15	1.62
May,	1.55	2.47	4.40	3.39	3.18	1.85	4.24	2.74	5.04	2.29	0.71	2.90
June,	3.24	4.58	2.08	2.08	3.63	5.05	2.19	1.41	3.18	7.33	4.76	3.59
July,	4.62	2.59	5.35	4.51	5.08	4.05	3.57	1.78	5.08	3.81	4.99	4.12
August,	1.74	2.09	3.46	2.37	0.48	3.12	4.40	5.69	0.89	1.92	1.50	2.51
September,	3.80	1.80	1.36	5.62	3.77	4.69	2.91	1.33	3.25	2.06	1.80	2.95
October,	4.10	5.03	5.11	2.26	2.65	3.69	2.59	5.32	8.11	3.56	4.11	4.23
November,	2.32	1.63	0.57	4.00	2.88	2.13	2.26	2.69	1.77	3.59	2.90	2.43
December,	3.20	1.48	2.08	2.21	1.68	4.07	2.95	1.63	3.31	1.83	2.26	2.41
	33.85	26.75	31.21	36.04	29.66	38.55	31.38	26.35	37.51	31.83	28.82	31.82

By the above table it appears that the greatest amount of water in any one year was 38.55 inches in 1847, and the least 26.35 in., in 1849,—range 12.20 in. The greatest monthly amount was 8.11 inches, in October, 1850, and the least 0.41 in. in February, 1849—range 7.70 inches. The proportion of the water, which falls in snow, is about one-fifth of the whole amount. The greatest rain-storms in the eleven years embraced in the above table, were on the 10th of July, 1844, when there fell 4.07 inches in twenty-four hours, and on the 22d and 23d of June, 1851, when the amount was 5.16 inches in thirty-six hours.

FALL OF SNOW, AND DAYS OF SLEIGHING IN TEN SUCCESSIVE YEARS.
Continued from page 12, Part I.

MONTHS.	1842–43	1843–44	1844–45	1845–46	1846–47	1847–48	1848–49	1849–50	1850–51	1851–52
	Inches.	Inches.	Inches.	Inches.	Inches.	Inches.	Inches.	Inches.	Inches.	Inches.
October,	0	10	0	0	7	0	0	0	5	0
November,	9	3	2	0	5	4	3	1	1	24
December,	47	18	13	36	22	13	18	19	48	14
January,	7	16	20	9	24	11	10	16	11	24
February,	20	13	22	24	25	19	10	18	8	17
March,	38	18	12	4	22	20	7	12	14	22
April,	2	0	6	0	7	1	0	8	0	12
Total.	123	78	75	73	112	68	48	74	87	113
Sleighing,			35 d.	102 d.	40 d.	16 d.	30 d.	80 d.	61 d.	87 d.

ADVANCE OF SPRING FOR ELEVEN SUCCESSIVE YEARS.
Continued from page 13, Part I.

Years.	Robins seen.	Bluebirds seen.	Barn Swallows seen.	Currants Blossom.	Red Plums Blossom.	Plums and Cherries Blossom.	Crab Apples Blossom.	Common Apples Blossom.
1842.	March 13	March 13	May 2	May 11	May 14		May 27	May 29
1843.	April 12	April 12	" 3		" 15	May 17		" 26
1844.	March 21	March 25	April 25	April 25	Apr. 30	" 4	" 9	" 11
1845.	" 9	" 13	May 3			" 13		" 21
1846.	" 25	" 26	April 29	" 29	May 5	" 10	" 13	" 17
1847.	" 25	" 25	May 4	May 16	" 20	" 22	" 26	" 28
1748.	" 25	" 28	" 4	" 5	" 11	" 14	" 18	" 20
1849.	" 14	" 26	" 10	" 20	" 23	" 26	June 1	June 4
1850.	" 29	" 28	" 2	" 13	" 19	" 20	" 2	" 4
1851.	" 20	" 22	April 25	" 9	" 14	" 15	May 21	May 24
1852.	" 16	" 17	" 30	" 16	" 18	" 20	" 23	" 27

Of our migratory birds, the Bob-o-link, *Icterus agripennis*, is undoubtedly one of the most regular in its return in the spring. In my account of that Bird, Part I, p. 70, it is said to make its appearance in the latter part of May. But from observations since made, and from information derived from others, I am satisfied that its arrival in Vermont very seldom varies more than two or three days from the 12th of May.

Closing and Opening of Lake Champlain and Running of the Line Steamers.—
Continued from page 14, Part I.

YEAR.	Lake Closed.	Lake Opened.	Line boats commen'd running.	Line Boats stop'd.
1842	not clos.		April 13.	Nov. 29.
1843	Feb. 16.	Apr. 22.	April 27.	Nov. 30.
1844	Jan. 25.	Apr. 11.	April 19.	Nov. 29.
1845	Feb. 3.	Mar. 26.	April 9.	Nov. 29.
1846	Feb. 10.	Mar. 26.	April 13.	
1847	Feb. 15.	Apr. 23.	May 3*.	Dec. 2.
1848	Feb. 13.	Feb. 26.	April 8.	Dec. 2.
1849	Feb. 7.	Mar. 23.	April 10.	Dec. 8.
1850	not clos		April 15.	Dec. 9.
1851	Feb. 1.	Mar. 12.	April 7.	Nov. 28.
1852	Jan. 18.	Apr. 19.	May 3.	Dec. 13.

The closing and opening of Lake Champlain have reference to the broadest part of the lake opposite to Burlington. With the exception of 1835, this part of the lake has never become frozen entirely over earlier than the 15th of January, within the last thirty-six years. The mean time of closing for that period would fall on the 1st day of February. During four of the years it did not close at all. The narrower parts of the lake usually become frozen over so as to interrupt navigation, through its entire length, early in December, and most of the bays become covered with ice about the same time.

* Although the Line Boats commenced running so late as the 3d of May, they were for several days after that unable to proceed farther north than Plattsburgh, on account of the ice. It was not till the 6th that they were able to pass through the whole length of the lake, and then only by cutting through the ice for a distance of nearly six miles. The boats were not able to reach St. Albans till the 10th, and ice remained in many of the bays up to that time.

SHELBURNE BAY.

The following record, kindly furnished me by my friend Robert White, Esq., of Shelburne, exhibits the number of days, during which teams were able to pass upon the ice from Shelburne Harbor across the mouth of Shelburne Bay and the southeastern part of Burlington Bay, to Burlington, in each year since 1835.

Year.	Days passable.	Year.	Days passable.
In 1836,	76	In 1845,	12
1837,	81	1846,	36
1838,	47	1847,	67
1839,	61	1848,	16*
1840,	21	1849,	40
1841,	48	1850,	not passa.
1842,	24	1851,	46
1843,	56	1852,	82
1844,	67		

Lake Champlain Phenomena.

In Part I, page 14, something was said respecting the sudden disappearance of the ice from Lake Champlain in the spring of some years, and an attempt was made to account for the phenomenon, without having recourse to the absurd notion that the ice sinks. The explanations there given were founded, partly on observed facts, and partly on theoretic views. Additional observations have since been made, which, while they go to confirm the general theoretic principles, require some modifications of the results. It was there supposed that, when the general surface of the lake commenced freezing, the great body of the water below might be at a temperature $7°$ or $8°$ above the freezing point, and this, in accordance with the researches of Count Rumford, would doubtless be true were the waters gradually cooled down without agitation. But it is not found to be true in fact; and from recent observations it appears probable that, in consequence of their violent agitation by the cold winds which prevail in the early part of winter, the whole mass of water is cooled down very nearly to the freezing point before any ice is formed at the surface over the deeper parts of the lake, and that, after the waters are protected from the winds by a covering of ice, their temperature is gradually, but slowly, elevated by the reception of heat from the earth beneath. The following experiments show that the temperature of the water under the ice is, generally, some degrees above the freezing point, but not so much above as we had supposed.

On the 27th of March, 1844, when the lake had been covered with ice about eight weeks, at the distance of one-fourth of a mile from the shore, the temperature of the water was found to be, at the surface $32°$, at the depth of 6 feet $32\frac{1}{2}°$—at 12 feet $34\frac{1}{2}$ and at 25 feet $35\frac{1}{2}°$. On the 8th of March, 1852, when the lake had been frozen over 7 weeks, one-fourth of a mile from the shore, where the water was 28 feet deep, the temperature at the bottom was $34\frac{1}{2}°$, that at the surface being $32°$. On the 5th of April following, at the distance of one mile from the shore, the water being 82 feet deep, the temperature at the bottom was $34°$. At the distance of $2\frac{1}{2}$ miles from the shore, at an open crack where the water was 125 feet deep, the temperature at the bottom was $34\frac{1}{2}°$.

The sudden disappearance of the ice from Lake Champlain has been a subject of remark and speculation, from the first settlement of the country. But to a person, who carefully observes the circumstances, there will not appear any thing in the phenomenon either mysterious or very wonderful. In order to its occurrence, the temperature of the great body of water must be some degrees above the freezing point, the ice must be reduced to the *honey-comb* structure, or brought into a condition in which it will easily separate into minute divisions, and there must be a wind sufficiently strong to produce considerable agitation of the water.

In addition to theoretic objections to the popular notion that the ice sinks, when it disappears suddenly, persons of observation, who live near the lake, have occular proof that it does not sink. The ice, while yet spreading over the entire surface of the lake, is seen to be gradually wasting as spring advances and to become less firm, till, at length, it is so far disintegrated that a stick may be thrust through it, while it is yet from 6 to 12 inches thick. This disintegration is sometimes carried so far, before the general icy covering is broken up, that the ice has little more solidity or tenacity than snow saturated with water. In this state of things, a strong wind soon produces rents in the ice,—the waters, before pent up and quiet, are thrown into violent agitation, and the slightly cohering masses are actually seen falling to pieces and dissolving on the surface of the water. But it is never seen sinking, nor was any ever seen lying at the bottom after it had sunk.

Some have supposed that the sudden absorption of so large an amount of caloric, as would be required for the liquefaction of the ice, would produce severe frost in the neighborhood of the lake, which is not found to be true in fact. But this difficulty is removed by the consideration, that the heat employed in melting the ice, is derived

---

* The mouth of Shelburne Bay only.

rather from the water than from the atmosphere, and that the surface of the lake, in contact with the atmosphere, after the ice is all melted, is still warmer than the icy covering was before.

There is another phenomenon connected with the freezing of Lake Champlain, which is of some interest. At Rouse's Point, where the lake passes into Canada, and where it narrows down into the form of a river and some current is perceptible, it becomes strongly frozen over long before the broad lake closes; but very soon after the broader and deeper parts of the lake become covered with ice, the ice begins to fail at this place and in a measure disappears, even while the cold is severe and ice is forming in other places. To many, this phenomenon has appeared somewhat mysterious; but its explanation may, probably, be found in the circumstance that the lake at Rouse's Point is quite narrow and shallow and that the water which passes there, before the broad lake freezes, is the surface water and consequently the coldest water of the lake. This cold water, passing the Point in a shallow, scarcely perceptible stream, is soon cooled down and congealed at the surface, and the ice usually becomes strong here before the main body of the lake is frozen over. But soon after the broad lake closes over, the ice begins to waste at the Point and usually fails here soon after it becomes good elsewhere. This failure of the ice here, is owing to the circumstance that, after the lake is covered with ice, the water passing off here is no longer the cold surface water, but the warmer water lying below. It is this warmer water by its motion, though moderate, under the ice and in contact with it, which causes the ice to fail here, while it is increasing in other places.

In February, 1851, there was an occurrence in Windmill Bay, on the west side of Alburgh, which is worthy of note; the lake and bay being at that time covered with ice. On Saturday, Feb. 15, the wind blew quite hard from the south, and the snow thawed so that water ran in the roads. Saturday evening the wind came suddenly round to the west and blew for a short time with great violence. In the morning of the 16th something unusual was observed in the bay, and on going to it, it was found that the ice had been ruptured for the space of five or six rods each way, and that there were two immense blocks of ice lying upon the firm ice at some distance from the opening made by the rupture. The largest of these blocks was 39½ feet long, with an average width of about 26 feet. The other was thirty-eight feet long and 20 wide, and their thickness was 17 inches. The nearest of these blocks was 7½ rods from the opening and they were both the same side up as when they were lying upon the water. The depth of the water at the opening was 17 feet, and the sides of the blocks matched, in part, the margin of the opening.

Respecting the cause and manner of this occurrence, there were various conjectures; many supposing that it must have been effected by the exertion of some sudden force or explosion from beneath. But as a fall from the least elevation must have inevitably broken such masses of ice into innumerable fragments, it is evident that it was not thrown out by a force acting upward, but by a lateral force, which caused the masses to slide upon the surface of the undisturbed ice, and to be thus removed from their bed without being broken. The cause of this lateral pressure was probably the wind. While the wind was blowing from the south, a crack might have been opened and these large fragments loosened. When the wind came round and blew violently from the west, this crack might have closed suddenly and the broken pieces, not returning exactly to their former position, might have been thrown out with a force sufficient to cause them to slide to the position in which they were found, without being broken.

QUADRUPEDS OF VERMONT.
Additional to Part I, Chapter II.

To our previous list of Mammalia, we now add two living species, and two extinct fossil species. They are the following:

Ves. noveboracensis, N. Y. Bat.
Mus leucopus, White bellied Mouse.
Eleph. primogenius? Fossil Elephant.
Beluga vermontana, Fossil Whale.

Besides these, we have made additions to our account, of the following:

Felis concolor, Panther.
Phoca vetulina, Seal.
Sciurus hudsonius, Red Squirrel.

NEW YORK BAT.
Vespertilio noveboracensis.—LINN.
DESCRIPTION.—Head small; nose point-

ed. Ears broad, rather small; targus club-shaped. Interfemoral membrane broader than long, including the entire tail. This membrane is hairy above, but two-thirds naked beneath. Hind feet with five sub-equal toes, of which the outer are shortest. Brachial membrane naked above, excepting near the body and at the base of the phalanges: beneath, the hair extends farther from the body, and the patch at the base of the phalanges much more extensive. General color of the fur above, tawny red—beneath, the same, but much lighter. A whitish patch on the sides of the body at the base of the wings, most conspicuous on the under side. The brachial membrane is dark brown, beautifully reticulated with lighter color. Length of the specimen before me $4\frac{1}{4}$ inches, spread of the wings 12 inches.

HISTORY.—This Bat is less common in Vermont than several other species, and Vermont is probably near the limit of its northern range. According to Dr. DeKay, this is the most common species in the state of New York. Its range is from Massachusetts to the Rocky Mountains, and south through Pennsylvania. This Bat, from its red or ferruginous color, is very commonly called the *Red Bat*, and is figured under that name in Wilson's Ornithology. With the exception of the Hoary Bat, this is the largest bat found in Vermont, and in its measurements it scarcely falls short of the Hoary Bat, but its form is more slender.

For the specimen from which the preceding description is made, I am indebted to my friend C. S. Paine, of Randolph.

PANTHER, or CATAMOUNT, (Part I.-37.)
Felis concolor.—LINNÆUS.

DESCRIPTION.— Color of the face, head and all the upper parts of the body dark gray, slightly brushed with red. Interior of the ears, under side of the body and tail, and inner side of the legs grayish ash ; between the hind legs and beneath the tail tawny white. Exterior of the ears, bottoms of the feet and extremity of the tail black. Also a black patch on each side of the nose, from which the whiskers proceed, and the two connected together by a brownish band over the nose. Chin, lower lip and part of the upper lip clear white. Nose naked, of a brownish copper color, and narrowly margined with white hairs. Whiskers $2\frac{1}{2}$ inches long, white, intermingled with a few black hairs. Eyes oblique, with a whitish spot above and a little in front of each, and a smaller one below. Irides orange. Claws completely retractile, one inch long, very strong and sharp. of a pearly white color, having a blood red tinge on the under side near the base.

Dental Formula—Inscisors $\frac{6}{6}$ canines $\frac{1\ 1}{1\ 1}$, molars $\frac{4}{3}\ \frac{4}{3} = 30$. Teeth all clear white, perfectly sound, exhibiting no marks of wear. Incisors small, outer ones largest. Canines conical and strong, projecting 1.1 inch beyond the gum. The carniverous molars project $\frac{3}{8}$ths of an inch. Posterior molars in the upper jaw not fully developed. The weight of the specimen before me, which is a male, is 86 pounds.

Length, from the nose to the root of the tail, 48 inches. Length of the tail (vertebræ 29.5, skin and hair beyond 1.5) 31. Total length 79 in., or 6 ft. 7 in. Length of facial line, from nose to occiput, 10. Width of the head between the ears, posteriorly, 4.5, anteriorly 6, between the eyes 2.5. Height of the rounded ear 3.5. Length of the humerus 8, fore-arm 9, thigh 11, leg 12. Circumference of the wrist 7.5, fore paw 7.5, ankle 6, hind foot 7. Height at the shoulders 26, at the rump 27. Girt of the neck 16 inches, just behind the fore legs 27 inches.

HISTORY.—The Panther here described, was caught on the western slope of the Green Mountains, in the town of Manchester, Bennington county, on the 5th day of February, 1850. It was taken, by a Mr. Burritt, in a trap set for a bear. Being caught by one of its paws only, and being quite ferocious, it was not deemed prudent to attempt to secure him alive, and he was killed by shooting him through the body. It was purchased by the Hon. L. Sargeant and a few others in Manchester, who, with a public spirit and zeal for the advancement of science truly commendable and worthy of imitation, presented it to the Museum of the University of Vermont, where its skin and skeleton are now preserved. In taking off the skin, the head, neck and inner sides of the fore legs were found very much filled with Hedge-Hog quills, which, in many cases, had passed entirely through the skin and were deeply embedded in the flesh. The trap, in which it was caught, had not been visited for some time previous, and, from appearances, it was supposed to have been several days in the trap, when found ; and when shot it bled very profusely. Its weight was very much diminished by both these circumstances, and it was the general opinion, that, when first caught, its weight was not less than 100 pounds.

The teeth of this Panther were all perfectly sound and white, showing no marks of wear, and as the posterior molars in the upper jaw were not fully developed, there can be no doubt that it was a young animal, probably about two years old.

The Panther, above described, is the last and the only one which has been, to my knowledge, killed in Vermont for many years; and as the animal is now exceedingly scarce, and there may never be another obtained, within the state, for any of our museums, I have thought it advisable to be thus minute in its description and history, notwithstanding the full general account given in Part I—p. 37.

SEAL.
Phoca vitulina.—LINNÆUS.

In Part I, page 38, of my History of Vermont, will be found some account of a Seal captured on the ice on Lake Champlain in 1810. Another Seal was killed upon the ice between Burlington and Port Kent, on the 23d of February, 1846. Mr. Tabor, of Keeseville, and Messrs. Morse and Field, of Peru, were crossing over in sleighs, when they discovered it crawling upon the ice, and, attacking it with the but end of their whips, they succeeded in killing it, and brought it on shore at Burlington, where it was purchased by Morton Cole, Esq., and presented to the University of Vermont, where its skin and skeleton are now preserved. Before it was skinned I noted down the following particulars :

Total length of the Seal 50 inches; thickness just behind the fore legs 12 inches; weight 70 pounds. Length of the fore paw 7, nails 1¼, width 4 ; hind paw, length 8, nail 1, width 11, measured along the margin of the web, with the foot spread. Tail 3.5 inches long and 2 broad at the base ; hair on the tail reversed, forming a crown at the extremity. Nose truncated and somewhat notched, being 2.5 inches across the extremity. Whiskers numerous, and nearly white ; four erect, stiff and nearly white bristles, situated above and a little behind each eye. Distance between the eyes 2¼ inches.

The specimen was a female, having two abdominal mammæ situated thus (. $\cdot^{\text{navel}}_{\text{teats.}}$) The teats appeared rather like cavities than protuberances, and she was doubtless a female which had never suckled young.

Dental Formula—Incisors $\frac{6}{7}$ canines $\frac{1-1}{1-1}$, molars $\frac{5-5}{5-5} = 34$.

Lower incisors quite small. Upper incisors larger, (the two outward ones largest,) overlapping the lower ones, when the mouth is shut. Canines rather large and hooking inward. The molars are placed obliquely in the jaw; that nearest the canines smallest, and increasing backward in size and in the number of their sharp pointed tubercles. Its dentition resembles very closely that of the common cat.

Its hair was short, stiff, thick and even. Color of the hair brown olive and tawny white, forming a beautiful dark spotted marbling, lighter and more tawny on the belly. Base of all the hairs on the hind feet brown olive, with the tips slightly brushed with white, giving them a hoary appearance. Hair on the fore feet obscurely mottled.

At the time the above mentioned seal was taken, the lake, with the exception of a few cracks, was entirely covered with ice.

WHITE BELLIED JUMPING MOUSE.
Mus leucopus.—RICHARDSON.

DESCRIPTION—.Head moderately large, with the nose pointed. Eyes medium size. Ears large, rounded above, and naked, with the exception of a short down, which is whitish, along the margin. Auditory opening rather large. Whiskers turned backward, in part, longer than the head, some of the hairs black and some white. Fore feet with four claws and a rudimentary thumb, without nail. Hind feet with five toes, having feeble curved claws, nearly concealed by long white hairs. Tail slender, and slightly tapering. Incisors yellow. Fur fine, and rather long. *Color* above reddish brown, darkest along the back. The reddish brown extends downward on the shoulders and on the outside of the thighs, forming a band. All the under parts, from the chin to the extremity of the tail, including the feet and nails, pure white, excepting a narrow band of reddish brown under the base of the tail. Color of the fur, plumbeous, at its base. Length of the specimen before me, which is a male, measuring from the snout to the extremity of the tail, 7 inches ; head 1, body 2.7, tail 3.3, fore feet, 0.45, hind feet, 0.8, whiskers 1.5.

HISTORY.—This Mouse is a very delicate and beautiful little animal. It is exceedingly active, often leaping to considerable distances in the manner of the Deer-Mouse, but it has nothing of that Kangaroo form, or disproportion between the fore and hind legs, which exists in that species. It is most common in forests and wooded places, but frequents, also, meadows and cultivated fields, particularly where grain and grass-seed abound. It also enters barns and houses in quest of food and shelter. Two or three have been taken in a trap, in my own cellar, during the past year, and they are frequently brought in by cats, in the village of Burlington. It is found on both sides of the Green Mountains. I lately received two specimens from my friend, C. S. Paine, which were taken in Randolph. It is found in all the northern states, and as far north as Hudson's Bay.

WHITE SQUIRREL.

Sciurus hudsonius, (Albino).—P. I., p. 46.

November 11, 1850, I obtained an individual of the species commonly called the *Red Squirrel*, or *Chickaree*, which was entirely white. It was shot, in the top of a large tree, near the railroad bridge, between Burlington and Colchester. There were two of these white squirrels in company, but only one of them was captured. This one was a male, and, although its form was slender and delicate, it had every appearance of having been healthy and active. Its entire length, from the nose to the extremity of the hairs of the tail, 12.5 inches—to the extremity of the vertebræ of the tail, 11, to the root of the tail, 6; length of the head 2. Color of the hair entirely of snowy whiteness. Nails white, with a slight carnation tinge. Eyes nearly transparent, with a slightly smoky aspect, but in the dead animal, they exhibited scarcely any of that redness, which is regarded as the characteristic of *albinos*.

FOSSIL ELEPHANT.

Elephas primogenius?—BLUMENBACH.

It is a remarkable fact that, in making the Rutland and Burlington Rail Road, which extends from Burlington to Bellows Falls, two of the most interesting fossils ever found in New England, were brought to light. These were the remains of an Elephant and a Whale: the former were found in Mount Holly, in 1848, and the latter in Charlotte, in 1849.

The Rutland and Burlington Railroad crosses the ridge of the Green Mountains, in the township of Mount Holly, at an elevation of 1415 feet above the level of the ocean, and the bones of the fossil Elephant were found at that height. In order that their true position may be understood, and a knowledge of it preserved, the accompanying rude map has been prepared. The map embraces an area of about 35 acres, lying at the summit level of the Green Mountains, over which the railroad passes.

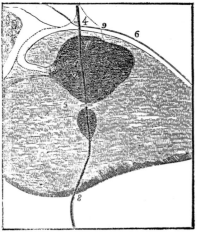

4. Station House.   6, Road.
Dotted lines, outlets of muck bed.   7. Division of water.

More than half of this area consists of a solid mass of rock, elevated considerably above the ground on each side, and only slightly covered with soil, or earth, excepting the cavities indicated, which are filled with vegetable muck. The line on the map, marked 5, denotes the ridge, which, previous to making the railroad, divided the waters flowing into Connecticut river from those falling into Lake Champlain. The cut, for the railroad, through this mass of rock, (from 4 to 8 on the map,) is about 180 rods in length, and from 12 to 35 feet deep. The muck beds are formed in basins excavated out of the rock. The larger basin appears to have been originally filled with water, and to have been a favorite resort for beaver, a large proportion of the materials which formed the lower part of the muck, consisting of billets of wood, about 18 inches long, which had been cut off at both ends, drawn into the water and divested of the bark, by the beavers, for food. When first taken out, the marks of teeth upon the wood were as distinct as if they were the work of yesterday. At 3, the outlet of the basin, the beavers had constructed a regular dam for the purpose of deepening the waters within. But at the time the excavation for the railroad was made, the basin had become entirely filled with vegetable matter, which was in parts 15 feet deep, and its surface was a swamp, on which plants, shrubs, and small trees were growing. The billets of wood, which the beavers had brought in, were, many of them, three inches in diameter, and were of several kinds, as ash, willow and alder.

FOSSIL ELEPHANT.

These, together with numerous cones of black spruce and white pine, in a good state of preservation, were embedded in a thick vegetable mucilage, nearly resembling clay in color, but which, when cut in cakes and taken in the hand, would shake and tremble like a mass of jelly. A cake of this mucilage, when dried, was much lighter than cork, and was diminished to about one-eighth of its original bulk. The mucilage was undoubtedly produced by the solution of leaves and wood, which had steeped for ages in that basin of cold water, from which there was not a sufficient flow to carry it off. The billets of wood, when taken out from the bottom of the muck, appeared plump and fresh, as if they had been recently pealed, but were very soft, and in drying, they lost full five-sixths of their bulk.

In making the excavation for the railroad, through the muck-bed above described, in the latter part of the summer of 1848, the workmen found, at the bottom of the bed, resting upon gravel, which separated it from the rock below, a huge tooth, the place of which is indicated on the map and Cut by 1. The depth of the muck at that place was 11 feet. Soon afterwards, one of the tusks was found, at 2, about 80 feet from the place of the tooth, above mentioned, which was a grinder. Subsequently the other tusk, and several of the bones of the animal were found near the same place. These bones and teeth were submitted to the inspection of Prof. Agassiz, of Cambridge University, who pronounced them to be the remains of an extinct species of Elephant. The Directors of the Rutland and Burlington Rail Road, to whom they belong, design to have them placed in the Museum of the University of Vermont, for preservation, and for the illustration of our fossil geology.

The form of the cut through the rocks and the muck, and the position of the fossils, may be seen in the accompanying section.

1. Grinder. 3. Original dividing ridge.
2. Tusks.   4. Present division of water.

The grinder is in an excellent state of preservation, and weighed 8 pounds, and the length of its grinding surface is about 8 inches. The tusks are somewhat decayed, and one of them badly broken. The chord, drawn in a straight line from the base to the point, of the most perfect tusk, measures 60 inches, and the longest perpendicular, let fall from that to the inner curve, of the tusk, measures 19 inches. The length of the tusk, measured along the curve on the outer surface, is 80 inches,

FOSSIL WHALE.

and its greatest circumference, 12 inches. The circumference has diminished very much since the tusk was taken from the muck bed, on account of shrinkage in drying, and several longitudinal cracks have been formed in it, extending through its whole length, and it was found necessary to wind it with wire to prevent it from splitting to pieces.

These are believed to be the only fossil remains ever found in New England, which have been, with certainty, ascertained to belong to an Elephant. Remains of Elephants have been found in several of the southern and western states, and very recently some fine specimens have been dug up in Ohio.

I have prefixed to this account the specific name of the Mammoth, or fossil Elephant of Europe, but have little doubt that ours is a distinct species, and I am happy in knowing that one of our best comparative anatomists is now investigating this very subject.

FOSSIL WHALE.
Beluga vermontana.—THOMPSON.

As many rare fossils are rendered nearly valueless by the want of an accurate knowledge of their localities, and of the circumstances in which they were found, I have deemed the above mentioned fossil, which is undoubtedly the most interesting of the organic remains yet found in Vermont, of sufficient importance to justify a minute history of its discovery and position, and the introduction of a small map of the locality. The discovery of this fossil took place in August, 1849. While widening an excavation for the Rutland and Burlington Rail Way, in the township of Charlotte, the workmen struck upon a quantity of bones, which were embedded in the clay at the depth of about eight feet below the natural surface of the ground. Some of the Irishmen remarked that they were the bones of a dead horse buried there; but little notice, however, was taken of them, till the overseers observed something peculiar in the form of several of the bones, and were, thereby, induced to examine them more carefully. It was soon found that the bones discovered, belonged to the anterior portion of the skeleton of some unknown animal, the head of which had already been broken into fragments, by the workmen, and many of the fragments carried away with the earth, which had been removed. On carefully removing more of the clay, a number of vertebræ were found, extending in a line obliquely into the bank, and, apparently arranged in the order in which they existed in the living animal. These vertebræ were taken out, and, together with the sternum, fragments of the head, ribs, &c., were forwarded to Burlington, and, by the kindness of Messrs. Jackson & Boardman, engineers on the railroad, were placed in my hands.

By a careful examination of these bones, I found that they belonged to some animal, with whose skeleton I was not acquainted, and that there were wanting, in order to complete the skeleton, the greater part of the head, all of the teeth, a considerable number of vertebræ and ribs, and the bones of the limbs. I was at first in some doubt, whether the animal belonged to the whale family or to the saurian; but this doubt was soon removed, by a careful examination of the caudal vertebræ. These were found to have their articulating surfaces convex, and rounded in such a manner as to allow of a very extensive vertical motion of the tail, but of a very limited lateral motion. This arrangement plainly indicated that the movements of the animal in the water, were effected by means of a horizontal caudal fin, and that it, therefore, belonged to the family of *Cetacea*, or Whales.

The manner in which these caudal vertebræ move upon each other may be seen in the cut, where Fig. 1 represents the 13th, 14th and 15th vertebræ of the tail,—*a*, as they appear viewed from above—*b*, as seen laterally*.

After having carefully removed from the bones, I had received, the adhesive clay, in order to prevent their crumbling by exposure to the air, and secure their preservation, I saturated them with a thin solution of animal glue, and then proceeded to Charlotte in order to recover, if possible, the bones, which were missing. By spending several days in the search, I succeeded in obtaining most of the anterior portion of the head, nine of the teeth, and thirteen additional vertebræ, together with the bones of one forearm, several chevron bones, and portions of ribs. From the portions of the head, which I obtained, and the fragments previously received, I was able to reconstruct so much of the upper and anterior portion of the head, as to exhibit distinctly its *spiracles*, or blow-holes, showing unequivocally that it belonged to the Whale family. My next object was to ascertain, if possible, whether it was a living, or an extinct, species of this family. Being without specimens for comparison, my only reliance for aid was Cuvier's great work on Fossil Bones. By a comparison of the Fossil Whale with the descriptions and figures in that work, it was found to resemble the living rather than the extinct types, and that the osteology of the head was very like that of the *Beluga leucas*, or small northern White Whale.*

Having collected together all the bones and fragments of the Fossil, within my reach, I proceeded with them to Cambridge, Mass., and submitted them to the inspection of Prof. Agassiz, who confirmed the opinion I had formed respecting them, and, for two days, very kindly lent me his aid, and his great skill and knowledge of the subject, in their collocation and arrangement. Having, all together, more than four-fifths of the bones of the skeleton, he was able, from the number, position and size of these, to determine the number, position and size of those, which were missing, and thus to determine the size and form of the whole animal.

The head of the skeleton, as already remarked, was broken into a great number of pieces, and only a portion of the fragments recovered; but enough to determine its entire length and general form. Fig. 2 represents the head, as reconstructed out of the fragments, viewed from above; and fig. 3, a side view, with the lower jaw dropped a little below its true place. The entire length of the head is 21.2 inches. The maxillary bone on the left side is mostly wanting, but on the right side, it is entire, so far as to embrace the alveolar margin, which is 6.85 inches in length, and perforated for 8 teeth. The corresponding alveolar margin of the lower jaw measures 5.5 inches, and is perforated for 7 teeth. Hence it appears that there were 16 teeth in the upper jaw and 14 in the lower, making 30 in the whole.

The teeth are all of one kind, being conical, with flat or rounded crowns, much worn, but, in their substance, very dense and firm. They are from one to two inches in length, with a diameter of half an inch. Fig. 4 represents their different forms and sizes. Only nine of the teeth were recovered, and none of those were in their places when found; but, that they were in their places, up to the time the bones were first discovered, is evident, from the fact, that, while every other cavity in the bones was filled with clay, the alveoli were all empty.

Of the vertebræ, 41 were secured, of which four were cervical, eleven dorsal, ten lumbar, and sixteen caudal. Three cervical vertebræ, the first, fifth and sixth, are evidently missing, which, with the four obtained, make *seven*, the usual number. These vertebræ are all free, not being soldered together, as in the common dolphin,

---

* The fractions after the number of the figure, when introduced in the accompanying cuts, denote the linear proportion of the cut to the object, which it represents.

* Cuvier's Osse. Foss., Vol. V, page 299 and Plate XXII, fig. 5 and 6,—Paris edition, 1825.

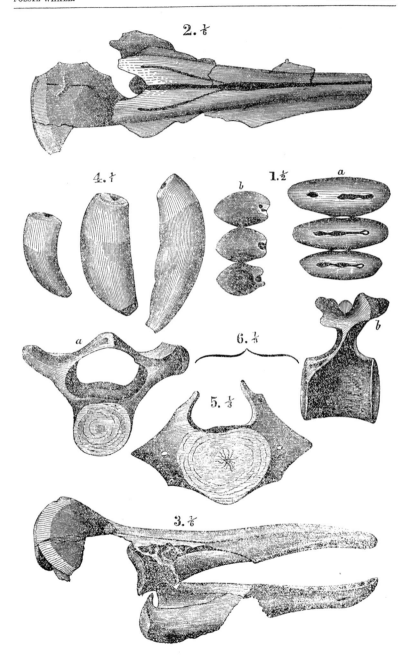

and some other cetaceans. Fig. 5 represents the third cervical vertebra.

The second and twelfth dorsal vertebræ are missing, the whole number being thirteen. Fig. 6, represents the seventh dorsal vertebra—*a*, as seen from behind—*b*, as seen laterally.

The lumbar vertebræ amount to twelve, of which the sixth and twelfth are missing. Fig. 7, represents the seventh lumbar vertebra. These vertebræ all have the same general form, but the lateral winged processes are more decayed and broken in some of them, than in the one here represented.

The eleventh and seventeenth caudal vertebræ are missing, and perhaps a nineteenth and twentieth, making the, probable, whole number, twenty. Fig. 8, represents the fourth caudal vertebra. The form of those nearer the extremity of the tail may be seen in fig. 1.

From these statements, it will be seen, that the whole number of vertebræ in the skeleton was 52, eleven of which are missing. Two of the missing vertebræ were known to have been taken away, after the bones were discovered. Articulating surfaces, at the meeting of the caudal vertebræ, indicate five chevron bones, of which the fourth only is wanting. Fig. 9, represents the second chevron bone.

The total length of the vertebral column, due allowance being made for the eleven missing vertebræ, and 17 inches for the aggregate thickness of the 51 intervertebral cartilages, is one hundred and thirty-seven inches. Of this length, the cervical vertebræ make 10 inches, the dorsal 40, the lumbar 48, and the caudal 39. The lumbar vertebræ are largest, having an average length of about 4 inches, with a diameter of 3 inches. The total length of the animal, including the head and caudal fin, must have been about 168 inches, or 14 feet.

Fig. 10, is the *hyoid* bone, and Fig 11, the *sternum*, both of which are large and strong, in proportion to the size of the skeleton. The former measures 8.5 inches in a right line, from point to point, and the latter is 15 inches long, from 3½ to 7 wide, and on an average about one inch thick.

The ribs are considerably decayed and broken. The longest entire rib measures just 24 inches along the curve. Fig. 12, represents the anterior rib, on one side. It is very strong, consisting of two portions, of nearly equal length, of solid bone.

Fig. 13, represents the scapula, the humerus and the bones of the fore-arm of the left fin, in their connexion. The scapula and the ulna of the right side were recovered, but all the other bones of the paddles are wanting. The height of the scapula is 7 inches; the length of the humerus 5, and of the fore-arm 4 inches.

I was able to obtain the following measurements of the head, which admit of direct comparison with a part of the measurements, given by Cuvier, of the head of *Beluga leucas*:

	B. vermontana.	*B. leucas.*
Length of the head, from the occipital condyles to the end of the snout,	21.2 inch.	20.9 inch.
" of one side of the lower jaw,	16.5 "	16.5 "
" of alveolar margin, "	8.2 "	7.8 "
" of the symphysis, "	3.1 "	3.1 "

Between these measurements, it will be seen that there is a very close agreement; but they disagree in their dental formulæ, as expressed below:

$$\text{Dental Formulæ,} \quad B.\ vermontana. \ \frac{8}{7}\ \frac{8}{7} = 30: \quad B.\ leucas. \ \frac{9}{9}\ \frac{9}{9} = 36.$$

They also differ in the relative width of the maxillary and intermaxillary bones, as developed on the upper side of the snout, and also in the outlines of the head.

Since the above measurements and comparisons were made, I have had an opportunity to examine the bones of three heads of *B. leucas*, in the Hunterian Museum, in London, and an entire skeleton of the animal in the collection of Prof. Agassiz, at Cambridge, Mass. On account of the absence of Prof. Agassiz, when I visited Cambridge, a minute comparison of my fossil bones, with the corresponding bones of his skeleton, was not gone into, but a sufficient number of bones was compared, to leave little doubt that they belong to different species of the same genus. I have, therefore, described my Beluga under the specific name of *vermontana*, which I gave it, provisionally, in my first account of the fossil*.

LOCALITY.—In order to prevent any doubt, hereafter, in regard to the precise place in which these fossil bones were found, I have here introduced a little map of the township of Charlotte, on which I have marked the locality by a black ▬. The township is six miles square, and bounded on the west by Lake Champlain. The single lines denote the principal roads passing through the township. The railroad passes through it, from north to south, nearly parallel to the lake shore, and at an average distance of 1½ mile from it. The distance between the two roads, which cross the railroad, one on the north and the other on the south side of the locality, is about 80 rods; the distance to the locality, from the north road, being perhaps 25 rods, and from the south road, 55 rods. The northern road crosses the railway on a bridge, over the excavation, elevated about 16 feet above

---

* Silliman's Journal of Science, Vol. IX, p. 256.

NATURAL HISTORY. 239

MAP OF CHARLOTTE.   FOSSIL WHALE.

MAP OF CHARLOTTE.

the track: the southern road crosses on a level with the track. The accompanying cut exhibits a section along the east side of the excavation, in which the bones were found. The surface of the ground slopes to the south, and, to the depth of four feet, consists principally of sand, showing no signs of stratification. Next, below this, is a mixture of sand and clay, finely and regularly stratified, for a depth of 2½ feet, below which is a vast bed of fine blue clay, in which were observed no signs of stratification, and which appears to have been, previous to the deposit of the stratified sand and clay above it, an extensive quagmire.

SECTION.

A   D   B   c

A and C denote the points where the two roads, above mentioned, cross the railroad; A the northern road, and C the southern, and the line A C the distance between the roads. From C to B, the railroad track is on the level of the natural surface of the earth; and from B to A at the bottom of the excavation in the clay bed. D indicates the point in the line of the road, where the fossil bones were found.

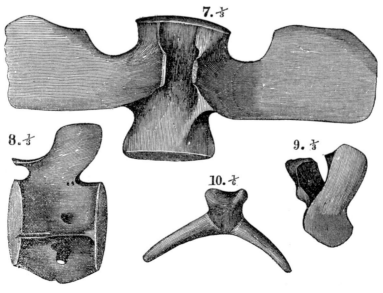

7. ⅛

8. ⅛   9. ⅛

10. ⅛

The fossil bones were embedded in this clay, at an average depth below its surface of nearly two feet. The head of the skeleton was towards the northwest, was lowest, and was nearly on a level with the railway, while the posterior parts extended obliquely into the bank, towards the southeast. In the blue clay, with the bones, were found some vegetable remains, and also specimens of *Nucula* and *Saxicava*. At the surface of the blue clay were great numbers of *Mytilus edulis* and *Sanguinolaria fusca*, and the latter were scattered through the stratified sand and clay above. The locality, as ascertained by the railroad survey, is 60 feet above the mean level of Lake Champlain, and 150 above the ocean.*

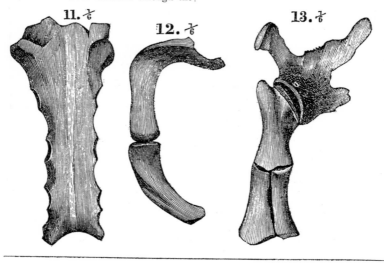

BIRDS OF VERMONT.

Additional to Part I, Chapter III.

To our list of Birds given in Part I, page 53, we now add the following species:

Tyrannus cooperi,	Olive sided King Bird.
Muscicapa traillii,	Traill's Flycatcher.
" *ruticilla*,	American Redstart.
" *pusilla*,	Green Blackcap Warbler.
Vireo gilvus,	Warbling Vireo.
Merula olivacea,	Olivebacked Thrush.
Sylvia striata,	Black-poll Warbler.
" *ruficapilla*,	Red-poll Warbler.
" *pardolina*,	Canada Warbler.
" *parus*,	Hemlock Warbler.
" *philadelphia*,	Mourning Warbler.
" *americana*,	Particolored Warbler.
Fringilla borealis,	Mealy Redpoll.
Coccoborus ludovicianus,	Rosebrested Grosbeak.
Tanagra rubra,	Scarlet Tanager.
Picus pileatus,	Crested Woodpecker.
Ardea minor,	American Bittern.
Tatanus melanoleucas,	Greater Yellow Shanks.
Tringa semipalmata,	Semipalmated Sandpiper.
Colymbus septentrionalis,	Red Throated Loon.

We also make additions to our former account, of the following:

Falco chrysaetas,	Golden Eagle.
Cypselus pelasgius,	Chimney Swallow.
Columba migratoria,	Passenger Pigeon.

THE GOLDEN EAGLE.

Falco chrysaetas.—LINN.

In May, 1845, two eagles of this species were observed flying near the summit of a high hill, in Pittsfield, in this State. One of these was shot and wounded. It flew about half a mile and pitched down into a thick forest, but could not then be found. About a week afterwards, it was discovered and captured. It was confined in a stable, fed on meat, and kept there more than a year. It was then sent to Middlebury, to Prof. C. B. Adams, who, on the 23d of Oct., 1846, sent it to me, at Burlington. I kept it in an open cage, or coop, in the corner of my yard, through the winter, and watched its conduct with much care. It was a female, and was, in her disposition, very savage; and during the 7 or 8 months I kept her alive, I made very little progress towards taming her. She would strike

* In 1847, a portion of the skeleton of a whale, was found in the same kind of clay, as that in which the bones were found in Charlotte, in the vicinity of Montreal. It was found about 15 feet below the surface, in digging clay for making bricks, and was about 100 feet above the level of the St. Lawrence. The portion found, consisted of 19 consecutive vertebræ, which measured, all together, when arranged in their order, 4 1-2 feet. About one-third of the vertebræ were caudal, the other two-thirds sacral and lumbar. These fossil bones were carried to London, by Mr. Logan, Provincial Geologist, where I had an opportunity of comparing with them some of the vertebræ of *B. vermontana*, at the Museum of the Geological Survey of Great Britain, and I have little doubt that they are identical in species.

with her feet with the quickness of a cat; and, after having had a piece of meat snatched from my hand so suddenly that I scarcely had a glimpse of the claws that took it, I thought it most prudent to keep my hands beyond her reach. She was most fond of meat when first killed, but, if hungry, she would not reject it after it had become putrid. When a hen, or dove, was killed and thrown into her cage, she would suddenly pounce upon it, striking in her claws with great force. She would then stretch up her neck and look around, as if exulting over the victory she had achieved. Before eating a particle of the fowl, she would take it to her roost, and holding it with one foot, she would pick it, with her beak, as cleanly as if it were to be cooked.

The length of this specimen was 33 inches, the spread of her wings 72, the folded wing 24, tarsus 3.5, tibia, 6.5, tail, consisting of 12 feathers, 10.5. It weighed, when killed, on the 19th of May, 1847, 10.5 pounds, the greater part of the weight being made up of the muscles of the wings and legs.

NOTE.—The specimen, belonging to the Museum of the University of Vermont, which I have described in Part I, page 60, without a name, I have satisfied myself to be an old female Bald Eagle.

THE OLIVE-SIDED KING-BIRD.

Tyrannus cooperi.—NUTTALL.

DESCRIPTION.—General color above, dark olive, becoming dusky brown on the head, wings and tail. Chin, throat, belly, and under tail coverts, white, tinged with light greenish yellow. Secondaries edged with white, and the wing coverts tipped with gray, giving the appearance of two obscure bars on the wings. Breast and sides of the belly, brownish, with an irregular yellowish white band from the throat down the breast to the belly. Legs and feet black. Upper mandible, blackish horn color; lower, yellowish, darker at the point. Irides hazel. Bill stout and broad. Second quill longest, first and third equal. Tail emarginate, extends one inch beyond the folded wings. Length, 6.5 inches; spread, 12.5.

HISTORY.—This species was first distinguished from the Wood Pewee, which it much resembles, by William Cooper, in 1829, and was described and named, in honor of its discoverer, by Mr. Nuttall. Its range is from Texas to the 53d parallel of latitude, and from New England to Oregon. It is a rare bird in New England, but numbers of them spend the summer and rear their young here. For the specimen, from which the above description was made, I was indebted to my friend C. S. Paine, who shot it in Randolph, in this state. The nest of this King Bird is usually built in the top of an evergreen, from 30 to 50 feet from the ground, and resembles, somewhat, that of the common King Bird. The eggs, four in number, are of a yellowish cream-color, thinly sprinkled with dark brown and purple spots. A nest, found by Mr. Paine, was on a horizontal branch of a tall hemlock, standing alone in a pasture, near the border of woods. The nest, containing three eggs, was composed of twigs, moss, and a few blades of grass. It was very flat, and slovenly put together. This bird manifests much uneasiness and anger, when its nest is approached, erecting its crest, and becoming very clamorous. These birds are known to breed, in the same locality, several years in succession.

TRAILL'S FLY-CATCHER.

Muscicapa traillii.—AUD.

DESCRIPTION.—Color of the head and body above, dark glossy olive green; circle round the eye and streak towards the bill, pale yellow. Wings, dark hair brown; secondaries and wing coverts edged with dull white, forming two bars across the wings. Bill, blackish above, flesh-colored beneath. Chin and throat yellowish white; breast, ashy brown; belly, and under tail coverts, pale sulphur yellow. Legs black. Tail emarginate. Length, 5.75; spread 8.75.

HISTORY.—This species bears a very strong resemblance to the *M. flaviventris.* It is quite a common bird at some places along the east side of the Green Mountains, in Vermont, particularly along the second branch of White River, in Bethel and Randolph, where, I am informed by my friend, Paine, it rears its young in large numbers. Its nest is usually built in a low bush, by the side of a stream, from one to four feet from the ground. The nest is composed, outwardly, of wild grass and wool, and lined with very fine grass and weeds. It is very snugly put together, and nearly two inches deep. The eggs, usually three, are of a yellowish white color, sparsely sprinkled with light umber toward the larger end.

THE AMERICAN REDSTART.

Muscicapa ruticilla.—LINNÆUS.

DESCRIPTION.—Upper parts, bill, chin and breast, black; sides of the breast, base of the primaries and of the tail feathers, excepting the two middle ones, fine reddish orange, sometimes approaching scarlet. Belly white. *Female* and *young* olive brown above, head cinereous; beneath, yellowish white. Sides of the breast, base of the quills and tail feathers, yellow, where they are orange in the male. Bill and legs smoky olive. Notch in the bill small. Second and third primaries subequal and longest. Length, 5 inches; spread, 6.5.

HISTORY.—This beautiful little bird is found in all parts of the United States, and extends its summer migrations as far north as the 62d parallel of latitude. It arrives in Vermont about the middle of May. It is a shy and retiring bird, confining itself to the forests and groves. It builds its nest, usually, between the forked branches of a small tree, or sapling, 10 or 12 feet from the ground. The nest is very securely and neatly built, being made of fibres and shreds of bark, very firmly agglutinated together. The eggs, 3 or 4 in number, are of a light cream color, thickly sprinkled with different shades of yellowish brown, particularly towards the large end.

THE GREEN BLACK-CAP WARBLER.
Muscicapa pusilla.—WILSON.

DESCRIPTION.—Crown, glossy black; back, rump, and upper tail coverts, olive green. Frontlet, line over the eye, and all beneath, bright lemon yellow. Tail and wings, hair brown, the feathers having their outer vanes edged with yellow olive. Bill, brown; legs, flesh color. No part white. Second and third quills longest, subequal. Tail long and rounded, reaching more than one inch beyond the tips of the folded wings. *Female* and *young* without the black crown. Length, 4.5 inches; spread, 6.5.

HISTORY.—This species appears to be very widely diffused, being found in Labrador, in latitude 58° north, where it breeds, and as far westward as Columbia river. According to DeKay, it arrives in New York early in May, but is in that state exceedingly rare. Their nests are built on low bushes, in which they lay about four grayish eggs, which are sprinkled with reddish dots, in a circle around the larger end.

THE WARBLING VIREO.
Vireo gilvus.—BONAPARTE.

DESCRIPTION.—Pale greenish olive. Head and upper part of the neck, dark ash, approaching to brown. Line over and beneath the eye, and extending from the eye to the nostril, light ash. Wings and tail, hair brown, the feathers edged with greenish gray. Upper mandible dark horn color, the lower one lighter. Chin and under parts dull greenish white. Sides of the body and beneath the wings, dull greenish yellow. Legs, bluish brown. Length, 5 inches; spread, 8.

HISTORY.—This Vireo, though plain and unadorned in plumage, is one of the most musical of the feathered warblers. Its range is through the whole extent of the United States, from the Atlantic to the Pacific. The nest of the Warbling Vireo is usually pendulous, and placed in the very summits of the loftiest trees. Nuttall has found them elevated to the airy height of more than 100 feet from the ground. The nest is made of the fibres of weeds and shreds of bark, and lined with grass. The eggs are from 4 to 6, white, with confluent spots and thread-like lines towards the larger end. My friend, C. S. Paine, who kindly favored me with a specimen of this bird, and its nest, which was built in the top of a lofty elm, near his dwelling, in Randolph, assures me that the favorite resort of the Warbling Vireo is among the trees and bushes, growing by the side of ponds and streams.

THE OLIVE-BACKED THRUSH.
Merula olivacea.—GIRAUD.

DESCRIPTION.—General color above, yellowish olive brown; breast and throat buff, spotted with black; belly, soiled white, spotted with reddish brown. Bill, brown, short and robust. Legs, flesh color, line from the bill to the eye slightly rufous. Outer webs of the primaries, yellowish. Lower tail coverts, white. Second primary longest, first and third nearly equal.—Length, 6.5 inches; spread, 11. Length of the tarsus 1 inch. Tail extends 1 inch beyond the folded wings.

HISTORY.—This species was long regarded as a variety of the Hermit Thrush, *M. solitaria.* It was first shown to be a distinct species by Mr. DeRhane, and was first described and named by Mr. Giraud, in his Birds of Long Island. This bird rears its young in Vermont, and is not very rare. Its nest, which is built on the ground in the woods, is composed of leaves and vines, and lined with fine roots. It has 4 light blue eggs, but somewhat darker than those of the Hermit Thrush, or very nearly the color of those of the Robin. It probably rears two broods in a season, since my friend Paine assures me, that he has found their nests, containing eggs, in the months of June, July and August.

THE BLACK-POLL WARBLER.
Sylvia striata —WILSON.

DESCRIPTION.—Head deep glossy black; back mottled with black, white and dark ash. Cheeks, collar round the neck, and under parts generally, white, largely spotted with black on the breast and sides; line of black spots from the chin towards the shoulders. Two white bars on the wings. Primaries brown, edged on their outer webs with greenish yellow. Tail, wood brown, the two outer feathers on each side having large white spots on their inner webs. Bill, dark horn color. Tail emarginate, reaching three-fourths of an inch beyond the folded wings. Legs, flesh color. *Female* and *young* dull yellowish olive, streaked with black and gray. Length 5 inches; spread, 8.

HISTORY.—The Blackpoll Warbler is pretty generally diffused over the United States, and has been observed as far north as the 54th parallel of latitude. Audubon found the nest of this species in Labrador, built in the forked branches of a fir tree, about three feet from the ground. It was formed of mosses and lichens, lined first with coarse dried grass, then with fine moss, and lastly with feathers. The nest contained 4 eggs, but he has given no description of them. It probably breeds in Vermont, but I am not aware that its nest has ever been found here.

THE RED-POLL WARBLER.
Sylvia ruficapilla.—LATHAM.

DESCRIPTION.—General aspect brownish olive, streaked with dusky brown; crown dark rufous. Line over the eye, and all beneath, yellow. The two lateral tail feathers with large spots of white on their inner webs, extending to their tips. The yellow on the breast streaked and spotted with bay. Legs and bill dusky brown. The first three quills nearly equal, second longest. Tail slightly notched, and reaches one inch beyond the folded wings. *Female* without the rufous crown, and having the spots on the breast brown instead of bay. In the young male the crown is spotted with bay, and the breast yellowish brown. Length, 4.75; spread, 7.5.

HISTORY.—The history of this little warbler appears to be very little known. I have two specimens, a male and a female, from which the above description is drawn. They were both shot by my friend Paine, in Orange county, in 1848, one on the 20th of April and the other in September. It has been observed, according to DeKay, from Mexico to the 55th degree of north latitude. Whether it breeds or not in Vermont, I have not been able to ascertain.

THE CANADA WARBLER.
Sylvia pardalina.—BONAPARTE.

DESCRIPTION.—All the upper parts bluish ash, with central parts of the feathers on the head, black, giving it a dark spotted appearance. Wings and tail brown, edged with grayish. Line under the eye descending down the side of the throat towards the shoulders, black. Spot in the forehead, a broad line towards the eye, and all beneath, bright lemon yellow. A broad rounded band of black spots across the breast, forming a sort of collar. Under tail coverts white, tinged with yellow. Upper mandible brownish: the lower mandible, the legs and feet, flesh color. Second and third primaries subequal, longest. Tail long, rounded, reaching 1.2 beyond the tips of the folded wings. The female is greenish above, and all its markings less distinct. Length, 5 inches; spread, 8.5.

HISTORY—This is a rare species, being only occasionally met with in Vermont. It breeds, according to Audubon, in Pennsylvania, Maine, and the British Provinces, and if so, it doubtless breeds in Vermont, though I am not aware that its nest has been found here. It is said to range as far north as the 55th degree of latitude. The nest is usually built in a low evergreen. The eggs, about five in number, are white, with a few dots of brownish red.

THE HEMLOCK WARBLER.
Sylvia parus.—WILSON.

DESCRIPTION.— Color above greenish yellow, striped with dusky; bill, wings and tail brownish black; two white bars on the wings; quills edged with greenish. Line over the eye, throat and neck yellow; beneath, yellow, streaked with dusky on the breast and sides; under tail coverts white; patches of white on the inner webs of the two outer tail feathers; legs and under mandible greenish yellow. First quill longest; tail emarginate. Length, 5.25; spread, 8.5

HISTORY.—This bird resides, for the most part, in thick Hemlock forests, and hence it has derived its name. Its nest, according to Audubon, is usually built in a hemlock or spruce, at a considerable elevation from the ground, and is composed of slender twigs and lichens, and lined with hair and feathers. The specimen above described was shot in Randolph, and the bird, no doubt, breeds here.

THE MOURNING WARBLER.
Sylvia philadelphia.—WILSON.

DESCRIPTION.—Head and sides of the neck bluish slate; upper parts of the body, wings and tail, dark yellowish olive-green; space before the eye, and frontlet, black.

PARTI-COLORED WARBLER. MEALY REDPOLL. ROSE-BREASTED GROSBEAK. SCARLET TANAGER.

Chin, throat and sides of the neck bluish gray. Breast black, with numerous fine crescent-shaped blue-gray lines. Beneath bright lustrous yellow. Bill smoky horn color; legs flesh color. In the female and the young, the throat and breast are buff, the latter much the darkest, and all the upper parts are a greenish olive. Length, 5 inches; spread, 7.5.

HISTORY.—The Mourning Warbler derives its name from its peculiar melancholy notes. The specimen, from which the above description is chiefly drawn, was shot by my friend, C. S. Paine, in Randolph, on the 4th of July. It was a male, had with it a mate and a brood of young ones, just able to fly. This warbler is a rare bird, and is of shy and solitary habits. Its range, so far as at present ascertained, is between the 23d and 47th parallels of latitude.

THE PARTI-COLORED WARBLER.
Sylvia americana.—LATHAM.

DESCRIPTION,—Color pale blue above, with a large golden umber spot on the back. Upper mandible black; lower, yellowish. Chin, throat and lower part of the breast, bright yellow. A blackish collar, bordered below with umber, mixed with yellow. Sides, under the edges of the folded wings, spotted with bay. Belly bluish white. Two white bars on the wings; and outer tail feathers largely spotted with white, on their inner webs. Wings and tail brown, the quills and feathers edged with light blue, on their outer webs. Legs and feet fuliginous. Three first quills nearly equal. *Female* without the dark collar on the breast. Length, 4.5, spread, 6.4 inches.

HISTORY.—This very beautiful little warbler ranges from Mexico to the 46th parallel of latitude, and is very common in the western states. It arrives in New England about the beginning of May. Its nest, according to Audubon, is built in the upright forks of small trees, and is composed principally of lichens, lined with downy substances. The eggs, about 4, are white, with a few reddish dots near the larger end.

THE MEALY REDPOLL.
Fringilla borealis.—SAVI.

DESCRIPTION.—Above dusky, streaked with yellowish white and rusty. Wings and tail, hair-brown, the feathers edged and tipped with yellowish white. Rump whitish. Crown dark rich crimson. Frontlet, lores and throat black. Beneath, grayish white, streaked with dusky. Legs, feet and nails black. Cheeks, sides of the body and posterior part of the rump, in the male, pale carmine. First primary longest, second and third nearly equal. Bill yellow, brownish towards the point; very acute, upper mandible longest. Hind nail long as the toe. Length, 5.5 inches; spread, 9.

HISTORY.—This species, though very rare, is quite widely diffused, being found in Maine, New Jersey and Oregon. The specimen from which the above description was made, was shot in Randolph, in the winter of 1850. They appeared there in flocks, and fed upon the seeds of weeds, which projected through the snow, in the open fields. They were not seen in the forests. Its notes were very much like those of the common yellow bird, *F. tristis.* In appearance it very closely resembles the Lesser Redpoll, *F. linaria;*—so closely that there is some difficulty in distinguishing them. It is, however, somewhat larger, and its colors a little lighter, particularly on the rump.

THE ROSE-BREASTED GROSBEAK.
Coccoborus ludovicianus.—LINNÆUS.

DESCRIPTION.—Head, chin and upper parts mostly black, varied with white on the wings and rump. Tail and wings brownish, with a broad white bar across the quills of the latter, and a narrower one on the wing coverts. Breast and under wing coverts carmine, or bright rose color. Beneath, yellowish white. Bill, cream color; legs and feet grayish brown. *Female* brown above, spotted with dull white on the wings; three yellowish white bands on the head, one passing from the bill over the crown to the occiput, and one passing along each side of the head, just over the eye. Feathers on the breast yellowish, with a brown central streak; under wing coverts sulphur yellow; no rose color. Bill brown horn color. Tail slightly emarginate. Bill notched near the point. Second quill longest. Length, 8 inches; spread, 13.

HISTORY.—The range of this bird is said to be from Texas to the 56th parallel of latitude. Though not numerous in Vermont, they are frequently met with and rear their young here. Its nest is usually built in thick forests, at a considerable height from the ground, and composed of twigs and lined with grass. The eggs are 4 or 5, bluish and spotted with brown.

THE SCARLET TANAGER.
Tanagra rubra.—LINNÆUS.

DESCRIPTION.—The bill robust, rather short, compressed towards the point, and acute. The second quill longest. Tail slightly forked. In the *male,* the plumage is of a brilliant scarlet, excepting the wings and tail, which are black, and the under wing coverts, which are yellow. Bill and legs brownish horn color. *Female* and *young* dull green, or brownish yellow. Wings and tail, brown, with the feathers edged with greenish. Color of the bill and legs lighter than in the male. Length, 6.6 inches; spread, 10.5.

HISTORY.—This bird, on account of the bright red color of the male, is sometimes called the *Fire Bird*. It is also known in many places as the *Blackwinged Red Bird*. It rears its young in Vermont, but is said to extend its summer migrations northward, as far as the 69th parallel of latitude. Its nest is usually built on the horizontal branch of a forest tree, 10 or 15 feet from the ground. It is composed of sticks, weeds and vines, nicely put together, and lined with finer materials. The eggs are usually 4, of a dull blue color, spotted with different shades of brown. It is a shy bird, occupying retired places, and manifests great solicitude for the safety of its young. One of the nests of this bird, found by my indefatigable friend, Paine, in Randolph, was on the branch of a maple, in the skirt of a forest, was 10 feet from the ground, and composed of hemlock twigs, laced and bound together with fibrous weeds and strings. It was 1.5 inch deep, and contained three eggs. The male bird showed much uneasiness when the discoverer approached the nest.

THE CRESTED WOODPECKER.
Picus pileatus.—LINNÆUS.

DESCRIPTION.—General color black. Chin white, with a rusty white stripe over the eye, and another from the nostril extending backward along the side of the neck to the base of the wings, which are, on the under side, of a delicate straw color. Vanes of the basal part of the wing feathers, white on the upper side, but nearly concealed by the wing coverts, when the wing is closed. Crest and mustachios, in the male, bright yellowish carmine red; crown variegated with black and golden yellow. Irides bright orange; bill and claws dark horn color, the bill a little lighter below, sharply ridged above and on the sides; with the mandibles, which are of equal length, brought to vertical cutting edges at their points. Tongue slender, protractile and barbed towards the point. Tail wedge-shaped; feathers 12, stiff and pointed, central ones longest. Length of the specimen here described, which was a female, 18 inches; spread, 28; from the point of the bill to the feathers 2.4;to the top of the crest 4.5. Length of the folded wing, 9.5,—tail, 7, reaching three inches beyond the tip of the folded wing.

HISTORY.—For the specimen here described I was indebted to Mr. Austin Isham, of Williston, who shot it near Shelburne pond, on the 10th of November, 1851. It was a female, and on skinning and dissecting it, I found in its craw more than 100 flat, jointed worms. They were, most of them, entire, about an inch long, and of a yellowish white color; such, in short, as are very common between the bark and wood of old trees. The gizzard contained parts of worms, and a large quantity of the fragments of ants and coleopterous insects, but no gravel.

Though no where numerous, this Woodpecker is found in all parts of the United States and as far north as the 63d parallel of latitude. In Vermont it has been very generally called the Woodcock. It is a very restless and retired bird, confining himself chiefly to the depths of the forests, and hence he is much more frequently heard than seen. In the early part of spring, as is well known to those employed at that season in the manufacture of maple sugar, his loud cackle and the sound of his powerful blows upon the old trees, are heard, reverberating through the naked forests, to a great distance. Like the other woodpeckers, it builds its nest in a cavity, hollowed out of an old tree, and lays about 6 purely white eggs.

CHIMNEY SWALLOW,—(Part 1—98.)
Cypselus pelasgius.—TEM.

In our account of this bird, we spoke of its habit, when the country was new, of resorting in immense numbers to hollow trees, in spring and autumn, and that there were many trees in this state, which were, on that account, extensively known as *swallow trees*. Many of these trees had, probably, been resorted to by thousands of birds, year after year, for centuries. The consequence would naturally be, that the hollow, in which they roosted, would be gradually filled up from the bottom, by the excrement, cast off feathers, exuviæ of insects, and rotten wood; and trees have been often found in this condition, long after the swallows had ceased to resort to them; and even after they had been blown down, and had become rotten by lying. One of this kind, in Ohio, is described in Harris' Journal, and quoted in Wilson's Ornithology. The tree was a sycamore, five feet in diameter, which had been blown down, and whose immense hollow was found filled, for the space of 15 feet, with a "mass of decayed feathers, with an admixture of brownish dust and the exuvia of various insects."

The remains of a tree of this description were found in this state, in Middlebury, so lately as the spring of 1852. The tree had been blown down, and had, nearly all, rotted away, leaving little besides the cylindrical mass, which had filled its hollow. The length of this mass was about seven feet, and its diameter 15 inches. Of the materials, which composed it, about one half consisted of the feathers of the Chimney Swallow, being, for the most part, wing and tail-feathers. The other half was made up of exuvia of insects, mostly fragments and eggs of the large wood-ant, and a brown substance, probably derived from the decayed wood of the interior of the tree.

This discovery at Middlebury, though interesting, would not have been regarded as very remarkable, if the materials, which had filled the hollow of the tree, had been promiscuously and disorderly mingled together. Such a jumbled mass would be what we should expect to find in a hollow tree which had been, for centuries, perhaps, the roosting place of myriads of Swallows. But this is not the case. In their general arrangement, the larger feathers have nearly all their quills pointing outward, while their plumes, or ends on which their webs are arranged, point inward. This arrangement might perhaps have arisen from the nesting of small quadrupeds in the hollow, making the feathers their bed. But this is not the most remarkable circumstance connected with the subject. In various parts of the mass, are found, in some cases, all the primary feathers of the wing; in others, all the feathers of the tail, lying together in contact, and in precisely the same order and position, in which they are found in the living swallow. In a lump of the materials, measuring not more than 7 inches by 5, and less than 3 inches thick, five wings and two tails were plainly seen, with their feathers arranged as above mentioned, and, in one of the wings, all the secondary quills were also arranged in their true position with regard to the primaries.

Now, we cannot conceive it possible that these feathers could be shed by living birds, and be thus deposited. We may suppose that the birds died there, and that their flesh had been removed by decay, or by insects, without deranging the feathers. But in that case, what has become of the skeletons? I do not learn that a bone, beak, or claw, has been found in any part of the whole mass. What, then, has become of these? They could hardly have been removed by violent means, without disturbing the feathers. But, if done quietly, what did it? What insect would devour the bones, and beaks, and claws, and not meddle with the quills? Or would the formic, or any other acid, which might be generated within the mass, dissolve the former without affecting the latter? These are questions, to which the *savans* have not yet returned any satisfactory response.

A specimen, from the above mentioned feathery mass, was obtained, in May, 1852, by Mr. J. A. Jameson, Tutor in the University of Vermont, and presented, by him, to the Museum of that Institution, to be preserved as a relic of primeval Vermont.

PASSENGER PIGEON.—(Part I, p. 100.)

Columba migratoria.—LINN.

Having learned that Pigeons had appeared and reared their young in large numbers, in the spring of 1849, in several towns on the Green Mountains, particularly in Fayston and Warren, in Washington county, and being desirous in case they should return there the next spring, to visit the localities, for the purpose of observing the habits of the Pigeons, and securing some of their eggs for specimens, I addressed a note of inquiry to Jacob Boyce, Esq., of Fayston. To this note I received the following reply:

FAYSTON, June 28, 1850.

MR. THOMPSON:

Sir,—I have received yours of the 10th inst., requesting information about Pigeons. They are not here the present season. Last year they came here early in April, and commenced building their nests by the middle of that month; and they left here with their young, about the middle of June. Their nests extended over a territory of, at least, 2,000 acres. Above the height of 25 feet from the ground, the tops of the trees were covered with nests. Some large birches had from 100 to 125 nests on a tree. The nests consisted of bunches of sticks, placed in the crotches of the limbs. They laid only two eggs in a nest, and raised only one brood. There might have been any quantity of eggs obtained from the nests; and great numbers of eggs rolled out of the nests and lay scattered on the ground, but I do not know that any of the eggs were preserved.

Respectfully yours,
JACOB BOYCE.

AMERICAN BITTERN.

Ardea minor.—WILSON.

DESCRIPTION.—General color yellowish ferruginous, mottled and sprinkled with dark brown. Crown dusky reddish brown. Chin and throat white, with reddish brown stripe. From the angle of the mouth a brownish black stripe proceeds downward, becoming broader on the side of the neck, and turning upwards towards the back side, where it is lost. The quills are also brownish black. Feathers of the neck and breast have their central part along the shaft dark yellow, sprinkled thickly with brown, broadly margined with tawny cream color. Dorsal plumage dark umber brown, with the feathers edged and spotted with yellowish brown and tawny white. Plumage about the vent and inside of the thighs, ochre-yellow. Legs, feet and nails greenish olive-brown. Bill dark greenish horn color, longer than the head, straight beneath, moderately arched above, stout, pointed, serrated on both mandibles, and, on the upper, notched towards the point. Tibia bare nearly an inch above the joint. Middle toe longest, pectinated. Hind nail longest. Feathers on the back of the head and neck loose and elongated. Tail small, rounded, and of 10 feathers. Length of the specimen before me, which is a female, 25 inches. Bill, along the gape, 4, along the ridge, 2.6 ; neck 11 ; folded wing 10 ; tail 3 ; tarsus 3 ; longest toe 3 ; longest nail 1.2.

HISTORY.—The specimen of American Bittern described above, was presented to me by my friend, N. A. Tucker, Esq. It was shot by him in his garden, in Burlington village, where it had alighted, on the 30th of April, 1845. It was a female, and contained several eggs, which were somewhat enlarged. About the first of June, Prof. J. Torrey found the nest of one of these birds in a swamp, in the east part of Burlington. It was made on the ground, of sticks and grass, was very shallow, and contained 6 eggs. The eggs were of a dark bluish brown clay color, and contained young, which were considerably advanced.

This bird is called by a great variety of names, but is most generally known in Vermont by the name of *Stake Driver.* This name is given it, on account of the resemblance of the sound, it makes in the breeding season, to that made by a smart blow and its echo, in driving a stake into the ground, resembling somewhat the uncouth syllables of *'pump-au-gah.* It is a sly, solitary bird, and feeds on mice, aquatic reptiles and the larger insects, and though not often seen, its sound is not unfrequently heard during the summer, proceeding from the depths of the swamps, in various parts of the state. Its range, according to DeKay, is between the 38th and 58th parallels of latitude.

THE GREATER YELLOW-SHANKS.

Totanus melanoleucas.—GEMLIN.

DESCRIPTION.—Color of the upper parts brown, spotted with black and white. Bill, black; rump and tail dusky white, barred with brown. Throat, belly, and under tail coverts, white. Legs and feet yellow. A small black spot before the angle of the eye. Shaft of the first primary white. Length, 13 inches; folded wing, 7.25; bill, along the ridge, 2.1; under mandible shorter, and both cylindrical towards the point. Tarsus 2.5 inches long; middle toe to the nail 1.5. A short web between the inner and middle toes.

HISTORY.—This bird appears in Vermont in the latter part of May, proceeding northward, where it is found in the summer up to the 60th degree of latitude. Some of them, however, remain in Vermont through the summer, and breed here. It builds its nest, according to Nuttall, in a tuft of rank grass, on the border of a creek or bog, and lays 4 eggs of a dingy white color, marked with spots of dark brown. The eggs are said to be remarkably large for the size of the bird. Perhaps its most common vulgar name is that of *Tell-Tale.*

THE SEMI-PALMATED SANDPIPER.

Tringa semipalmata.—WILSON.

DESCRIPTION.—The bill is shorter than the head, straight, enlarged and flattened towards the end, and acutely pointed at the tip. Tibia one-fourth naked; tarsus compressed and of the length of the bill. Hind toe short and small. First quill longest. Tail pointed, reaching beyond the folded wings; middle feathers longest. The color of the bill is black; the legs dark dusky olive. General color above grayish ash, thickly streaked and spotted with dusky brown, while the feathers are edged with light gray and rufous. Frontlet and line over the eye, light gray. All beneath, white, excepting the breast and lower front of the neck, which are gray, with brownish spots and streaks. Length, 6 inches; folded wing, 3.7; bill and tarsus each 0.8; middle toe, which is longest, including the nail, 0 8.

HISTORY.—This little Sandpiper ranges through all parts of the United States. It appears in Vermont in May, and remains here till autumn, and undoubtedly breeds here, although I have not seen its nest. According to Nuttall, it makes its nest, early in June, of withered grass, and lays 4 or 5 eggs, which are white, spotted with brown. For the specimen above described I am indebted to Mr. C. S. Paine, of Randolph, who shot it in the fall of 1850.

RED-THROATED LOON.

Colymbus septentrionalis.

DESCRIPTION.—Color of the head and upper parts of the neck, deep ash. Chin and sides of the mouth, white. Sides of the throat and neck, white, spotted, or striped with ash. Upper parts brownish, spotted with white, the feathers usually having a white spot on each side toward the point. White beneath, with a brownish transverse band across the vent. Wings brownish black; second quill longest, first nearly equal. Tarsus much compressed, with a slight web along the edge, black on the outside and whitish on the inner side. Outward side of the feet, and a part of the web on the inner side, blackish. Bill bluish black, lightest towards the point, narrow and pointed; upper mandible longest and a little curved; lower, incurved on the sides, acute at the tip and grooved beneath. Tongue pointed, with a fringe at the base directed inward. Eye moderate; irides dark purple; pupils black. Length of the specimen above described, which is evidently a young fowl, 24 inches; bill, along the ridge, 1.7, beneath, 1.9, along the gap, 2.9; folded wing, 10.6; tarsus, 3.2; longest toe, (the outer,) 3.2, nail 0.3. Tail very short and rounded, reaching two inches beyond the folded wings. Adults have the head lead color, the upper parts blackish, the belly white, and a reddish stripe along the throat and neck.

HISTORY.—The Loon above described was shot in Burlington Bay, on the 1st day of November, 1846. It is very rare in Vermont, in comparison with the *C. glacialis*, or Great Northern Diver, described in Part I, p. 111. They are common in the northern parts of both continents, and rear their young in the neighborhood of fresh water lakes. They lay their eggs, 2 in number, on a small quantity of down, or other soft materials, near the edge of the water. They are of a pale oil-green color, and are nearly 3 inches long and 1¾ in thickness. This fowl is called, in England, the Sprat Loon, by the fishermen. It is known in some places by the name of Scape Grace.

NOTE.—In Part I, Chap. III, we have described 141 species of Vermont Birds; and we have in this Appendix described 20 additional species, making the whole number described 161 species. And even this number falls very considerably short of the whole number of species found in the state. I have specimens of several species, which are not here described, on account of doubts with regard to their proper names. And it is well known that we have a considerable number of ducks and other water fowl, which spend some time with us, in spring and fall, in their annual migrations north and south. The Swan, *Cygnus americanus*, is occasionally met with, even in the small ponds in the interior of the state. My friend, Dr. Ariel Hunton, of Hydepark, informs me that a Swan was shot in Mud pond, in Cambridge, by Mr. Eliel Page, in 1841. It was very large, said to be six feet high, to spread its wings eight feet, and to weigh 57 pounds. These statements are doubtless exagerated, particularly the last. The length of the American Swan is usually stated at about 5 feet, and spread 7 feet.

REPTILES OF VERMONT.

Additional to Part I, Chapter IV.

Although we are well satisfied that we have a considerable number of species of reptiles, which are not embraced in our list in Part I, page 113, we shall here add only the two following:

Emys geographica, Geographic Tortoise.
Trionyx ferox, Soft-shelled Tortoise.

GEOGRAPHIC TORTOISE.

Emys geographica.—LE SUEUR.

DESCRIPTION.—Shell oval, rather depressed, smooth, widely emarginate in front, serrated behind, and deeply notched over the tail. Vertebral plates slightly carinate, the first hexagonal, rounded in front—the three following somewhat larger, subequal, and hexagonal. The two intermediate lateral plates largest, and pentagonal—the posterior rhomboidal. Marginal plates 25, the three first on each side subequal, with a nearly equal margin,—the three following restricted, with their outward margins turned upwards; the seventh slightly turned upwards and widening posteriorly. The five remaining ones on each side are two toothed on their outer margins, the bidentations becoming more distinct to the last. Sternum deeply notched behind, and slightly before—scapular plates small, triangular—brachial plates truncate, triangular; third pair of plates narrow, with their exterior edges projecting laterally and backward, and joining the 4th and 5th marginal plates at their junction,—the fourth pair largest and joining the fifth marginal plate and a small intermediate one; five pairs of trapeziums, with the longest of the parallel sides outward; caudal plates rounded posteriorly, with the two straight sides forming an acute angle. Head moderately large; edges of the jaws very sharp. Legs rather long; upper sides of the fore legs covered with flat roundish scales, largest on the outer margin; fore feet armed with five sharp incurved claws; hind feet broad, palmate, covered with flat scales towards the posterior margin, and armed with five claws, longer but less curved than on the fore feet. *Tail* conical, pointed, and reaching 1¼ inch beyond the shell. *Color*, greenish brown, with meandering yellow lines, crossing one another in various directions. Under side of the marginal plates greenish yellow, with numerous and somewhat regular brown markings. *Sternum*, yellowish flesh-color. Head, neck and legs, beautifully striped with brownish and yellow. Jaws of a uniform yellowish umber,—a yellow spot on each side of the head, back of the eye. *Eyes* yellow, with a horizontal black stripe. Sutures, at the junction of the plates above, a little elevated. Length of the shell, 10 inches; breadth, 8.5; length of the head, 2.75; width, 1.8; between the orbits, 0.5; width of the palmated hind foot, 2.4; length of the tail, from the attachment of the vertebræ, 3.3; beyond the shell, 1.25.

HISTORY.—The specimen here described, was taken in Colchester, near the mouth of Winooski river, on the 28th of May, 1846. It was a female, containing 14 mature eggs in the oviduct, with about the same number, considerably developed, and innumerable small ones, in the ovaries. She was crawling very fast over the sandy plain, when taken, and was evidently in search of a suitable place for depositing her eggs. The form of the mature eggs, was that of an ellipsoid, with one end a trifle larger than the other, and they differed not sensibly in size, being 1.4 inch long, 0.9 thick, and having their greatest circumference 3.7 in., and least, 2.9. The oviduct, containing the mature eggs, was taken from the abdomen, cut into three pieces, and laid aside, and, in the course of ten minutes, by repeated visible contractions, or throes, all the eggs were expelled from it.

Another female of this species was taken, June 10th, 1846, near Clay Point, in Colchester. She was sitting over a hole she had excavated in the sand, in the act of depositing her eggs, and made no effort to escape when approached. Her oviduct was filled with mature eggs. I learn that in ploughing the sandy lands near this Point, nests of this tortoise, containing from 12 to 20 eggs, have been frequently laid open.

The chief habitat of this species is in the states at the south-west, and I was not aware of its existence in New England, at the time of the publication of my History of Vermont, in 1842. Since that time, I have found that it is quite common all along the eastern shore of Lake Champlain. It has not, however, to my knowledge, been found any where else in New England; and, for the present, this may be regarded as its eastern limit.

The dimensions of the shell of this species, given by Dr. DeKay, are: length, 6.5 in.; breadth, 5, and height 3. Most of those observed in this vicinity have been from 7 to 10 inches long, and from 6 to 8.5 broad. Their flesh is said to be a very palatable article of food.

GENUS TRIONYX.—*Geoffroy*.

Generic Characters.—Shell without plates, and, together with the sternum, cartilaginous, and extending over the edges into a flexible margin. Feet palmated, with three sharp claws. A corneous beak, covered with fleshy lips. Nose produced. Vent near the extremity of the tail.

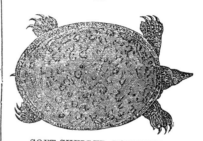

SOFT-SHELLED TORTOISE.

Trionyx ferox.—GMELIN.

DESCRIPTION.—General color of the shell brownish olive, above, with ocellated spots, formed mostly by a circular arrangement of black dots, and with a border formed of black dots around the margin of the shell. The spots are usually about the size of a dime. In dried specimens, the color is nearly black, and the spots very obscure. All the under parts dull white, or light flesh-color. A bright yellow line, edged on both sides with black, extends from the snout to each eye, and from the eyes backwards, till lost in the marbling of the neck. Irides bright yellow, crossed by a black medial stripe. Upper side of the legs varied with black and yellow. Form orbicular; shell bony in the central part, with the margin cartilaginous, soft and flexible. Head long and pointed, snout projecting beyond the jaws, with large open nostrils at the extremity. Jaws horny, with the lips fleshy and revolute. Upper side of the legs, next the margin of the shell, covered with horny scales. Five toes on each foot, three of which have well developed claws, the others are enveloped in a web, forming paddles for swimming. Tail projects less than an inch beyond the shell, with the vent near the extremity

SOFT-SHELLED TORTOISE.

Dimensions:—Length of the shell, 13 inches; breadth, 10¾; height, 3½; head, 2¼; head and neck, 7.

HISTORY.—The existence of this species of Tortoise in our waters, was not even suspected by me at the time of the publication of my History of Vermont, in 1842. My first specimen of it, I obtained on the 10th of August, 1844. It was caught on a fish-hook, in the river Lamoille, in Milton, by Mr. Joseph Dupau, to whom I have been indebted for many interesting specimens of reptiles and fishes. Since that time I have obtained several other specimens, which were taken in or near the mouth of Winooski river, in Burlington. Most of these I kept alive for some weeks, but I could induce none of them to take any food; and, although it might be inferred, from the name, *ferox*, that they were of a ferocious disposition, I could never cause either of them to bite at a stick, notwithstanding I frequently provoked them for that purpose.

The specimen, which furnished the materials for the preceding description, was taken in a seine, at the mouth of Winooski river, on the 6th of May, 1848. I kept it alive till about the middle of June, when I killed and skinned it. It was a female, and her ovary contained 29 eggs, enlarged to near the size of a musket ball, besides innumerable minute eggs.

The *Trionyx ferox*, though common in the western waters, has not, to my knowledge, been found any where in New England, excepting the western parts of Vermont, along the margin of Lake Champlain; but it would seem, from the dimensions given by naturalists, that it attains a larger size here than at the west, where it is more common. The shells of three specimens, taken in this vicinity, measured as follows, viz :

	Length.	Breadth.	Height.
1.	11 inches;	9.5 inches.	2.2 inches.
2.	13 "	10.75 "	3.2 "
3.	13.5 "	11 "	3.5 "

The dimension of the shell of this species, given by Dr. DeKay, Zoology of New York, Part III, p. 6, are : length, 5.3 in.; breadth, 5 in.; height, 1.4.

This species resembles the sea-turtle, in its structure and habits, much more than our other tortoises. It leads a more aquatic life, and, probably seldom, if ever, crawls out upon the land, except for the purpose of depositing its eggs. On account of the shortness of its legs, and the great width of the shell, it travels upon land with much difficulty, especially where the surface is uneven, or covered with vegetation. Its flesh is esteemed a wholesome and nutritious article of food.

FISHES OF VERMONT.

FISHES OF VERMONT.

Additional to Part I, Chapter v.

To our list of Vermont Fishes, given in Part I, page 128, we now add the following species :

Lucio-perca canadensis,	Ground Pike Perch.
Boleosoma tassellatum,	Darter.
Cottus gobioides,	Little Stargazer.
Leuciscus atromaculatus,	Small scaled Dace.
Esox nobilior,	Masquallonge.
Salmoperca pellucida,	Trout Perch.
Coregonus clupeiformis,	Herring Salmon.
Amia ocellicauda,	Bowfin.

THE GROUND PIKE-PERCH.

Lucio-perca canadensis.—SMITH.

L. grisea, DEKAY, Zoology of New York, Part IV., page 19.

DESCRIPTION.—General form elongated, cylindrical, and proportionally more slender along the abdomen than in the common Pike-Perch, but the head and opercules resemble that species very closely. The preoperculum is finely serrated on the posterior margin, and more coarsely below. There are also fine serratures on the lower margins of the preoperculum and suboperculum, near their junction. Instead of a single ridge proceeding from the upper anterior angle of the operculum, and terminating backward in a single spine, there are usually several ridges radiating thence, and often terminating in the opposite margin in very acute spines. Humeral bones armed posteriorly with several spines. Vent midway between the ventrals and the caudal fin. The anal fin commences under the fourth ray of the second dorsal. The first dorsal begins in a vertical line, passing through the base of the ventrals. Caudal forked.

General *color* grayish or brownish yellow, or orange, lightest beneath. First dorsal sprinkled with roundish black spots of the size of a small pea, usually arranged in two or three rows, nearly parallel to the line of the back, but without the black patch on the posterior part, which forms a conspicuous mark in the *L. americana*. Second dorsal, and the caudal, barred with black, or brown. Usual length, 13 inches.

Rays, B. 7, D. 13–1|18, P. 13, V. 1|5, A. 12, C. $17\frac{5}{5}$.

HISTORY.—When the Natural, Civil and Statistical History of Vermont was published, I was well satisfied that the species here described was distinct from the *L. americana*, but was not so clear whether it was a species already described, or not. The difference between this species and the *L. americana* is so obvious, that they are instantly distinguished, even when there is no difference in size ; but while the latter species often exceeds two feet in length, and weighs five or six pounds, the *L. canadensis* seldom, if ever, exceeds 14 inches in length, or half a pound in weight. It is much less common in Lake Champlain than the *L. americana*, but is frequently taken in company with it. It usually swims very near the bottom of the water, and hence it has received the name of *Ground Pike*, (Pike-Perch). As an article of food, this species is held in the same high esteem, as the Common Pike-Perch.

GENUS BOLEOSOMA.—*De Kay.*
Generic Characters.—Two dorsal fins. Opercule scaly, with a single spine. Preopercule smooth on the margin. Six branchial rays. Nape depressed, contracted.

THE DARTER.
Boleosoma tassellatum.—DEKAY.

DESCRIPTION.—A small fish, with a row of quadrated black spots, about seven in number, along the dorsal ridge, occupying about one-half of the space. A row of lozenge-shaped black spots, a little smaller than those on the back and about the same in number, along the lateral line, on each side of the body. General color of the remaining parts brownish yellow, approaching to white on the belly. Eyes moveable in their sockets; pupils black, surrounded by a golden line, which fades outward into a gray iris. Fins yellowish white, with faint brownish bars on the dorsals and caudal fin. Body cylindrical, covered with rough scales. Head rather small; eyes large and projecting; nape depressed.

HISTORY.—The habits of this fish are quite peculiar. It moves not from place to place by an even labored motion, like other fishes, but proceeds by sudden leaps, or darts, impelling itself forward by its tail and pectoral fins, which it moves as a bird does its wings. It remains suspended in the water no longer than it keeps its pectoral fins in rapid motion. When the motion of its fins ceases, the fish sinks, at once, to the bottom, showing that its specific gravity is greater than water, owing, doubtless, to its want of a swimming bladder. When it reaches the bottom, it alights upon its stiff ventral fins, upon which it stands on the bottom, balanced, with its head elevated, as a bird stands on its feet. I kept several specimens of this fish alive, in a vessel of water, for some time, for the purpose of watching their motions and learning their habits. They were very uneasy, and seemed extremely anxious to escape from their confinement. Aided by their caudal and pectoral fins, in giving them an impulse upwards, and by their ventrals in climbing and adhering, they would often raise themselves up the perpendicular side of the vessel, entirely above the surface of the water, excepting only the caudal fin. Another peculiarity of this fish, is its power of bending its neck and moving its head without moving the body, in which respect it equals many of the reptiles. This fish is entitled to the name of *Darter*, both from its sudden motion, and from its having the general form of a dart.

THE LITTLE STAR-GAZER.
Cottus gobioides.—GIRARD.

DESCRIPTION.—*Color*, blackish on the back, mottled with light yellow; abdomen white; lower parts of the sides and under parts posterior to the vent, yellow—all the under parts finely sprinkled with black specs. *Fins;* first dorsal blackish, edged with red; all the others barred with brown and yellow, except the ventrals, which are white, close together, and a little behind the pectorals. *Teeth* sharp and fine, like velvet pile, on both jaws. *Tongue* large and fleshy, with a patch of teeth. *Head* large, broad, and a little flattened, with the eyes, which are large, on the upper side and near together. A sharp, stout spine on the preoperculum. *Lips* rather fleshy, and the upper one a little protractile.— *Body* thick forward, tapering very fast towards the tail. Lateral line nearest the back, consisting of a furrow with the edges a little raised. Caudal fin rather large, and nearly even. Pectorals very large, and rounded. Ventrals long and narrow. Vent anterior to the middle. Skin thickly covered with mucus. No scales.

Total length, 4 inches; to the commencement of the first dorsal fin, 1.2; to the vent, 1.7. Width of the head, 0.95.

Fin Rays, B. 6, D. 7—17, P. 14, A. 12, C. 15.

HISTORY.—For the specimen here described, I was indebted to the kindness of Mr. R. Colberth. He caught it, while fishing for trout, in a branch of the river Lamoille, in Johnson. This fish usually lies still at the bottom, or concealed under the stones in the streams, and seldom moves, except when disturbed, and then its motions are sluggish and labored. It is called, in some places, the Slow Fish.

It probably derived the name of Star-Gazer, from the favorable position of its

SMALL-SCALED DACE.

eyes for looking upwards, they being placed very near the top of the head. It seldom exceeds 4 inches in length.

Mr. Girard ascertained our Cottus to be an undescribed species from the identical specimen, which I have described here, and he gave it the name of *C. gobioides*, from its strong resemblance to the *Cottus gobio* of Europe. His description of the fish, accompanied by a beautifully engraved figure of it, is contained in his valuable Monograph of the Cottus Family of Fishes, published in the second volume of Smithsonian Contributions to Knowledge.

SMALL-SCALED DACE.

Leuciscus atromaculatus.—MITCH.

DESCRIPTION.—Color of the head and back dark olive-brown; sides lighter, often with bronzy reflections, and passing into a whitish flesh-color on the belly. Dorsal, caudal and outer margins of the pectoral fins, brownish; inner margins of the pectorals, the ventrals and the anal fin, dull orange. Eyes rather small; pupil black, surrounded by a fine golden line; iris brown. Scales small and crowded, as far backward as the ventrals. Lateral line begins at the top of the gill opening, bends rapidly downward over 11 scales, and then runs a straight course to the tail, passing over, in the whole, 60 scales. Tail lunated. Pectoral and ventral fins rounded. A squarish distinct black spot on the anterior part of the base of the dorsal fin.

Length of the specimen described, 6.5 inches; to the pectoral fins, 1.5; dorsal 3; vent, 3.75; anal, 3.9; to base of caudal, 5.5; width between the eyes, 0.6; head, 0.8.

Rays, D. 8, P. 16, V. 8, A. 8, C. 19.

HISTORY.—This is one of the most common fishes of this genus in the western part of Vermont. It abounds almost every where, both in the rivers and small streams. Its insipidity and small size prevent its being sought as an article of food ; but, as it takes the hook with great readiness, it affords the boys an opportunity to indulge in the cruel sport of catching them for mere amusement. They are also caught to be used as bait in taking larger fishes.

MASQUALLONGE.
Esox nobilior.—THOMPSON.

Esox estor, Richardson Fauna Boreali, Part III,–127.
Esox estor, Herbert's Frank Forester's Fish and Fishing.

MASQUALLONGE.

DESCRIPTION.—Back nearly black; sides bluish gray, mostly covered with irregular roundish dark-brown spots, usually about an inch in diameter, and often confluent, with a few meandering yellowish lines. Belly grayish white, with ruddy tinges. Fins dark brown; pectorals ruddy. Eyes moderately large; pupils black, surrounded by a bright yellow ring, which fades outward over the irides into grayish orange. Length of the specimen before me, 48 inches, from the tip of the under jaw, which is longest, to the extremity of the tail ; to the anterior nostril, 4 inches ; to the orbit, 5 ; to the nape, or beginning of the scales, 8 ; to posterior edge of the preoperculum, 8.5; do. of the operculum, 10.5; the beginning of ventral fins, 24.5; do. dorsal, 31.5 ; do. anal, 32.5 ; to the centre of the base of the caudal, 42.75; width between the orbits, 3. Fins : dorsal, length, 5; height 4,—pectorals, length, 1.7; height 5,—ventrals, length, 1.5; height, 4,—anal, length, 4; height 4 5,—caudal, 4 across the base; longest rays, 6.25. Lower part of the cheek, in front of the preoperculum, naked. Face nearly flat between the orbits.

Rays, D. ²18, P. 16, V. ¹12, A. ³16, C. 19⅗.

HISTORY.—This fish has, till lately, been confounded with the *Esox estor*, or Common Pike, or Lake Pickerel. When my description of the *E. estor* was published, in 1842, I doubted the existence of this species in our lake, but since that time my mind has changed on the subject. In May, 1847, I received from my friend, the Hon. A. G. Whittemore,* of Milton, a fish caught near the mouth of the river Lamoille, which the fishermen called Masquallonge. It was 26 inches long, and weighed about 6 pounds. Upon examining it, I was fully satisfied that it was of a species distinct from *E. estor*, and, as I could find no description of it under any other name, I made out a description and gave it the name of *Esox nobilior*.

In April, 1848, I received another specimen from the same source, which weighed 19 lbs., and was 41½ inches long. In May, 1849, two specimens were brought along, both caught near the mouth of the Lamoille, one of which weighed 40, and the other 27 pounds. I purchased the latter, and from it the preceding description is chiefly drawn.

Believing this species to attain a larger size, and to be a more excellent fish for the table, than any other epecies of the Pike

---

* Since the above was written, I have received intelligence of the death of my esteemed friend, Albert G. Whittemore, Esq. He was accidentally killed at Zanesville, Ohio, on the 10th of November, 1852, aged 55 years; where he was engaged as contractor on the rail road from that place to Wheeling. He was a gentleman of intelligence and enterprise, and of many estimable qualities as a man and a citizen.

family, found in the United States, I have given it the specific name of *nobilior*. It is a fish, which is eagerly sought, and commands the highest price in market, but it is rare in Lake Champlain, compared with *E. estor*, or Common Pike. Very good figures of both of these species are given in Frank Forester,s Fish and Fishing, but both under wrong names; the *E. nobilior* being figured under the name of *E. estor*, and the *E. estor* under that of *E. lucioides*.

The vulgar name, MASQUALLONGE, appears to have been given by the early French settlers of Canada to the Pikes and Pickerels generally, it being a term, or phrase, descriptive of the whole family, *Masque*, signifying face or visage, and *allonge*, lengthened,—they all having lengthened, or elongated heads. In modern times this name, Masquallonge, has been confined, by the fishermen, to the species here described, while the other species bear the vulgar name of Pike, or Pickerel. The methods of spelling this Canadian-French name, have been almost as numerous as the authors, who have used it, as may be seen by the following list:

Maskallonge,	Le Sueur.
Masquinongy,	Dr. Mitchell,
Maskinonge,	Dr. Richardson.
Muskallonge,	Dr. Kirtland,
Muskellunge,	Dr. DeKay.

The oldest forms of this name, it will be seen, approach nearest, both in spelling and pronunciation, to the phrase *Masque allonge*, which we have supposed to be its origin, and, therefore, afford presumptive proof of the correctness of our supposition.

This fish may usually be distinguished from the Common Pike by its dark circular markings, and its more robust proportions. Its head is proportionally shorter, the face flatter and less grooved, and the width across the eyes and upper jaw greater than in the *estor*. But, perhaps, the mark by which it may be most readily distinguished is on the cheek, the lower half of the cheek in the *E. nobilior*, in front of the preoperculum, being naked, or without scales, while in the *E. estor* the whole cheek is covered with scales. The difference in the general aspect of the two species may be seen by comparing the figure of the *E. estor* below, with the *E. nobilior* at the head of this article.

Esox estor.—LESUEUR.

The specimen here described was a female, with her abdomen filled with eggs, contained in two ovaries, which extended nearly the whole length of the cavity. This fish abounds much more in the streams and smaller lakes in Canada than in Lake Champlain.

GENUS SALMOPERCA.—*Thompson.*

Generic Characters.—Two dorsal fins, the first supported by flexible rays, and the second adipose, as in the trouts. Opercules smooth. A band of fine teeth in each jaw. Scales with serrated edges, as in the perches.

TROUT-PERCH.

Salmoperca pellucida.—THOMPSON.

Percopsis guttatus, Agassiz Lake Superior, p. 284, and Plate I., fig. 1 and 2.

DESCRIPTION.—General color, light brownish yellow, with longitudinal rows of brown spots, about one-tenth of an inch in diameter, usually one row along the dorsal line, and two rows on each side between this and the lateral line. A broad satin stripe embraces the lateral line. Belly white. Fins and flesh translucent–the vertebral column, the contents of the abdomen, and portions of the head, only appearing opaque, when held towards the light. Fins all large, in proportion to the size of the fish. The rays of the pectorals reach backward half of their length beyond the ventrals, which are attached near the middle of the abdomen, and under a point a little anterior to the first dorsal, and reach backward to the vent. The anal fin has its first ray short and spinous. Caudal fin forked. Nostrils and eyes large; irides yellow. A depression on the head, between the orbits, divided longitudinally by a long ridge. Scales rather large and rough, having finely serrated edges. Length, from 3 to 5 inches. The following are the measurements of one out of three living specimens before me, when the above description was made: Total length, 3.9 inches; to the pectoral fin, 1; to ventral, 1.45; first dorsal, 1.5; anal, 2.1; adipose, 2.6; central base of caudal, 3.2. Fin rays, B. 6, D. ^{2}10—0, P. 13, V. 8, A.1|7, C. 18_6^4.

HISTORY.—The first knowledge I had of this fish was in the summer of 1841, when I found a specimen of it, 5 inches long, which was dead, and had been drifted up by the waves on the lake shore, in Burlington. On examining it, I found it to possess the adipose and abdominal fins of the trouts, but, in its teeth, gill covers and particularly in its hard serrated scales, to bear considerable resemblance to the perch family. After searching all the books within my reach, without finding it described, I concluded that it might be new, both in genus and species, and accordingly, in allusion to the above mentioned properties, I described it in my journal under the provisional generic name of SALMOPERCA. A notice of this fish was omitted in my History of Vermont, published in 1842, because I had then only one specimen, and, upon that one, with my little experience, I did not think it prudent to found a new genus and species. When Prof. Agassiz was at Burlington, in 1847, I submitted the above mentioned specimen to his inspection, having at that time obtained no others. At first sight, he thought it might be a young fish of the salmon family, but, upon further examination, he said it was not a salmon, nor any other fish with which he was acquainted.

During the summer of 1847, I found three other specimens of this fish, dead, on the lake shore. One of these I took with me to Boston, in September, to the meeting of the Association of American Geologists and Naturalists, and put it into the hands of my friend D. H. Storer, M. D., with a request that he would ascertain what it was, and let me know.

In May, 1849, I obtained from Winooski river a number of living specimens, which I kept alive for some time; and, observing the great translucency of the living fish, when held up towards the light, I gave it the specific name of *pellucida*, having previously called it, in my journal, *eoceta*, from its wing-like pectoral fins.

About this time I noticed, in the proceedings of the Boston Society of Natural History, that Prof. Agassiz had laid before the Society an account of a new genus of fishes discovered by him in Lake Superior, which he proposed to call PERCOPSIS. Suspecting, from the brief description given of it, that it was identical with my SALMOPERCA, I wrote to Dr. Storer and inquired of him, if the specimens from Lake Superior, presented to the Society by Prof. Agassiz, were like the one I put into his hands in 1847. He wrote me that he could not say—that the specimen went out of his hands soon after he received it, and he had not seen it since.

In Prof. Agassiz Lake Superior, page 248, I find an account of his genus PERCOPSIS, and his species *P. guttatus*, and I have no doubt that it is identical with my *Salmoperca pellucida*. Still, I have thought it best to let it remain, in this Appendix, under the name I had given.

HERRING SALMON.

Coregonus clupeiformis.—MITCH.

Coregonus artedi.—LESUEUR.

Argyrosomus clupeiformis.—AGASSIZ Lake Superior, p. 339.

DESCRIPTION.—Color of the back bluish brown; sides lighter, with silvery reflections; belly white. Gill covers and cheeks, with silvery and cupreous reflections. Head small, pointed and somewhat flattened above; under jaw longest; mouth small, without teeth; eyes large, round—irides silvery yellow. Scales large and circular. Lateral line distinct, nearly straight, and passes over 72 scales; 13 rows of scales between the first dorsal and the ventral fin—a long slender bract at the base of the ventrals. Pectoral fins long and pointed; ventrals under the anterior part of the dorsal, and triangular; first dorsal nearly midway between the point of the lower jaw and the extremity of the caudal fin; second dorsal adipose and over the posterior part of the anal, and triangular; caudal forked.

Length, total, 14 inches; to the posterior edge of the operculum, 2.4; to the beginning of the dorsal fin, 6; to the ventrals, 6.2; to the vent, 9; to the anal, 9.3; to the adipose, 10.2; to the central base of the caudal, 12; greatest depth in front of the first dorsal, 2.5; thickness, 1.4. Length of the longest fin rays: first dorsal, 1.6; Pectoral, 1.5; Ventral, 1.4; Anal, 1, and Caudal 2.

Rays, B. 8, D. ³10,—0, P. 4, V. ⸺11, A. ⸺11, C. $18\frac{6}{5}$.

HISTORY. –This fish is only occasionally met with in Lake Champlain, but they sometimes appear here in myriads. In the spring of 1847, they were, for a short time, taken at Burlington, in very large numbers; as many as 200 being taken at one haul of the seine. In some years none at all are taken here. The specimen from which the preceding description is made was taken in 1848, and I learned of only two others being taken that season. It resembles, somewhat, the Lake Shad, *C. albus*, but is a rounder fish, having much less depth in proportion to its length. It is much esteemed as an article of food. It is common in Lake Ontario and Lake Erie, and is called in many places the Shad Salmon.

GENUS AMIA.—*Linnæus.*

Generic Characters.—Small paved teeth behind the conical ones. Head flattened, naked, with conspicuous sutures. Twelve flat gill-rays. A large buckler between the branches of the lower jaw. Dorsal long. Anal short. Air-bladder cellular, like the lungs of reptiles.

THE BOWFIN.
Amia ocellicauda.—RICHARDSON.
Amia occidentalis.—DEKAY.

DESCRIPTION.–General color above, brown, waved with dull bronzy yellow, approaching to white on the belly, and having the sides sprinkled with yellowish white spots. Pectoral, ventral and anal fins, brownish; dorsal and caudal with alternate bars of brown and brownish white. A large and conspicuous black spot near the upper part of the tail, at the base of the 4th, 5th, 6th and 7th rays of the caudal fin. Head without scales, covered with scabrous bony plates; opercules bony, with membranous edges. Gill-rays flat. Cartilaginous buckler between the branches of the lower jaw. Two short cirri on the upper lip. Eyes moderate, deeply sunken. Jaws broad, rounded and even. A row of sharp conical teeth in each jaw, paved behind with short blunt teeth. Scales large and thin. Lateral line distinct, nearly straight, nearest the back, on the anterior part of the body, crossing 70 scales, which are smaller than those adjacent. Attachment of the caudal fin oblique—caudal rounded. Total length of the specimen before me, 19.2 inches; from the snout to the upper side of the gill-opening, 4; to the beginning of the dorsal, 6.8; to the ventrals, 9; to the anal, 11.5; to the lower edge of the caudal. 15; depth behind the pectorals, 3.6. Width of the head, 3; back of the pectorals, 2.6. Distance between the eyes, and from the orbits to the end of the snout, 1.3 each; between the cirri, 0.6. Length of the dorsal fin, 8.7; height, 1.2,—commences midway between the pectorals and ventrals, and reaches almost to the tail.

Fin Rays, D. 48, P. 17, V. 7, A. ?8, C.21.

HISTORY.—This fish abounds upon the muddy bottoms and the marshy coves of the southern part of Lake Champlain. It is very plentiful in the vicinity of Whitehall, and also about the mouth of Otter Creek. From its partiality to muddy bottoms, it has acquired, in many places, the name of Mud Fish. From its resemblance in form to the Ling, it is called in some places the *Scaled Ling*. But its more common appellation in Vermont, is that of *Bowfin*. It attains to considerable size, frequently exceeding two feet in length, and weighing 10 or 12 pounds; but its flesh is soft and ill flavored, very little esteemed as an article of food.

BOTANY OF VERMONT.

Additional to Part I., Chapter VII.

In the first edition of my Gazetteer of Vermont, published in 1824, I gave a simple catalogue of the plants then known to be indigenous, in this state. The materials for that catalogue were derived, principally, from a list of plants growing in the vicinity of Middlebury, prepared by Dr. Edward James, and published, in 1821, in Prof. Frederick Hall's statistical account of Middlebury. The additions to this list were mostly furnished by Dr. William Paddock, Prof. of Botany in the University of Vermont. At that time, very little attention had been given to the scientific botany of the state, and the whole number of plants contained in my catalogue was only 569.

Between 1824 and the publication of my general history of Vermont, in 1842, our state was explored by several eminent botanists from abroad, and by a number of enthusiastic disciples of Linnæus, raised up in our midst, by whose united labors our list of known indigenous plants was greatly enlarged. While engaged in collecting together these scattered materials, for the purpose of making my Catalogue as complete as possible, in the work I was preparing for publication, I was so fortunate as to become acquainted with the late Wm. Oakes, Esq., of Ipswich, Mass. He was at that time engaged in investigating the botany of the western part of Vermont, and he very generously undertook, for me, the systematic arrangement of a complete Catalogue of Vermont plants. I, therefore, put into his hands my former catalogue and all the additional materials, I had accumulated, and the full and beautifully arranged Catalogue in Part I, Chapter VII, is the result of his labor. That Catalogue contains 929 species of Vermont plants, and is an honorable memorial of its

BOTANY OF VERMONT. CATALOGUE OF PLANTS.

Author, both of his kindness as a friend, and of his zeal and accuracy as a botanist*.

Since the publication of the Catalogue above mentioned, the number of known Vermont plants has been considerably increased, and we have doubtless many more species to reward the labors of botanists. By the kindness of several friends, I am enabled to add to the previous list 105 species, making in the whole 1034.

For the arrangement of these additional species, and for the identification of a large number of them, I am indebted to the kindness of my friend, Prof. Joseph Torrey, D. D., of the University of Vermont.

* It is my painful duty here to record the death of my esteemed friend, William Oakes, Esq. He was drowned on the 31st of July, 1848, while passing from Boston to East Boston, under circumstances which left it doubtful, whether by accident, or in a temporary fit of insanity, to which he was subject. He was 49 years of age.

CATALOGUE OF VERMONT PLANTS.

Continued from page 177, *Part I.*

CLASS I. EXOGENOUS OR DICOTYLEDONOUS PLANTS.

ORDER RANUNCULACEÆ.

Clematis, *Linn.*
    viorna, Wildn. Found at Castleton, by *Mrs. J. Carr.* A very rare species. June, July.
Anemone, *Haller.*
    aconytifolia, Mx. Castleton. *Mrs. J. C.*
Ranunculus, *L.*
    fascicularis, Muhl. Low grounds. Burlington, *T**. Brattleborough, *C. C. Frost.*

ORDER MAGNOLIACEÆ.

Liriodendron, *L.*
    tulipifera, L. A tree not rare in the southern part of this state, fifty years ago. Some large specimens are still left in Bennington county, valley of the Hoosic river. *Mrs. J. C.*

ORDER CRUCIFERÆ.

Nasturtium, *R. Br.*
    hispidum, D. C. Low grounds, Burlington. Also found in Brattleborough, *C. C. F.*
Arabis, *L.*
    lyrata, L. Mountain-garden. Willoughby lake. *C. C. F.* May.
    canadensis, L. Rocks below Winooski Falls; Colchester. *T.* June.
Cardamine, *L.*
    rotundifolia, Mx. In Vermont, locality not specified. *Dr. Robbins.*
    Virginica, Mx. Hill-sides, Vt. *A. Wood.* June.
Sisymbrium, *All.*
    thaliana, Gay. Rocks and sandy fields, Vermont. *A. Wood.* May.
Draba, *L.*
    verna, L. Willoughby lake. *A. W.* May.
Erysimum, *L.*
    cheiranthoides, L. Brattleborough, *C. C. F.*
Isatis, *L.*
    tinctoria. Banks and islands of Winooski river; Burlington, *T.* Probably introduced.

ORDER VIOLACEÆ.

Viola, *L.*
    Selkirkii, Goldie. Rich cedar swamps. Grand Isle. *T.*
    pedata, L. Brattleborough. *C. C. F.* April, May.

ORDER HYPERICACEÆ.

Hypericum, *L.*
    sarothra, Mx. Brattleborough, *C. C. F.* July, August.

* The authority to which *T.* refers in this catalogue, is *Prof. Joseph Torrey*, of the University of Vermont.

Order CARYOPHYLLACEÆ.

Saponaria, *L.*
  officinalis, L. Brattleborough, *F.* Banks of Castleton river. *Mrs. J. C.*
Silene, *L.*
  inflata, Smith. Brattleborough, *F.* Castleton, *Mrs. J. C.* June.
Sagina, *L.*
  procumbens, L. Brattleborough, *F.* June.

Order PORTULACEÆ.

Claytonia, *L.*
  Virginica, L. Intervale lands in Colchester. Quite distinct from the species *Caroliniana*, Mx., *T.* April, May.

Order ACERACEÆ.

Negundo, *Moench.*
  aceroides, Moench. Abundant in some localities on the banks of Winooski river, Burlington and Colchester, *T.* April.

Order RHAMNACEÆ.

Rhamnus, *L.*
  catharticus, L. Bethel, *R. Green.* July.

Order LEGUMINOSÆ.

Astragalus, *L.*
  Canadensis, L. Burlington, near Red Rocks, *R. Benedict.* July.
Hedysarum, *L.*
  boreale, Nutt. Willoughby lake. *A. Wood.* June, July.

Order ROSACEÆ.

Sanguisorba, *L.*
  Canadensis, L. Brattleborough, *C. C. F.* July.
Rubus, *Tourn.*
  Idæus. Cambridge, *Dr. Robbins.* June.

Order ONAGRACEÆ.

Epilobium, *L.*
  molle, Torr. Intervals, Burlington, *T.* June.
Oenothera, *L.*
  fruticosa, L. Willoughby lake, *C. C. F.* August.

Order MELASTOMACEÆ.

Rhexia, *L.*
  Virginica, L. Brattleborough, *C. C. F.*

Order SAXIFRAGACEÆ.

Saxifraga, *L.*
  oppositifolia. Willoughby lake, *A. Wood.*
  aizoides. Willoughby lake, "

Order UMBELLIFERÆ.

Sium, *L.*
  lineare, Mx. Burlington, *T.* June and July.

Order COMPOSITÆ.

Diplopappus, *Cass.*
  linarifolius, Hook. Brattleborough, *C. C. F.* September.
Sericocarpus, *Nees.*
  solidagineus, Nees. Brattleborough, " July.
Aster, *L.*
  radula, Ait. Brattleborough, " July.
  elodes, T. & G. Brattleborough, " July.
  cyaneus, (var.) Brattleborough, " July.
  amethystinus, Nutt. Brattleborough, " August.
  sagittifolius, Ell. Brattleborough, " September.

Solidago, *L.*
   rigida, L. Burlington, *T.* August and September.
   stricta, Ait. Burlington, *T.* August and September.
   serotina, Willd. Burlington, *T.* August and September.
   patula, Muhl. Burlington, *T.* August and September.
   thyrsoides, Meyer. Willoughby lake, *A. W.* Mansfield mountain, *W.*
   Muhlenbergii, T. & G Brattleborough, *C. C. F.* August.
   corymbosus. Willoughby lake, *C. C. F.* August.
Artemisia, *L.*
   Canadensis, Michx. Willoughby lake, *A. W.*
Rudbeckia, *L.*
   hirta, L. Brattleborough, *C. C. F.* July.
Helianthus, *L.*
   frondosus, (var.) Brattleborough, *C. C. F.* July.
   trachelifolius, Willd. Brattleborough, *C. C. F.* August.
Hieracium, *L.*
   Gronovii, Tourn. Colchester, *T.* Brattleborough, *C. C. F.* July.
   scabrum, Michx. Brattleborough, *C. C F.* August.
Cirsium, *Tourn.*
   horridulum, Michx. Brattleborough, *C. C. F.* July.
Cichorium, *Tourn.*
   Intybus, L. Burlington, in the lanes, *Mrs. A. P. Judd.* August.

ORDER LOBELIACEÆ.

Lobelia, *L.*
   Dortmanna, L. Willoughby lake, *A. W.* July.

ORDER ERICACEÆ.

Andromeda, *L.*
   ligustrina, Muhl. Brattleborough, *C. C. F.* June.

ORDER AQUIFORDELELIACEÆ.

Prinos, *L.*
   laevigata, L. Mouth of the Winooski, Burlington, *T.* June.

ORDER ASCLEPIADACEÆ.

Asclepias, *L.*
   purpurascens, L. Brattleborough, *C. C, F.* June.
   variegata, L. Brattleborough, " July.
   verticillata, L. Brattleborough, " July.

ORDER BORAGINACEÆ.

Myosotis, *L.*
   stricta, Link. Brattleborough, *C. C. F.* June.
Symphytum, *L.*
   officinale, L. Pownal, *T.* Introduced. July.

ORDER LABIATÆ.

Pycnanthemum, *Michx.*
   linifolium, Pursh. Brattleborough, *C. C. F.* July.
   aristatum. Michx Brattleborough, " August.
Trichostema, *L*
   dichotoma, Brattleborough, " August.

ORDER SCROPHULARIACEÆ

Verbascum, *L.*
   blattaria, L. Burlington, *T.* Brattleborough, *C. C. F.* July.
Ilysanthus, *Rafinesque.*
   gratioloides, Benth. Brattleborough, *C. C. F.* July.
Pentstemon, *L'Her.*
   laevigatum, Soland. Burlington, Red Rocks, *T.* August.

ORDER LENTIBULACEÆ.

Utricularia, *L.*
   inflata, Walt. Brattleborough, *C. C. F.* August.

ORDER PRIMULACEÆ.

Primula, *L.*
   Mistassinica, Michx. Willoughby lake, *A. Wood.*

Order PLANTAGINCÆ.

Plantago, *L.*
  lanceolata, L. Burlington, *T.* July.
  Virginica, L. Brattleborough, *C. C. F.* July.

Order POLYGONACEÆ.

Polygonum, *L.*
  erectum, L. Burlington, *T.* Brattleborough, *C. C. F.* July.
  punctatum, Ell. Brattleborough, *C. C. F.* July.
Rumex, *L.*
  sanguineus, L. Brattleborough, *C. C. F.* July.
  aquaticus, L. Brattleborough, " July.

Sub-Order MYRICEÆ.

Myrica, *L.*
  gale, L. Wells, border of the pond, *T.* July.

CLASS II. GYMNOSPERMS.

Order CONIFERÆ.

Cupressus, *Tourn.*
  thyoides, L. Willoughby lake, *C. C. F.* May.
Juniperus, *L.*
  Sabinus, L. West Rutland, *Mrs. J. C.*

CLASS III. ENDOGENS OR MONOCOTYLEDONS.

Order AMARYLLIDACEÆ.

Hypoxis, *L.*
  erecta, L. Brattleborough, *C. C. F.* June.

Order ALISMACEÆ.

Sagittaria, *L.*
  lancifolia, (var.) Burlington, *T.* Brattleborough, *C. C. F.* July.
  natans? Brattleborough, *C. C. F.* July.

Order XYRIDACEÆ.

Xyris, *L.*
  Caroliniana, Walt. Brattleborough, *C. C. F.* August.

Order RESTIACEÆ.

Eriocaulon, *L.*
  decangulare, Michx. Willoughby lake, *A. W.*

Order PODOSTEMACEÆ.

Podostemon, *Michx.*
  ceratophyllum, Michx. Brattleborough, *C. C. F.* July.

Order FLUVIALES.

Potamogeton, *L.*
  praelongus, Wolff. Willoughby lake, *A. W.* July.
  oblongus. Brattleborough, *C. C. F.* July.
  pulcher. Brattleborough, *C. C. F.* July.
  hybridus, Michx. Brattleborough, *C. C. F.* July.
  spiralis. Brattleborough, *C. C. F.* July.

Order CYPERACEÆ.

Rhyncospora, *Vahl.*
  alba, Vahl. Burlington, *T.* August.

Carex, *Micheli.*
  lanuginosa, Michx. Burlington, *T.* July and August.
  folliculata, L. Burlington, *T.* July and August.
  angustata, (Boot.) Burlington, *T.* July and August.
  filiformis, Linn. Burlington, *T.* July and August.
  striata, Michx. Burlington, *T.* July and August.
  dioica, L. Burlington, *T.* July and August.
  scirpoidea, Schk. Willoughby lake. July and August.

ORDER GRAMINEÆ.

Koeleria, *Pers.*
  Pennsylvanica, D. C. Burlington, *T.* July.
Oryzopsis, *Michx.*
  melanocarpa, Muhl. Willoughby lake.
Aira, *L.*
  atropurpurea, Wahl. Mansfield mountain, *T.* August.
Lolium, *L.*
  perenne, L. Willoughby lake, *C. C. F.*
Muhlenbergia, *Schreb.*
  sylvatica, T. & G. Willoughby lake, *C. C. F.*

CLASS IV. ACROGENS.

ORDER FILICES.

Isoetes, *L.*
  lacustris, L. Brattleboro, *C. C. F.*
Woodsia, *R. Brown.*
  glabella. Willoughby lake.

GEOLOGY OF VERMONT.

Geological Survey.

In my Preface to the Natural, Civil and Statistical History of Vermont, it was stated that Chapter VIII, Part I., remained to be written, after a Geological Survey of the state should be effected. Little did I then think that ten years would be suffered to pass away, and so desirable a work remain unperformed. But such is the fact ; and I am, therefore, yet under no obligation to redeem my pledge, to write that chapter. But since, within those ten years, a Geological Survey of the state was *begun*, and since, through that beginning, and other means, important geological facts have been brought to light, I shall here give a brief history of the labors, which have been performed, and a brief sketch of the knowledge of our geology which has been acquired.

The first state Geological Survey, prosecuted under legislative authority, was, I think, authorized by North Carolina, in 1823. In 1824, the legislature of South Carolina authorized a geological survey ; and in 1830 provision was made for a geological survey of Massachusetts, under the authority of that state. The execution of the survey of Massachusetts was committed to Professor, (now President), Hitchcock, of Amherst College, and was prosecuted with so much ability and success, that most of the other states followed the example, and authorized surveys.

In the execution of these surveys, and in the publication of the results, the state of New York has, by far, outdone any of the other states. The plan of the New York survey embraced, not only the Geology and Mineralogy of the state, but also the Botany and Zoology ; and ample provision was made for carrying out that plan. The corps of surveyors embraced four distinguished geologists, one mineralogist, one palæontologist, one botanist and one zoologist, with their respective assistants. Arrangements were made for commencing the work in 1836, and, after five years of incessant labor, in 1842, several volumes of the Final Reports were in readiness for publication, which, with other volumes afterwards prepared, have since been published. These reports are published in large quarto form, on excellent paper, and fully illustrated with excellent engravings. Eighteen volumes have been published, five of which are devoted to zoology, four to geology, four to agriculture, one to mineralogy, two to organic remains, and two to botany.

The subject of a Geological Survey of Vermont was first brought before the legislature of the state in 1836*. In 1837 the subject was referred to the committee on education, in behalf of which, Professor Eaton submitted to the Senate a very able Report, accompanied by several important documents. The report closed by recommending the passage of a resolution, ordering the report and documents to be printed and circulated among the people of the state, and by expressing the belief that, upon due consideration, the popular voice would be in favor of providing for the survey at the next session of the legislature. In 1838, the subject was again taken up, discussed and dismissed, without any provision being made for commencing the survey; and nearly the same process, with the same result, was repeated at each succeeding session of the legislature down to the year 1844, when a bill, authorizing a Geological Survey of the state, was finally passed, in the Senate, by 20 yeas to 7 nays, and in the House, by 96 yeas to 92 nays, and received the Governor's approval.

This act authorized and directed the Governor to appoint a competent State Geologist, who should have power, with the Governor's approbation, to appoint the necessary assistants, fix the amount of their compensation, and direct their labors. It made it "the duty of the State Geologist, as soon as practicable, to commence and prosecute a geological and mineralogical survey of the state, embracing therein a full and scientific examination and description of its rocks, soils, metals and minerals," and report to the Governor, annually, on the 1st day of October, the progress of the work. For the purpose of carrying the provisions of this act into effect, the sum of $2,000 annually, for the term of three years, was appropriated.

His Excellency, William Slade, Esq., being Governor, upon him devolved the appointment of the State Geologist, and the arrangements, on the part of the state, for carrying the contemplated survey into effect. After some time spent in deliberation and inquiry, he finally commissioned Charles B. Adams, at that time Professor of Chemistry and Natural History in Middlebury College, the State Geologist, who was to enter upon his duties on the 1st day of March, 1845†.

In arranging the details of the survey, it was provided that, so far as should be found practicable, eight suites of specimens of all the rocks and minerals should be collected, trimmed and ticketed. These specimens, when the material admitted, were to be three inches square, and from one to two inches thick. The destination of these suites of specimens were as follows:—one, (and the best, where there was a choice,) for a state collection at Montpelier; one for the University of Vermont; one for Middlebury College; one for Norwich University; one for each of the Medical Colleges, at Castleton and Woodstock; one for the Troy Conference Academy, at Poultney, and one to be the property of the State Geologist. With the approbation of the Governor, the State Geologist appointed the Rev. S. R. Hall and Z. Thompson, general assistants in the field labor, and Dr. S. P. Lathrop, assistant in the depot of specimens, and in occasional field services. The field labors were commenced as soon as the advancement of the season would permit, which was early in May, and were prosecuted during the summer with unremitted diligence. The labors of the general assistants were confined to the northern half of the state ; and, during their four months' services, they together, or separately, visited and explored, more or less thoroughly, about 110 townships. The State Geologist, with Dr. Lathrop and other occasional assistants, labored, for the most part, in the southern half of the state. During the season, about 6,000 specimens were collected and forwarded to the depot, in Middlebury. These were mostly trimmed, ticketed and catalogued, in the course of the following winter.

During the years 1846 and 1847, the business of the Survey was diligently prosecuted by the State Geologist, and the assistants were employed, for several months in each summer, in field labors. At each session of the legislature reports were made to the Governor, of the progress of the work ; and these annual reports were published and circulated among the people.

---

* See Part II., page 104.

† Since the above was written, I have received the painful intelligence of the death of my esteemed friend, Prof. Charles B. Adams. He died of fever, on the 19th of January, 1853, on the island of St. Thomas, W. I., whither he had gone for the double object of improving his health and furthering himself in his favorite pursuits of Natural History. In the death of Prof. Adams, the scientific world has lost a most indefatigable and successful laborer. During the last ten years, few individuals have done more than he did, for the advancement of the natural sciences. By his contributions to Conchology, and his minute investigation of the geographical distribution of mollusks, he has erected to himself an honorable monument ; and, although removed by a mysterious Providence, in the prime of life, and in the midst of his usefulness, his name will long be cherished by his personal friends, and will be handed down to future generations, deeply engraved upon the records of science.

But they were by no means intended to exhibit the entire results of each years' labors, but merely to indicate the advancement of the survey, and to furnish such general information as would enable the people of the state rightly to understand, and duly to appreciate those results, when they should be collected and systematised in a Final Report.

Before the close of the third year, for which provision had been made by the act of the legislature authorizing the survey, the State Geologist was appointed a Professor in the College at Amherst, Mass. Believing that the remaining field labors, for the completion of the survey, would be finished during the next season, and that he should derive much aid in the preparation of the Final Report, from the collections and library at Amherst, he deemed it his duty to accept the professorship offered him ; but he did it, with the expectation that he would not be required to enter fully upon the duties of the professorship, until he had completed the survey and prepared his Final Report, and that our legislature would make the appropriations necessary for those purposes.

At this time, only a part of the specimens, collected from the various sections of the state, had been trimmed, ticketed and sent to the institutions for which they were designed. The remainder, embracing those which had been ticketed for the state cabinet, were in the depot at Middlebury. Anticipating legislative provision at the next session, for the completion of the work as above mentioned, the State Geologist directed his assistant, at Burlington, to obtain, at that place, a suitable room, or rooms, to serve as a depot for the tools, fixtures and untrimmed specimens ; which being done, the articles, amounting to several tons, were forwarded from Middlebury and placed in it. In doing this, he reserved the principal fossils and the specimens ticketed for the state cabinet, which he, soon afterwards, took with him to Amherst, that they might be at hand, for examination and reference, while preparing his final Report.

At the session of the legislature, in Oct. 1847, the subject of the survey was taken up, but no appropriation was made, either for its continuance, or for the preparation of a Final Report.

In 1848 the subject was again taken up, but with no better success, and all that was done in relation to it, was the passage of a resolution, directing the Governor to employ some person to get back into the state, the materials and manuscripts, belonging to the Survey, and place them in the charge of the State Librarian, at Montpelier. That duty the author of this work had the honor of discharging, in the summer of 1849, and his report to the Governor, was published in the Appendix of the House Journal, for that year. Since 1849, the subject of the Survey has, once or twice, been called up in the legislature, but nothing further has been done. The untrimmed and unticketed specimens are lying, packed in boxes, at Burlington, with a portion of the tools and fixtures ; and the remainder are in charge of the State Librarian at Montpelier, and all these are fast losing their value.

The Geological Survey of the state, having been suspended before the examinations were completed; and the results of the labors performed, having never been collected together and systematically arranged, a full and satisfactory account of our Geology cannot yet be expected ; and all that will now be attempted, is a hasty sketch of the general geological features of the state. There are important scientific questions, which an accurate knowledge of the geology of Vermont would, doubtless, very much aid in solving, but the acquisition of this knowledge will require much additional patient investigation and research ; and the discussion of these questions, would require more space than could be afforded to the subject in this Appendix.

CHAMPLAIN ROCKS.

We shall begin our sketch of Vermont Geology at the western border of the state, and, proceeding eastwardly, give some general account of the different rock-formations in their order.

The rocks which occupy the lowest parts of the valley of Lake Champlain belong to that division of the Palæozoic rocks, denominated, by the New York geologists, the *Champlain Group*. Beginning with the oldest and most westerly, these rocks are arranged in the following order :

1. Potsdam Sandstone.
2. Calciferous Sandstone.
3. Chazy, or Isle la Motte Limestone*.
4. Trenton Limestone.
5. Utica Slate and Hudson River Shales.

The *Potsdam Sandstone* is largely developed at several places on the west, or New York, side of Lake Champlain, but is no where found *in situ*, within the limits of Vermont. The remarkable Chasm, through which the river Ausable passes, near Port Kent, is in this rock.

Calciferous Sandrock.—This, the second member of the Champlain group, appears on the Vermont side of the lake, but very sparingly. It is seen at the base of the

---

* This division embraces the Chazy and Bird's eye limestone, and Isle la Motte marble of the New York Geologist.

uplift of Snake mountain, in Addison county, and in a few other places.

Chazy, or Isle la Motte Limestone.—This is the most important member of the Champlain group, and the oldest, which is in much force in Vermont. This rock forms the principal part of the Isle la Motte, the western part of Grand Isle and the eastern shore of the lake, from Charlotte southward. It usually lies in thick, even-bedded strata, dipping, for the most part, slightly towards the east or northeast. It is of a close, compact texture, easily broken into regular blocks, and easily sawed, or hammered, and yet sufficiently strong to serve as the very best of building stone. It constitutes, in many places, the shore of the lake, and is in a position highly favorable for quarrying, and for transportation by water. Quarries of this limestone have been opened in various places, and it is extensively used for building and other purposes. Some of the best of these quarries are on the Isle la Motte; and among these, Fisk's quarry, on the west side of that island, is probably the most interesting and valuable. This quarry rises directly on the lake shore, and lies but a few rods from the usual line of steamboat navigation through the lake; and the shore is here so bold that the largest vessels on the lake may safely approach it within a few feet, and a very good landing is constructed. The quarry presents a working breast, rising about 35 feet above the lake. The strata vary somewhat in their aspect, but they are, in general, of a bluish gray color. The thickness of the strata, varies from eight inches to five or six feet, and each stratum preserves its thickness with great uniformity. The general dip of the strata is about 4° towards the north east.

Other excellent quarries have been opened on this island, of which Hill's quarry, and the Black Marble quarry, on the east side, are the most important. The Isle la Motte limestone, obtained at these quarries, and at others along the shore of the lake, is already extensively used in the construction of buildings and rail road bridges, and considerable quantities are sawed for hearths, or for being polished as marble. The black marble takes a very fine polish, and some of it is exceedingly beautiful. The surfaces of the natural seams and fractures of the strata of this marble, are frequently covered with a black, often iridescent, glazing, resembling the surface of anthracite, and it is probably carbonaceous.

The Isle la Motte limestone abounds in fossils, among which, species of Maclurea, orthoceras and corals are conspicuous, being seen in the worn and weathered surface of the rocks, in great numbers.

Maclurea magna.

This cut represents the Maclurea magna, as it appears on the worn surfaces of the rocks, on the Isle la Motte, and at most places where the rock is found. The Maclureas are spiral shells, resembling in form our little fresh water shell called the Planorbis, but they grow to a very large size. When in the surface of the rock, and about half worn away, they frequently present a spiral coil, eight or ten inches in diameter, sometimes having so much resemblance to serpents coiled up, that the early settlers in the valley of Lake Champlain, regarded them as *petrified snakes*.

Orthoceras.

This cut represents the general form of the Orthocerata, as they appear in the weathered surfaces of the Isle la Motte limestones. The number of species found in this formation is very great, and the number and magnitude of the individuals, accumulated at some localities, is remarkably so. At some places on the Isle la Motte the rocks, for rods in extent, and several feet in thickness, seem to be made up almost wholly of Orthoceras, closely packed together in a limestone cement. Some of these are 18 or 20 inches long and 6 or 8 inches in diameter at the larger end. The interior of these shells is usually filled with calcareous spar, but they are sometimes found empty.

Columnaria alveolata.

Several species of coral are found in this limestone. Some of these have a structure resembling that of honey comb, and hence they have been supposed, by persons ignorant of geology, to be honey-comb petrified. Like the coral reefs, which are now in the process of formation in many parts of the ocean, they are the work of minute insects called *zoophytes*.

The *Trenton Limestone.*—This lies next in the ascending series. It occupies only a small extent of territory in Vermont, but is every where recognized by its characteristic fossils. From near the south end of the lake it extends northward as far as

Charlotte, showing itself in the uplifts, at various places. It appears again in South Hero, and extends northward, through the western part of Grand Isle, and constitutes the south eastern and highest parts of the Isle la Motte. It also caps some of the elevations near the Medicinal Spring, in Highgate.

This rock is sufficiently compact and firm, in some places, to serve as a building stone, but it is, for the most part, thin bedded and shaly, and of very little value, excepting that it forms the basis of a good soil.

The species of fossils in the Trenton Limestone are exceedingly numerous. In the single genus, orthis, they amount to no less than seventeen, which are peculiar to this rock; and in many other genera the species are nearly as numerous. In Grand Isle, this limestone is rather thick bedded, is of a light gray color, and almost entirely made up of shells of the orthis. This stone, when the edge of the stratified mass was exposed to the heat, was found, unlike most limestone, to withstand the action of the fire, and, on that account, it was much used by the early settlers, for the construction of fire-places, on which account it is still distinguished by the name of *fire stone*.

 This figure exhibits the general form of an Orthis.

Utica Slate and Hudson River Shales.—Still higher, and to the eastward of the Trenton Limestone, lie a series of black slates. Some of these slates are rather thick bedded, are quite calcareous, and break with conchoidal fracture, and, lying immediately above the Trenton Limestone, are in some cases, with difficulty, distinguished from it. In other parts, the slaty laminæ are quite regular, and readily separated. But far the greater part of it appears to be crushed and broken into wedge-shaped masses, interspersed with seams of calcareous spar. In many places, these wedge-shaped shaly masses are covered with glazing, giving them the lustre and appearance of anthracite. So strong is this resemblance to coal, that many have supposed that there must be coal beneath it, and considerable excavations have been made in it with the vain hope of finding it.

These shales are the only rock in the place, in Alburgh; they form nearly the whole of North Hero, the eastern half of Grand Isle, Rock Dunder, Juniper Island, and most of the other small islands; and it forms the bank of the lake, along the east side, throughout almost its entire length. With the exception of that portion of them which lies next to the Trenton Limestone, these shales are totally useless as a building stone. They, however, disintegrate into a black, rich soil, and are a valuable material for making roads.

Trilobite.

Trilobites are occasionally met with, particularly in the older portion, which has been sometimes separated from the other shales, under the name of Utica Slate, as well as in the Isle la Motte Limestone, before described.

Graptolites are common in some few places, but as a whole, these shales are quite barren of fossils. The above cut will furnish some idea of the general appearance of graptolites. It pretty nearly represents *Graptolites amplexicaule*, found in the Trenton Limestone.

Graptolites.

Red Sandrock.—The next series of rocks, lying above and to the eastward of shales, has been generally known in the neighborhood, as the Red Sandrock formation. This rock extends from south to north nearly the whole length of Lake Champlain. It makes its appearance in uplifts, presenting mural precipices towards the west, with a dip from 5° to 30° towards the east. Its western limit is marked by a series of considerable hills, which are at some little distance from the lake shore at the south and in the northern part of Franklin county; but from Shelburne to St. Albans Bay, it lies, for the most part, along the shore of the lake. Sugar Loaf and Glebe Hill, in Charlotte, Red Rocks and Lonerock points, in Burlington, and Mallet's Head, in Colchester, belong to the same line of uplifts. The accompanying cut represents a section passing through the uplift at Lonerock Point, where the thick bedded sandrock is seen resting on black glossy shales.

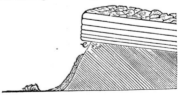

The shale, here, has been washed out from under the sandrock, large masses of which have broken off by their weight and fallen into the lake. These, excepting one, are covered, when the lake is high, but that one is seen at all times, and from all points of view, to stand prominently out of the water; hence the name, *Lone-rock*

Point. We are aware that it has been generally called *Sharpshins*, but we think that it is quite time that vulgar name was discarded.

Towards the northern part of this uplift there commences a bed of dove-colored limestone, between the shale and the sandrock, which appears with increasing thickness at Mallet's Head and St. Albans Bay; and at Swanton is quarried for marble. From Swanton it continues northward into Canada.

One of these uplifts, that of Snake Mountain, affords a fine exhibition of all the members of the Champlain Group of rocks, which we have been describing, as may be seen by the section below, which is copied from Prof. Adams' Second Annual Report on the Geology of Vermont, p. 168.

a. Red Sandrock, forming the summit of the mountain, with an easterly dip of 20°.
b. Debris from the Red Sandrock.
c. Hudson River Shales, considerably covered with drift and debris.
d. Utica Slate.
e. Trenton Limestone.
f. Isle la Motte Limestone.
g. Calciferous Sandstone.
h. Clay.

The rocks, which constitute the Red Sandrock series, differ very much in color, and in composition, or lithological character. The lower strata are, in many places, considerably calcareous, and thick-bedded, with the planes of stratification so much obliterated, as to give them the appearance of an igneous, or unstratified rock. The color of this portion is often gray, or variegated with different shades of brownish red and yellowish white; and parts of it are sufficiently calcareous to admit of being polished, and make a very compact and beautiful variegated marble. The best specimens of this have been found in boulders in connexion with the drift.

The middle portion of this series is almost entirely siliceous, and, through a great part of its extent from south to north, is of a dark reddish brown color; and it is the color of this portion which has given the name of Red Sandrock to the series. But in some places, this middle portion is nearly destitute of coloring matter, appearing as a light gray stratified quartz rock. In some places it is so purely siliceous as to be suitable for the manufacture of glass. Cases also occur, where a stratum of pure white quartz intervenes between strata which are highly colored. There is an example of this in Willard's quarry, in Burlington.

The colored strata of this sandrock furnish a very durable and beautiful stone for foundations and underpinnings of buildings, and, though somewhat refractory and difficult to work, has been very much used for that purpose. The foundations of the greater part of the buildings in Burlington, are of this material.

From the middle portion of the Red Sandrock series, the strata become more and more calcareous, in proceeding upward and eastward, till they, at length, become in many places a very pure limestone. This limestone is, generally, of a bluish color; but in some places, particularly in the eastern part of Shelburne, its color is pure white. Portions of this limestone make the very best of quicklime, which is largely manufactured from it, not only for use in the neighborhood, but for transportation into the interior of the state, and to places where no good limestone exists*. At Penniman's quarry and kilns, which are by the side of the railroad above Winooski Falls, Messrs. Penniman & Catlin manufactured, in 1852, about 67,000 bushels of quicklime, and others, in the neighborhood, manufactured about 40,000 bushels, making over 100,000 bushels, the principal part of which was sent by railroad into the central and eastern parts of the state, and to other parts of New England.

Some portions of the Red Sandrock series are very regularly and handsomely stratified, but other portions are much disturbed and broken, or bent and folded. One of the most interesting plications in this rock, which have been noticed, is in Monkton. The south end of this plication is represented in the figure below.

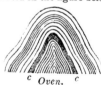
c Oven. *c*

The white spaces between the curved lines represent the edges of the strata. These are all of the ordinary sandstone. The broad dark stratum, *c c*, is argillaceous slate, having the laminæ nearly perpendicular to the plane of deposit. The upper portion of this, which is left wholly dark, has been removed, forming the cavity called *the Oven.*

* The good qualities of the quicklime manufactured from this stone is not only attested by those who have used it in this country, but has been fully acknowledged, by competent judges, abroad. In 1851, U. H. Penniman, Esq., sent out a cask of his lime for exhibition at the World's Fair, in London. This lime was examined, by a Jury, appointed for that purpose, under the royal commission, and this jury awarded him a Prize Medal and Certificate, as testimonials of its superior excellence, placing it in the first rank in competition with the world.

| TACONIC ROCKS. | ROOFING SLATE. | STOCKBRIDGE LIMESTONE. |

The portion represented in the cut, is about 30 feet broad at the base.

The general strike of the Red Sandrock formation is about N. 20 E., and the dip varies from 4° to 25° or more. Its width, from east to west, is very variable, but will average, perhaps, five miles.

This rock is very barren in fossils, and those found are very obscure, consisting of fucoidal layers, and fragments of crinoidea and trilobites. Marks of rain-drops, and wave and ripple marks are very common, and well defined. The fragments of trilobites have been found most abundant in this rock in Highgate, but they are there so much decayed, and so obscure, that it is very difficult to determine the species.

TACONIC ROCKS.

Under this name have been embraced the rocks in the southern half of the state, which lie to the eastward of the Champlain group, and to the westward of the main ridge of the Green Mountains. They occupy a large part of the counties of Bennington, Rutland and Addison. They derive their name from a range of high lands, which extend from the western part of Massachusetts into Vermont, and which are called the *Taconic Mountains.* The true geological position and character of these rocks is not yet well settled. While some regard them as primary, and others as metamorphic silurian rocks, Prof. Emmons, and some others, have maintained that they are a distinct group of palæozoic rocks, which are older than the Potsdam Sandstone, which is the oldest member of the Champlain group.

The Taconic group of rocks consists of Roofing Slates, Sparry Limestone, Magnesian Slates, Stockbridge Limestone and Granular Quartz.

Roofing Slate.—The roofing slate of this formation is found principally in the western part of Rutland county, particularly in the towns of Castleton, Poultney and Fairhaven. Some sixteen or eighteen slate quarries have already been opened in these towns, many of which yield slate of a very superior quality. There are two principal varieties of the slate, one of which is of greenish color, and the other reddish brown. Several of the quarries have been opened very recently, and have not yet yielded a large amount. The yield of all the quarries above mentioned, in 1852, was estimated to exceed 10,000 squares, and the annual yield will doubtless go on increasing, from year to year, indefinitely. It already finds its way, not only to Boston and New York, but to Buffalo, Cleveland, and other cities at the west.

Sparry Limestone.—This rock stretches through the western parts of the counties of Addison and Rutland. It is divided and checked by numerous beds of calcarious spar. Its color is bluish, or gray of different shades.

Magnesian Slates.—These slates lie to the eastward of the roofing slates and sparry limestone, and to the westward of the Stockbridge limestone. They sometimes alternate with the latter, as the two former do with each other. They are most fully developed in the northwestern part of Bennington county, and southwestern part of Rutland county.

The magnesian slates are usually of a light grayish color, and often of a greenish hue. They, in many places, are easily split into broad flat masses, the surfaces of which often have a pearly lustre, and an oily feel. But for the most part, these slates are largely filled and checked with veins and seams of white quartz.

Stockbridge Limestone.—In an economical view, the Taconic group probably furnishes the two most valuable rocks in the state, excepting only the Isle la Motte Limestone; and these two are, the roofing slate, already mentioned, and the Stockbridge Limestone.

Commencing at the south line of the state, in Pownal, the Stockbridge Limestone forms a belt, which extends northward through the counties of Bennington, Rutland and Addison, as far as the town of Monkton. This belt is, on an average, nearly five miles wide, having the Magnesian slate on the west, and a range of Granular Quartz on the east. To the northward of Bennington county this rock occupies, for the most part, the valley of Otter Creek.

This range of limestone furnishes, through almost its entire extent, an abundance of excellent marble. Its color is generally light, varying from dove color to the purest white. Some portions of it are of a light flesh-color, and others are beautifully variegated; and at several places a very good statuary marble is found. Stephenson's statue of the wounded Indian, which was exhibited at the World's Fair, in London, was made from Vermont marble, obtained, I think, from a quarry in Rutland.

Marble quarries, at various places in the Stockbridge Limestone, have been more or less worked for many years. The principal of these are in the towns of Dorset, Rutland, Pittsford, Brandon and Middlebury. The great expense of transportation, for a long time prevented these quarries from being extensively worked, but the construction of railroads, along the whole line of this formation in Vermont, has opened easy outlets for the marble, and already largely increased the marble business. We have not room in this Appendix to go into particulars with regard to the yield of the various quarries. There is no doubt but that the marble business is destined to be one of the most important resources of the state.

Granular Quartz.—This forms a narrow range, or belt, extending from the south line of the state to the northern part of Addison county, between the Stockbridge limestone on the west, and the rocks of the Green Mountains on the east. This range is quite irregular, and in some places not easily traced. It is mostly of a dark gray, or brownish, color, and is very barren in minerals, containing only occasionally crystals of sulphuret of iron and schorl.

The Taconic rocks, generally, contain few interesting minerals; and the fossils which have been found in them are very few and obscure. These rocks have, for the most part, a steep eastern or southeastern dip.

All the rocks, in place, in Vermont, lying to the eastward of the Champlain and Taconic groups, already mentioned, and occupying more than three-fourths of the state, have, till recently, been regarded as primary; but facts are daily coming to light which render it probable that the greater part of them belong to the palæozoic, or silurian series, and that they have been changed, and have had their fossils nearly all obliterated by heat. We shall not enter at all into the discussion of the geological age of these rocks, but confine ourselves to a hasty general description of them.

In a former work* I have regarded these rocks as primary, and have described them as belonging to two grand divisions, which are distinguished from each other by very obvious characteristics. The first of these divisions, lying next eastward of the rocks already described, and constituting the main central body of the Green Mountains, was denominated the *Telcose slate formation*, or division, from the general prevalence of that rock, particularly in the northern portion of it. The other division, extending eastward from this to Connecticut river, was called the *Calcario-mica slate formation*. The line between these formations is, for the most part, well defined and easily traced, from south to north, through the whole length of the state. From the south line of the state, in Halifax, it runs in a direction nearly north through the towns of Newfane, Cavendish, Bridgewater and Bethel, to Northfield, and thence a little east of north through Montpelier,

---

* Geography and Geology of Vermont, for Schools.

Calais, Craftsbury and Irasburgh, to Memphremagog lake.

TALCOSE SLATE DIVISION.

This division, which constitutes the central portion of the Green Mountains, varies in width from about 14 miles, in the south part of the state, to 30 miles in the northern, and it embraces the loftiest mountain summits in the state.

The rocks of this division, though, generally, more or less talcose, vary considerably, in their aspect and composition. Beginning in the northerly part of the state, with the rocks next eastward of the Red sandrock formation, we find them shaly, very quartzose, and with very little talc or mica in their composition. They have a dip of about 40° to the east, and in some places the beds, or strata, are a fine conglomerate, the rounded pebbles being, for the most part, quite minute. In some parts the rocks have a greenish, or chloritic hue, and are so thick bedded and compact, as to make a very good building stone. This is particularly the case in the towns of Jericho and Westford.

In proceeding eastward the dip of these rocks increases rapidly, till it becomes vertical along the western foot of the Green Mountains, forming a synclinal axes. The line of this axis passes through the towns of Berkshire, Enosburgh, Bakersfield, Cambridge, Underhill and Jericho. To the eastward of this line the dip continues nearly vertical for several miles, being sometimes to the west, and, at others, to the east, forming a succession of synclinal and anticlinal axis. The dip then becomes uniformly westward, and continues so through the eastern part of the formation.

Interstratified with the Talcose Slate, we frequently find well characterized clay and mica slates; and in many places along the slopes of the Green Mountains, the talc and mica enter into the composition of the same slates in such equal proportions as to make it difficult to say which name more properly belongs to it. In Berkshire and Enosburgh there are extensive beds of well characterized clay slate, portions of which may hereafter be found suitable for roofing. A little further east, in Richford, is a narrow range of plumbaginous slate, which has been traced southward as far as Huntington, in the south eastern part of Chittenden county. In Cambridge, it is found sufficiently soft and black to form a tolerable substitute for black lead.

In many places along the western slope of the Green Mountains, the rocks lie in thick beds, or strata, each stratum splitting with nearly equal facility in all directions, and approaching to gneis in appearance and composition; and it has been proposed to denominate the rocks, which constitute this great axis of the Green Mountains, *Green Mountain Gneis.*

The rocks embraced in our Talcose Slate Division, in the southern part of the state, are much less characteristic, than in the northern, and the different varieties of rock are much more broken and jumbled. No true granite or gneis have been observed in this formation, in the northern half, but both these rocks show themselves in the southern half, in various places. With the exception of a few small patches at the south, and three or four thin beds of saccharoid limestone, at the north, there are no rocks which contain any sensible amount of lime, in the whole territory embraced in what we have called the talcose division, and which constitutes about one-third of the whole surface of the state. Quartz is the great mineral element of this formation, for, besides forming the principal part of the various slates, shales, &c., it is almost every where infused and spread through them in great abundance, in the form of seams and veins. The color of these seams and veins is usually yellowish, white, or hyaline.

Gold Formation.—It has been known for a great number of years, that we have, in Vermont, a formation agreeing, in almost all respects, with the gold formation in the southern states, and in many other parts of the world; and it is a well known fact, that native gold was found here more than twenty-five years ago. The statement, which we published in a note on page 127, Part III., respecting a lump of gold picked up in Newfane, and weighing 8½ ounces, was extensively circulated in the newspapers soon after it was found. Our statement was derived from Gen. Martin Field, who had the lump in his possession. It was a fact well known to us, when our History and Gazetteer were published, that gold had been found in small quantities in the township of Somerset, by washing the alluvial gravel; but believing then, as we do now, that the success of Vermonters, in digging for gold, will be best secured by observing the Quaker's directions, *never to dig for it more than plough deep*, we took no pains to give prominence to these facts.

What we here call the Gold Formation constitutes a part of what we have been describing under the name of the Talcose Division. It forms a narrow and irregular belt, extending along near the eastern margin of the great division, above mentioned, and reaching through the entire length of the state. Beginning at the line of Massachusetts, in Whitingham, it extends northward, through the western part of Windham county, through Ludlow, Bridgewater and Rochester, in Windsor county; through Roxbury, Moretown and Waterbury, in Washington county, and thence through Morristown, Eden, Lowell and Troy, to the north line of the state. The rocks, which mark the line of this formation, are talcose slate, steatite and serpentine, accompanied by magnetic, specular, chromic and titaniferous iron, also sulphuret and hydrous peroxide of iron. At some places, beautiful specimens of rock crystal occur, many of which are traversed in various directions by hair-like crystals of rutile, rendering them exceedingly interesting to mineralogists. The fine specimens of this kind which have been found in the drift in the valley of the Connecticut, probably had their origin in this formation. Although, long since, aware of the fact that the formation, in which gold was found in Windham county, extended through the whole length of the state, we had no knowledge that gold existed in Vermont to the northward of that county, previous to the fall of 1852, when gold was discovered in Bridgewater, Windsor county, by a Mr. Kennedy, and the discovery made known to the public by Prof. O. P. Hubbard, of Dartmouth College. The gold is found there in seams of quartz, and also, in alluvial gravel. Sufficient time and opportunity for examination have not yet been had, since the discovery was made, to determine its value. Some specimens of the gold, which we have seen in the quartz, though small, were exceedingly fine and beautiful.

In the neighborhood of the gold in Bridgewater, very fine specimens of galena, or sulphuret of lead, are also found, but we are not informed with regard to its extent: but as Bridgewater is our native town, we hope ere long to have occular view of the revelations, which are being made there.

Although the formation, (in which gold is found) may be traced through the entire length of the state, it is not to be expected that gold will be found through its whole extent; nor is it, at present, at all certain that the *placers*, where gold has already been found, will yield gold enough to pay for working. This same gold formation, which passes through Vermont, has been traced from the north line of the state at Troy, nearly 200 miles into Canada. It passes along a little to the westward of Memphremagog lake to Orford, near Sherbrooke, and thence takes a more northeasterly course to the neighborhood of Quebec. Gold was found, in this formation, along the river Chaudiere, as early as 1834, and the discovery was announced in Silliman's Journal in April, 1835. From that time gold was collected there, in small quantities, up to the time of the discovery of gold in California; amounting in the whole to only a few hundred dollars. Since the geological survey of Canada has been in progress, more attention has been given to the subject, and it is found that the auriferous district is quite extensive. During the last three or four years the search for gold has been prosecuted more extensively, and the yield has amounted to several thousand dollars. In Ascot, near Sherbrooke, gold has been found in veins, associated with copper pyrites in a quartz gague; and it is reported that a lump of gold, weighing 14 oz., was obtained in that vicinity in the fall of 1852.

The steatite, or soapstone, and the serpentine, which we have mentioned, as indicating the line of the gold formation, are, probably, destined to be of quite as much economical value to the state, as the gold itself. The steatite is abundant, and is, in many places, of a very good quality. It has been quarried at Grafton, Bridgewater, Bethel, Moretown, Waterville, and, perhaps, a few other places. The serpentine is largely developed at Cavendish and Ludlow, at Roxbury, and at Lowell and Troy. Much of this serpentine is compact and firm, beautifully variegated with every shade of green, from the lightest tints to an almost perfect black; and, as it admits of a high polish, and is unaffected by heat and acids, it forms a most valuable ornamental marble. It has long been used, to a limited extent, in some of the neighborhoods where it is found, for fire-places, centre-tables, &c., and the opening of rail roads, through these several localities, will, probably, be the means of bringing this valuable marble extensively into use.

CALCARIO-MICA SLATE DIVISION.

Under this general name, we embrace all the territory of Vermont, not included in the divisions already described, with the exception of a few tracts of granite. It has been called the *calcario-mica slate* formation, or division, from the fact, that it consists, to a very considerable extent, of impure limestone, interstratified with argillaceous and mica slate. These three constitute the principal rocks, but they, in many parts, run into several other varieties of slate. Through the central part of Orleans county, and in Caledonia county, are extensive ranges of what might properly be called hornblende slate. In the northern part of Essex county, extending into Canada, is a range of siliceous slate; and in the southern part of that county there is a considerable development of chlorite slate. The western portion of this formation, from Barnard northerly to lake Memphremagog, is mostly clay slate. This slate constitutes a large proportion of that beautiful and fertile swell of land extending from Winooski to White River, through the towns of Berlin, Williamstown, Brookfield and Randolph. It is also largely developed in the north part of Montpelier, and in Calais, Craftsbury and Coventry. At Berlin, this slate has been found to answer very well for roofing; and it is not improbable that good roofing slate will be obtained from some of the other localities, which we have mentioned.

Clay slate also exists, in large quantities, along the Passumpsic and the west bank of the Connecticut river, in the counties of Orange and Caledonia, and also in the southeastern part of Windham county. At the latter place, it extends through the towns of Guilford, Brattleborough and Dummerston. The slate here is found to be very suitable for roofing, and has been, more or less, quarried for that purpose for many years.

The mica slate of this division is not, in general, very well characterized as mica slate. Indeed, the slates, or shales, of this division, appear to be a combination, or jumble, of almost all the known varieties, sometimes exhibiting a predominance of one kind and sometimes of another; and, again, we find the materials of three or four different varieties combined in a single stratum. There are, however, some small tracts, to which the above remarks are not applicable. This is the case with some parts of Windham and Windsor counties, where mica slate is found, well characterized, and forming a valuable and beautiful material for flagging.

The limestone of this division is, every where, very impure, containing a very large proportion of siliceous sand. It is burned in several places for quicklime, but the lime is nowhere of a good quality. It is made to answer in mortar for stone work, where better is not to be had; and it is usefully applied, in agriculture, as a fertilizer, to soils deficient in lime. But the lime made from the beds of shell-marl, which abound in this division, though that is not of the best quality, is much preferable to the above, both for the purposes of masonry and agriculture.

The color of this limestone, where unaffected by the weather, is of a bluish shade, and the stone is very compact and homogenious, splitting, or breaking, with nearly equal facility, in all directions. Where long exposed to the weather, it is recognized at once by its rust-colored, rotten surface. This rotten surface consists of the siliceous sand, which remains after the lime, which

| TERTIARY FORMATION. | WHITE CLAY. | YELLOW OCHRE. | IRON ORE. | BROWN COAL. |

had cemented it together, has been dissolved and washed out. In Hardwick, Berlin, and some other places, this blue silicious limestone is regularly arranged in parallel strata, showing very distinctly planes of deposit. But it more commonly occurs in irregular beds of unequal thickness, in the different varieties of slates.

Throughout nearly the whole of what we have called the *talcose* division of the state, the waters are soft and very pure, but those of the calcario-mica slate division are, on the contrary, hard, being, in general, strongly impregnated with lime. But the reason is obvious; for, in the former case, there is no lime, excepting what exists in the materials of the drift, while in the latter, besides the lime in the drift, the blue silicious limestone of the formation is diffused through every part, and being, by exposure, readily disintegrated and dissolved, keeps the waters of the neighborhood constantly impregnated with lime.

The principal metalic ores found in this division, are iron and copper *pyrites*, or the sulphurets of iron and copper. These, and particularly the sulphuret of iron, are found, though, for the most part, sparingly, throughout the whole division, usually in the form of small yellow cubes, which are not unfrequently mistaken for gold. The most extensive deposits of pyrites are at Strafford, Corinth, Woodbury and Brighton. That at Strafford is fully described, together with an account of the manufacture of copperas from it, in our description of that town in Part III., page 167. The veins of pyrites at Corinth consist of the sulphurets of copper and iron in nearly equal proportions. In Thetford there is a small vein of galena or sulphuret of lead.

In the different parts of this great division of the state, there is found a considerable variety of interesting minerals, most of which are mentioned in Part III., under the names of the towns in which they are found.

Tertiary Formation.

It has been generally supposed, until within a very few years, that no geological formation existed in Vermont, of an age intermediate, between the lower silurian and the drift, or post-tertiary. In other words, it was supposed that the corboniferous series, and the secondary and tertiary formations, were entirely wanting. Still it has been long known that there was a series of deposits along the western foot of the Green Mountains, the geological age of which was extremely doubtful, and it was not till the discovery of the deposit of Brown Coal in Brandon, in 1848, that the uncertainty was in any degree removed.

The deposits above mentioned, commence in the south part of the state, at Bennington, and, extending northward, have been traced as far as Milton, in Chittenden county; and, probably, will be traced still further north into Canada. The material in these deposits, which first brought them into notice, was the brown oxide of iron, or brown hematite. This iron ore has been known and worked at Bennington, Pittsford, Brandon and Monkton, for a great number of years.

It was also early noticed that there were beds of a beautiful white clay, along the same line, generally in the vicinity of the brown iron ore. The nature of this clay was little understood, but being found to answer as a substitute for whiting, it was, for a while, considerably used in making putty for setting glass. Hence these beds of clay became known as *putty beds*. During the war with Great Britain, in 1812, one of these beds in Monkton was examined by Prof. J. Muzzy, who published an account of it, with an analysis of the clay, in the "Repository," a monthly periodical, published at Middlebury. He showed it to be *kaolin*, or porcelain clay; and efforts were made, about that time, to get up a manufactory of porcelain ware.

Subsequently, associated, for the most part, with the beds of brown hematite, were found, not only extensive beds of pure yellow ochre, but large quantities of the ores of manganese, both of which are articles of much economical value; and at some localities in the same connexion, were also found beds of very pure white quartz sand. The deposits, above mentioned, along the western foot of the Green Mountains, have been, as already remarked, known for many years. But in addition to these, in sinking shafts in the iron ore-bed, in Brandon, about 1848, a deposit of Lignite, or Brown Coal, was discovered, which has thrown some light upon the geological age of the deposits above described.

Of all the localities, to which we have referred, that at Brandon is the most interesting, not only on account of the Brown Coal, but on account of having all the other materials in conjunction with it. We have here, in the area of a few acres, the following substances, which are of economical value:

1. Pure white quartz sand.
2. Beautiful white and stained kaolin, or porcelain clay.
3. Yellow ochre.
4. Brown hematitic iron ore.
5. Ores of manganese.
6. Brown Coal.

The two first, in the above list, make their appearance at, or very near, the surface;

COAL AND IRON AT BRANDON.

and the coal may also be traced to the surface. But the great bulk of the clay, iron, manganese and coal, is buried at a considerable depth beneath the drift, which consists principally of pebbles, gravel and ochrey earth.

In the area above mentioned, there have been sunk, principally for obtaining the iron ore, five shafts, to depths varying from 100 to 130 feet. From these shafts, at depths of 80 or 90 feet, drifts have been sent off in various directions, connecting the different shafts, and various galleries have also been formed by the removal of the ore. By the shafts and drifts, the iron, clay and coal have been passed through in various directions, and something has been learned respecting their relative position and extent. The locality was visited during the summer of 1852, by a number of distinguished geologists, among whom were Dr. Hitchcock, President of Amherst College, Sir Charles Lyell, Prof. James Hall, of Albany, and Mr. Foster, United States Geologist; and the conclusion seems to be, that the formation, embracing the hematitic iron ore, the manganese, the kaolin and the coal, are of the same geological age, as the brown coal of Europe, and, therefore, belong to the tertiary period.

The extent of the brown coal at Brandon, is not yet ascertained. It shows itself at, or very near, the surface of the ground, and has been found at the depth of 90 feet. It seems to descend somewhat obliquely, by the side of the kaolin, in a columnar form, about twenty feet wide and fourteen feet thick. The carbonaceous materials are of a dark brown color, approaching to black. Some portions of them are very completely converted into coal, while, in other parts, the woody structure and the form of the trees are clearly seen. Scattered in this mass of materials, for the most part near the surface, are found many varieties of seeds or fruits, which vary in size from that of a fig to that of less than a barley-corn. These fruits were at first supposed to be butternuts, walnuts, chestnuts, hazelnuts, &c., such as are now indigenous in Vermont, but a very slight examination suffices to show that they are unlike any vegetation now growing in our country.

President Hitchcock, in an interesting article* on the deposit of brown coal at Brandon, has figured about twenty species of the fruits found in it, and his figures, for the most part, agree very well with specimens of the fruit obtained by myself from the coal. To furnish some idea of these fruits, I give, in the next column, figures of a few of such of the fruits as I have in my possession.

* Silliman's Journal of Science, Vol. XV.,—p. 95.

FIGURES OF FRUIT.

The Brandon coal contains a considerable amount of earthy matter, but it burns readily, even when first taken from the bed; and is employed, almost exclusively, for fuel in driving the steam engine, by which the iron ore is raised and the water pumped from the mine.

As the hematitic iron ore, kaolin, manganese, &c., which occur at Brandon in conjunction with the coal, are found at numerous other places in Vermont, along the western foot of the Green Mountains, it is, also, highly probable that at some of these places, coal will likewise be found.

The conclusion to which President Hitchcock has arrived, from his examination of the subject, is, that the formation of which the Brandon deposit is a type, belongs to the tertiary period, and that it extends through the entire length of the United States, from Canada to Alabama.

IGNEOUS ROCKS.

The only unstratified igneous rocks in Vermont, which occupy any considerable extent of territory, are granite and serpentine. The fields of granite are nearly all included in the calcario-mica slate division of the state. The granite appears, every where, to have been forced up from beneath, in a melted state, between the strata and beds of slate and limestone, sometimes in small isolated elevations; but for the most part in long narrow ranges, extending north and south, in accordance with the strike of the outcrop of the strata. This is particularly observable in the eastern part of the counties of Orleans and Washington, and in the western part of Caledonia county.

The most extensive tracts of granite, and the only ones, which have much width from east to west, are in Essex county, and in the southwestern part of Caledonia county, and the adjacent parts of the counties of Washington and Orange. It was from the southwestern part of this last tract, in Barre, that the granite was obtained of which the State House was built. Further south, in the counties of Windsor and Windham, there are many isolated patches of granite and gneis, but with the exception of Ascutney mountain, they are of quite limited extent. In numerous places, granite is seen traversing the other rocks, in the form of dikes, veins and seams. This is particularly observable in Marshfield and Woodbury; and this fact, and, also, the fact that fragments of clay slate are there found, embedded in the granite, make it certain that the granite has been in a melted state since the formation of the slate.

Granite boulders are scattered, more or less abundantly, over the whole of this division of the state. In the northeastern part, they are exceedingly numerous, and many of them are of very great magnitude. From a single granite boulder, in Greensborough, the material for a good sized stone house, including the walls of the cellar, were obtained, without using it all. Another isolated boulder in that town, is 41 feet long, 22 feet high, and, in the widest part, 25 feet wide, and is calculated to weigh more than a thousand tons. About half a mile from this large boulder, there are two smaller granite boulders, about 80 feet apart, so nicely balanced, on other granite rocks, as to be easily rocked by a push with the hand, and hence they have acquired the name of *the rocking stones.*

Rocking Stones.

The accompanying rude cut will serve to show their relative positions. They are both considerably elevated above the surrounding country. The one at the right hand in the figure is 9 feet high, 12 feet long, and weighs about 70 tons. It rests upon another mass of granite about 16 feet high. The other rocking stone, at the left, is 8 feet high and 11 long, weighing about 40 tons.

The granite of this division, though generally good, and, much of it, of a superior quality for building purposes, exhibits, nevertheless, several varieties. Perhaps the most remarkable of these, is that found in place in Craftsbury and Northfield, and which has, sometimes, been called *Nodular Granite.* The granite is of the ordinary character, with the exception of having flattened balls of black mica, about one inch in diameter, scattered through it, like plums in a pudding. These balls, or concretions, are composed of concentric layers of black mica, separated from each other by extremely thin layers of pure white quartz. In some portions of this granite the balls, or nodules, constitute quite one-half of the entire mass, while, in other portions, they are scattered very sparingly, often several inches asunder, in all directions. The only locality, beside those just mentioned, where this granite is found in place, is just over the north line of Vermont, in Stanstead, C. E. Boulders of it are scattered, sparingly, over a great part of the surface of the counties of Orleans and Caledonia.

The *serpentine* has been already mentioned in our account of the talcose division, as occurring along the line of the gold formation. Some of those tracts are quite extensive, forming hills of considerable elevation. This is the case in Cavendish, Lowell and Troy. In the serpentine in Lowell, fine specimens of asbestus and of different varieties of amianthus, are common. In Troy, it contains a large irregular bed or vein of iron ore. The ore appears well, and extensive works were erected for manufacturing it into iron; but the difficulty of working it, on account of the titanic acid it contains, and the cost of transportation, rendered the business unprofitable, and the works were, therefore, abandoned and have gone to decay. The following is the result of the analysis of this titaniferous iron ore, by Mr. Olmsted:

Peroxide of iron, - - - - 81.20
Protoxide " - - - - - 13.37
Titanic acid, - - - - - 4.10
Silica, - - - - - - - 1.33

100.00
Metalic iron, - - - - - 66.62

Chromic iron is also met with in many places in the serpentine of this neighborhood. In Jay, there are veins of it two feet wide. The ore is of good quality, and might easily be obtained to any amount. Its analysis, by Mr. Hunt, gave the following result:

Green oxide of chromium, - - 49.90
Protoxide of iron, - - - - - 48.96
Alumina, with traces of silica, &c., 1.14

100.00
One hundred parts of this ore will yield 191 parts of chromate of lead, or chrome yellow.

TRAP AND PORPHYRY.

These are found in Vermont only in the form of dikes, or intrusive beds among the other rocks. Trap dikes are met with in all parts of the state, but they are much more common in some parts than in others.

In the central part of the state, in the talcose slate formation, they are exceedingly rare. They are more common in the eastern part of the state; but abound most of all in the vicinity of Lake Champlain, and, particularly, in the neighborhood of Burlington.

The strike of the various stratified rocks in Vermont is, generally, from a little west of south to a little east of north, while the trap dikes, for the most part, cut through these rocks in a direction nearly east and west. The width of these dikes varies from three or four inches to five or six feet. The width of the greater part differs but little from three feet. They sometimes cut through the rocks quite obliquely, both to the strike and the horizon, but are more commonly nearly perpendicular to both. In some cases, the same dike may be traced for several miles, in nearly a straight line, across the outcrop of the strata. In other cases they will terminate suddenly, and commence anew, at some little distance to the right or left, and then proceed onward in the same direction as before. Faults of this kind are of frequent occurrence in the numerous trap dikes, which exist in the black shales along the eastern shore of Lake Champlain. The accompanying cut represents one of these, as seen in the bank of the lake at Clay Point, in Colchester.

Trap Dike.

The fault is an offset of about three feet. The dike is in black slate. The part of the bank above it is sand. At Hubbell's Fall, in Winooski-river, two faults may be seen, in the same dike, in the bottom of the river.

Some of these dikes are very compact and homogenious. Some have a concretionary structure, and, by exposure to the weather, separate into spheroidal masses. Others again exhibit signs of a columnar structure; and still others contain numerous light colored crystals, giving it an amygdaloidal character. An interesting dike of this character may be seen in a small island in Lake Champlain, a little to the northward of Colchester point.

The Porphyry Dikes are mostly confined to the southwestern part of Chittenden county. Like the trap dikes, they have, in general, an easterly and westerly course, but they are much more irregular in their direction, and much less uniform in width. In some places they seem to bilge up in large rounded masses, crowding and crushing the slate all around. The color of these dikes and intrusive masses, varies from a dark chocolate brown to a light cream color. In some cases, the embedded crystals are very numerous; in others, they are rare; and in others still, no crystals are seen, but they appear to consist of a homogeneous mass of feldspathic mineral.

No part of the state, which has been examined, so much abounds in dikes, both trap and feldspathic, as the northwestern part of Shelburne. Pottier's point is crossed by a dozen, at least. At Nash's point, the two kinds of dike are seen together, in circumstances to afford a clear indication of their relative ages. Their positions may be understood by the accompanying cut.

Porphyry Dikes.

The cut, which represents the western side of the point, exhibits a perpendicular face of porphyry, about 11 feet high and some rods in length, resting upon black slate, and covered above by about 2 ft. of black slate and soil. Cutting through the slate, in an easterly direction, beneath the porphyry, are two parallel trap dikes, about eight feet apart, and each about one foot wide. Portions of these trap dikes are also found in the slate overlying the porphyry. These facts make it certain that the trap dikes existed in the slate before the porphyry was thrown up, and that they were broken off, and parts of them lifted up with the slate by the intrusion of the porphyry. The more recent origin of the porphyry is also inferred from the fact that it is frequently found to have flowed laterally between the strata of the rocks, while the trap is never found to have done so, showing that the latter was formed under a much greater incumbent pressure than the former. Trap has been no where found in Vermont in the condition of an overlying mass.

The only purely igneous rock, observed any where in Vermont, on the west side of the mountains, in any other form than that of dikes, is in Charlotte. It there forms the hill, south of the Four Corners, and presents a surface of a number of acres. It is, in appearance, intermediate between common trap and porphyry, and most of it is exceedingly hard. The position of this hill may be seen on the map of Charlotte, given on page 19, it being the hill indicated on the map nearest the locality of the fossil whale.

Superficial Deposits.
Drift Scratches.

The rocks which have been briefly described, with the exception of our tertiary formation, are all fixed in the places in which they are found, and form the solid foundation of our territory. The surface of these rocks, where exposed to view, are every where found to be ground, or worn down by some agency, frequently having their surfaces finely polished, and crossed by numerous parallel striæ, or scratches. These striæ, or scratches, lying in the same direction, in which the loose materials, resting upon the solid rocks, have been transported, are supposed to have been produced by the movement of these materials; and, as the materials have received the name of Drift, the striæ, or grooves, are called *Drift Scratches*. The general direction of these scratches, and of the transportation of drift materials, is towards a point a little to the east of south, but varies in different parts of the state, somewhat in conformity to the directions of the valleys and the ranges of mountains.

The smoothing and striation of the surfaces of the rocks are most conspicuous, when the earthy materials are first removed from them. In some varieties of rocks they are seen, in a great measure obliterated by exposure to the weather. These polished and striated surfaces are found, not only in the bottoms of the lowest valleys, but upon the tops of the highest mountains in the state. Mount Washington, in New Hampshire, appears to be the only point in New England, which was not reached by the agency which produced them. The rocks, which form the summit of that mountain, are all sharply angular, exhibiting no appearance of having been worn, or scratched.

Drift Materials.—The smoothed and striated solid rocks, which we have been describing, are, in nearly all parts of the state, covered by a deposit, very variable in thickness, and consisting of boulders, pebbles, gravel, sand and clay, variously and irregularly blended together, without any distinct signs of stratification. These materials are, for the most part, different from the rocks on which they rest; and, as they are usually accompanied by evidence that they have been transported in a south-easterly direction, they have received the general name of *Drift*. In some places the drift consists entirely of sand; in other, of clay; in others, of gravel; but these are usually of small extent. The materials are more commonly mixed, but in very different proportions in different places. It is quite common to find them in the condition of what is called *hard-pan*, wherein sand, clay, gravel and pebble are so completely bedded together as to make it extremely difficult to penetrate them.

The proofs that the drift has been transported in a southerly direction, and never in a northerly direction, are very abundant. The large boulders of sienite and other rocks scattered through the valley of Lake Champlain, have evidently been brought from beyond the 45th parallel of latitude, there being no rocks of their kinds within the limits of the state. Boulders of Trenton and Isle la Motte limestone are found scattered along the western slope of the Green Mountains, and resting on talcose slate, far to the southeastward of the quarries from which they must have been derived. Boulders of the red sandrock, which is found in place only near the lake shore, are met with in masses, of several tons, in Williston and Richmond, and high up on the sides of the mountains, and even to the eastward of these mountains. To the eastward of Camel's Hump, in Duxbury, there is a boulder, which weighs about three tons, and which very clearly came from the lower strata of the red sandrock formation, near the level of Lake Champlain. It is now 20 miles from the nearest part of that formation, and rests upon talcose slate, at an elevation of about 700 feet above the level of the lake.

The transportation of boulders of what we have called Nodular Granite, has already been mentioned. They are found, of large size, in Waterford, Ryegate, and other places, 30 miles from any locality, where the rock is in place. Instances might be mentioned, where boulders of this rock have been transported over deep vallies, and lodged near the summit of the elevation on the opposite side.

Lawrencian Deposit.

Throughout the valley of the St. Lawrence, the valley of Lake Champlain, and around Lake Ontario up to the Falls of Niagara, there is found a regularly stratified formation of sands and clays, to which has been given the name of the *Lawrencian Deposit*. The thickest parts of this formation, in the valley of Lake Champlain, is about 200 feet, and the highest part of it is, at least, 400 feet above the level of the ocean. The southern portion of it consists chiefly of clay, while in Franklin county and the northern part of Chittenden county, sand predominates, particularly at the surface.

It is clearly a marine deposit, being well filled with remains of marine bivalve molluscs and other animals; and, as nearly, or quite, all of these remains belong to existing species, it is plain that it belongs

NATURAL HISTORY.

FOSSIL SHELLS. **FOSSIL SPONGE.** **FOSSIL WOOD.**

to a very recent geological period, or is what is called a Pleistocene, or post tertiary formation. These remains are mostly shells of molluscs, with a few remains of whales, seals and fishes; and they are nearly all identical in species with those living on the coast of New England, and in the Gulf of St. Lawrence.

Nearly an entire skeleton of a whale was found in this formation, in Charlotte, as mentioned on page 15. A number of the lumbar and caudal vertebræ of a whale, probably of the same species, have also been found in a similar clay bed, on the Island of Montreal, as already stated in a note on page 20. The species of fossil shells, found in this formation, are quite numerous, but the most common in the Champlain valley are the following:

Sanguinolaria. fusca.—This is the most common and abundant species. It is met with, in hundreds of places, along the banks of the lake and streams, in digging wells, in making excavations for roads and railroads, and in cultivating the lands. It is met with at the distance of several miles from the lake, and often more than 200 feet above it.

Saxicava rugosa.—This species is quite common, but is not so generally diffused as the preceding. The shell, being thick and strong, is often found in a very good state of preservation.

Mya arenaria.—This is the largest of the fossil shells found here. There are fewer localities of this than of the two preceding species, but at some of these the individuals are so multiplied as to be exceedingly numerous; and they are often well preserved.

Nucula portlandica.—This shell is found low in the blue clay, but is not abundant.

Mytilus edulis has been found only at a few localities; but in some cases the individuals are quite numerous. They are seldom well preserved.

The *Sanguinolaria fusca* is a littoral mollusc, which lives and propagates only in the sweep of the tide. This fact throws light upon the progress of subsidence of the St. Lawrence and Champlain vallies, by which the ocean, (from which the Lawrencian deposit took place,) was admitted into them. In various strata we have this mollusc embedded in the position in which it buries itself, when alive, and where it had evidently propagated, with the two valves united, and the epidermis undisturbed. These strata must, therefore, have been at the surface of the ocean, when the animals were buried. But we find them thus bedded in strata more than 60 feet apart, in vertical height, showing clearly that the subsidence was a very gradual one.

Fossil Sponge.—While digging a well in Alburgh, about four years ago, at the depth of 11 feet, the workmen came upon a horizontal stratum of what appeared to be mats of hair. It was in quite compact clay, was about two inches thick, and extended over nearly the whole bottom of the excavation. It excited much curiosity, but very little of it was saved. Having obtained a small quantity of it, I sent it in a letter to my friend, Prof. J. D. Dana, who pronounced it, both upon his own authority and that of Prof. Bailey, of West Point, to be *Fossil Marine Sponge.*

Fossil Wood.—It is not uncommon, in the vicinity of Lake Champlain, to find wood, and other vegetable matter, buried at various depths in the earth, in places and under circumstances, in which we should little expect it. There have been several cases of this kind in the village of Burlington, which I shall here mention, and all of which, with one exception, have fallen under my own observation; and that one is well attested.

The first of these cases was in 1835. That year the Hon. Alvan Foote, who resides, about 40 rods, directly north of the University, dug a well near his residence. The surface of the ground, at the place, was originally covered with a heavy growth of timber, and large boulders were thickly scattered over it. In digging the well, the first four feet were loose earth and gravel. The next 20 feet were what is commonly called *hardpan*, consisting of pebbles, gravel, sand and clay, very solidly compacted together. Next came a sandy earth, which could be shoveled without being loosened with a pickaxe, for about 4 feet, when the workmen, to their astonishment, broke into a hollow cavity, extending across the bottom of the well.

Upon examination, the cavity was found to have been occupied by a large tree, supposed to be pine, parts of which were remaining, and quite sound. It had been embraced by the sand; but, a few inches lower down, a stratum of black carbonaceous matter was found, resembling muck. The natural surface of the ground, where the well was dug, was about 240 feet above Lake Champlain, and the tree was $29\frac{1}{2}$ feet below the surface of the ground.

The next case was in 1850. In making the excavation, on Pearl street, for the reservoir, connected with the aqueduct, which supplies the lower part of Burlington with water, at the depth of 13 feet from the surface of the ground, a large amount of wood, sticks and leaves were found embedded in clean gravel. The locality is about 200 feet above the lake, and the size of the excavation was 36 feet by 40. The surface of the ground sloped moderately towards the northwest, and was originally covered heavily with timber. The earth, after getting below the soil, was sand and gravel, which had been washed and assorted by water, and was lying in irregular beds, sloping steeply towards the northwest. The vegetable remains formed a mass in the gravel about two feet wide, one and a half foot deep, and 36 feet long, extending in a right line, and was, at first, mistaken for a rotten tree; but, on breaking it to pieces, it was found to consist of roots, limbs, bark, stems and leaves, snugly bedded together, and all of a dark brown color; some portions of it approaching, in appearance, to brown coal. Many of the sticks and roots were perfectly sound, and exhibited the structure of the wood completely, and are, I have little doubt, the American Larch, *Pinus peudula*.

In October, in laying the aqueduct pipes in the south part of the village, wood, resembling larch and oak, were found, at the depth of 10 feet beneath hardpan. And in April, 1852, in deepening the well at the Pearl Street House, which is midway between the two localities first mentioned, a piece of wood, ten inches long, six wide and three thick, was found below hardpan, 24 feet from the surface. The Pearl Street House is about 230 feet above the lake. Wood has also been found in the central part of the village, in the stratified sand and clay, 20 feet below the surface.

The question now arises—to what geological period does this fossil wood belong? The last mentioned certainly belongs to the post tertiary or Lawrencian, for the characteristic shells were found with it. In the other cases, the earth was unstratified, and the materials, which covered the wood, evidently belonged to the drift. But did the wood belong to the drift period?—or to the tertiary which preceded it?

To these last questions, I would answer, that, in my opinion, it belonged to neither. The wood, and materials associated with it, are totally unlike the lignite, and its associates, which constitute the tertiary at Brandon, and no one can for a moment regard them as belonging to the same period. But the wood is beneath or within, what are, evidently, drift materials. How can this be, unless the wood and drift are of the same age?

To answer this question, we are to consider that the elevation on which the University stands, was, at the close of the drift period, a high ridge of drift deposit, having a steep descent towards the northeast and northwest. Subsequently the whole Champlain valley subsided, the sea was let in, and this elevated ridge became more and more immersed, and the materials forming its steep declivities were gradually washed down and re-arranged by the action of the waves.

Previous to the burial of the tree first mentioned, there appears to have been a small marsh at the foot of the steep bank of drift. When the subsidence had let the sea in upon this marsh, the tree was floated in and lodged at the foot of the bank. The subsidence continued, and the action of the waves soon washed down the drift materials and covered the tree; and we have evidence that the valley continued to sink till the whole ridge was immersed, and the island disappeared. During this immersion, the materials continued to be washed down, and beaten and pressed together by the surf and weight of the water, until the wood became buried in the condition in which it is found, since the sea was emptied out by the upheaval of the valley; so that while the wood is buried in the drift, it has been buried by a re-arrangement of the drift materials, since the drift period.

Shell Marl.

The beds of shell marl in Vermont are considerably numerous; and some of the beds are quite extensive; but they are entirely confined to what we have called the calcario-mica slate division, on the east side of the Green Mountains, and to a small portion of the western border of the state. On that large central portion of the state, which we have called the talcose slate division, not a single marl-bed is known to exist. The marl, which constitutes these beds, has a general resemblance to pulverised chalk, and consists, essentially, of carbonate of lime, which has resulted from the partial decay and crumbling of innumerable fresh-water shells, with sometimes a slight intermixture of sand and clay. Though, when wet, like a bed of putty, and when dry, a pulverulent mass, still shells, more or less entire, are found to be scattered through all parts of it: and near the surface unbroken shells are often numerous.

These shells are, for the most part, of the same species, which are now found living in the ponds and streams of the neighborhood, and belong chiefly to the following genera, viz: Paludina, Limnæ, Physa, Planorbis, Pupa and Cyclas.

Marl beds exist in all the counties on the east side of the mountains, but are most numerous in Caledonia county. There are several in each of the towns of Barnet, Peacham and Danville. In Orleans county, and in the eastern part of Washington county, there are a few. The following section exhibits the thickness and association of one of these marl beds in Derby, with its overlying muck and underlying sand:

Muck, 4 feet.

Marl, 3 feet.

Sand.

Muck and Marl Bed.

The most valuable bed of marl known on the east side of the mountains, is in Williamstown. It is from 6 to 18 feet deep, and slightly covered with a dry soil. It is very pure carbonate of lime, and makes the best quicklime obtained in that part of the state. Its analysis, by Mr. Hunt, gives the following results:

Carbonate of Lime,	89.
Carbonate of Magnesia,	4.2
Silica, with traces of oxide of iron and alumina,	1.
Water and organic matter.	5.5=99.7.

The marl beds on the west side of the mountains are not numerous, but some of them are quite extensive. The most interesting beds are those in Monkton and Alburgh. That at Alburgh extends over 60 acres. Where examined, it was found to be from 6 to 9 feet deep, resting on fine blue clay, and covered by vegetable muck to the depth of five feet, upon which there had been a large growth of forest trees. Supposing the average depth of the marl to be only three feet, the aggregate amount would exceed 60,000 cords, and the muck resting upon it would probably exceed 100,000 cords.

An account of the marl in Monkton pond, and of the manner in which these marl-beds are formed, may be found in Prof. Adams' Second Annual Report on the Geology of Vermont, page 148.

Shell marl is valuable, both for the manufacture of quicklime, and as a fertilizer to be applied to the soil. To obtain good quicklime from it, it should be moistened, made into the form of bricks, and, after being dried, should be arranged and burned in kilns, for the expulsion of the carbonic acid, by fires placed beneath, as is done in the manufacture of common bricks. This is the course pursued at Williamstown, where our best marl lime is made.

The value of marl, as a fertilizer, depends upon the constituents of the soil to which it is applied. If the soil is already sufficiently supplied with lime, for the purposes of vegetation, the application of marl will produce no sensible effect; and this is generally found to be the condition of the soil in those neighborhoods in which marl-beds are found. Indeed, the connexion between the marl and the lime, in the soil and waters of the vicinity, is very obvious; for the pre-existence of the latter, is absolutely necessary, for the existence and multiplication of the molluscs, whose shells form the marl. If the lime did not exist in the water, there would be no material for the formation of the shells, and, therefore, the animals could not exist. Hence we learn the reason why there are no marl-beds on the talcose slate division of the state. There are, there, no limerocks, and only a very minute amount of lime in the soil and water, and hence there are hardly any land, or fresh-water shells. The soil, throughout the whole of that division, would undoubtedly be improved by the application of marl. The soil in the calcario-mica slate division, is, in general, well supplied with lime, by the decomposition of the blue siliceous limestone; and, in the western part of the state, by the marine fossil shells contained in it, and by the decomposition of the different limestones, which abound; excepting the sandy plains. These, though resting upon limestone, are very deficient in lime, and are greatly benefitted by the application of marl.

Vegetable Muck.

In all parts of the state are found deposits of muck, consisting of partially decomposed leaves and other vegetable matter. These deposits vary in extent from a few square rods to many acres, and are from a few inches to 15 and 20 feet in depth. When the country was new, most of these were bogs, many of which have since become dry, either by draining, or by exposure to the sun and winds, in consequence of the removal of the forests. They are, not unfrequently, found resting on beds of marl, as has been already mentioned.

The cavities, in which these beds, both of muck and marl, are found, were, undoubtedly, originally, little pools or ponds of water, which gradually became filled up with the shells of successive generations of molluscs, and vegetable matter, thus diminishing the size of the pond by surrounding it with a bog, or, what was more commonly the case, filling it entirely, leaving only a bog in its place.

This muck is a valuable manure for most soils, and nature has provided plentiful stores of it, in almost every part of the state. Some of our farmers have already learned its use, as a fertilizer, and profited by its application; and we trust that it will soon be more generally appreciated, and more extensively used. The value, both of the muck and marl, for the improvement of some soils, is thought to be much enhanced by applying them in conjunction.

INFUSORIAL SILICA.

Several deposits are met with in different parts of the state, which, in their situation and appearance, very much resemble shell-marl, but instead of being, like that substance, calcareous, are a fine siliceous earth. By examination, under the microscope, this earth is found to have originated from the flinty shell of infusorial animalcules, in the same manner that the marl was formed from the calcareous shells of molluscs, and hence it received the name of *Infusorial Silica.*

The most extensive deposit of infusorial silica, known in the state, is in Hosmer's pond, in the southwestern corner of Peacham. This pond is surrounded by granite hills, and covers about 250 acres. The infusorial deposit is thought to average about six inches in depth, over the bottom of about two-thirds of the pond. When taken out and dried in lumps, it is a very good substitute for chalk. When dried and pulverised, it resembles calcined magnesia; and, hence, the pond is called, sometimes *Chalk* pond, and sometimes *Magnesia* pond. There is another small deposit of infusorial silica, in Maidstone, in Essex county.

By the examination of specimens of the silica, from these deposits, by Prof. Bailey, of West Point, the shields of more than twenty distinct species of animalcules, were discovered in it. Some of these are so exceedingly minute, that, incredible as it may seem, it would require a million of them to make the bulk of a single mustard seed. By the labors of Ehrenberg, and others, these microscopic fossils have all been arranged, described and figured, so far as known, and many of the forms are exceedingly beautiful. I give, below, the figures of a few of the species found in Hosmer's pond. Their areas are magnified in the cuts, about 73,000 times.

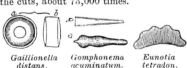

Gaillionella distans.   *Gomphonema acuminatum.*   *Eunotia tetradon.*

CLAY STONES.

Concretions of various kinds are found in Vermont, but the most common are those found in beds of clay, and generally known by the name of *Clay Stones.* These clay stones exhibit an almost infinite variety of forms. Many of them appear as if skilfully turned in a lathe, or beautifully carved by art; and hence they are every where regarded as objects of curiosity. Their most common form is that of a convex lens, or flattened sphere; but various forms are often blended together, in the most grotesque and fanciful manner. At some localities, they are found in the form of a perfect ring, like the ring of an ox-yoke, both in form and size. Those concretions, which abound in the Lawrencian formation, in the neighborhood of Lake Champlain, are generally cylindrical, having their longer axis nearly perpendicular to the stratification of the clay, and prolonged through several of the strata. These cylindrical concretions are all formed of concentric layers around the axis of the cylinder, which axis is a capillary opening, extending through its whole length. They appear as if they had been formed around fibrous roots, which had afterwards decayed out, leaving a small perforation, like a pith, extending through their whole length.* Localities of clay stones and other interesting concretions exist in various parts of the state, and are too numerous to be particularized.

---

* Having an opportunity, in 1851, to examine the Crag formations in the east part of England, I satisfied myself that the, so called, coprolites, which abound there, and are so highly prized and extensively used as a fertilizer, are, for the most part, at least, concretions formed in the same manner as those above named. They differ in the materials of which they are composed, but do not, apparently, differ in the manner, in which they are formed. While ours contain carbonate of lime, those found in the Crag, in England, are said to consist of 65 per cent. of the phosphate of lime, and hence the great value of the latter as a fertilizer. The clay in which those concretions were found, had probably abounded in fossil bones, and the decomposition of these bones furnished the lime, in the condition of a phosphate, for the formation of the concretions. Thousands of tons of these concretions are said to be, annually, separated from the Crag, and used as a manure.

POPULATION OF VERMONT.

There have now been seven complete enumerations of the inhabitants of Vermont, since the organization of her government. The result of six of these are given, by towns, in Part II., page 209. The result of the seventh is given below.

POPULATION OF VERMONT IN 1850.

Towns.	Pop.	Towns.	Pop.	Towns.	Pop.	Towns.	Pop.
Addison,	1279	Andover,	725	Averill,	7	Bakersfield,	1523
Albany,	1052	Arlington,	1084	Avery's Gr., F.C.	48	Baltimore,	124
Alburgh,	1568	Athens,	359	Buell's Gore,	18	Barnard,	1647

CIVIL HISTORY.

POPULATION OF VERMONT—SEVENTH CENSUS.

Towns.	Pop.	Towns.	Pop.	Towns.	Pop.	Towns.	Pop.
Barnet,	2521	Fairlee,	575	Montgomery,	1001	South Hero,	705
Barre,	1845	Fayston,	684	Montpelier,	2310	Springfield,	2762
Barton,	987	Ferdinand,	0	" East,	1447	Stamford,	833
Belvidere,	256	Ferrisburgh,	2075	Moretown,	1335	Starksborough,	1400
Bennington,	3923	Fletcher,	1084	Morgan,	486	Sterling,	233
Benson,	1305	Franklin,	1646	Morristown,	1441	Stockbridge,	1327
Berkshire,	1955	Georgia,	2686	Mt. Holly,	1534	Stow,	1771
Berlin,	1507	Glastenbury,	52	Mt. Tabor,	308	Strafford,	1540
Bethel,	1730	Glover,	1137	Newark,	434	Stratton,	286
Bloomfield,	244	Goshen,	486	Newbury,	2984	Sudbury,	794
Bolton,	602	Goshen Gore, nor.	183	Newfane,	1304	Sunderland,	479
Bradford,	1723	" " south,	32	Newhaven,	1663	Sutton,	1001
Bradleyvale,	107	Grafton,	1241	Newport,	748	Swanton,	2824
Braintree,	1228	Granby,	127	Northfield,	2922	Thetford,	2016
Brandon,	2835	Grand Isle,	666	North Hero,	730	Tinmouth,	717
Brattleborough,	3816	Granville,	603	Norton,	0	Tapsham,	1668
Bridgewater,	1311	Greensborough,	1008	Norwich,	1978	Townsend,	1354
Bridport,	1393	Groton,	895	Orange,	1007	Troy,	1008
Brighton,	193	Guildhall,	501	Orwell,	1470	Tunbridge,	1786
Bristol,	1344	Guilford,	1389	Panton,	559	Underhill,	1599
Brookfield,	1672	Halifax,	1133	Pawlet,	1843	Vergennes,	1378
Brookline,	285	Hancock,	430	Peacham,	1377	Vernon,	821
Brownington,	613	Hardwick,	1402	Peru,	567	Vershire,	1071
Brunswick,	119	Harris' Gore,	8	Pittsfield,	512	Victory,	168
Burke,	1103	Hartford,	2169	Pittsford,	2026	Waitsfield,	1021
Burling- (vil. 6110) ton, (town, 1475)	7585	Hartland,	2063	Plainfield,	808	Walden,	910
		Highgate,	2653	Plymouth,	1226	Wallingford,	1688
Cabot,	1356	Hinesburgh	1834	Pomfret,	1546	Waltham,	270
Calais,	1410	Holland,	669	Poultney,	2329	Wardsborough,	1125
Cambridge,	1849	Hubbardton,	701	Pownal,	1742	Warren,	962
Canaan,	471	Huntington,	885	Putney,	1425	Washington,	1348
Castleton,	3016	Hydepark,	1107	Randolph,	2666	Waterbury,	2352
Cavendish,	1576	Ira,	400	Reading,	1171	Waterford,	1412
Charleston,	1008	Irasburgh,	1034	Readsborough,	857	Waterville,	753
Charlotte,	1634	Isle la Motte,	476	Richford,	1074	Weathersfield,	1851
Chelsea,	1958	Jamaica,	1606	Richmond,	1453	Wells,	804
Chester,	2185	Jay,	371	Ripton,	567	Wenlock,	26
Chittenden,	675	Jericho,	1837	Rochester,	1493	West Fairlee,	696
Clarendon,	1477	Johnson,	1381	Rockingham,	2837	Westfield,	502
Colchester,	2575	Kirby,	509	Roxbury,	967	Westford,	1458
Concord,	1153	Landgrove,	337	Royalton,	1850	Westhaven,	718
Corinth,	1906	Leicester,	596	Rupert,	1101	Westminster,	1721
Cornwall,	1155	Lemington,	187	Rutland,	3715	Westmore,	152
Coventry,	867	Lewis,	0	Ryegate,	1606	Weston,	950
Craftsbury,	1223	Lincoln,	1057	St. Albans,	3567	West Windsor,	1002
Danby,	1535	Londonderry,	1274	St. George,	127	Weybridge,	804
Danville,	2577	Lowell,	637	St. Johnsbury,	2758	Wheelock,	855
Derby,	1750	Ludlow,	1619	Salem,	455	Whiting,	629
Dorset,	1700	Lunenburgh,	1123	Salisbury,	1027	Whitingham,	1380
Dover,	709	Lyndon,	1752	Sandgate,	850	Williamstown,	1452
Dummerston,	1645	Maidstone,	237	Searsburgh,	201	Williston,	1669
Duxbury,	845	Manchester,	1782	Shaftsbury,	1896	Wilmington,	1372
East Haven,	94	Marlborough,	896	Sharon,	1240	Windham,	763
Eden,	668	Marshfield,	1102	Sheffield,	797	Windsor,	1928
Elmore,	504	Mendon,	504	Shelburne,	1257	Winhall,	762
Enosburgh,	2009	Middlebury,	3517	Sheldon,	1814	Wolcott,	909
Essex,	2052	Middlesex,	1365	Sherburne,	578	Woodbury,	1070
Fairfax,	2111	Middletown,	875	Shoreham,	1601	Woodford,	423
Fairfield,	2591	Milton,	2451	Shrewsbury,	1268	Woodstock,	3041
Fair Haven,	902	Monkton,	1246	Somerset,	321	Worcester,	702

NOTE.—Montpelier and Windsor have each been divided into two towns, since the sixth census, and the town of Mansfield has been swallowed up by annexation to Stow. All these changes were made at the session of the legislature, in 1848.

APPENDIX TO THOMPSON'S VERMONT.

POPULATION BY COUNTIES.   LITERARY INSTITUTIONS.   PRODUCTIONS OF AGRICULTURE.

POPULATION BY COUNTIES.

Counties.	Population.
Addison,	26,549
Bennington,	18,589
Caledonia,	23,595
Chittenden,	29,036
Essex,	4,650
Franklin,	28,586
Grand Isle,	4,145
Lamoille,	10,872
Orange,	27,296
Orleans,	15,707
Rutland,	33,059
Washington,	24,654
Windham,	29,062
Windsor,	38,504
Total aggregate,	314,304
White Males,	159,748
White Females,	153,838
Total Whites,	313,586
Colored Males,	375
Colored Females,	343
Total Colored,	718
Total aggregate, as before,	314,304

LITERARY INSTITUTIONS.

	No·	Teachers.	Pupils.
Colleges,	5*	30	464
Public Schools,	2789	4,204	94,795
Academies and Private Schools,	95	272	6,231

LIBRARIES.

Public, (including School)	77	43,705
Private,	133	55,773

NOTE.—In most of the counties no private libraries were reported which contained less than 1,000 volumes.

NEWSPAPERS.

Whole No. 36. Whig, 14 ; Democratic, 7 ; Literary, &c., 15.

	No.	Circulation.	Ann. Issues.
Daily,	2	550	171,050
Semi-Weekly,	1	2,200	228,800
Weekly,	31	41,206	2,142,712
Monthly,	2	2,000	24,000
	36		2,566,562

* Two of them Medical Colleges.

PRODUCTIONS OF AGRICULTURE IN VERMONT,—SEVENTH CENSUS—1850.

COUNTIES.	Acres of Land in Farms. Improved.	Unimproved.	Cash Value of Farms.	Value of Farm Implements.	Horses.	Asses and Mules.	Milk Cows.	Working Oxen.	Other Cattle.	Sheep.	Swine.	Value of Live Stock.
Addison,	243,312	115,287	$7,799,257	$256,270	5,921	1	10,691	2,816	13,248	188,154	5,822	$1,283,608
Bennington,	138,065	85,760	3,338,756	131,194	3,344		6,667	1,983	7,402	71,294	5,162	662,281
Caledonia,	210,474	151,604	4,751,609	219,559	5,705	11	11,914	4,402	12,529	30,252	2,864	1,027,836
Chittenden,	177,707	104,454	5,624,439	217,343	4,897	17	12,790	2,324	3,448	57,184	6,492	898,732
Essex,	42,993	52,310	767,185	61,096	1,025		2,549	1,260	14,890	7,519	822	215,370
Franklin,	180,848	127,002	4,284,070	215,418	5,396	94	16,217	3,307	1,465	58,509	5,413	942,262
Grand Isle,	33,171	15,113	1,189,082	38,993	1,300		1,856	238	6,124	18,949	937	122,588
Lamoille,	76,083	76,070	1,851,471	107,505	2,032	39	5,511	2,225	13,564	15,193	2,476	422,671
Orange,	226,257	120,142	4,807,788	279,641	5,580	5	10,777	5,138	9,592	71,551	7,337	974,258
Orleans,	119,377	127,520	2,492,090	150,114	3,724	6	8,191	3,631	12,673	27,422	3,825	645,412
Rutland,	290,392	154,524	7,972,180	255,240	6,151		17,151	3,590	11,196	186,319	5,034	1,543,936
Washington,	165,654	120,239	3,905,385	165,179	4,140	15	11,507	3,922	17,744	32,355	5,507	873,766
Windham,	319,558	99,674	6,301,500	248,874	5,054	1	13,975	6,031	20,420	58,553	6,005	1,223,724
Windsor,	377,523	174,711	8,221,815	292,856	6,788	29	16,918	7,661		190,868	8,600	1,800,789
Total,	2,601,409	1,524,143	$63,367,227	$2,789,282	61,057	218	146,128	48,577	154,143	1,014,122	65,296	$12,643,228

NOTE.—By comparing the numbers in the above table with the returns of 1840, given in Part I., p. 52–57, it will be seen that, while the number of horses and cattle has remained nearly the same, there has been a very great diminution of the number of hogs and sheep. The number of sheep returned shows a diminution of more than 600,000.

PRODUCE OF THE YEAR ENDING JUNE 1, 1850.   SEVENTH CENSUS.

PRODUCE OF THE YEAR ENDING JUNE 1, 1850.

COUNTIES.	Wheat, Bushels.	Rye, Bushels.	Ind. Corn. Bushels.	Oats, Bushels.	Peas and Bea. Bu.	Potatoes, Bushels.	Barley, Bush'ls.	Wool, Pounds.
Addison,	103,434	20,096	175,478	211,385	26,355	318,421	149	622,594
Bennington,	6,973	17,270	150,920	177,511	3,150	200,013	3,003	221,679
Caledonia,	62,551	2,090	96,389	218,735	6,419	565,341	3,658	136,790
Chittenden,	36,491	25,566	198,598	184,752	10,390	383,113	682	185,215
Essex,	8,826	1,360	21,931	45,597	2,506	94,124	1,221	29,614
Franklin,	55,488	9,138	137,896	145,840	10,255	258,757	815	209,350
Grand Isle,	31,324	3,986	23,245	81,027	10,469	31,793	739	70,291
Lamoille,	14,466	6,663	66,017	90,434	4,351	278,252	629	49,053
Orange,	52,822	9,740	176,586	205.457	5,658	599,925	1,861	248,715
Orleans,	58,515	4,853	70,306	169,587	3,723	407,132	8,974	81,947
Rutland,	25,874	20,598	258,831	183,706	4,220	416,000	627	623,199
Washington,	30,580	10,567	133,477	208,554	4,954	446,551	865	152,843
Windham,	8,749	18,302	210,141	160,393	2,279	338,295	14,124	179,122
Windsor,	39,862	26,004	312,581	224,756	9,920	613,297	4,803	589,305
Aggregate,	535,955	176,233	2,032,396	2,307,734	104,649	4,951,014	42,150	3,400,717

PRODUCE OF THE YEAR ENDING JUNE 1, 1850.

COUNTIES.	Buckwh't Bushels.	Orchard Produce.	Wine, Gal's.	Pro. Mar. Garden.	Butter, Pounds.	Cheese, Pounds.	Hay, Tons.	Clo.S. Bush.	Gr. Se'd Bush'ls.
Addison,	15,659	$41,696	114	$	876,771	817,149	88,793	5	1,589
Bennington,	22,797	16,629	7	1,558	502,786	558,494	54,600		622
Caledonia,	14,380	26,094	47	355	1,206,272	121,602	59,449	179	2,991
Chittenden,	10,003	33,841	303	10,913	838,481	1,663,456	57,407	2	619
Essex,	15,400	4,523			292,615	122,321	14,972	38	923
Franklin,	10,095	19,429		107	1,399,455	1,196,660	78,619		1,050
Grand Isle,	12,140	11,223		12	93,225	26,793	6,980	1	300
Lamoille,	10,373	9,095	94		437,110	213,035	26,973	9	587
Orange,	28,942	23,980		270	869,042	428,876	70,549	206	609
Orleans,	15,305	5,920			645,160	68,092	45,288	39	1,798
Rutland,	12,051	38,457	19	537	1,120,814	1,930,047	103,950	1	773
Washington,	10,135	20,620		1,475	970,368	437,476	54,959	37	767
Windham,	7,531	19,139	15	581	1,144,653	469,728	84,749	76	392
Windsor,	25,006	44,609	60	3,045	1,741,228	667,105	118,865	167	1,916
Aggregate,	209,819	315,255	659	18,853	12,137,980	8,720,834	866,153	760	14,936

PRODUCE OF THE YEAR ENDING JUNE 1, 1850.

COUNTIES.	Hops, Pounds.	Flax, Pounds.	Fl'x S'd Bush.	Silk Co. Pounds.	Map. Sugar, Pounds.*	Molas. Gall's.	Wax and Hon. lbs.	Ho. Manufac. Val.	Ani. Slaughtered. Value.
Addison,	5,962	1,282	51	76	205,263	650	40,654	$9,648	$176,856
Bennington,	193	2,522	132		220,009	165	14,814	6,450	86,123
Caledonia,	1,422	2,365	113		854,820	364	22,863	40,343	135,537
Chittenden,		968	26	4	242,842	70	18,319	13,359	134,536
Essex,	28,250	855	11		145,041	129	3,855	22,044	37,020
Franklin,	1,610	1,052	33		684,511	36	20,536	26,247	141,682
Grand Isle,		331	8	30	32,665		4,866	3,449	19,967
Lamoille,	15,657	1,293	41		427,918	23	11,501	6,584	80,296
Orange,	23,827	3,752	158	15	532,156	674	12,438	27,346	160,430
Orleans,	77,605	660	140		656,883		6,461	16,422	86,672
Rutland,	162	986	22		492,664		37,370	12,620	184,251
Washington,	12,125	2,730	31		765,429	407	17,299	17,269	155,477
Windham,	41,510	518	10	1	470,934	1,360	7,255	13,321	189,095
Windsor,	79,700	1,538	163	142	618,222	2,119	31,191	52,608	273,394
Aggregate,	288,023	20,852	939	268	6,349,357	5,997	249,422	267,710	1,861,336

*By comparing the amount of Maple Sugar here given, with amount made in 1840, as stated in Part 1, page 210, it will be seen that the advance in the annual manufacture of this article, amounts to 1,701,423 pounds. There has also been a very great improvement in the quality of the sugar made as well as increase in quantity. Two *prize Medals* were awarded at the World's Fair, in London, in 1851, for Vermont Sugar, one to Mr. L. Dean, of Manchester, and the other to Mr. W. Barnes, of Rutland.

PRODUCTIONS OF INDUSTRY.   REAL AND PERSONAL ESTATE.   TAXES.   WAGES.   PAUPERISM.

PRODUCTIONS OF INDUSTRY IN THE YEAR ENDING JUNE 1, 1850.

COUNTIES.	No. of Estab.	Capital Invested.	Value of raw Material	No. of Hands. Male.	No. of Hands. Fema.	Monthly Wages. Male.	Monthly Wages. Female.	Value of Ann. Prod.
Addison,	161	$289,375	$360,069	523	74	$12,143	$ 704	$659,888
Bennington,	150	468,050	414,622	652	117	15,137	1,077	880,216
Caledonia,	243	444,180	399,427	742	74	18,633	869	799,053
Chittenden,	202	771,610	700,192	848	368	19,211	3,121	1,320,730
Essex,	32	31,250	23,589	55		1,235		48,794
Franklin,	112	147,710	126,879	364	30	7,133	247	285,697
Grand Isle,	9	13,100	1,790	47		675		15,600
Lamoille,	45	110,300	93,108	115	31	2,909	421	175,861
Orange,	85	171,045	110,774	226	27	5,122	244	219,165
Orleans,	68	64,450	60,148	116	9	2,188	73	119,036
Rutland,	276	828,975	490,507	1,280	99	30,703	957	1,284,756
Washington,	77	231,337	223,705	375	74	9,656	784	525,236
Windsor,	196	476,720	399,933	649	273	15,345	3,601	831,209
Windham,	193	953,275	767,809	902	375	25,976	5,365	1,405,729
Aggregate,	1,849	5,001,377	4,172,552	6,894	1,551	166,066	17,463	8,570,920

REAL AND PERSONAL ESTATE IN VERMONT.

Valuation of Real and Personal Estate, by Assessors,   -   -   - $71,671,651
Estimated true value of Real and Personal Estate,   -   -   -   -   92,205,049

TAXES.

General State Tax,   -   -   -   -   -   -   -   -   - $138,533
School Tax,   -   -   -   -   -   -   -   -   -   -   88,930
Poor Tax,   -   -   -   -   -   -   -   -   -   -   90,809
County, Town, &c., Taxes,   -   -   -   -   -   -   -   401,142

Total,   -   -   -   -   -   -   -   -   - $719,414

WAGES.

Average Monthly Wages of a Farm Hand,   -   -   -   - $13,00
    " to a Day Laborer, with Board,   -   -   -   -   0,72
    " to a Day Laborer, without Board,   -   -   -   -   0,97
    " Day Wages to a Carpenter, without Board,   -   -   -   1,44
Weekly Wages to a Female Domestic, with Board,   -   -   -   1,19
Price of Board to Laboring Men,   -   -   -   -   -   -   1,95

PAUPERISM.

	Native.	Foreign.	Total.
Whole No. of Paupers within the year ending June 1, 1850,	2043	1611	3654
Whole No. of Paupers on June 1, 1850,	1565	314	1879

CRIME.

	Native.	Foreign.	Total.
Whole No. of Criminals convicted within the year ending June 1, 1850,	34	45	79
Whole No. in Prison on June 1, 1850,	64	41	105

RAIL ROADS IN VERMONT.

At the time of the publication of our History of Vermont in 1842, we had neither canals nor rail roads within the state; but we ventured the opinion, (Part I, page 217) that Boston would in time be connected with Lake Champlain by a continuation of the Lowell and Concord rail road. At that time we little thought that the short period of ten years would witness the completion of a net-work of rail road over the whole country. Ten years ago the construction of a railway across the Green Mountains from the valley of the Connecticut to Lake Champlain, was very generally regarded as a chimerical notion, which would never be realized, and they who entertained it were looked upon as visionaries. But events have proved it otherwise. We have already two rail roads crossing the state from east to west, connecting these vallies; and, also a road in each of these vallies running north and south, through

RAIL ROADS IN VERMONT.

nearly the entire length of the state.

The first rail road commenced in this state was the Vermont Central, and the ground was first broken for the construction of that road in the spring of 1846 at Windsor. The Rutland and Burlington road was commenced in the spring of 1847, and both of these roads were opened from Connecticut river to Burlington in December, 1849.

The following table exhibits the names, the terminations, the lengths, and the times of opening the several rail roads, in operation in April, 1853.

NAMES.	TERMINATIONS.		LENGTH.	OPENED.
Atlantic and St. Lawrence,*	Bloomfield,	Norton,	34	1853
Conn. and Passumpsic Rivers,	White River,	St. Johnsbury,	61	1851
Rutland and Burlington,	Burlington,	Bellows Falls,	119	1849
Rutland and Washington,	Rutland,	Poultney,	18	1852
Rutland and Whitehall,	Castleton,	Whitehall,	12	1850
Vermont Central,	Burlington,	Windsor,	117	1849
Vermont and Canada,	Essex Junction,	Rouse's Point,	47	1850
Vermont and Massachusetts,	Brattleborough,	South Vernon,	10	1849
Vermont Valley,	Bellows Falls,	Brattleborough,	24	1851
Western Vermont,	Rutland,	N. Bennington,	51	1852
			493	

Several others are in contemplation within the state, and no great length of time will probably elapse before the Connecticut and Passumpsic Rivers road will be continued northward from St. Johnsbury to Canada Line. The effects which these roads have produced upon the towns through and near which they pass, are marked and obvious, but I have not room to particularize them.

*This is a section of the rail way designed to connect Portland, Me , with Montreal, C. E. It is now opened (April, 1853,) from Portland to Island Pond in Brighton and from Montreal to Sherbrooke. The intermediate portion from Sherbrooke to Island Pond is nearly ready for the rails and is expected to be opened in the course of a few months. The length here given is only an estimate from the Map.

MAGNETIC TELEGRAPH.

The Magnetic Telegraph, which seems to be essential to the safe management of rail roads, sprang into being very soon after the time when railroads themselves had their origin; and they were introduced simultaneously into Vermont. The first line of telegraph in Vermont, forms a part of the *Troy and Canada Junction Line*, and was commenced in 1847. It was opened for communication, from Troy to Burlington, on the 2d of Feb. 1848, and was soon after carried through to Montreal. This line enters the state at Bennington, passes thro' Manchester, Rutland, Castleton, Whitehall, Orwell, Brandon, Middlebury, Vergennes, Burlington and St. Albans, and leaves the state at Highgate. The length of this line, within the state, is 200 miles.

The Northern Telegraph Line connects Boston with Rutland. Proceeding from Boston by way of Fitchburg and Keene it enters the state at Bellows Falls and follows the line of the Rutland and Burlington rail road through Chester and Ludlow to Rutland. Length within the state 50 miles.

Vermont and Boston Telegraph Line.— Proceeding from Boston by way of Lowell and Concord, this line enters the state at White River Junction, and, after going to Woodstock and back, 20 miles, follows the line of the Central rail road, passing thro' South Royalton, West Randolph, Northfield Montpelier, Waterbury, and Essex Junction to Burlington. From Burlington it follows the Vermont and Canada rail road through St. Albans and Swanton to Rouse's Point, where it leaves the state, and proceeds in two branches, one to Montreal and the other to Ogdensburgh. Connected with this line and crossing it at White River Junction, the same company have a line along the valley of the Connecticut, reaching from St. Johnsbury to Springfield, Mass. From St. Johnsbury it follows the rail road through Newbury and Bradford to Norwich, where it crosses over to Hanover and back, and then proceeds down to White River Junction. From the Junction it proceeds to Windsor, crosses over to Claremont, N. H., then back to Weathersfield Bow, thence to Springfield—then by way of Charlestown bridge to Charlestown, and down the Sullivan rail road to Bellows Falls. From Bellows Falls it proceeds down the Connecticut on the Vermont side thro' Brattleborough into Massachusetts. The whole length of telegraph line belonging to this company is about 700 miles, of which more than 300 are in Vermont. The whole length of telegraph wire in the state is little less than 600 miles, and the cost of building, including appurtenances and patent privileges has been about $215 per mile.

INDEX

TO THE APPENDIX TO THOMPSON'S VERMONT.

	Page.
Acres of Land or Farms,	280
Adams' C. B., - - -	261
Agricultural Productions,	280
Amia ocellicauda, - -	255
Animalcules, - - -	278
Ardea minor, - - - -	246
Boleosoma tassellatum,	251
Beluga vermontana, - -	235
Birds of Vermont, - -	240
Bittern, - - - - - -	246
Botany, - - - - - -	255
Bowfin, - - - - - -	255
Brandon Ores, - - -	270
Brown Coal, - - - -	270
Calciferous Sandstone,	262
Catalogue of Plants, - -	256
Champlain Rocks, - -	262
Charlotte, Map of - -	239
Chazy Limestone, - -	263
Chromic Iron, - - - -	272
Clay Stones, - - -	278
Climate and Meteorology,	227
Coccoborus ludovicianus,	244
Colleges, - - - - -	280
Columba migratoria, -	246
Colymbus septentrionalis,	248
Coregonus clupeiformis,	254
Corals, - - - - - -	263
Cottus gobioides, - -	251
Crime, - - - - - -	282
Cypselus pelasgius, -	245
Dace, small-scaled, - -	252
Darter, - - - - - -	251
Dikes, - - - - - -	273
Drift, - - - - - -	274
Eagle, Golden - - - -	240
Elephant, Fossil - -	234
Elephas primogenius, -	234
Emys geographica, -	248
Essox nobilior, - - -	252
Essox estor, - - - -	253
Falco chrysaetos, - - -	240
Felis concolor, - - -	232
Fishes, - - - - - -	250
Flycatcher, Traill's -	241
Fossil Seeds, - - - -	271
Fringilla borealis, - -	244
Geological Survey, - -	260
Geology of Vermont, -	260
Gold Formation, - - -	268
Grain, - - - - - -	280
Granite, - - - - -	272
Grosbeak, Rose-breasted,	244
Heights, - - - - -	226
Hematite, - - - -	270
Herring Salmon, - - -	254
Hudson River Shale, -	264
Igneous Rocks, - - -	271
Infusorial Silica, - -	278
Isle la Motte Limestone,	263
Jumping Mouse, - -	233
Kaolin, - - - - - -	270
King-bird, Olive-sided,	241
Lawrencian Deposit, - -	274
Leuciscus atromaculatus,	252
Libraries, - - - - -	280
Lime, - - - - - -	265
Limestone, Blue, - - -	269
Literary Institutions, -	280
Live Stock, - - - -	280
Loon, Red-throated -	248
Lucioperca Canadensis, -	250
Maclurea magna, - -	263
Magnesian Slate, - -	266
Manufactures, - - -	282
Marble, - - - - -	266
Marl, Shell - - - -	276
Masquallonge, - - -	252
Merula alivacea, - -	242
Muck, Vegetable - - -	277
Muscicapa traillii, - -	241
" ruticilla, - -	241
" pusilla, - -	242
Mus leucopus, - - -	233
Mya arenaria, - - -	275
Mytilus edulis, - - -	275
News Papers, - - -	280
Nucula portlandica, - -	275
Oakes, Wm., - - -	256
Orthis. - - - - - -	264
Orthoceras, - - - -	263
Oven, - - - - - -	265
Paupers, - - - - -	282
Phenomena, L. Champla.,	230
Phoca vitulina, - - -	233
Pickerel, - - - - -	253
Picus pileatus, - - -	245
Pigeon, Passenger - -	246
Pike Perch, - - - -	250
Population by towns, -	278
" by counties,	280
Porphyry, - - - - -	272
Products of Industry, -	282
Quadrupeds of Vermont,	231
Quartz, Granular - -	267
Rail Roads, - - - -	282
Redpoll, Mealy - - -	244
Red Sandrock, - - -	264
Redstart, - - - - -	241
Reptiles of Vermont, -	248
Rocking Stones, - - -	272
Roofing Slate, - - -	266
Salmoperca pellucida, -	253
Sandpiper, semipalmated,	247
Sanguinolaria fusca, - -	275
Saxicava rugosa, - -	275
Schools, - - - - -	280
Sciurus hudsonius, -	234
Serpentine, - - - -	272
Snake Mountain, - -	265
Snow, fall of - - - -	229
Sparry Limestone, - -	266
Sponge, Fossil - - -	275
Spring, Advance of -	229
Squirrel, White - - -	234
Stargazer, - - - -	251
Stockbridge Limestone, -	266
Superficial Deposits, -	274
Swallow Tree, - - - -	245
Sylvia striata, - - -	242
" ruficapilla, - -	243
" pardolina, - -	243
" parus, - - - -	243
" philadelphia, -	243
" americana, - -	244
Taconic Rocks, - - -	266
Talcose Division, - - -	267
Tanager, Scarlet - -	244
Telegraph, - - - -	283
Temperature, Burlington,	227
" Newbury, -	228
" Extremes - -	227
Tertiary Formation, -	270
Titaniferous Iron Ore, -	272
Topography, - - -	225
Tortoise, Geographic - -	248
Tortoise, Soft-shelled -	249
Totanus melanoleucas, -	247
Trap Dikes, - - - -	272
Trenton Limestone, - -	263
Trilobites, - - - -	264
Tringa semi-palmata, -	247
Trionyx ferox, - - -	249
Trout Perch, - - - -	253
Tyrannus cooperi, - -	241
Utica Slate, - - - -	264
Vespertilio novobaracen.,	231
Vireo gilvus, - - - -	242
Warbler, Green black-cap	242
" Black-poll - -	242
" Red-poll - -	243
" Canada - - -	243
" Hemlock, - -	243
" Mourning -	243
" Parti-colored -	244
Warbling Vireo, - -	242
Water, Fall of - - -	228
Whale, Fossil - - -	235
Wood, Fossil - - -	275
Woodpecker, Crested -	245
Yellow Shanks, - - -	247

INDEX

TO PART FIRST, OR NATURAL HISTORY.

[G. stands for Genus. For an Index to the Genera of Plants see page 207.]

Acanthopterygii	129	Bunting Savanna.	87	Domestic Fowls,	111
Acipenser, G.	149	——— Snow,	86	Dove, Carolina,	100
Alasmodonta, G.	165	Butcher Bird,	75	Duck, Wood,	109
Alburgh Springs	8	Butternut,	215	——— Mallard,	109
Alcedo, G.	96	Buttonwood,	214	——— Dusky,	110
Alosa, G.	144	Caledris, G	102	——— Bluewing.	110
Ammocœtes, G.	150	Caprimulgus, G	99	Eagles,	59
Anas, G.	109	Cartilag. Fishes,	148	Earthquakes,	16
Anchor Ice,	15	Carp Sucker,	133	Eel, Black,	148
Ancylus, G.	164	Carp Family,	133	——— Common,	147
Angle Worm,	170	Cat,	52	——— Silver,	148
Anodonta, G.	164	Catalog. Quadru.,	24	Eel-pout,	147
Annulata,	169	——— Birds,	57	Elk,	50
Anser, G.	108	——— Reptiles,	113	Elm, White,	214
Anthus, G	86	——— Fishes,	128	——— Red,	215
Appear. of Birds,	13	——— Plants,	173	Emberiza, G	86
Appendix (shells)	169	Catamount,	37	Emys, G	113
Arachnides, G	170	Catastomus, G	133	Emysaurus, G	114
Arbor Vitae,	218	Cat Bird,	78	Ermine,	31
Arctomys, G	44	Cat Fish,	139	Esox, G	137
Ardea, G	103	Cattle,	51	Etheostoma,	132
Area of Vermont,	2	Caves,	8	Extent of Vt.,	2
Arvicola, G.	41	Cedar Bird,	74	Face of country,	3
Ash,	211	Cedar, Red,	218	Falco, G	58
Ass,	53	——— White,	218	Finches,	90
Astacus Bartonii,	170	Certhia, G	95	Fisher Martin,	32
Aurora Borealis,	18	Centrarchus, G	131	Fishes,	127
Bass, Black,	131	Champlain, Lake,	5	Fly-catchers,	75
——— Rock,	131	Cherry,	209	Fox, Black,	36
Basswood,	209	Cherry Bird,	74	——— Cross,	35
Bat, Carolina,	25	Chestnut,	213	——— Red,	35
——— Hoary,	25	Chickadee,	73	——— Sampson,	36
——— Say's,	25	Chickaree,	46	Fringilla, G	87
——— Silver-haired,	26	Class. of Animals,	23	Frog, Bull,	119
Batrachia,	119	Climate,	9	——— Black,	121
Bays,	6	Climates compar'd	20	——— Horicon,	121
Bear,	28	Clytus pictus,	172	——— Leopard,	120
Beaver,	38	Coccyzus, G	92	——— Pickerel,	120
Beech,	212	Coluber, G	115	——— Spring,	120
Beetle,	172	Columba, G	100	——— Woods,	121
Bill-Fish,	145	Colymbus, G	111	——— Tree,	122
Birch,	213	Corvina, G	132	Fulica, G	106
Birds,	56	Corvus, G	71	Gadidæ,	146
Black Bird, Cow,	69	Cougar,	37	Gallina. Birds,	100
——— Crow,	70	Counties,	2	Gar Fishes,	145
——— Red-wing.,	68	Crane, Whoop'g,	103	Geology,	222
——— Rusty,	71	Crawfish,	170	Goldfinch,	89
Blue Bird,	85	Crow,	71	Goosander,	110
Bob-o-link,	70	Cross Bill, Com.	91	Goose, Wild,	108
Bombycilla, G	74	——— white wing.	92	Gos-hawk,	62
Boundary,	1	Crustacea,	170	Granivor'us Birds,	86
Borer, Locust,	172	Cuckoos,	92	Grosbeak, Pine,	91
Bos, G	54	Currant,	220	Grouse,	101
Botany,	173	Cyclas, G.	168	Grus, G	103
Bug, Cucumber,	172	Cypselus, G	98	Gull, Bonapart.	107
Bufo, G	123	Dace,	135	——— Herring,	108
Bull Frog,	119	Dark Days,	15	Halcyons,	96
Bull Pout,	138	Dobchick,	107	Hang Bird,	68
Bunting,baywing.	87	Dog,	52	Hare,	48

Hawk, Br.winged,	61
——— Cooper's,	62
——— Fish,	60
——— Gos,	62
——— Large-foot.	62
——— Marsh,	62
——— Pigeon,	63
——— Red-shoul.	60
——— Red-tailed,	63
——— Slate color.	61
Hedge-hog,	47
Height of lands,	3
Helix, G. snails,	58
Hemlock,	217
Heron, Blue,	103
——— Green,	104
——— Night,	103
Herring family,	144
Hickory,	215
Hiodon, G	144
Hirundo, G	97
Hog,	53
Hog Fish,	132
Hornbeam,	212
Horned pout,	139
Horse,	52
Horses in Vt.,	53
Horse Leech,	169
Humming bird,	96
Hydrargira fusca,	137
Hyla,	122
Hylodes Pickerin.	121
Ice,	14
Icterus, G	67
Indian summer,	16
Insectivor's birds,	74
Insects,	170
Iron-wood,	212
Islands,	6
Jay, blue,	72
——— Canada,	73
Jumping Mouse,	44
King Bird,	75
King Fisher,	96
Lakes,	5
Lamprey, Blue,	150
——— Mud,	150
Lanius, G	74
Larch,	217
Lark, Brown,	86
——— Meadow,	67
Latitude,	1
Larus, G	107
Lepisosteus, G	145
Limax, G	163
Limnæa, G	153
Ling,	146
Linnet, Pine,	89
——— Purple	91

INDEX TO NATURAL HISTORY.

Lizards,	115	Perch, Common,	129	Salmo, G	140	Tatler Solitary,	105
Lobe footed birds,	106	——— Pond,	130	Salmon,	140	——— Spotted,	105
Lobster, F. water,	170	Percidæ,	129	Salmonidæ,	140	Teal, Blue-wing.	110
Longe,	140	Perdix, G	101	Sanderling,	102	Tebenophorus, G	163
Longitude,	1	Petromyzon, G	150	Sassafras,	211	Temperature,	9
Loon,	111	Pewee,	76	Sauria,	115	Tetrao, G	101
Lota, G	146	Pheasant,	102	Saw-whet,	66	Thrasher,	78
Loxia, G	91	Philomycus, G	163	Scolopax, G	105	Thrush, Aquatic,	79
Lucio-Perca, G	129	Phœbe,	76	Seal,	38	——— Brown,	78
Lynx,	36	Physa, G	154	Seasons,	13, 20	——— G. crown.	80
——— Bay,	37	Pickerel,	138	Serpents,	115	——— Hermit,	80
Magnetic varia'ns,	19	Picus, G	93	Shad,	143	——— New York,	79
Maple, Red,	210	Pigeon,	100	——— Winter.	144	——— Wilson,	79
——— Sugar,	209	Pike,	137	Sheep,	55	Tip-up,	105
——— White,	210	Pike-perch,	130	Sheep-head,	133	Titmouse, bl'k cap,	73
Martin, Pine,	32	Pimelodus, G	138	Sheldrake,	110	——— Hud'n bay,	73
——— Purple,	97	Pine Grosbeak,	91	Shiner,	136	Toad, Common,	123
Medicinal Springs,	7	——— Linnet,	89	Shrew, Forster's,	26	——— Tree,	122
Melania, G	152	——— Martin,	32	——— Short tailed,	27	Tortoise, painted,	113
Meleagris, G	101	Pine,	215	Shrew-mole,	27	——— sculptur'd	114
Menobranchus G	126	Planorbis, G	154	Siluridæ,	138	——— snapping,	114
Mergus, G	110	Plover, Sand'ling	102	Sitta, G	94	Totanus, G	104
Metals,	223	——— Upland,	104	Situation,	1	Trochilus colubris,	96
Meteorology,	9	Podiceps, G	107	Skunk,	33	Troglodytes, G	84
Meteors,	16	Pomotis, G	130	Smelt,	142	Trout, Brook,	141
Minerals,	224	Ponds,	5	Smoky atmosph'e,	15	——— Salmon,	140
Mink,	31	Poplars,	174	Snake, Black,	117	Tupelo,	211
Minnow, Brook,	136	Porcupine,	47	——— Brown,	116	Turdus, G	78
Moles,	27, 28	Pout, Bull,	138	——— Chicken,	118	Turkey, Wild	101
Moose,	49	——— Horned,	139	——— Green,	117	Unio, G	166
Mouse,	43	Productions,	6	——— Rattle,	118	Valvata, G	152
——— Meadow,	41	Pupa, G	157	——— Ribband,	115	Verd-mont,	4
——— Jumping,	44	Pyrrhula, G	91	——— Ringed,	117	Vireos,	77
Mud Fish,	137	Quadrupeds in Vt.	23	——— Spott. neck,	116	Vitrina, G	162
Mulberry,	212	Quay Bird,	103	——— Striped,	115	Wading Birds,	102
Mule,	53	Quail,	101	Snipe,	105	Walnut,	220
Muræna, G	147	Quiscalus,	70	Snow,	12	Warblers, b. & w.	83
Muscicapa, G	75	Rabbit,	48	Snow Bird,	88	——— Blackburn's,	82
Musk Rat,	41	Raccoon,	29	Snow Bunting,	86	——— black-throat.	83
Namaycush,	140	Rail,	106	Soil,	6	——— cœrulean,	82
Name of Vermont,	4	Rain in Vt.,	12	Sparrow, Blue,	88	——— green,	81
Night Hawk,	99	Rallus, G	106	——— Chipping,	88	——— Maryland,	83
Norway Rat,	42	Rana, G	119	——— Field,	88	——— Nashville,	81
Nuthatches,	95	Rat, Black,	43	——— Song,	87	——— pine-creep'g	82
Oak, Red, White,	213	——— Brown,	42	——— Swamp,	89	——— spotted,	81
Oil-nut,	215	Raven,	72	——— Tree,	88	——— summer,	81
Omnivorous Birds,	67	Red-poll,	89	Spruces,	216	——— yellow crown.	80
Ophidia,	115	Regulus, G	83	Squirrel, Black,	45	——— yell. red poll,	80
Oriole, Baltimore,	68	Reptiles,	112	——— Flying,	47	——— worm-eating,	83
Osmerus, G	142	Rivers,	4	——— Gray,	45	Weasel,	30
Otter,	33	Robin,	79	——— Red,	46	Web-footed birds,	107
Owl, Barn,	67	——— Golden,	68	——— Striped,	46	Whip-poor-will,	99
——— Barred,	66	Rocks,	222	Starnosed Mole,	28	White Fish,	143
——— Cinereous,	65	Rumina. Animals,	49	Streams,	4	Winds,	10
——— Great-horn.,	65	Rusticola, G.	105	Sturgeons,	149	Winter Shad,	144
——— Hawk,	64	Sable,	32	Sturnus, G	67	Wolf,	34
——— Screech,	64	Salamanders,	123	Succinea, G	156	Wolverine,	30
——— Short-eared,	66	——— Glutinous,	125	Suckers,	133	Woodchuck,	44
——— Snowy,	64	——— Many-spot.,	123	Sun Fishes,	130	Woodcock,	106
Ox,	54	——— Red backed,	125	Swallow, Bank	98	Woodpeckers,	93
Paludina, G	151	——— Salmon col ,	124	——— Barn,	97	Wren, House,	84
Panther,	37	——— Symmetri'l,	123	——— Chimney,	98	——— Winter,	84
Partridge,	101	——— Tiger,	124	——— Cliff,	97	——— Wood,	85
——— Spruce,	102	——— Two-lined,	125	——— White bell.	98	Yellow Bird,	89
Parus, G	73	——— Violet col'd,	125	Sylvia, G	80	——— Summer,	81
Pekan,	32	Salamandra, G	123	Tamarack,	217	Yellow Throat,	83
Perca, G	129	Salia, G	85	Tatler, Bartram's	104	Yoke-toed Birds,	92

Waking to the Dream